S0-DXD-651

Future Trends in Superconductive Electronics

(Charlottesville, 1978)

AIP Conference Proceedings
Series Editor: Hugh C. Wolfe
No. 44

Future Trends in Superconductive Electronics

(Charlottesville, 1978)

Editors

B.S. Deaver, Jr.
University of Virginia

C.M. Falco
Argonne National Laboratory

J.H. Harris
National Science Foundation

S.A. Wolf
Naval Research Laboratory

American Institute of Physics

New York 1978

L.C. Catalog Card No. 78-66638
ISBN 0-88318-143-6
DOE CONF - 780339

CONFERENCE
ON
FUTURE TRENDS
IN
SUPERCONDUCTIVE ELECTRONICS

Charlottesville, Virginia

March 23 - 25, 1978

Sponsored by

Division of Basic Energy Sciences, Department of Energy
Division of Engineering, National Science Foundation
Division of Physical Sciences, Office of Naval Research

With Local Arrangements and Support by

Department of Physics
Department of Electrical Engineering
Center for Advanced Studies
Division of Continuing Education

University of Virginia, Charlottesville, Va.

ORGANIZING COMMITTEE

Bascom Deaver, Jr.	University of Virginia	Chairman
Charles Falco	Argonne National Laboratory	Organizer–SQUID Symposium
John Clarke	University of California	Organizer–SQUID Symposium
Richard Brandt	Office of Naval Research	
William Clinton	Department of Energy	
Jay Harris	National Science Foundation	
Robert Mattauch	University of Virginia	
Mark Wittels	Department of Energy	
Stuart Wolf	Naval Research Laboratory	
Warren Salsbury	University of Virginia	Conference Coordinator

PROGRAM COMMITTEE

Malcolm Beasley	Stanford University
Richard Brandt	Office of Naval Research
Robert Buhrman	Cornell University
John Clarke	University of California
William Clinton	Department of Energy
Bascom Deaver, Jr.	University of Virginia
Edgar Edelsack	Office of Naval Research
Charles Falco	Argonne National Laboratory
Jay Harris	National Science Foundation
Robert Hein	Naval Research Laboratory
Donald Langenberg	University of Pennsylvania
William Little	Stanford University
Robert Mattauch	University of Virginia
James Mercereau	California Institute of Technology
Paul Richards	University of California
Sidney Shapiro	University of Rochester
Arnold Silver	Aerospace Corporation
Theodore Van Duzer	University of California
Nancy Welker	Laboratory for Physical Sciences
Mark Wittels	Department of Energy
Stuart Wolf	Naval Research Laboratory

PREFACE

In 1967 a symposium entitled "Physics of Superconducting Devices", sponsored by the Office of Naval Research, was held in Charlottesville at the University of Virginia to exchange ideas on techniques and devices made possible by the unique properties of superconductors, particularly, but not exclusively, ones based on the Josephson effects. This meeting was extremely successful and included some of the earliest discussion of SQUIDs, magnetometers, the voltage standard, the noise thermometer, devices based on thin film bridges, and Josephson devices as electromagnetic detectors and as possible computer elements. (The proceedings of that symposium is available as document number AD661848 from the National Technical Information Service, Springfield, Virginia 22151.) At that time the field was largely the province of the low temperature physicist.

Now, eleven years later, SQUID magnetometers are commercial devices, enormous progress has been made toward practical high speed digital devices, and Josephson devices may be on the verge of competing usefully as detectors of electromagnetic radiation. It can be argued that superconducting devices have indeed emerged from the research laboratory and are becoming attractive candidates for use in electronic systems offering heretofore unavailable levels of precision, speed, sensitivity and low loss.

The Conference on Future Trends in Superconductive Electronics addressed the interface between those who study and develop devices and those who design and produce practical electronic systems. Evidence for the reality of this interface comes from the increasing number of companies developing some capability in superconducting technology and the number of universities beginning to offer courses in superconducting electronics to meet the perceived need for engineers and applied physicists with some training in this area.

The purpose of the conference was to bring together a select, international working group of physicists and engineers representing research, development, manufacturing and project management to examine the current status of superconductive electronics and to discuss the role of superconducting devices in electronic systems of the next ten years.

The program included two types of presentations. First, a group of speakers was invited to give a series of overview or summary talks. There were 35 of these talks divided into five sessions: 1 & 2. Symposium on SQUIDs and their Applications; 3. Metrology and Generation and Detection of Electromagnetic Signals; 4. Junctions, Microfabrication, Materials and Cryogenic Techniques; 5. Digital Systems.

In Session 4 two speakers described the techniques being used by the semiconductor industry to fabricate submicron devices. Subsequent speakers described progress in applying some of these techniques to fabricating superconducting devices. Also, there was a brief description of the plans for the National Submicron Facility, an NSF-funded facility at Cornell University that during the coming year will make available to the research community some of these state-of-the-art techniques.

For the second type of presentation, all participants were invited to contribute papers for poster displays on the same general topics as the summary talks. The two evenings of the Conference were devoted to viewing and discussion of these displays.

This Proceedings contains papers prepared by the speakers and papers describing most of the work presented as poster displays. The topics have been somewhat more finely subdivided and the contributed (poster) papers have been interspersed with the summary talks under the appropriate headings.

In keeping with the intent that this Proceedings be a useful resource book, it seems appropriate to call attention here to the comprehensive literature survey issued quarterly entitled "Superconducting Devices and Materials" prepared by N. A. Olien, M. Nisenoff, S. A. Wolf, D. B. Sullivan and E. A. Edelsack. The initial volume issued in 1967 covered work from 1959 to 1967 and the survey has been compiled regularly since that time. Copies are available from Cryogenic Data Center, National Bureau of Standards, Boulder, Colorado 80302.

We want specifically to acknowledge the indispensable encouragement and support provided by the Department of Energy, the National Science Foundation and the Office of Naval Research, the key role of the conference committees in assembling the program, and the outstanding job done by Warren Salsbury and his colleagues at the University of Virginia Conference Center. Particular credit is due to Edgar Edelsack for his insights, guidance and assistance, to Charles Falco and John Clarke for organizing the SQUID symposium for DOE, to Stuart Wolf and the reviewers who worked so diligently on the manuscripts, and most of all to all the people who took time to participate with such enthusiasm in yet another conference - but one that we hope will play some useful role in guiding the future of the rapidly evolving field of superconductive electronics.

Bascom S. Deaver, Jr.
Robert M. Mattauch

University of Virginia
Charlottesville, Virginia
April, 1978

TABLE OF CONTENTS

METROLOGY, OSCILLATORS AND HIGH Q DEVICES

DETECTION OF ELECTROMAGNETIC SIGNALS

JUNCTIONS: FABRICATION AND PROPERTIES

SUPERCONDUCTOR-SEMICONDUCTOR DEVICES

MATERIALS

CRYOGENIC TECHNIQUES

CHARLOTTESVILLE REVISITED

John Clarke
Department of Physics, University of California,
and Materials and Molecular Research Division
Lawrence Berkeley Laboratory, Berkeley, California 94720

The 1967 Charlottesville Symposium on the Physics of Superconducting Devices was the first to be devoted exclusively to this subject. It was also, incidentally, the first conference that I attended in the United States, and I found it an exhilarating and stimulating two days. For the first time, I was able to fully appreciate the tremendous impact that Brian Josephson had made on measurement technology. Needless to say, some of the devices described were rather crude by today's standards, for example, the SLUG, others weren't really practical devices, for example, the rf SQUID, while yet others weren't even off the drawing board, for example, the noise thermometer. Although I remember the conference as being predominantly a "Josephson Conference", in fact only 10 of the 22 papers presented dealt specifically with the Josephson effect. However, in the intervening 11 years the term "superconducting devices" has increasingly come to imply "Josephson devices". Thus, at the present conference 32 out of the 36 invited papers are concerned with the Josephson effect, and, of the remaining 4 papers, 2 are on refrigeration.

In these opening remarks, I should like to revisit Charlottesville 1967 to pick out the main areas of interest, and then try to give some kind of overview of how these areas have developed subsequently. The list I have come up with is rather long, and should be considered representative rather than exhaustive. Obviously, I cannot go into details, or even mention the names of the many people who have been involved in the work.

The first topic that I have picked out (quite at random!) is SQUIDs (Table I). At Charlottesville '67 there were two talks on the dc SQUID. This was already a practical instrument with which people had made measurements that would otherwise have been impossible. There were also two talks on the rf SQUID, a device that had not yet been used as an instrument, but that obviously had a promising future. It was quite clear that the SQUIDs were already vastly superior to any competing devices for measuring small changes in magnetic flux. In his concluding remarks at the '67 meeting, Mike Tinkham made the following statement: "In the sense that a device is defined as something which can be sold at a profit, the conference may have been premature." At the time, this statement was undoubtedly true. The only device that saves the present conference from the same remark is the SQUID: SQUIDs are indeed commercially available today, presumably at a profit to their manufacturers!

The SQUID is the most highly developed and widely applied Josephson device. There are basically two types: the dc SQUID,

ISSN: 0094-243X/78/001/$1.50

TABLE I: SQUIDS

Charlottesville: 1967
- Clarke: SLUG (DC SQUID)
- Beasley and Webb: DC SQUID
- Silver: RF SQUID
- Mercereau: RF SQUID

1978

DC SQUID: Thin-film device with shunted tunnel junctions. Increase in sensitivity likely with smaller area junctions.

RF SQUID: Commercially available. Point contact 20 or 30 MHz devices at limit of sensitivity. Increase in sensitivity with higher frequencies.

Magnetometers ($\sim 10^{-11}$ $GHz^{-\frac{1}{2}}$)

Gradiometers ($\sim 10^{-12}$ $Gcm^{-1}Hz^{-\frac{1}{2}}$)

χ-measurement ($\sim 10^{-12}$ emu cm^{-3})

Voltmeters ($\sim 10^{-15}$ $VHz^{-\frac{1}{2}}$)

Laboratory measurements
Geophysics: magnetotellurics, rock magnetometry
Biomedical: heart and brain waves
Low frequency communication
Standards: rf power measurements, voltage and current comparators
Gravity wave detectors
Quark hunting

which consists of two junctions on a superconducting ring, and the rf SQUID, which has a single junction on a superconducting ring. Although the dc SQUID preceded the rf SQUID, the latter has become by far the more popular--for the simple reason, I suspect, that it needs only one Josephson junction. In the late '60's and early '70's, when it was difficult enough to make a single junction let alone two junctions, this was a decided advantage. The commercial SQUIDs, which use point contact junctions and operate at 20 or 30 MHz, have reached the limits of their sensitivity, although improvements have been achieved by operating at higher frequencies. The dc SQUID, which uses shunted tunnel junctions, can be made more sensitive by decreasing the area, and hence the capacitance of the junctions, and is likely to become substantially more sensitive in the near future.

SQUIDs, almost always in conjunction with a superconducting flux transformer, are used to measure magnetic fields, magnetic field gradients, magnetic susceptibility, and voltage. Although the flux transformers have usually been made of wire, there may be a considerable future for thin-film integrated devices, for example, gradiometers. Quite apart from laboratory measurements, SQUIDs have been used in a wide variety of applications, for example, geophysics--magnetotellurics and rock magnetometry, biomedical applications--heart and brain waves, low-frequency communications, standards work--rf power measurement and voltage and current comparators, gravity wave detectors, and even quark hunting. SQUIDs are well tried and tested devices, and have already made a substantial impact on non-cryogenic fields. The future outlook for this field is extremely bright.

The next topic on my list is high-frequency (millimeter and submillimeter) detectors (Table II). There were two talks at Charlottesville '67, on the Josephson video (square law) detector, and on the Josephson mixer. These were not very practical devices, but a good deal of optimism was expressed about the future of this field. Subsequently, a wide variety of new devices have appeared, not all of them Josephson-based or even superconducting, and it is by no means clear which of them is going to be the best in any given application. In particular, it is not obvious that the Josephson-based device will emerge as the most sensitive, and it is necessary to keep a wary eye on the opposition.

The Josephson devices include the video detector, the heterodyne mixer with an external or internal local oscillator, and the parametric amplifier with external or internal pumping. The Josephson heterodyne mixer is the best of any presently available in the 36 to 500 GHz frequency range. However, it appears to be reaching the limit of its sensitivity, and it is still way above the single photon detection limit. Thus, although its present status represents a considerable achievement, it is not clear that the Josephson mixer ultimately will be the best available. The externally-pumped parametric amplifier shows considerable promise,

TABLE II: MILLIMETER AND SUBMILLIMETER DETECTORS

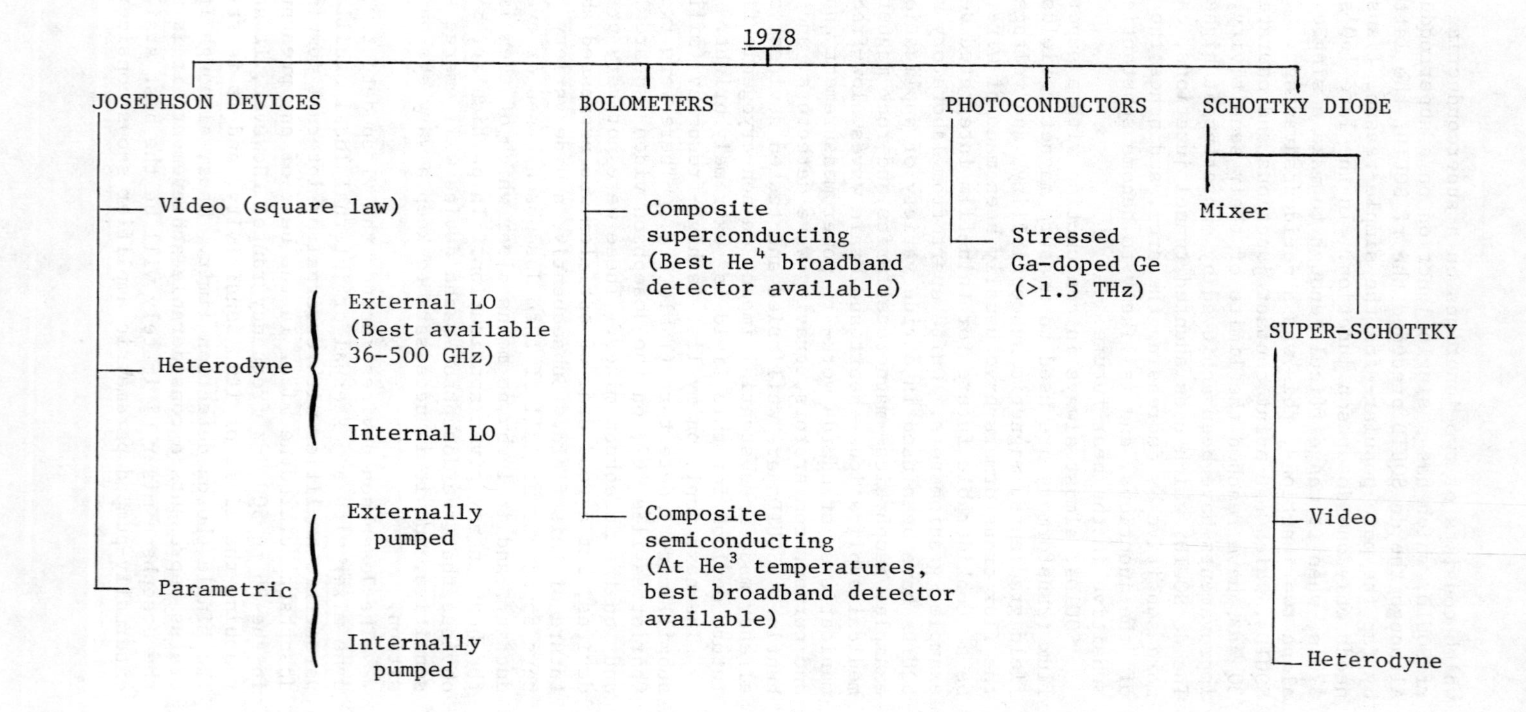

particularly the version with an array of small tunnel junctions, but it remains to be seen how well it operates at the higher frequencies, say above 50 GHz. At present, the best broadband detectors are composite bolometers. The composite bolometer consists of a sapphire substrate with a film to absorb the incoming radiation on one side, and a temperature sensitive element on the other side. The superconducting transition edge bolometer is the most sensitive in the He^4 range, while a He^3 - temperature semiconductor bolometer is the most sensitive of any presently available. Stressed Ga-doped Ge photoconductors have recently become available at frequencies as low as 1.5 THz, and are likely to be even better than bolometers at these higher frequencies. Finally, there are the Schottky and Super-Schottky detectors. The latter has promise as a heterodyne detector, but there have been difficulties in making it operate at frequencies much above X-Band.

Thus, this field remains in a state of flux, with no one device being clearly vastly superior to the others. One cannot say that Josephson devices have made a major impact on high frequency detection, although they may yet do so. It will be interesting to watch future developments.

Let me move on to metrology (Table III). In 1967 Taylor, Parker, and Langenberg had already made their famous measurement of e/h using $h\nu = 2eV$, and its impact on maintaining the standard volt and on the values of the fundamental constant had been established. Kamper proposed his noise thermometer. There were two papers on high-Q superconducting cavities.

This field has blossomed tremendously over the past 11 years. A variety of clever devices have been invented, mostly at national laboratories, particularly the NBS. The Josephson effect is routinely used to maintain and compare the standard volt in several national laboratories, and a commercial standard has just become available. A cryogenic voltage divider and a SQUID voltmeter enable one to make voltage comparisons to 0.1ppm or better. Efforts are now directed towards the use of series arrays of junctions to produce larger Josephson voltages. Noise thermometry is now a reality, in two versions. In the first, one measures the linewidth of the Josephson radiation emitted by a resistively shunted junction: The linewidth is proportional to the absolute temperature. In the other (commercially available) versions, the Johnson noise generated by a resistor is measured with a SQUID voltmeter. Both thermometers can be used down to a few mK. Another type of thermometry relies on a SQUID to measure the nuclear susceptibility of Cu ($\lesssim$ 1K). The NBS now sells superconducting fixed-point thermometers with an absolute accuracy of ±1mK (Cd to Pb), and I understand that higher and lower temperatures will be available soon. Another important development is the inductive current comparator, which uses a SQUID to compare currents to 1ppb. Rf SQUIDs are used to measure power levels and to calibrate rf attenuators. Josephson junction mixers are be-

TABLE III: METROLOGY

Charlottesville 1967:	Taylor, Parker, and Langenberg: e/h	
	Kamper: Noise Thermometer	
	McAshan	Superconducting
	Siegal, Domchick, and Arams	high-Q cavities

1978

$h\nu = 2eV$	Standard volt, e/h	
THERMOMETRY	Noise (mK range)	Linewidth: $\Delta f \propto T$.
		Voltmeter: $\overline{V^2} \propto T$.
	Nuclear χ of Cu (SQUID) $\lesssim$ 1K.	
	Fixed points: Superconducting transition (Cd-Pb) ± 1mK.	
COMPARATORS	Voltage: Cold resistive divider + SQUID (standard volt) ~ 0.1ppm.	
	Current: Superconducting inductive comparator + SQUID ~ 1ppb.	
RF POWER	Power Level (rf SQUID): ± 0.1dB, 0 - 1GHz.	
	Attenuator Calibration (rf SQUID): ± 0.002dB/60dB.	
FREQUENCY	Stable oscillator: Superconducting cavity, with $Q \sim 10^{11}$	
	Frequency synthesis: Josephson junction mixer	

coming important in frequency synthesis to simplify the frequency chains for accurate infrared frequency measurements. Finally, superconducting cavities with Q's of 10^{11} or more are the basis of high-stability oscillators.

Thus, there has been considerable growth in the standards field, and there is no doubt that routine standards' measurements with cryogenic devices will continue to develop in the future.

At Charlottesville '67, Matisoo described the Josephson junction switch, and suggested that it could be used in computers as a logic and memory element (Table IV). The device had current gain and a switching time that was less than 1ns. Although nobody had made junctions on a large scale 11 years ago, there was a widespread feeling that this idea heralded a new generation of computers. Subsequently, there has been an enormous effort in this field at IBM, and, more recently, a relatively small-scale effort at Bell Labs. The Josephson switch is very fast--I think the record is 24ps--and, more importantly, has an extremely low power dissipation. There are semiconductor devices that rival the Josephson devices in speed, but that dissipate perhaps 4 orders of magnitude more power. The low dissipation of the Josephson devices allows them to be packed very densely, thus greatly reducing the transit times between elements, and enabling one to take full advantage of the fast intrinsic switching times. It is this feature that gives the Josephson devices a prime advantage over their semiconducting competitors. All the Josephson circuits have used tunnel junctions, and have relied heavily on photolithographic techniques.

At IBM, the emphasis has been on switching the junctions between zero and non-zero voltage states by means of a magnetic field. The memory cells consist of two-junction interferometers. The non-destructive readout (NDRO) cell has a switching time of 80ps and a power-delay product of about 10^{-16}J. A single-flux quantum cell has also been developed (with destructive readout) with a switching time of 50ps, and a power-delay product of about 10^{-18}J. The logic circuits originally used single-junction cryotron of the type described by Matisoo: Adders, multipliers, and shift registers have all been successfully operated. More recently, a 3-junction interferometer has been introduced as an in-line gate, and this appears to be the fastest logic element available. Workers at Bell Labs have concentrated on current-switched junctions, and have successfully produced several devices, for example, a shift register, and a full adder.

More recently, there has been interest in signal-processing applications of Josephson junctions, for example, A-to-D converters and correlators. To me, computing is the most exciting of the Josephson fields, and one in which we are likely to see great progress in the future. I'm enough of an optimist to believe that we shall see a superconducting computer in operation in the not-too-distant future.

TABLE IV: COMPUTERS

Charlottesville 1967: Matisoo--Josephson junction switch--current gain, < 1ns.

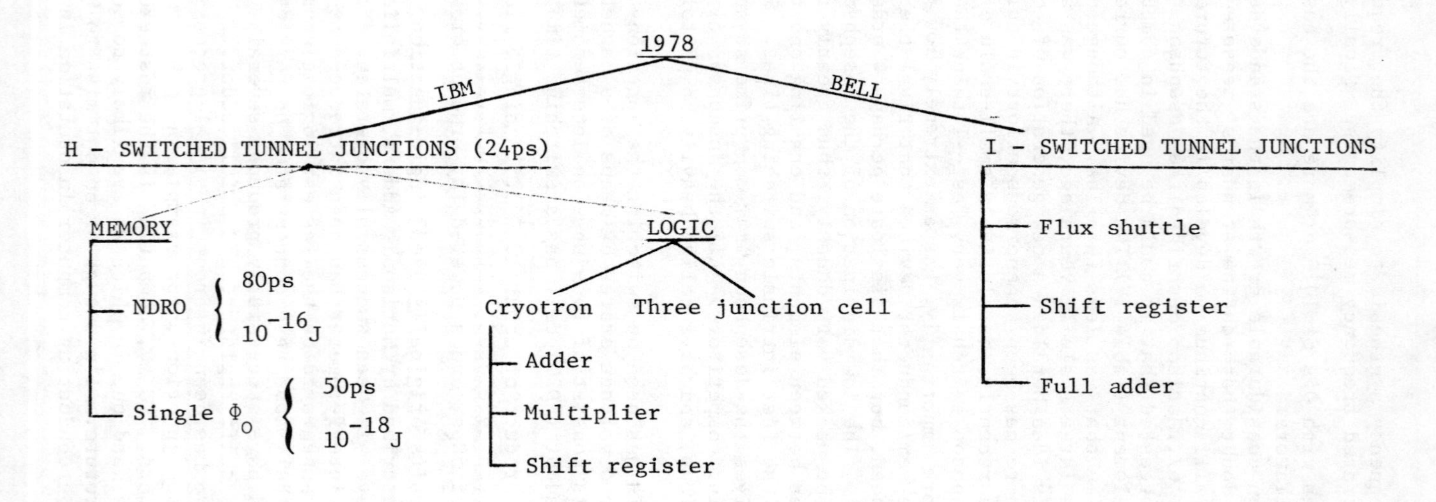

All circuits use Josephson tunnel junctions and photolithographic techniques.

ADVANTAGES
- High Speed
- Low power-delay product--high packing density and short transit times

Let me briefly touch on one other subject raised at Charlottesville '67, namely noise. Scalapino talked about current noise in SIS tunnel junctions, and Burgess discussed quantization and fluctuations in superconductors. There has been a good deal of work in this area subsequently, and theory and experiment are mostly in good accord, although there remain some details to be worked out.

I'm going to finish up with two areas that weren't discussed at Charlottesville '67. The first is junction fabrication: At that time, there had been little systematic investigation of methods of making junctions--we were grateful for anything that worked. In the intervening 11 years there has been an enormous effort in this area, and three principal types of junctions have emerged. The tunnel junction is the most reliable and reproducible, and is well-understood theoretically. Some very sophisticated circuits have been built for computing in which the tunnel junction has a hysteretic current-voltage characteristic and is used as a switch. Resistively shunted non-hysteretic tunnel junctions have been used in dc SQUIDs, and are likely to be used in the future in rf SQUIDs and high frequency detectors. The second type of junction is the microbridge--constant thickness, variable thickness, proximity effect, and granular. Electron-beam machining techniques have been used to make these bridges reproducibly, but, unfortunately, they are rather easy to burn out. Their theory is in a decidedly murky state, and GL or TDGL theories seem inadequate to properly explain the behavior. Additional factors that have been considered include heating (a dominant effect), non-linear quasiparticle conductance, quasiparticle relaxation processes (there are several to choose from!), gap relaxation processes, vortices, and non-uniform current distribution. It remains to be seen which of these factors are essential to a proper description. Microbridges have been used in SQUIDs and high frequency detectors, but their use as devices is not particularly widespread at present. The third and undoubtedly most popular junction is the point contact. The modern commercial version seems to be quite reliable, although a good deal of black magic remains in its manufacture. The point contact is widely used in SQUIDs, in high frequency detectors, and in metrology.

The future of junctions remains open. My own feeling is that we shall see a move away from the point contact towards thin film devices, particularly shunted tunnel junctions, for SQUIDs and high frequency detectors. To avoid hysteresis, one requires $2\pi I_c R^2 C \lesssim \Phi_0$, and at the same time one needs $I_c R$ to be as large as possible. These constraints imply that C, and hence the junction area, should be as small as possible. I suspect that we shall see a strong effort to fabricate and use small-area tunnel junctions. The future of the microbridge is uncertain. For many applications, one would like a relatively high resistance, say 100Ω, a value that is difficult to achieve. In my opinion, the undesirably strong temperature dependence of the critical current, the ease with which they can be destroyed, and their pronounced heating effects make microbridges of somewhat limited use as devices. However, the fact that microbridges have been operated well above He^4 temperatures may make them useful in certain applications where liquid He^4 is not available.

The final topic is another that was not discussed at Charlottesville '67, namely refrigeration. We were a bunch of low temperature physicists who took cryogenic techniques very much for granted, and didn't need to talk about them. However, when we tried to operate experiments outside our low-temperature laboratories, we soon found that the traditional liquid-nitrogen-shielded glass cryostats were unsuitable. The immediate problem was solved with the advent of the commercially available fiberglass cryostat that is rugged, portable, needs no liquid nitrogen, and consumes one liter or less of liquid helium per day. This cryostat has made possible the field work that we shall hear about at this conference. Nevertheless, there is a critical need for a closed-cycle refrigerator that is reliable, low-noise, and operable in remote areas. There has been some progress in this area using the Stirling cycle, but I have a feeling that the widespread acceptance of superconducting devices depends strongly on the availability of suitable refrigerators.

To conclude: I have tried to give my impressions of the main areas of research in superconducting devices over the past 11 years. To me, at least, this has been a very exciting time that has seen a great deal of first-class research. I have every reason to believe that this trend will continue. I look forward to this conference which promises to be as exciting as the previous one. At the same time, if Bascom Deaver can be persuaded to organize Charlottesville 1989, I wonder what that conference will have in store for us?

RF SQUIDs: THE STATE OF THE SCIENCE*

R. P. Giffard
Department of Physics, Stanford University
Stanford, Ca. 94305

ABSTRACT

The theory of rf SQUID operation, including the effects of fluctuations, is discussed. The theory is in agreement with the large signal and small signal behavior of real devices under optimum conditions. The noise levels of many existing rf SQUID systems are substantially above those predicted on the basis of amplifier noise temperatures. All rf SQUIDs seem to exhibit excess noise of unknown origin at low input frequencies. The limitations of devices and future directions of development are considered.

INTRODUCTION

The use of superconducting weak links in rf-biased circuitry follows directly from Josephson's observation[1] in 1964 that the behavior of a working junction would correspond phenomenologically to that of an inductance whose value was a function of current. Since a Josephson junction could be virtually lossless, and non-linearity would occur at exceedingly small signal levels, such a device would constitute a unique, parametrically variable, circuit element. Silver and Zimmerman[2] in 1967 gave the first detailed analysis of the static and dynamic behavior of a superconducting ring containing a single Josephson junction. They then demonstrated that the response of a device at 30 MHz showed parametric effects and a dissipative response periodic in applied flux. Although the use of point-contacts was, and still is, socially unacceptable in some places, this paper marks the state of a decade of exciting scientific investigation which has given birth to an almost unique detector of magnetic flux, the rf-biased magnetometer or SQUID. The development of useful devices, and the detailed analysis of their properties, have provided valuable testing grounds for the electrodynamics of superconducting circuits and the physics of Josephson junctions.

There have naturally been many reviews of the properties of Josephson junctions or, more generally, weak links and their applications. An up-to-date and useful group is contained in the proceedings of the 1976 NATO Advanced Study Institute on Small Scale Applications of Superconductivity.[3]

This paper contains a brief review of the appropriate theory, followed by a comparison with the performance of existing devices. Finally, the limitations of present SQUIDs and some possible future developments are discussed.

*Work supported by ONR under N00014-76-C-0848 and NSF under PHY76-80105.

ISSN: 0094-243X/78/011/$1.50

BASIC THEORY

An rf-biased SQUID consists essentially of a single weak link connected in a closed circuit with a superconducting inductor. Because of the Josephson effect, the total impedance of the circuit is modulated by any magnetic flux which is externally applied, the period of the modulation being equal to the superconducting flux quantum, $\Phi_o = 2.07 \times 10^{-15}$ Wb. The impedance changes are usually read out at some suitable rf frequency f by their effect on an inductively coupled circuit which matches the device to an amplifier.

The impedance level of the device is determined by the requirement that the weak-link coupling energy, $\Phi_o i_c$, where i_c is the weak-link critical current, should be considerably greater than the thermal energy $k_B T_d$ at the device temperature T_d. The inductive impedance of the weak link at zero average current $Z_j = \Phi_o f / i_c$ must therefore satisfy the relationship

$$Z_j << \Phi_o^2 f / k_B T_d. \tag{1}$$

At the low rf frequencies typically used in the past, about 30 MHz, Z_j must be much less than 1 Ω, and a large impedance transformation is required. At microwave frequencies, Z_j can approach the impedance level of conventional transmission lines and wideband matching circuits can be used.

The loading of the weak link by the superconducting circuit can be expressed in terms of a parameter $\beta = 2\pi L_s i_c / \Phi_o$, where L_s is the total series inductance of the SQUID. Silver and Zimmerman[2] pointed out that different phenomena would be obtained for the conditions $\beta < 1$, and $\beta > 1$. For $\beta < 1$, the response of the device can be developed using conventional circuit theory. The impedance modulation is reactive,[4] and successful device operation requires the critical current to be precisely controlled. When $\beta > 1$, the magnetization curve of the SQUID is hysteretic and multi-valued, and the behavior of the circuit must be worked out explicitly using piece-wise linear methods. Modulation of the resistive component of the device impedance is now observed with an amplitude which is relatively independent of the weak link properties. Although important work continues

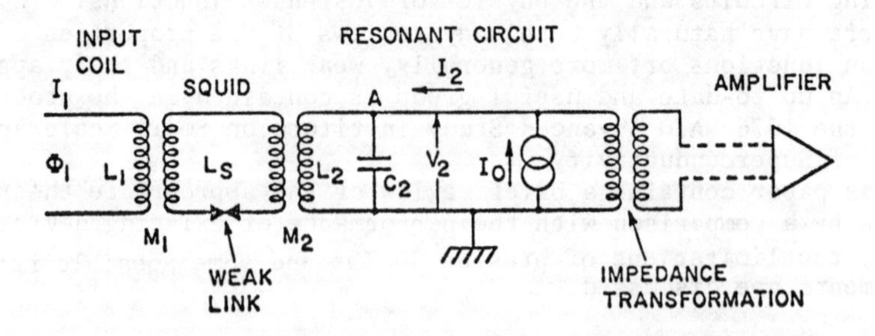

Fig. 1. Circuit arrangement for rf SQUID operation.

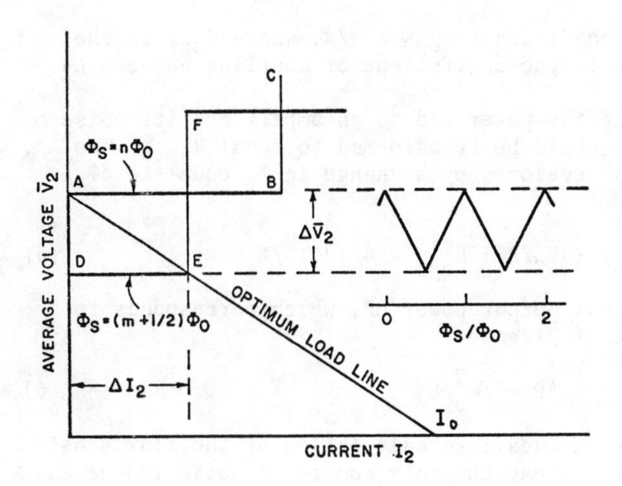

Fig. 2. Output characteristic of ideal SQUID at zero temperature.

to be done with $\beta < 1$ devices, most practical SQUIDs operate in the hysteretic régime.

Several circuit arrangements have been used for rf-biased SQUID operation, most of which are electrically equivalent to that shown in figure 1. The SQUID is coupled with mutual inductance M_2 to a circuit L_2C_2 which is excited at its loaded resonance frequency f. Although the SQUID is sometimes used as a direct detector of flux, input signals are usually applied through an input coil L_1. Zimmerman, Thiene and Harding[5] first considered the behavior of this circuit with a hysteretic SQUID. Their theory, which gives a good account of many properties of real devices, is now seen to be correct in the limit of a perfect weak link at zero temperature. The idealized relationship between the average rms voltage $\overline{V}_2$ appearing at point A in figure 1 and the rms current I_2 is shown in figure 2 for integer and half-integer values of the average normalized magnetic flux in the SQUID Φ_s/Φ_o. It is assumed that the circuit is dissipationless, although some high-frequency loss mechanism is required to damp out transients. The inset shows the relationship between $\overline{V}_2$ and Φ_s for any value of I_2 smaller than ΔI_2. The significant quantities $\Delta\overline{V}_2$ and ΔI_2 are independent of i_c and given by

$$\Delta\overline{V}_2 = \pi f L_2 \Phi_o / 2^{\frac{1}{2}} M_2, \qquad (2)$$

and

$$\Delta I_2 = 2^{\frac{1}{2}} \Phi_o M_2 / \pi L_2 L_s. \qquad (3)$$

Since the circuit has zero differential impedance, but is nonlinear, there is an optimum load line through A and E in figure 2, and the corresponding optimum drive current I_o allows the full modulation depth $\Delta\overline{V}_2$ to be obtained on-load. The optimum load resistance R_{opt} is thus given by

$$R_{opt} = \Delta\overline{V}_2/\Delta I_2 = \pi^2 f L_2^2 L_s / 2M_2^2. \qquad (4)$$

This corresponds to the condition $\kappa_2^2 Q_{opt} = \pi/4$, where Q_{opt} is the optimum loaded Q, and κ_2 is the coefficient of coupling between L_2 and L_s.

In order to maximize the power fed to an amplifier, its noise matched input impedance should be transformed to equal R_{opt}. The signal power δP which is developed by a change in Φ_s equal to $\delta\Phi_s$, is then given by

$$\delta P = (\delta\Phi_s)^2(\partial\overline{V}_2/\partial\Phi)^2 R_{opt}^{-1} = 4f(\delta\Phi)^2/L_s. \tag{5}$$

The maximum possible signal output power ΔP, which corresponds to a flux change of $\frac{1}{2}\Phi_o$, is thus given by

$$\Delta P = f\Phi_o^2/L_s. \tag{6}$$

Equation (5) allows an idealized calculation of the flux sensitivity to be made, assuming that the only source of noise is the amplifier. Since the smallest power which can be detected in a bandwidth of 1 Hz by an amplifier of noise temperature T_N is equal to $k_B T_N$, the smallest flux change Φ_n detectable at a 1 Hz rate should be given by

$$(\Phi_n)^2/L_s = k_B T_N/4f. \tag{7}$$

INTRINSIC FLUCTUATIONS

At the usual operating temperatures of SQUIDs, thermal fluctuations must be taken into account. Using the resistively-shunted junction model for the weak link at temperature T_d, Kurkijärvi[6,7] and Kurkijärvi and Webb[8] have adapted the Zimmerman, Thiene and Harding theory to account for fluctuations. The most important predictions are (i) that the output characteristic shown in figure 2 should have a non-zero impedance R_i, where

$$R_i = R_{opt}\pi^{-1}\beta^{1/3}\left(\frac{\beta + \pi/2}{\beta - \pi/2}\right)\left(\frac{3\pi^2 k_B T_d L_s}{2^{\frac{1}{2}}\Phi_o^2}\right)^{2/3}, \tag{8}$$

and (ii) that the output voltage $\overline{V}_2$ should show random noise with a spectral density $S_V(f)$ given by

$$S_V(f) \approx \left(\frac{\alpha^2\Phi_o^2 R_{opt}}{2L_s}\right)(1 - \beta^{-2})^{1/3}\left(\frac{\beta - \pi/2}{\beta + \pi/2}\right)^2. \tag{9}$$

where $\alpha = R_i/R_{opt}$ is a device-dependent step slope parameter introduced by Jackel and Buhrman.[9] Equation (8) shows that because of the fluctuations the optimum load resistance is reduced to a value R'_{opt} given by

$$R'_{opt} = (1 - \alpha)R_{opt}. \tag{10}$$

The signal power level predicted by equation (5) is consequently reduced to $(1 - \alpha)$ of its ideal value, and the maximum power level ΔP is reduced somewhat more due to noise rounding of the corners of the characteristic at A and E. The flux noise level due to the amplifier predicted by equation (7) must be increased by a factor $(1 - \alpha)^{-1}$.

EQUIVALENT CIRCUIT

A new comprehensive model of rf SQUID operation has recently been introduced by Ehnholm.[10] The model, based on the work of Kurkijärvi and Webb, takes account of the interaction between the SQUID and the input coil L_1 and follows both reactive and resistive components of $\bar{V}_2$. A novel prediction of the model is the reverse transfer of signals and noise from the output of the device to its input. An understanding of this phenomenon is necessary before accurate calculations can be made of the input impedance, the effect of source impedance on the output impedance, and the optimum device noise temperature. The reverse transfer phenomenon also indicates two sources of noise in rf SQUID systems which had not previously been understood.

A SQUID device is almost invariably used in a small signal mode, usually within a feedback loop. The important device parameters are then coefficients connecting small changes in the output amplitudes V_2 and I_2 with the input signals which cause them. The state of the input circuit is most conveniently expressed in terms of the input current I_1 and the input flux $\Phi_1 = \int V_1 dt$, where V_1 is the input voltage. All possible linear relationships between the small-signal dependent quantities $\delta\Phi_1$ and δV_2 and the variables δI_1 and δI_2 may conveniently be expressed by the equation

$$\begin{pmatrix} \delta\Phi_1 \\ \delta V_2 \end{pmatrix} = \begin{pmatrix} t_{11} & t_{12} \\ t_{21} & t_{22} \end{pmatrix} \begin{pmatrix} \delta I_1 \\ \delta I_2 \end{pmatrix}, \tag{11}$$

where t_{11} - t_{22} are four constant, small-signal parameters given by the equations

$$t_{11} = \partial\Phi_1/\partial I_1, \qquad t_{12} = \partial\Phi_1/\partial I_2$$
$$t_{21} = \partial V_2/\partial I_1, \qquad t_{22} = \partial V_2/\partial I_2 \,. \tag{12}$$

The Ehnholm model allows values of the four parameters t_{11} - t_{22} to be calculated to the accuracy with which the SQUID and circuit constants are known. At an average bias flux of $(n \pm 1/4)\Phi_0$, the parameters t_{11}, t_{22} and t_{21} correspond approximately to the quantities L_1, R_i and $\mp \pi f L_2 M_1/M_2$ respectively, as calculated using the Kurkijärvi and Webb analysis. The reverse transfer parameter t_{12} had not previously been calculated and is approximately equal to $\pm \pi L_2 M_1/4M_2$.

A slightly simplified small-signal equivalent circuit for the components of the SQUID system to the left of point A in figure 1 is

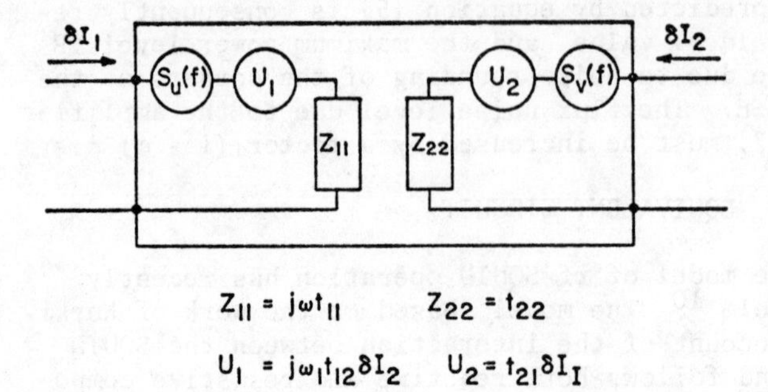

Fig. 3. Simplified small signal equivalent circuit of rf SQUID.

shown in figure 3. The voltage generator $S_V(f)$ represents the intrinsic noise given by equation (9). The generator $S_u(f)$ corresponds to voltage fluctuations induced in the input circuit by intrinsic flux fluctuations in the SQUID. Since, under some conditions, Z_{11} can have a negative real part,[10] it follows that the device may exhibit spontaneous quantum noise.[11] Since $S_V(f)$ and $S_u(f)$ arise in part from the same physical process, they will in general be correlated. If the equivalent circuit is combined with an appropriately characterized source and amplifier, the overall noise level of the SQUID circuit may be evaluated using conventional noise analysis.[12]

The Ehnholm model suggests two sources of noise which are less easy to characterize. Changes in I_2 will produce an effect equivalent to a changing input signal even when subcarrier modulation[13] is

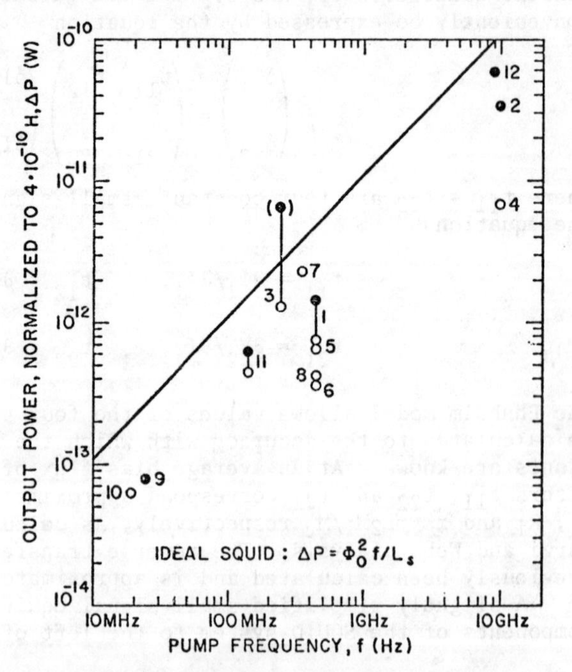

Fig. 4. Output power data for a number of well-documented SQUIDs. Solid circles - measured power; open circles - data deduced from noise measurements. Circles 1-12 denote data described in references 14-25 respectively.

used. Since the optimum excitation current I_o is proportional to the weak link critical current, changes in i_c will have a complementary effect, and fluctuations will produce extra noise.

It is interesting to compare the predicted performance of the rf-biased SQUID with the Manley-Rowe values for conventional parametric amplifiers. In an ideal reactive parametric amplifier with the same input and output frequencies as the SQUID, the forward and reverse transfer parameters would be related by $4j\pi f t_{12} = t_{21}$. The values calculated using the Ehnholm model are related by $3.7j\pi f t_{12} = t_{21}$, showing that although the SQUID functions by means of a varying hysteretic property, its overall behavior is remarkably similar.

PERFORMANCE OF EXISTING DEVICES

Figure 4 shows output power data for a number of well-documented rf SQUIDs.[14-25] Solid circles show the change in available power ΔP corresponding to a $\frac{1}{2}\Phi_o$ change in the average SQUID flux plotted against the working frequency f. In order to compensate for variations in SQUID inductance, each power has been normalized to 4×10^{-10}H by plotting the quantity $\Delta P(L_s/4\ 10^{-10}\ \mathrm{H})$. The solid line corresponds to ideal, zero temperature, matched performance as expressed by equation (6). The data clearly establishes the expected dependence of output power on frequency and shows no loss of performance at frequencies up to 10 GHz. This is consistent with the fact that, with the exception of the point labelled 4, all the devices employ niobium point-contact weak links which are expected to function up to considerably higher frequencies. Most devices fall between one half and one order of magnitude below ideal behavior. The decreased performance is easily accounted for by the fluctuation effects discussed above and possible departures from ideal matching.

Jackel and Buhrman[9] have carefully compared values of α obtained with SQUIDs containing oxidized niobium point-contacts and tin microbridge weak links with the values predicted by equation (8). Good agreement was found in some cases, but many point-contacts and all the microbridges gave significantly different results. These differences can be attributed[9] to resistive effects and departures of the individual current-phase relationships of the weak links from the ideal Josephson sinusoidal form.

The performance of a commercial point-contact SQUID[26] has recently been compared in detail[22] with theory at an operating frequency of 24 MHz. The value of α was found to be correctly predicted by equation (8), and the measured small-signal parameters defined by equation (12) agreed with values predicted by the Ehnholm model to within the experimental error of a few percent. It thus appears that the small-signal behavior of rf SQUIDs with good weak links agrees closely with current theory.

Jackel and Buhrman[9] have emphasized that the output noise level $S_V(f)$ due to intrinsic fluctuations in the SQUID may be predicted using equation (9) with measured values of α. Since the matched noise power output will be equal to $S_V(f)/(1 - \alpha)R_{opt}$ per unit bandwidth, it is smaller than amplifier noise unless $(\alpha\Phi_o)^2/2L_s(1 - \alpha) \gtrsim k_B T_N$. This inequality is not usually satisfied in practice, and the intrinsic noise level is, with some possible exceptions,[27] not observed.

Since the noise level achieved in practical SQUID systems is determined by amplifier noise, it should be calculable through equation (7), taking account of the fluctuation-dependent effects described above. The open circles in figure 4 demonstrate the performance of SQUIDs with known noise levels. A normalized effective power $\Delta P = k_B T_N(\Phi_o/2\Phi_n)^2(L/4.10^{-10}\ \mathrm{H})$ is plotted against the working frequency f. Equation (7) shows that for an ideal SQUID system with $\alpha \ll 1$, these points should also fall on the line $\Delta P = (\Phi_o)^2 f/L_s$. In fact, most SQUID systems show rather poorer performance, presumably due to incorrect matching, significant values of α, or an elevation of the amplifier noise temperature by dissipation in the tuned circuit.

The useful sensitivity of a SQUID system must be referred to the terminals of the input coil L_1. For all applications except those in which L_1 is part of a resonant circuit, the most useful quantity is the so-called energy noise level $A(\omega)$ given by

$$A(\omega) = \tfrac{1}{2} L_1 S_i(\omega), \tag{13}$$

where $S_i(\omega)$ is the spectral intensity of the total noise level at frequency ω, referred to current in the input coil. Equation (7) shows that the value of $A(\omega)$ has an ideal minimum value

$$A(\omega)\Big|_{min} = k_B T_N/8(1 - \alpha)\kappa_1^2 f, \tag{14}$$

where κ_1 is the coefficient of coupling between L_1 and L_s. Measured $A(\omega)$ data is only available for a small number of the SQUID systems shown in figure 4. Commercial SQUID systems operating between 20 and 30 MHz have been shown[3] to have $A(\omega)$ values of about $5 \times 10^{-29}\ \mathrm{J\ Hz^{-1}}$, within a factor of about 2 of the value which can be estimated with equation (14). Pierce, Opfer and Rorden[17] have reported a value of $2 \times 10^{-30}\ \mathrm{J\ Hz^{-1}}$ for a 10 GHz SQUID system using a parametric amplifier with a noise temperature of 100 K. All systems show an increase in $A(\omega)$ at low input frequencies, generally taking the form of an additional ω^{-1} spectrum.

For applications in which the input coil is resonated, a more complete noise calculation based on the equivalent circuit shown in figure 3 is required.

FUTURE DEVELOPMENTS

Reliable and well-characterized rf-biased SQUID systems are now in widespread use for a variety of magnetic measurements. A number of demanding applications would nonetheless benefit from more sensitivity, better performance at extremely low frequencies, and wider closed-loop frequency response.

Figure 4 shows clearly that the output power of rf SQUIDs increases with frequency as expected from equation (6), indicating that

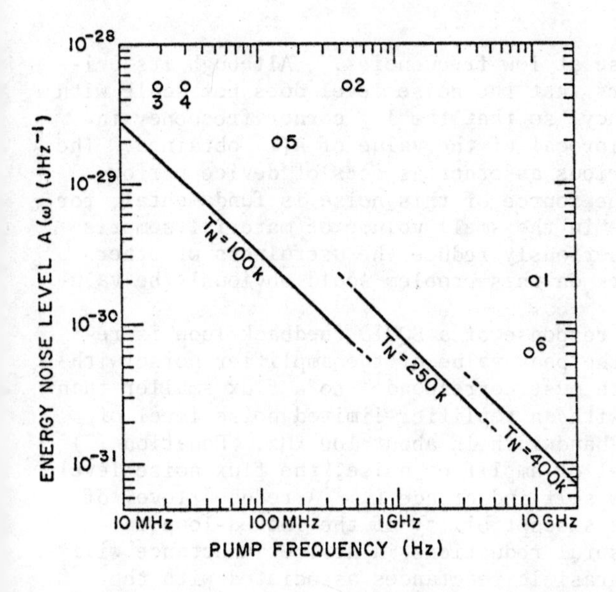

Fig. 5. Energy noise level. Straight lines: practical limit for various amplifier noise temperatures with $(1-\alpha)(\kappa_1)^2 = 0.5$. Circles: noise data as follows; 1 ref. 17, 2 ref. 19, 3 and 4 ref. 3, 5 ref. 24 and 6 ref. 25.

the high frequency limit discussed by Buhrman and Jackel[28] has not yet been reached. The potential sensitivity therefore depends mainly on the quantity f/T_N characterizing the sensing amplifiers. The performance of present commercial instruments is characteristic of T_N values of about 100 K at 20-30 MHz. Room temperature solid state amplifiers are now capable of achieving noise temperatures of 100 K up to about 0.5 GHz, 250 K at 4 GHz, and 400 K at 10 GHz. The solid lines in figure 5 show the performance which could be obtained with such amplifiers on the somewhat optimistic assumption that devices for which $(1 - \alpha)(\kappa_1)^2 = 0.5$ can be made. The data points show $A(\omega)$ values for existing systems. It is clear that unless some unsuspected noise source is present, considerably improved performance should be attainable with optimal SQUID design and correct amplifier noise matching. An attractive frequency range would be between 400 and 500 MHz, where high-performance low-cost amplifiers are available.

Refrigerated solid-state devices, parametric amplifiers of various degrees of sophistication, and masers offer further reductions in noise temperature, with an eventual quantum limit at $k_BT_N \approx hf$. It appears from equation (9) that intrinsic noise will become important for values of T_N below about 30 K, suggesting the use of the highest possible frequency. Kanter and Vernon[29] have recently observed SQUID action at 100 GHz, demonstrating that the niobium point-contact weak link operates well at this frequency. Intrinsic noise can in principle be decreased by cooling the SQUID, although at high frequencies dissipation may heat the weak link above the temperature of its surroundings. A return to non-hysteretic SQUID operation would eliminate intrinsic noise but would require more sophisticated detection systems for the resulting phase-modulated output signals.

A limitation of existing rf SQUID systems in some applications is the level of excess noise at low frequencies.[3] Although its origin is not known, it appears that the noise level does not scale with the SQUID operating frequency, so that the $1/f$ corner frequency increases roughly as the reciprocal of the value of $A(\omega)$ obtained. The noise thus becomes more serious as other aspects of device performance are increased. If the source of this noise is fundamental, for example the intrinsic noise in the small volume of material comprising the weak link, it may seriously reduce the usefulness of other noise reductions. More work on this problem would obviously be valuable.

The maximum frequency response of a SQUID feedback loop is restricted by the fact that the peak value of the amplifier noise within the closed-loop bandwidth must correspond[17] to a flux smaller than about $\Phi_o/10$. For a SQUID with an amplifier-limited noise level of $10^{-4}\ \Phi_o\ Hz^{-1}$, this maximum bandwidth is about 100 KHz. Equation (7) shows that for a given level of amplifier noise, the flux noise level is smaller for SQUID of low self-inductance L_s. A reduced level of flux noise also reduces the susceptibility of the locked-loop to transient interference. Useful reductions in device inductance will eventually be limited by parasitic reactances associated with the weak link structure, and by the fact that a good input coupling coefficient is increasingly difficult to achieve. In direct field-sensing SQUIDs good use has, however, been made of a parallel arrangement of several inductors.[30]

Shunted tunnel junctions, point-contacts, metallic microbridges, and various other weak links have been used successfully in rf SQUIDs at low frequencies. Significant parasitic or kinetic inductance is undesirable, as it contributes to the total device inductance, reducing the maximum possible value of the input coupling coefficient κ_1, and apparently increasing the level of intrinsic noise.[9] At microwave frequencies the requirements are more demanding, and clean, adjustable, niobium point-contacts seem to be most widely used. The development of faster tunnel junctions, good thin-film microbridges[31] or permanently adjusted, clean, point-contacts would help to improve the reliability of high performance SQUID systems.

Although the behavior of microwave SQUIDs[15,17,29] seems very similar to that observed at low frequencies, existing models of operation have not yet been extended to the interesting régime in which the device dimensions are comparable with the pump wavelengths.

ACKNOWLEDGMENT

It is a pleasure to acknowledge many useful discussions with M. R. Beasley and with J. N. Hollenhorst who also contributed unpublished microwave SQUID data.

REFERENCES

1. B. D. Josephson, Rev. Mod. Phys. 36, 216 (1964).
2. A. H. Silver and J. E. Zimmerman, Phys. Rev. 157, 317 (1967).

3. J. Clarke, in *Superconductor Applications: SQUIDs and Machines*, B. Schwartz and S. Foner, eds. (Plenum, New York, 1977), p. 67.
4. P. K. Hansma, J. Appl. Phys. 44, 4191 (1973).
5. J. E. Zimmerman, P. Thiene and J. T. Harding, J. Appl. Phys. 41, 1572 (1970).
6. J. Kurkijärvi, Phys. Rev. B 6, 832 (1972).
7. J. Kurkijärvi, J. Appl. Phys. 44, 3729 (1973).
8. J. Kurkijärvi and W. W. Webb, in *Proc. Appl. Superconductivity Conf.*, Annapolis, Maryland, May 1-3, 1972, IEEE Pub. No. 72 CHO 682-5-TABSC (IEEE, New York, 1972).
9. L. D. Jackel and R. A. Buhrman, J. Low Temp. Phys. 19, 201 (1975).
10. G. Ehnholm, J. Low Temp. Phys. 29, 1 (1977).
11. W. Louisell, *Radiation and Noise in Quantum Electronics* (McGraw-Hill, New York, 1964), Chap. 7, p. 277.
12. See for example: Haus et al., Proc. IRE 48, 69 (1960).
13. R. P. Giffard, J. C. Gallop and B. W. Petley, Prog. Quant. Electr. 4, 301 (1976).
14. L. R. Corruccini, Rev. Sci. Instr. 44, 1256 (1973).
15. R. A. Kamper, M. B. Simmonds, C. A. Hoer and R. T. Adair, U. S. National Bureau of Standards Technical Note 643 (1973).
16. D. Pascal and M. Sauzade, J. Appl. Phys. 45, 3085 (1975).
17. J. M. Pierce, J. E. Opfer and L. H. Rorden, IEEE Trans. Mag. MAG-10, 599 (1974)
18. J. Ahopelto, P. J. Karp, T. E. Katila, R. Lukander and P. Mäki-pää, in *Proceedings of the 14th Int. Conf. on Low Temp. Physics*, Otaniemi, Finland, August 14-20, 1975, M. Krusius and M. Vuorio, eds. (North-Holland, Amsterdam, 1975), Vol. 4, p. 262.
19. T. D. Clark and L. D. Jackel, Rev. Sci. Instr. 46, 1249 (1975).
20. D. Duret, P. Bernard and D. Zenatti, Rev. Sci. Instr. 46, 474 (1975).
21. L. D. Jackel, J. M. W. Stoeckenius and R. W. Epworth, J. Appl. Phys. 49, 471 (1978).
22. R. P. Giffard and J. N. Hollenhorst, to be published in Appl. Phys. Lett., May 15, 1978.
23. S.H.E. Corporation, San Diego, California, SQUID System TSQ-330X.
24. R. P. Giffard and M. B. Simmonds, unpublished.
25. R. P. Giffard and J. N. Hollenhorst, unpublished.
26. Toroidal Point-Contact SQUID Model TSQ, S.H.E. Corporation, San Diego, California.
27. M. R. Gaerttner, in *Proceedings of the Int. Mag. Conf.*, Toronto, 1974, unpublished.
28. R. A. Buhrman and L. D. Jackel, IEEE Trans. Mag. MAG-13, 879 (1977).
29. H. Kanter and F. L. Vernon, IEEE Trans. Mag. MAG-13, 389 (1977).
30. J. E. Zimmerman, J. Appl. Phys. 42, 4483 (1972).
31. S. S. Pei and J. E. Lukens, Appl. Phys. Lett. 26, 480 (1975).

PRESENT STATUS AND FUTURE OF THE dc SQUID

M. B. Ketchen
IBM Thomas J. Watson Research Center,
Yorktown Heights, New York 10598
and
J. Clarke and W. M. Goubau
Department of Physics, University of California and
Materials and Molecular Research Division,
Lawrence Berkeley Laboratory,
Berkeley, California 94720

ABSTRACT

A simple theory of the tunnel junction dc SQUID shows that the energy resolution per Hz with respect to an input coil is of order $4k_BT(\pi LC)^{1/2}/\alpha^2$, where C is the capacitance of each junction, and α is the coupling coefficient between the input coil and the SQUID inductance L. The dc SQUID with junction areas of about 10^4 μm^2 has a typical energy resolution per Hz of 7 x 10^{-30} JHz^{-1}. If the junction area is reduced to 10^2 μm^2 or 1 μm^2, the energy resolution per Hz is expected to improve by one or two orders of magnitude respectively. Further improvement should result from lowering the operating temperature.

INTRODUCTION

The first Superconducting QUantum Interference Device was the tunnel junction dc SQUID of Jaklevic et. al.[1] (1964). This prototype was followed by a variety of other dc SQUIDs, but subsequently the rf SQUID[2] became the more popular device. The rf SQUIDs now commercially available are superior to the first generation of dc SQUIDs. However, a recently developed tunnel junction dc SQUID[3] with 10^4 μm^2 junctions is more sensitive than the commercially available rf SQUIDs. In this paper we briefly review the theory of the dc SQUID and describe the fabrication and performance of the present tunnel junction version. Projections for the sensitivities of dc SQUIDs with 10^2 μm^2 and $1\mu m^2$ junctions are made.

THEORY

We give a simple approximate theory of the dc SQUID that allows us to project improvements. The dc SQUID consists of two Josephson junctions incorporated in a superconducting ring of inductance L. Each junction is resistively shunted so that its current-voltage characteristic is non-hysteretic;[4] that is, $\beta_c = 2\pi I_o R^2 C/\phi_o \lesssim 1$ where I_o, R

ISSN: 0094-243X/78/022/$1.50

and C are the critical current, shunt resistance and capacitance per junction, and ϕ_o = h/2e. When the SQUID is biased with a current I_B somewhat greater than the maximum critical current $2I_o$, the voltage is periodic in the applied flux ϕ with period ϕ_o. Tesche and Clarke[5] have done a detailed computer analysis of the dc SQUID and concluded that the optimum performance is achieved when $\beta = 2LI_o/\phi_o \approx 1$. If $\beta = 1$, the maximum change of the SQUID's critical current in response to an applied flux is $\Delta I_c \approx \phi_o/2L \approx I_o$, and the corresponding change in the voltage across the SQUID is $\Delta V \approx \Delta I_c(R/2) \approx \phi_o R/4L$. The flux-voltage transfer function is given by

$$\left(\frac{\partial V}{\partial \phi}\right)_{I_B} \approx \frac{\Delta V}{\phi_o/2} \approx \frac{R}{2L} \quad . \tag{1}$$

When the SQUID is modulated with an ac flux and the voltage across it is lock-in detected, the value of $(\partial V/\partial \phi)_{I_B}$ is close to that given by Eq. (1).

The intrinsic noise voltage across the SQUID is on the order of the Johnson noise voltage of the shunts, and has a spectral density

$$S_v \approx 4k_B T(R/2) \quad . \tag{2}$$

The intrinsic flux noise of the SQUID has a spectral density

$$S_\phi = S_v/(\partial V/\partial \phi)^2_{I_B} \approx 8k_B T L^2/R \quad . \tag{3}$$

For many purposes, a convenient figure of merit is the energy resolution per Hz with respect to an input coil, $\varepsilon = S_\phi/2\alpha^2 L$, where α is the coupling constant between the SQUID and the input coil. From Eq. (3)

$$\varepsilon \approx \frac{4k_B T}{\alpha^2 R/L} \approx \frac{4k_B T(\pi LC)^{1/2}}{\alpha^2} \quad , \tag{4}$$

where we have used $2LI_o \approx 2\pi I_o R^2 C \approx \phi_o$. Thus $\alpha^2\varepsilon$ is on the order of $k_B T$ divided by the resonant frequency of the ring, $[2\pi(LC/2)]^{-1/2}$ This frequency is also approximately the Josephson frequency at the optimum bias voltage ($\sim I_o R/\phi_o$). The analysis of Tesche and Clarke indicates that more carefully computed values of S_ϕ and ε are larger than those estimated in Eqs. (3) and (4) by about a factor of four. However, these equations enable us to make estimates of performance for SQUIDs with different parameters provided the noise is dominated by thermal noise in the shunts. This requires that both the preamplifier noise and 1/f noise be negligible.

A PRACTICAL dc SQUID

The dc SQUID of Clarke et. al.[3] is shown in Fig. 1. The films are sputtered or evaporated through machined masks, and the Nb strips are thermally oxidized to form the tunnel barriers. The SQUID is coated with a thin layer of Duco cement, and a superconducting ground plane (not shown in Fig. 1) is evaporated over the slit in the cylinder and the narrow strips to lower the stray inductance. Important parameters of the device are: cylinder diameter ≈ 3 mm, junction area ≈ 10^4 μm^2, C ≈ 600 pF, R ≈ 0.6 Ω, L ≈ 1 nH, $I_o \approx 1$ μA, and $\alpha^2 \approx 0.37$. The residual parasitic inductance ($\lesssim$ 0.5 nH) is largely responsible for the relatively low value of α^2.

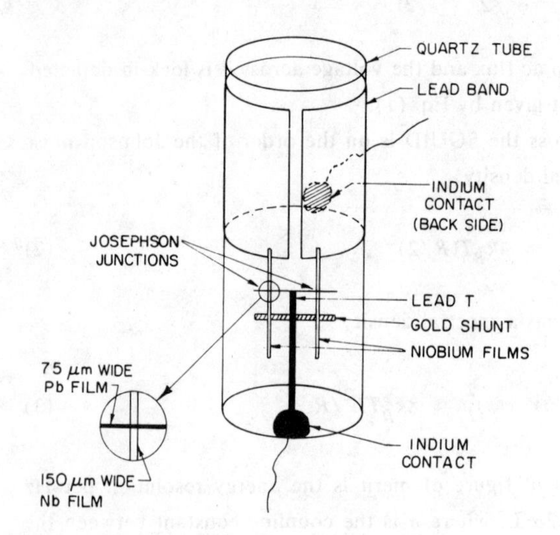

Fig. 1: Configuration of tunnel junction dc SQUID (3 mm cylinder diameter).

The SQUID is flux-modulated at 100 kHz, and the voltage across it is enhanced by a cooled superconducting tank circuit or resonant transformer that optimally couples the SQUID to a room temperature FET preamplifier (noise temperature ~ 1 K). After amplification, the signal is lock-in detected, and coupled back as a flux to the SQUID, which thus operates as a null detector in a flux-locked loop. The measured spectral density of the flux noise of a typical dc SQUID in a superconducting shield at 4.2 K is shown in

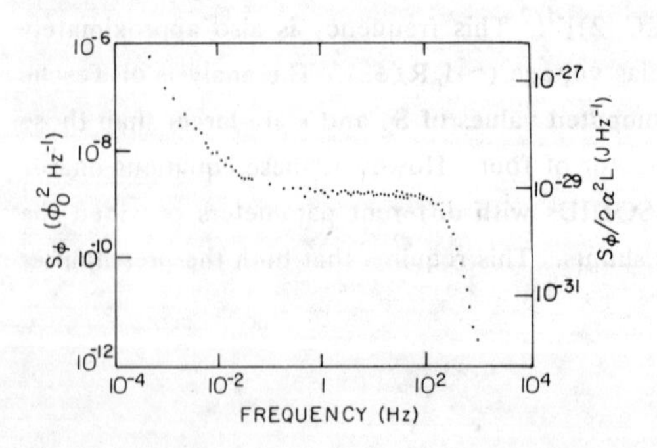

Fig. 2: Noise power spectrum of tunnel junction dc SQUID. The axis on the right specifies ε with respect to a 24-turn input coil.

Fig. 2. In the white noise region, $S_\phi^{1/2} \approx 3.5 \times 10^{-5}\ \phi_o \mathrm{Hz}^{-1/2}$, about a factor of two greater than the predictions of Tesche and Clarke. The corresponding energy resolution with respect to a 24-turn input coil is about $7 \times 10^{-30}\ \mathrm{JHz}^{-1}$. The roll-off above 200 Hz is due to filtering in the electronics. With resonant transformer coupling the white noise region has been found to extend out to at least 40 kHz. Below about 2×10^{-2} Hz the spectral density varies as 1/f. The drift rate of the SQUID in a temperature regulated helium bath has been reduced to $\lesssim 2 \times 10^{-5}\ \phi_o \mathrm{h}^{-1}$. The dynamic range is about $\pm\ 3 \times 10^6$ in a 1 Hz bandwidth, and the slewing rate using a resonant coupling transformer is about $2 \times 10^5\ \phi_o \mathrm{s}^{-1}$ at 1 kHz.

Planar dc SQUIDs have also been constructed and incorporated into thin-film gradiometers.[6]. These SQUIDs have an inductance of about 2.5 nH and a correspondingly poorer flux resolution, about $10^{-4}\ \phi_o \mathrm{Hz}^{-1/2}$. The best thin-film gradiometer has a sensitivity of approximately $2 \times 10^{-13}\ \mathrm{Tm}^{-1}\ \mathrm{Hz}^{-1/2}$.

PROJECTIONS FOR IMPROVED SENSITIVITY

If one maintains the SQUID at 4.2 K, as is likely for most applications, Eq. (4) suggests that one can improve the sensitivity by reducing L and C and by increasing α. Decreasing the diameter of the SQUID from 3 mm to 2 mm and adopting the more compact configuration of Fig. 3, one should be able to reduce L to ~ 0.3 nH. Because of the lowered parasitic inductance, α^2 would increase to perhaps ~ 0.6. To keep $\beta \approx \beta_c \approx 1$, I_o would be increased to ~ 3 μA and R reduced to ~ 0.35 Ω. These modifications

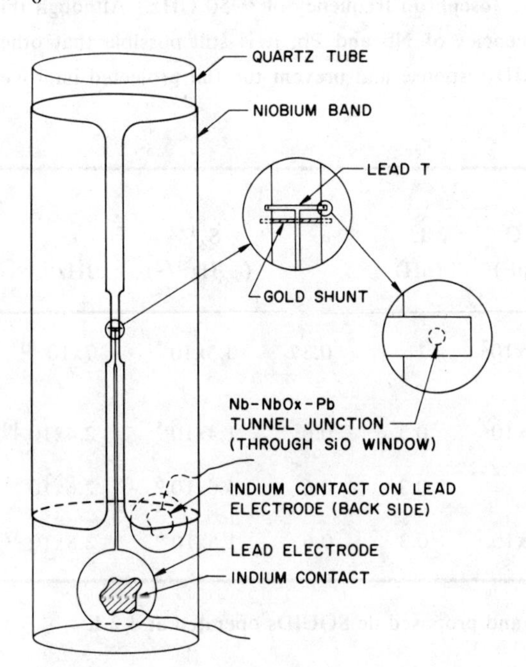

Fig. 3: Suggested configuration of dc SQUID with higher sensitivity (2 mm cylinder diameter).

would reduce ε by about a factor of 3. We feel that a further reduction in L would greatly increase problems of fabrication and probably result in a lower value of α^2.

The largest potential improvement in sensitivity is likely to be gained by a reduction in C. The present SQUID has junctions of relatively large area (~ 10^4 μm^2), and low critical current density (~ 10^2 Am^{-2}). With more sophisticated masks or with straightforward photolithographic techniques one could achieve junction areas of 10^2 μm^2. With more refined photolithography or E-beam resist techniques 1 μm^2 junctions are possible. The critical current density would have to be increased to 3 x 10^4 Am^{-2} and 3 x 10^6 Am^{-2} respectively in order to maintain $I_o \sim 3$ μA as is optimum for the low inductance SQUID design. A critical current density of 3 x 10^4 Am^{-2} is attainable with thermal oxidation provided that the Nb is not contaminated with resist. A value of 3 x 10^6 Am^{-2} is likely to require sputter cleaning and oxidation.[7] The increase in critical current density requires a corresponding reduction in barrier thickness. From the work of Basavaiah and Greiner[8] we estimate that C will be decreased by a factor of 7.3 x 10^3 when the area is reduced by a factor of 10^4.

In Table I we give the values of important parameters for the present dc SQUID and projected values for lower inductance and smaller junction area devices. These values are consistent with $\beta \approx \beta_c \approx 1$. We have assumed that the low noise temperature of the preamplifier is maintained as the SQUID resistance is increased by appropriate choice of the matching tank circuit or resonant transformer. A SQUID with 1 μm^2 junctions is seen to have an expected energy resolution of about 3 x 10^{-32} $JHz^{-1/2}$. The bias voltage would be about 100 μV with a corresponding Josephson frequency of ~ 50 GHz. Although this frequency is far below the gap frequencies of Nb and Pb, it is still possible that other relaxation processes may reduce SQUID response and prevent the full projected improvements from being achieved.

	Junction Area (μm^2)	I_o (μA)	R (Ω)	C (pF)	L (nH)	α^2	$S_\phi^{1/2}$ ($\phi_o Hz^{-1/2}$)	ε JHz^{-1}
Present	10^4	1	0.6	$6x10^2$	1	0.37	$3.5x10^{-5}$	$7x10^{-30}$
Proposed	10^4	3	0.35	$6x10^2$	0.3	0.6	$1.4x10^{-5}$	$2.4x10^{-30}$
	10^2	3	3.2	7	0.3	0.6	$4.6x10^{-6}$	$2.6x10^{-31}$
	1	3	30	$8x10^{-2}$	0.3	0.6	$1.5x10^{-6}$	$2.8x10^{-32}$

Table I Parameters of present and proposed dc SQUIDs operated at 4.2 K

One also expects a further improvement in performance from cooling the SQUID below 4.2 K, provided that a preamplifier with a sufficiently low noise temperature is available (for example, a second SQUID). At a low enough temperature the thermal voltage noise of the shunt resistors should become less than the shot noise voltage across the junctions. The shot noise has a spectral density $S_v^{(s)} \approx 2e(2I_o)(R/2)^2 = eI_oR^2$. In the limit $k_BT << eI_oR/2$

$$\varepsilon^{(s)} = \frac{S_v^{(s)}}{2\alpha^2 L(\partial V/\partial\phi)^2_{I_B}} \approx \frac{h}{2\alpha^2}, \qquad (6)$$

where we have used $2LI_o \approx h/2e$. The temperature at which this limit is expected to be reached is on the order of 100 mK.

CONCLUSIONS

The energy resolution per Hz, ε, of present tunnel junction dc SQUIDs is typically 7×10^{-30} JHz^{-1}. With a streamlining of the SQUID design and a four order of magnitude reduction in junction area (to ~ 1 μm^2), one should be able to reduce ε to ~ 3×10^{-32} JHz^{-1} at 4.2 K. By lowering the temperature to ~ 100 mK, one may be able to achieve a shot noise limited condition with ε on the order of Planck's constant.

REFERENCES

1. R. C. Jaklevic, J. Lambe, A. H. Silver and J. E. Mercereau, Phys. Rev. Lett. 12, 159 (1964).
2. J. E. Zimmerman, P. Thiene and J. T. Harding, J. Appl. Phys. 41, 1572 (1970); J. E. Mercereau, Rev. Phys. Appl. 5, 13 (1970), and M. Nisenoff, Rev. Phys. Appl. 5, 21 (1970).
3. J. Clarke, W. M. Goubau and M. B. Ketchen, J. Low Temp. Phys. 25, 99 (1976).
4. W. C. Stewart, Appl. Phys. Lett. 12, 277 (1968); D. E. McCumber, J. Appl. Phys. 39, 3113 (1968).
5. C. D. Tesche and J. Clarke, J. Low Temp. Phys. 29, 301 (1977).
6. M. B. Ketchen, W. M. Goubau, J. Clarke and G. B. Donaldson, J. Appl. Phys. (June 1978).
7. J. H. Greiner, J. Appl. Phys. 42, 5151 (1971).
8. S. Basavaiah and J. H. Greiner, J. Appl. Phys. 47, 4201 (1976).

PROPERTIES OF HIGH TEMPERATURE SQUIDs*

Charles M. Falco and C. T. Wu†

Argonne National Laboratory, Argonne, Illinois 60439

ABSTRACT

We review the present status of weak links and dc and rf biased SQUIDs made with high temperature superconductors. A method for producing reliable, reproducible devices using Nb_3Sn is outlined and comments are made on directions future work should take.

INTRODUCTION

There are a wide variety of techniques available for producing weak links suitable for use as elements in both dc and rf biased superconducting quantum interference devices (SQUIDs) operating at or below 4.2 K. However, there have been only a few reports of SQUIDs operating above 10 K[1-5] where closed-cycle refrigeration systems are available. This is due both to the difficulty in producing high-T_c superconductors and to the difficulty in producing weak links in these materials--which may have critical-current densities as high as 10^7 A/cm^2.

Recently, several groups have succeeded in fabricating weak links and SQUIDs operating in the temperature region 10-17 K. These developments have important implications for their possible future use in conjunction with closed-cycle refrigeration systems. This paper reviews the fabrication and properties of present high temperature SQUIDs as well as outlines a reproducible, reliable method for producing rf SQUIDs using Nb_3Sn thin films. Also discussed are several directions future work should take in order to produce devices useable for various possible applications. Finally, we speculate on the future importance of high temperature SQUIDs.

DC SQUIDs

One of the major requirements of a weak link in a SQUID, either dc or rf, is that it carry a critical current $I_c \cong \Phi_o/L$ where Φ_o is the magnetic flux quantum and L is the inductance of the device. SQUIDs can be made to operate with a large I_c if they are fabricated in a low inductance geometry.

Palmer, Notarys, and Mercereau[1] noted that the T_c of NbN and Nb_3Sn showed appreciable thickness dependence below 1000 Å. By photoresist masking and argon ion beam etching, they were able to reduce the thickness and thereby locally introduce a reduction

*Work supported by the U. S. Department of Energy.
†Present address: IBM Thomas J. Watson Research Center, Yorktown Heights, NY 10598.

ISSN: 0094-243X/78/028/$1.50

in the pair density in a well-defined region. Using weak links made in this manner, they reported the operation of dc SQUIDs at temperatures up to 17 K.

The area enclosed by these devices was 21 x 54 μm indicating an inductance of approximately 10-100 pH and therefore a critical current of 20-200 μA. While these small area devices exhibit quantum interference effects, their small inductance makes it difficult to couple in external signals and consequently limits their usefulness for a number of applications.

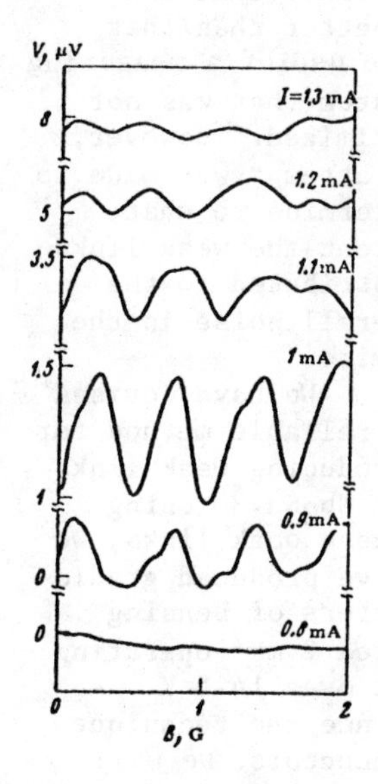

Fig. 1. Voltage-field characteristics of dc SQUID at 13.96 K for different biasing currents (after Golovashkin et al.).

By mechanically scribing 5000-7000 Å thick Nb_3Sn films, in order to form 1-5 μm wide by 1-2 μm long weak links, Golovashkin et al.[3,4] were able to fabricate dc SQUIDs operating up to 16.5 K (see Fig. 1). The area enclosed by these interferometers, and hence the inductance, was a factor of 5-10 smaller than that of Pàlmer et. al.[1] This enabled the devices to operate with correspondingly larger critical currents, and consequently quantum interference behavior was seen over a wide temperature range (more than 10 K). However, the low inductance limits the usefulness for practical applications.

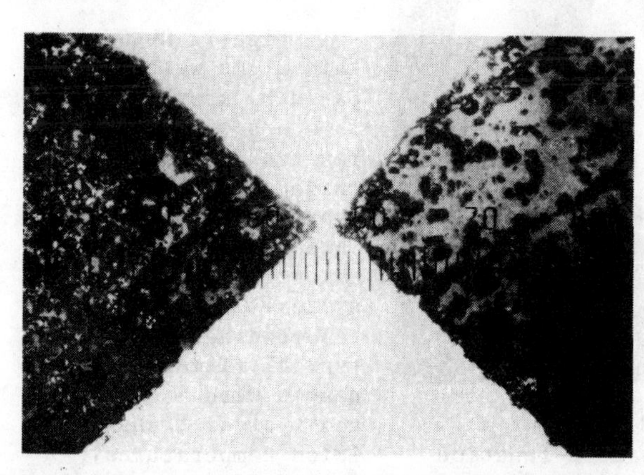

Fig. 2. A photograph of a planar SQUID. A close-up view of the NbN bridge, where the scale indicates 3μ m/div. (after Fujita et al.).

RF SQUIDs

Fujita et al[2] reported a technique for sputtering expitaxial single crystal NbN films with T_c's over 15 K using cleaved single crystal MgO substrates. By subsequent photoresist masking and sputter etching, they were able to produce well-defined weak links as is shown in Fig. 2. By incorporating these weak links in the 2-hole geometry shown in Fig. 3 which has a sensing area of 2 mm^2, they were able to make a gradiometer with a flux resolution $10^{-3}\Phi_o/Hz^{1/2}$. The highest operating temperature obtained was ∿ 15 K with an operating range of approximately 0.5 K. The signal from one of their devices is shown in Fig. 4. They estimate that the intrinsic flux resolution was a factor of 10 better than that measured due to inadequate shielding and to the use of a measuring system that was not optimized. However, no attempt was made to determine to what extent the weak links contributed to the overall noise in the device.

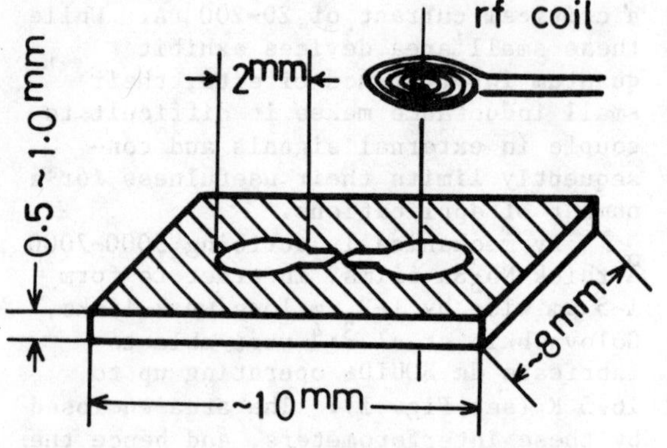

Fig. 3. A schematic drawing of a NbN planar SQUID coupled to an rf coil, (after Fujita et al.).

We have devised a reliable method for producing weak links in Nb_3Sn.[5] Using these weak links, we have produced gradiometers of sensing area 2 mm^2 operating at over 14.5 K. Since the technique should be applicable to other high T_c superconductors, we will describe it in some detail (see also Ref. 6).

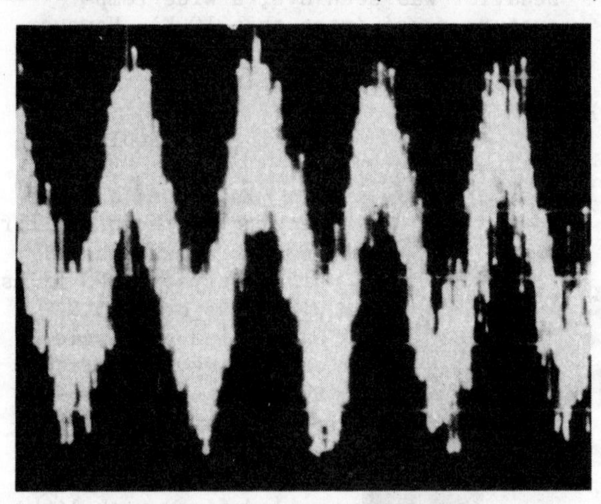

Fig. 4. An oscilloscope display of a periodic signal vertically as a function of magnetic flux in a NbN planar SQUID (after Fujita et al.).

RADIATION DAMAGE PRODUCED Nb_3Sn WEAK LINKS

High-T_c Nb_3Sn thin films were prepared by sputtering from a stoichiometric-reacted pressed-powder target onto sapphire substrates using a Varian high-rate magnetron sputtering system. For the work reported here, typical film thicknesses were ∿ 500 Å, and typical transition temperatures were ∿ 17.5 K. The T_c seems to be slightly thickness dependent, since the transition temperatures for our thicker films (∿ 1 μm) are typically ∿ 17.8 K. The typical resistivity ratio

$\rho_{300}/\rho_{17.5} = 2.4$ is smaller than the values we have obtained with thicker films (3.5–5.4). The preferred orientation and lattice parameter determined by x-ray diffraction analysis were <210> and 5.291 Å, respectively, for our thin Nb_3Sn films.

The six steps we use for the fabrication of Nb_3Sn thin-film weak links are as follows: (1) 500 Å Nb_3Sn is deposited on a polished sapphire substrate; (2) a bridge of width 5–30 μm and length 200 μm including four pads for the four-point probe measurement is etched using acid (see inset to Fig. 5); (3) a Cu film ∿ 4000 Å thick is deposited over the entire substrate; (4) a sharp razor blade is used to cut an opening of 0.5–5 μm in the Cu over the bridge; (5) the entire system is irradiated with a controlled dosage of α particles of energy 50 keV; (6) the Cu is etched away. Figure 5 shows the resistance-vs-temperature measurements of one of our weak links. The major portion of the film still has the same T_c after radiation damage, indicating that the Cu film was sufficiently thick to protect the underlying Nb_3Sn from damage. Figure 6 shows the critical-current variation of this weak link as a function of temperature ($I_c^{2/3}$ versus T). It is clear that over quite a large temperature range, there is a linear relation between $I_c^{2/3}$ and temperature.

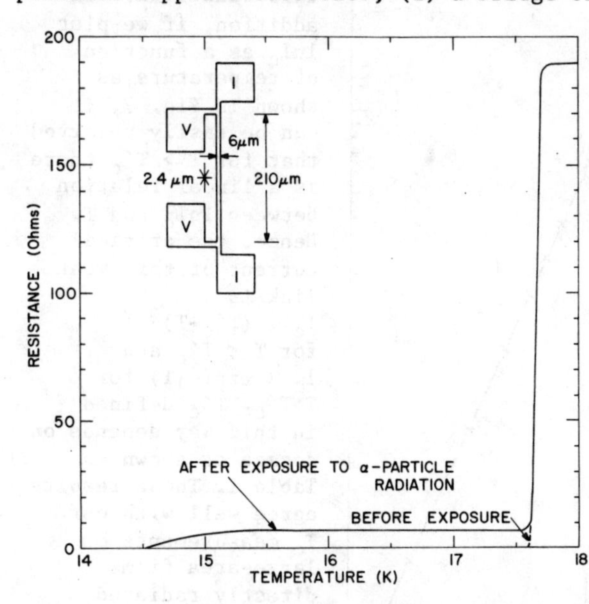

Figure 5. Resistance measurement as a function of temperature for a weak link.

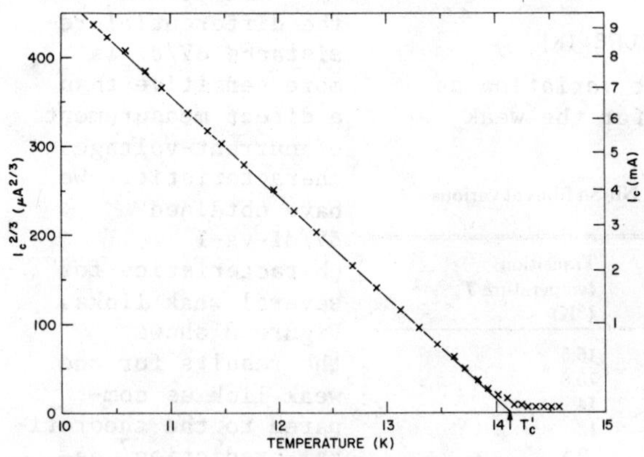

Figure 6. Critical-current variation as a function of temperature for the weak link of Figure 5.

If we define the intrinsic transition

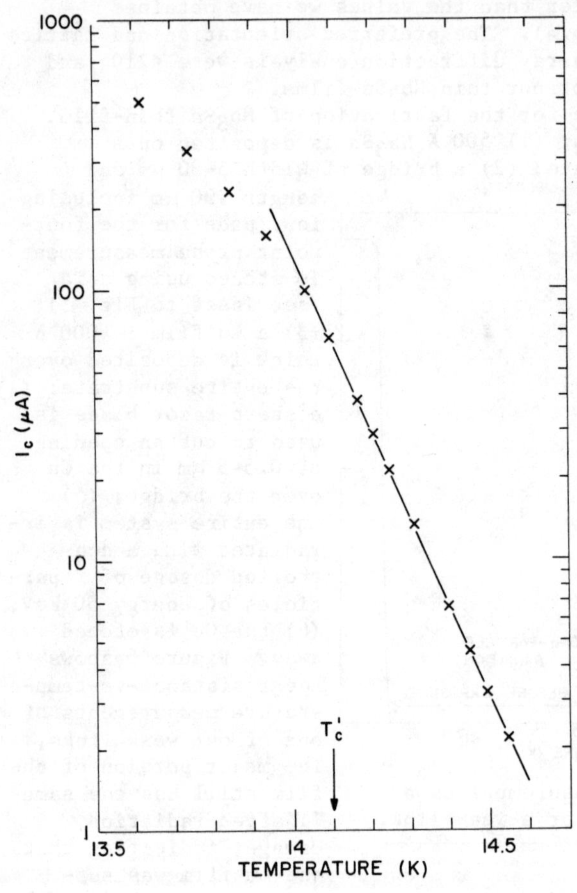

Figure 7. Critical-current variation as a function of temperature for the weak link of Fig. 2.

TABLE I. Transition temperature of Nb_3Sn films at various dosages.

Dosage (ions/cm²)	Transition temperature T_c' (°K)
1.2×10^{14}	16.5
2.4×10^{14}	15.8
7×10^{14}	14
1×10^{15}	13
1.5×10^{15}	9.5
5×10^{15}	2.5 ~ 3.0

temperature of the irradiated region T'_c as the intersection of the extrapolation to the temperature axis, then T'_c is 14.1 K for this weak link. In addition, if we plot $\ln I_c$ as a function of temperature as shown in Fig. 7, it can be easily observed that for $T > T'_c$ there is a linear relation between $\ln I_c$ and T. Hence, the critical current of this weak link is $I_c \propto (T'_c - T)^{3/2}$ for $T < T'_c$ and $I_c \propto \exp(-\gamma T)$ for $T > T'_c$. T'_c defined in this way depends on dosage as shown in Table I. These results agree well with our T_c measurements on large-area films directly radiated without Cu protection.

Measurements of the differential resistance dV/dI is more sensitive than a direct measurement of current-voltage characteristic. We have obtained dV/dI-vs-I characteristics for several weak links. Figure 8 shows the results for one weak link as compared to the theoretical prediction[7] assuming two different effective-noise temperatures. It has been shown[8] at lower temperatures with samples immersed directly in liquid He that a measuring system such as ours allows effective-noise temperatures equal to the substrate temperature to be obtained. If this is

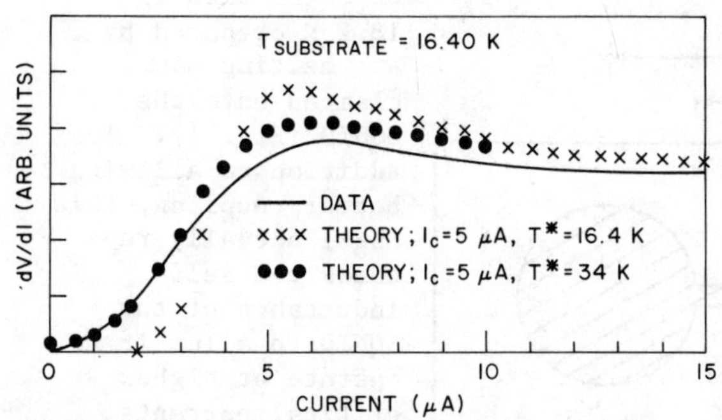

Figure 8. Differential resistance dV/dI as a function of I. Theoretical points are calculated from the theory of Ref. 7 assuming two different noise temperatures T*.

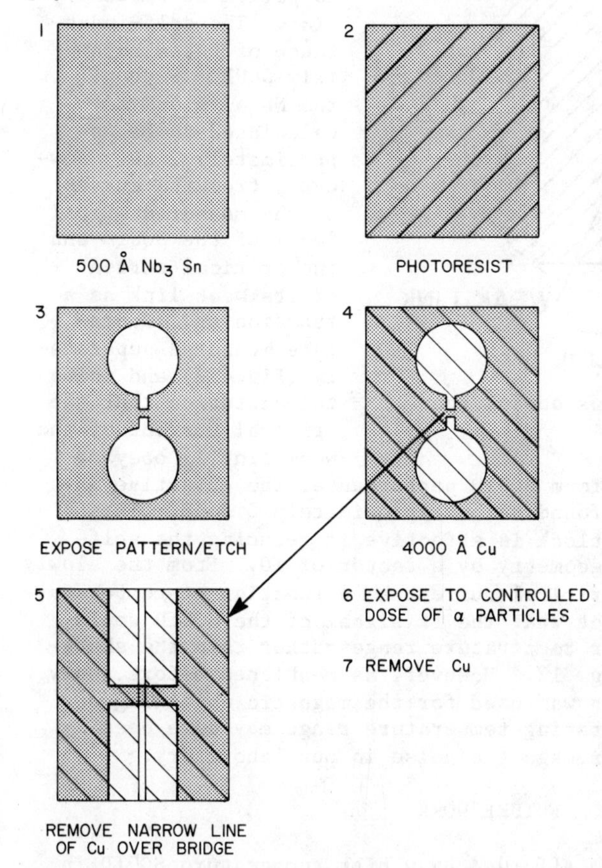

Figure 9. Steps for the fabrication of Nb_3Sn SQUIDs.

the case here, it indicates a much higher intrinsic noise in these weak links than in those made in low-T_c materials. However, it will be necessary to study Nb_3Sn weak links made to operate with low critical currents at 4.2 K in order to determine whether the large noise temperature is caused by the absence of direct contact with liquid helium or by large intrinsic noise in these weak links.

Nb_3Sn SQUIDS

There are six steps in the fabrication of Nb_3Sn SQUIDs as shown in Fig. 9. (1) Approximately 500 Å Nb_3Sn is deposited on a sapphire substrate; (2) the film is coated with photoresist, and (3) a two-hole pattern is etched using standard photolithographic techniques. (4) A protective Cu layer is deposited and (5) a narrow line is cut through the Cu layer over the bridge region. (6) The whole film is subjected to a controlled dose of α particles at 50 keV. (7) The Cu layer is then etched away after the irradiation. The detailed dimensions of the final SQUID are shown in Fig. 10. In order to couple external signals into the SQUID more effectively, a large block

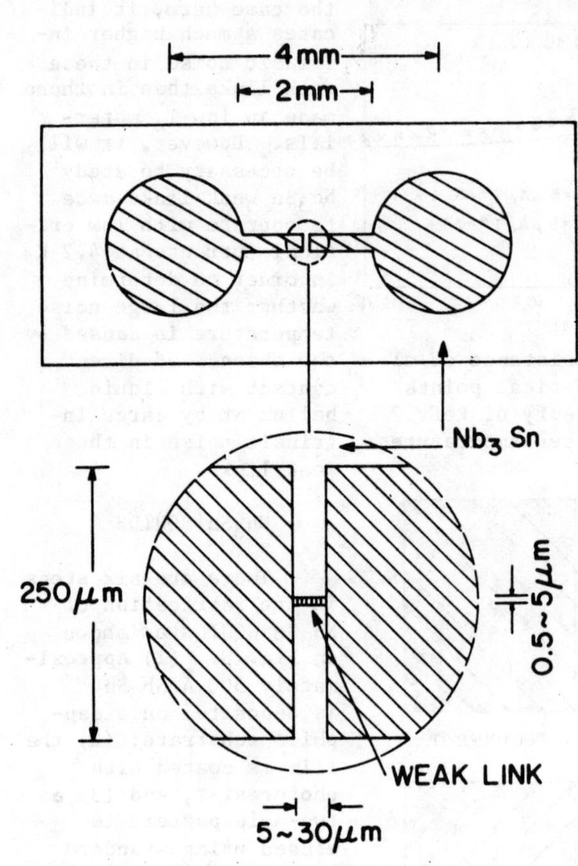

Figure 10. Detailed dimensions of the Nb_3Sn SQUIDs.

of Nb_3Al with $T_c \sim$ 18.2 K prepared by arc melting was clamped onto the SQUID (Fig.11). In addition to allowing better coupling, this Nb_3Al actually reduces the self-inductance of the SQUID to allow it to operate at higher critical currents. This block was polished with fine emery paper; however, no attempt was made to polish it optically flat. The self-inductance of these thin-film SQUIDs without the Nb_3Al block is calculated to be approximately 2 nH. However, from the curves of the measured signal level of the SQUID and the critical current of its weak link as a function of temperature measured separately (Fig. 12) and using the criterion that the critical current of the weak link I_c obeys a relation $I_c \sim \Phi_o/L$ for optimum SQUID performance, the effective inductance of the SQUID is found to be approximately 0.3 nH. This indicates that the Nb_3Al block is effective in reducing the self-inductance of the planar geometry by a factor of 10. From the slowly varying behavior of the critical current as a function of temperature (Fig. 12), one would expect that the rf signal of the SQUID would vary smoothly over a wider temperature range rather than the sharp-peak behavior shown in Fig. 12. However, as mentioned before, only a single-layer μ-metal can was used for the magnetic shielding of the SQUID so that the operating temperature range may have been limited by external electromagnetic noise in our laboratory.

FUTURE WORK

The ease of operation afforded by a high temperature SQUID in conjunction with a closed cycle refrigerator makes further work in this area especially valuable. All work to date has produced

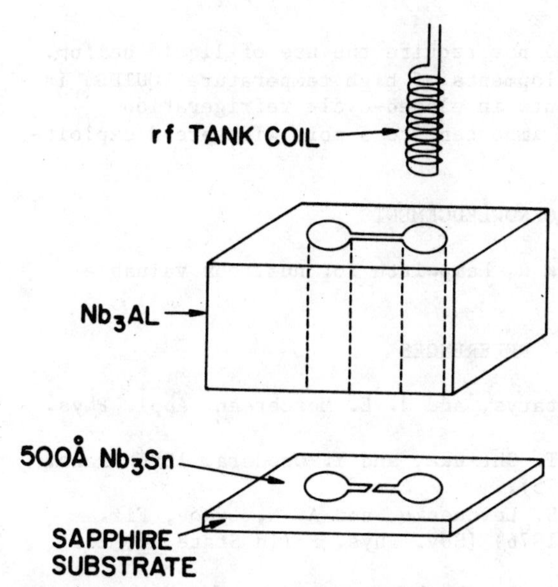

Figure 11. Schematic drawing of the planar SQUID coupled to an rf tank coil shielded with a Nb_3Al block.

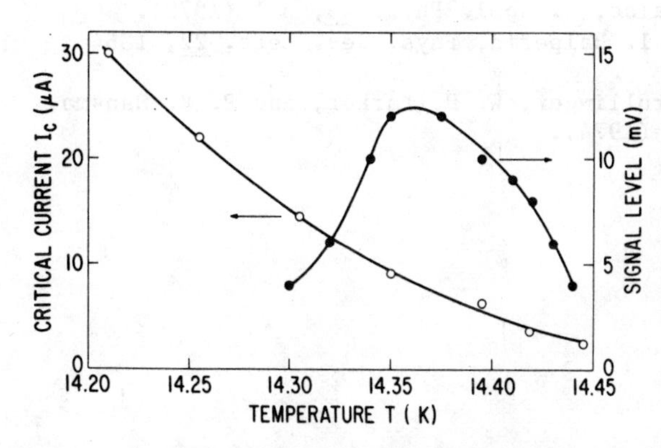

Figure 12. Measured signal level of a SQUID and the corresponding critical current as a function of temperature of its weak link.

devices which are either limited in sensing area or in operating temperature range. In addition, it is likely that the present weak links exhibit non-sinusoidal current-phase relations, leading to excess noise and a reduction in overall sensitivity.

For many applications, a reduced SQUID sensitivity can be tolerated. Here, effort should be made to find techniques for producing weak links operating over a wide temperature range; even if they exhibit non-ideal Josephson behavior. In the opposite extreme, it is also necessary to have high temperature SQUIDs with maximum achievable sensitivity within the limits that operation at high temperatures impose. Here, ideal Josephson weak links are required even if they only operate over a narrow temperature range. For convenience in temperature regulation, operation at the hydrogen triple-point temperature 13.96 K would be desirable.

Applications for SQUIDS reported to date in such orders as biomedicine, geophysics, rock magnetism, and communication systems have been almost exclusively initiated or developed by low temperature physicists. It seems clear that in order for there to be widespread use of these developments, or for there to be additional applications in areas, it will be necessary to provide complete stand-alone systems that do not require the use of liquid helium. Consequently, further developments of high temperature SQUIDS, in conjunction with developments in closed-cycle refrigeration systems, will certainly be important to a more widespread exploitation of SQUID technology.

ACKNOWLEDGEMENTS

We would like to thank R. Kampwirth for numerous valuable discussions.

REFERENCES

1. D. W. Palmer, H. A. Notarys, and J. E. Mercereau, Appl. Phys. Lett 25, 527 (1974).
2. T. Fujita, S. Kosaka, T. Ohtsuka, and Y. Onodera, IEEE Trans. Magn. MAG-11, 739 (1975).
3. A. I. Golovashkin, I. S. Levchenko, and A. N. Lykov, Fiz. Tverd. Tela 18, 3642 (1976) [Sov. Phys. Solid State 18, 2121 (1976)].
4. A. I. Golovashkin, I. S. Levchenko, A. N. Lykov, and L. I. Makhashvili, Pis′ma Zh. Eksp. Teor. Fiz. 24, 565 (1976) [JETP Letters 24, 522 (1976)].
5. C. T. Wu and C. M. Falco, Appl. Phys. Lett. 30, 609 (1977).
6. C. T. Wu and C. M. Falco, J. Appl. Phys. 49, 361 (1978).
7. V. Ambegaokar and B. I. Halperin, Phys. Rev. Lett. 22, 1364 (1969).
8. C. M. Falco, S. E. Trullinger, W. H. Parker, and P. K. Hansma, Phys. Rev. B10, 1864 (1974).

THE APPLICATION OF SQUIDS TO LOW TEMPERATURE TRANSPORT PHYSICS

J. C. Garland
The Ohio State University, Columbus, Ohio 43210

ABSTRACT

In a little over ten years, the sensitivity of low temperature transport experiments has increased from 10^{-9}V to 10^{-15}V. This increase in resolution, made possible by the development of SQUID voltmeters, has spawned several new avenues of research, such as the study of the interface resistance between superconducting and normal boundaries, and the study of transport effects in inhomogeneous superconductors near the percolation threshold. In addition, other research areas have become revitalized, as previously well established theories were found unable to withstand the scrutiny of ultra-high resolution measurements. The study of thermoelectricity in pure metals has developed rapidly during the decade, and new experimental techniques have been devised with resolution comparable to the best resistivity measurements.

I. INTRODUCTION

The rapid development of SQUIDs into routine laboratory instruments has been responsible for an enormous increase in the sensitivity and resolution of low temperature transport measurements. One need not have been a practicing physicist too long to be able to remember when the moving coil galvanometer was the mainstay of the low temperature laboratory. This paper will review some of the advances in transport physics which have been made possible by a sensitivity increase of approximately six orders of magnitude in a little over ten years. In keeping with the spirit of this conference, I will stress applications and measurement techniques rather than basic physical problems.

As an experimental objective, we might hope that ultimately the resolution of our measurements would be limited by the Johnson noise of the sample being studied and not by the inherent noise of the measuring apparatus. Figure 1 shows how close we have come to meeting that objective with a typical d.c. SQUID wired as a low impedance voltmeter. In this figure, I have assumed an apparatus noise level of 10^{-14}V and a bandwidth of 1 Hz; by careful attention to signal coil design, vibration isolation, and r.f. screening, it is possible to reduce the instrument noise level considerably, but this value probably represents a reasonable estimate for day-to-day operation in an experiment designed for general purpose usage. From the figure, we see that the Johnson noise in a 1 micro-ohm resistance at liquid helium temperatures can be easily resolved. This kind of sensitivity means that virtually all low temperature studies of transport effects in alloys, semi-metals, and

ISSN: 0094-243X/78/037/$1.50

semiconductors can, if necessary, routinely attain the Johnson noise limit. Furthermore, the price of this sensitivity is not

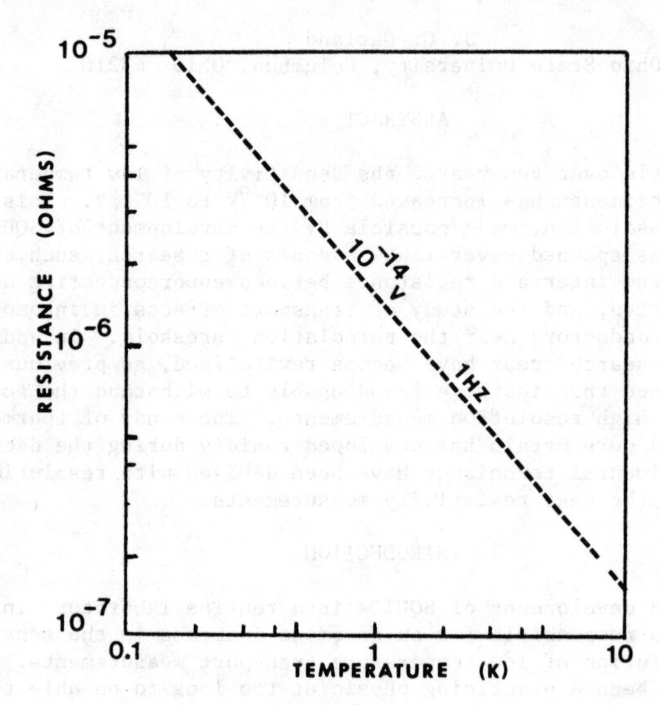

Fig. 1. Temperature dependence of the resistance required to produce a Johnson noise voltage of 10^{-14}V in a 1 Hz bandwidth. In an experiment whose instrument noise level is 10^{-14}V, resistance measurements above the curve will be Johnson noise limited; those below the curve will be instrument noise limited.

exorbitant. Although a commercial SQUID instrumentation system may cost several thousand dollars, a Clarke SLUG voltmeter[1] may be built by an industrious graduate student having access only to an oscilloscope and two hundred dollars worth of electronic components.

Before becoming too self-congratulatory, however, we should keep in mind that many of the important problems of transport physics pertain not to alloys and semiconductors, but to very pure metals and to metals containing superconducting constituents. In these highly conducting materials, we are usually well away from the Johnson noise limit. To illustrate this point, and to set the stage for subsequent discussion, let us consider a specific example.

A bar of high purity zone-refined aluminum with a residual resistance ratio of 25,000 will have a mean-free-path for impurity scattering of about 0.5mm. For electron-phonon scattering at 4.2K the mean-free-path will be about 5cm, and for electron-electron scattering it will be anywhere from 5cm to 20cm (depending on whose calculations one chooses to believe). It is obvious from these numbers that if we want to measure the transport properties of bulk aluminum (that is, aluminum whose transport coefficients are not seriously influenced by scattering from sample boundaries) we must use samples with diameters of at least several millimeters. The electrical resistance of this sample at 4.2K will be about 10^{-9} ohms, which is about a thousand times less than the contact resistance of the current wires joined to the ends of the sample. To minimize Joule heating effects at the ends of the sample and elsewhere in the cryostat, we must keep the sample current below about 100mA so that maximum signal voltages will be about 10^{-10}V. If we are to measure this voltage with 0.1 percent resolution, our voltmeter must have a minimum sensitivity of 10^{-13}V.

The thermal conductivity of our hypothetical aluminum specimen poses analogous experimental difficulties. The thermal conductivity of 25,000 residual resistance ratio aluminum at 4.2K is about 10^2 greater than the room temperature thermal conductivity, which makes the aluminum a far better thermal conductor than anything else in the cryostat. Should one attempt to develop a temperature difference greater than a few mK along the specimen (in order to measure the thermal conductivity or thermoelectric power), one is likely to find that everything else in the dewar is heating up and that large temperature gradients are appearing everywhere except along the sample.

As a final introductory remark, we note that one of the characteristics of studying ultra-low resistance materials is that the time scale for measurements is restricted by the inductance of the voltage probes connected to the sample. In this respect, the experimental problems are not too unlike those accompanying the design of electronic circuits at UHF frequencies. In each case one must take great pains to insure that circuit performance is not degraded by the stray inductance of interconnections.

II. GENERAL TRANSPORT EQUATIONS

In metals, the flow of electric current $\vec{J}$ and heat current $\vec{J}_Q$ can be described quite generally by two coupled linear transport equations

$$\vec{J} = \sigma\vec{E} + \frac{L}{T}(-\vec{\nabla}T) \tag{1}$$

$$\vec{J}_Q = L\vec{E} + \kappa(-\vec{\nabla}T). \tag{2}$$

In these equations, σ is the electrical conductivity, κ is the thermal conductivity, and L is a coefficient which characterizes thermoelectric conduction. (I have adopted a slightly unconventional definition for thermal conductivity. One ordinarily measures the thermal conductivity by electrically open-circuiting the sample; in Equation 2, however, κ is the thermal conductivity defined under conditions of zero electric field and not zero current. The difference between the two definitions is a small term of order $(k_BT/\mu)^2$

which is ignorable in most circumstances.)

There are two points to note about Equations 1 and 2. First, the three transport coefficients are scalars only in the absence of a magnetic field (and, strictly speaking, in metals with cubic symmetry). In strong magnetic fields the coefficients become highly anisotropic tensors. For example, in the limit of an infinitely strong magnetic field, the conductivity of simple metals becomes essentially one-dimensional; the metal conducts only along the direction of the magnetic field and is an insulator in other directions. Second, the three transport coefficients are not always independent of one another. Under conditions of elastic scattering, the thermal and electrical conductivities obey the Wiedemann-Franz law

$$\frac{\kappa}{\sigma T} = L_o \qquad (3)$$

where the Lorenz number $L_o = 2.44 \times 10^{-8} V^2 K^{-2}$. Under more general scattering conditions, the ratio $\kappa/\sigma T$ is smaller than L_o. Such relationships are frequently useful at very low temperatures where elastic impurity scattering is the dominant relaxation mode for transport currents. Let us now turn to some recent developments in measurements of transport coefficients in metals at low temperatures. I will not address the formidable problems of measuring thermal conductivities of very pure metals since these measurements ordinarily do not involve SQUIDs.

III. ELECTRICAL CONDUCTIVITY

Inasmuch as physicists have been measuring the electrical conductivity of metals at low temperatures ever since Kamerlingh Onnes liquified helium in 1908, one might reasonably wonder whether there is still anything of fundamental interest left to learn. Surprisingly, yes. The enormous improvement in sensitivity of superconducting voltmeters over conventional voltmeters has, in fact, opened several entirely new areas of experimental research. I cite as one example the physics of superconducting-normal interfaces. Measurement of an unusual boundary resistance between superconducting and normal metals in the past few years has shed new light on the nature of chemical potentials in superconductors and on the conversion of normal electrons into superconducting electrons.[2]

As another example, consider the emerging subject of inhomogeneous superconductivity.[3] Filaments of type II superconductor are frequently embedded in a normal metal matrix in order to minimize dissipative effects arising from the motion of flux lines. Below a critical concentration of superconductor, the composite material retains a remnant resistivity which is orders of magnitude below that of the normal metal constituent. This resistivity depends in a complicated way on the geometry of the superconducting filaments and on proximity effect coupling between filaments. It has recently been suggested that the onset of perfect conductivity in these systems may represent a class of critical phenomena with critical exponents depending only on the dimensionality of the system.[4] The role that SQUID instrumentation is playing in the investigations of this new area can be inferred from Figure 2, which shows some recent data of C. Lobb and M. Tinkham on the resistivity of a superconducting niobium-copper composite known as Tsuei wire; the region of interest ($R < 10^{-7}$ ohms) would be completely inaccessible by conventional electronics.

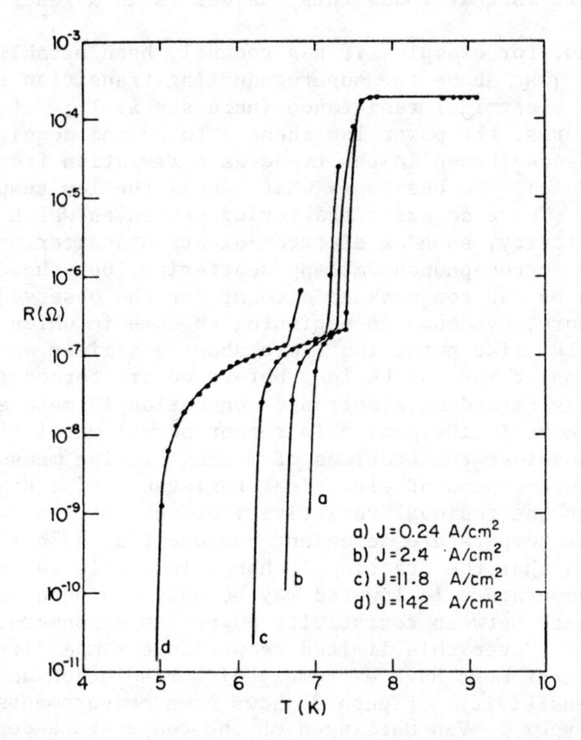

Fig. 2. Resistive transitions in a Tsuei wire composite containing 7% Nb filaments in a Cu matrix. The different curves correspond to measurements taken with different current densities. Magnetic impurities were added to the wire to suppress the proximity effect and thus narrow the transition. (C. Lobb and M. Tinkham, ref. 3).

As a final example, let us consider the temperature dependent electrical resistivity of the simple metals. This subject was ostensibly laid to rest in the 1950's when it appeared that the Bloch-Grüneisen law adequately explained existing electrical resistance data of most metals below Θ_D. According to this law, at low temperatures the resistance of simple metals (eg, potassium or aluminum) should vary as

$$\rho(T) = \rho_o + AT^5, \qquad (4)$$

the AT^5 term coming from the scattering of conduction electrons off of a Debye phonon spectrum, and the ρ_o term arising from elastic electron-impurity scattering. More sophisticated calculations of the electron-phonon resistivity in the 1960's and 1970's, which allowed for band structure effects, Umklapp processes, and anisotropic phonon spectra, suggested that power law dependences other than T^5 might be expected. Accordingly, during the past ten years, experimentalists have begun taking another look at the temperature dependent resistivity of these simple metals using high resolution SQUID instrumentation. The result is that today this subject is in a general state of turmoil.

In aluminum, for example, it has recently been established[5] that at temperatures just above the superconducting transition temperature (1.2K) the electrical resistance increases as T^2; at slightly higher temperatures, the power law changes to T^3 and acquires a weak impurity dependence (known in the trade as a deviation from Matthiessen's Rule). No one knows what causes the low temperature T^2 resistivity. There do exist scattering processes which produce a quadratic resistivity, such as electron-electron scattering and certain kinds of electron-phonon Umklapp scattering, but these effects are believed to be far too weak to account for the observed aluminum data. Furthermore, evidence is beginning to come in which suggests that other metals, like potassium[6], are showing similar anomalous resistivities, so it may not be long before we are forced to rethink many of our ideas regarding electronic conduction in metals.

What role have SQUIDs played in recent normal metal transport experiments? Consider the problems of making precise measurements of the temperature-dependent electrical resistance of a high purity metal. Although the residual resistivity of the metal may be 10^{-9} ohm-cm, the temperature dependent component at 4.2K may be only 10^{-11} ohm-cm, so that the fractional change in sample resistance as the specimen temperature is lowered may be only one percent. If we are to distinguish between resistivity power law exponents, for example T^2 vs. T^3, over this limited temperature range, it is clear that our experiment must have extremely high resolution as well as high sensitivity. Figure 3 shows some measurements taken in my laboratory by D. J. Van Harlingen of the temperature-dependent resistivity of aluminum. The purpose of the measurements was to distinguish between T^2 and T^3 resistivity between 1K and 4K.

The apparatus for these measurements (Figure 4) was built in 1970 using a Clarke SLUG as a null detector in a potentiometric configuration, germanium resistance thermometers for measuring temperature differences, and a five place digital voltmeter for monitoring current through the sample and the standard resistor. Although this apparatus, which was capable of measuring the thermoelectric coefficient and thermal conductivity as well as electrical resistivity, was relatively sophisticated when it was built, it was not capable of the high resolution now needed for precise low temperature transport measurements.

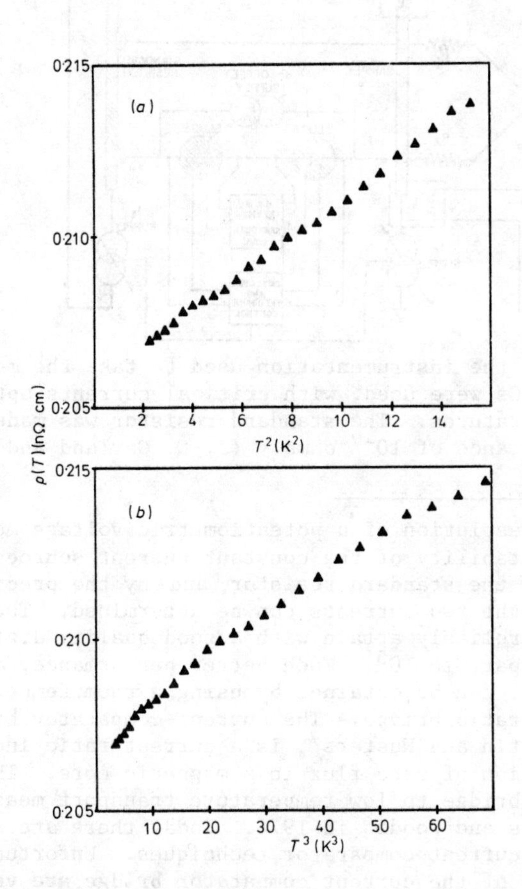

Fig. 3. The temperature-dependent electrical resistivity of aluminum plotted (a) as a function of T^2 and (b) as a function of T^3, illustrating the superiority of the T^2 fit. (J. C. Garland and D. J. Van Harlingen, ref. 5)

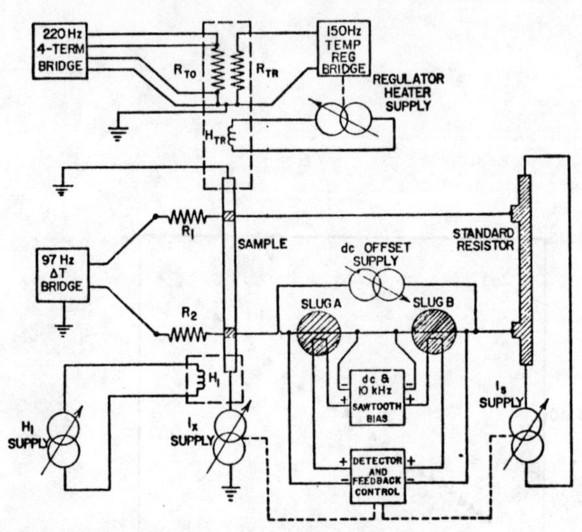

Fig. 4. Diagram of the instrumentation used to take the measurements of Fig. 3. Two SLUGs were used, with critical currents optimized for different temperatures. The standard resistor was made of OFHC copper with a resistance of 10^{-7} ohms. (J. C. Garland and D. J. Van Harlingen, ref. 5)

The ultimate resolution of a potentiometric voltage measurement is limited by the stability of the constant current sources which feed the sample and the standard resistor and by the precision with which the ratio of the two currents can be determined. The highest resolution one can reliably attain with a good quality digital voltmeter is about one part in 10^5. Much better performance, approaching one part in 10^7, can be obtained by using a room temperature current-comparator ratio bridge. The current-comparator bridge, developed by MacMartin and Kusters[7], is a current ratio indicator based on the detection of zero flux in a magnetic core. The first application of the bridge to low temperature transport measurements was made by Rowlands and Woods[8] in 1976. Today there are several laboratories using current comparator techniques. Unfortunately, commercial versions of the current comparator bridge are very expensive, costing in excess of $10,000; it is to be hoped that more economical ways of making precise current ratio measurements can be developed, inasmuch as there is today a real need for additional high-resolution low temperature measurements of the resistivity of metals.

IV. THERMOELECTRICITY

It is well known that of the three basic transport coefficients the thermoelectric coefficient L is most sensitive to the details of electronic scattering. It is also the coefficient which, in pure metals, has been the least investigated; the reason for this lack of information on an obviously interesting quantity is not hard to discern if one looks at the numbers.

The coefficient L has, to my knowledge, never been measured directly, since to do so would require driving an electric current with a temperature gradient while at the same time insuring that no thermoelectric fields were present (a difficult but not impossible task). Instead, it has been customary to measure L in combination with the electrical conductivity. The thermoelectric power is defined as the ratio $L/\sigma T$, which by setting the left hand side of Equation 1 to zero can be seen to be given by

$$S = \frac{L}{\sigma T} = \left(\frac{E}{\nabla T}\right)_{J=0} \qquad (5)$$

In pure metals at liquid helium temperatures S is typically 10^{-7} to 10^{-8} V/K and decreases to zero as the temperature is lowered. Since, for reasons stated previously, it is difficult to sustain a temperature difference of more than a few mK across a pure metal, the thermoelectric voltages one must measure are only 10^{-10} to 10^{-11}V. In order to achieve a one percent measurement resolution, a voltmeter sensitivity of at least a picovolt is required. Quite obviously, the development of SQUID voltmeters has provided access to an area of research which was almost totally inaccessible to other methods; as might be expected, there has been a great deal of experimental research into thermoelectricity in just the past few years which would have been impossible a decade ago.

Even given the great voltage sensitivity now available to low temperature experimentalists, the measurement of thermoelectric powers still requires knowledge of very small temperature differences. Germanium resistance thermometers in a differential bridge configuration can resolve temperature differences of about 10μK. This resolution limits the precision of a thermoelectric power measurement of a typical pure metal at 4.2K to about one percent -- not too bad, perhaps, but nowhere near the precision with which conductivities can be measured. Furthermore, the situation becomes much worse at lower temperatures, not only because the thermoelectric power decreases to zero at T=0, but also because one is forced to reduce the heat load on the cooling system (dilution refrigerator, ^{3}He cryostat, etc.). In short, it is nearly impossible to measure the thermoelectric power of a pure metal below 1K, even using SQUID instrumentation.

As a way around the experimental difficulties associated with thermopower measurements, I will cite as a final example a new approach to thermoelectricity which has found increasing use in the past few years. I suggested this approach originally in 1973 and am discussing it here both because I am naturally familiar with it and also because it depends fundamentally on SQUID instrumentation; the method cannot work with non-superconducting instruments[9].

According to this new approach, neither voltage differences nor temperature differences are measured; instead, the method involves the measurements of two currents, a heat current J_Q and electric current J, while a SQUID is used as a superconducting galvanometer. We do not obtain the thermoelectric power from this method but instead a related thermoelectric function G. This function G exhibits the same sensitivity to scattering as the thermopower S, but it does not drop to zero at T=0 and can be measured with orders-of-magnitude greater resolution than the thermopower. G is defined as the ratio of two transport coefficients:

$$G \equiv \frac{L}{\kappa T} \qquad (6)$$

This expression can be compared with the analogous expression for the thermoelectric power in Equation 5, and it should be noted that each expression is proportional to the thermoelectric coefficient L. If we divide the two transport equations (Equations 1 and 2) by each other, setting $\vec{E} = 0$, we obtain

$$G = \left(\frac{J}{J_Q}\right)_{\vec{E}=0} \qquad (7)$$

Thus it is possible to make measurements of G by applying a heat current J_Q and an electric current J in just the proportion required to null out the electric field within the sample. Before discussing how this measurement might be performed, I might mention some interesting properties of the G coefficient. First, G is measured when there is an electric current in the metal but no electric field. (By "electric field" I actually mean "effective electric field", which is the negative gradient of the electrochemical potential.) As far as the source of electric current is concerned, therefore, the metal appears to be a perfect conductor. In fact, there is no need for the current source to supply any energy to the sample because there is no Joule heating; the only source of energy comes from the process of heat conduction down the sample. The energy ordinarily dissipated as Joule heat has not vanished, of course, but appears as an additional load on the heat reservoir which supplies the heat current. Second, there is a reversible Thomson heat associated with measurements of G which is proportional to $\vec{J}\cdot\vec{\nabla}T$. This heat is negative for free electron metals and, when the details are worked out[10], is found to have the same magnitude but opposite sign as the Joule heat. Thus we see that in simple metals the Joule heat associated with a measurement of G is just swallowed by the Thomson heat. As a final remark I note that, unlike the thermoelectric power, G is not required to vanish at T=0; thermodynamic considerations only require GT=0 at T=0, so that G may actually diverge at low temperatures so long as the power law dependence is weaker than T^{-1}. In a free electron metal, G is expected to have a temperature-independent value of $3e/2\mu$.

Figure 5a shows a simplified schematic representation of an experiment to measure G in pure metals. One end of a sample is anchored to a heat reservoir at temperature T_0 while the other end is connected to a heater. When the heater is turned on, a heat current flows through the sample and sets up an electric field given by $S\vec{\nabla}T$. This electric field is short circuited by a superconducting wire containing a SQUID galvanometer. The current through the wire ideally would be equal to the open circuit thermoelectric voltage $S\Delta T$ divided by the sample resistance, although in practice the current is reduced substantially by the contact resistance between the superconducting loop and the sample. The galvanometer "deflection" is amplified by a current amplifier and used to control a constant current source. This current source supplies a feedback current J to the sample which restores the SQUID galvanometer to equilibrium. The quantities which are measured in the experiment are therefore the feedback current J and the heat current J_Q, their ratio giving the G coefficient. Note that the SQUID is used only as a null detector and that it is not necessary to measure any temperature gradient in the experiment. Figure 5b shows early measurements of G on high purity tungsten taken in my laboratory.[9] Some of these measurements have recently been extended to millikelvin temperatures by Pratt, et al. at Michigan State University[11]; this group found that magnetic impurities in tungsten can have a profound influence on the temperature dependence of G below 100mK.

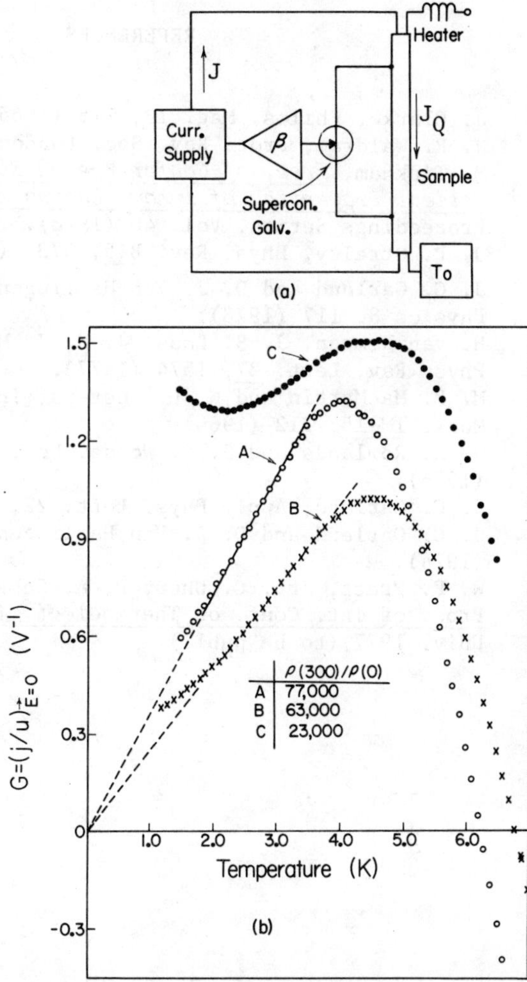

Fig. 5. (a) Schematic diagram of the potentiometric circuit used to measure the thermoelectric coefficient G. (b) Temperature variation of G for high purity tungsten. (J. C. Garland, ref. 9)

REFERENCES

1. J. Clarke, Philos. Mag. 13, 115 (1965).
2. J. R. Waldram, Proc. Roy. Soc. London A345, 231 (1975).
3. M. Tinkham, *Proc. of Conference on Electrical Transport and Optical Properties of Inhomogeneous Media,* APS Conference Proceedings Series, Vol. 40 (1978).
4. J. P. Straley, Phys. Rev. B15, 5733 (1977).
5. J. C. Garland and D. J. Van Harlingen, J. Phys. F: Metal Physics 8, 117 (1978).
6. H. van Kempen, J. S. Lass, J. H. J. M. Ribot and P. Wyder, Phys. Rev. Lett. 37, 1574 (1977).
7. M. P. MacMartin and N. L. Kusters, IEEE Trans. on Instr. and Meas. IM-15, 212 (1966).
8. J. A. Rowlands and S. B. Woods, Rev. Sci. Instrum. 47, 795 (1976).
9. J. C. Garland, Appl. Phys. Lett. 22, 203 (1973).
10. J. C. Garland and D. J. Van Harlingen, Phys. Rev. B10, 4825 (1974).
11. W. P. Pratt, Jr., C. Uher, P. A. Schroeder, and J. Bass, Proc. of Int. Conf. on Thermoelect. in Metals, Mich. State Univ. 1977 (to be publ.)

RECENT ADVANCES IN SQUID NMR TECHNIQUES*

R. A. Webb
Argonne National Laboratory, Argonne, Illinois 60439

ABSTRACT

A discussion of the techniques used for performing a variety of magnetic resonance experiments using a SQUID is presented. In particular, measurements of the nuclear magnetism of liquid ^{3}He in the temperature range 1.2 K to 2 mK in magnetic fields of 3.7 to 309 Oe using pulsed NMR, cw NMR, adiabatic fast passage, and static susceptibility using the same detection system will be discussed.

INTRODUCTION

In the last ten years a great deal of progress has been made in the experimental understanding of superconducting devices based on the Josephson effect.[1] This in turn has led to a wide variety of applications, primarily in the measurement of weak magnetic phemomena. One area of importance that has not received a great deal of attention is in the use of a SQUID for the detection of nuclear magnetic resonance (NMR) signals. The ability to isolate and study a single nuclear spin species combined with the sensitivity of the SQUID should allow new and interesting magnetic phenomena to be investigated under experimental conditions that previously made a conventional NMR experiment impossible.

In this paper a discussion is presented of the techniques used for performing pulsed NMR, cw NMR, adiabatic fast passage as well as static susceptibility measurements using the same SQUID-based detection system in magnetic fields as low as 3.7 Oe in the temperature range 4.2 to 0.002 K. SQUID NMR is not new. One of the first NMR uses of a superconducting weak link was by Silver et al.[2] who observed the rf absorption in the ^{59}Co nuclear spin system. Both the adiabatic fast passage and cw NMR measurements reported to date[3,4] used a SQUID to detect the change in the average value of the z-component of magnetization as the frequency of a transverse H_1 field was swept through resonance. Only recently have experiments using pulsed SQUID NMR been reported[5,6] even though the advantage of this technique over conventional methods is the most apparent. Following a single rf pulse, both a complete record of the relaxation process and the value of susceptibility are obtained, independent of dephasing times. A detailed description of SQUID NMR techniques can be found elsewhere.[7]

*Work performed under the auspices of the U.S. Department of Energy.

ISSN: 0094-243X/78/049/$1.50

APPARATUS

The measurements were made in an adiabatic demagnetization cell using Cerous Magnesium Nitrate (CMN) as the low temperature refrigerant. This cell is shown schematically in Fig. 1. It was thermally connected to the mixing chamber of a dilution refrigerator via a Zn heat switch. The ^{3}He measured was contained in a 3 mm i.d. tower located above the main CMN. The static field H_o, parallel to the axis of the tower, was trapped in a 10.9 mm i.d. Nb tube. A 42-turn rf saddle coil, wound on a diameter of 5 mm using 0.076 mm Nb wire, provided the transverse rf field, H_1. The change in the ^{3}He magnetization was sensed by a 3 mm long Nb pickup coil wound astatically on a 4.3 mm diameter. The pickup coil was connected to the superconducting input coil of a two-hole symmetric SQUID operated in the flux-locked loop[8] configuration. Temperatures were determined, with the aid of a second SQUID, by 17 Hz mutual inductance measurements[8] on 10 mg of CMN located in a second magnetically shielded tower.

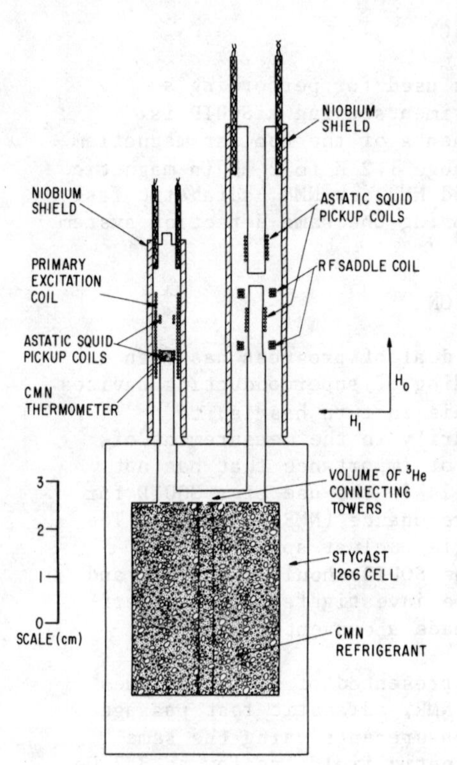

Fig. 1. Schematic drawing of the demagnetization cell used in this work.

In order to achieve the optimum signal-to-noise ratio, the H_1 field must be kept from coupling to the SQUID. In the earlier SQUID NMR experiments[5,6] the H_1 field was shielded from the SQUID detection coils by a thin, high conductivity metallic shield. The disadvantage of this technique is that the pickup coil must be located outside the eddy current shield and results in a lower detection sensitivity. Also the L/R time constant of the shield may severely limit the frequency response of the system. Furthermore, the Johnson noise in this shield will increase the background noise of the system, and the magnetic impurities in the shield may interfere with the static susceptibility measurements. For ultra-low temperature use, there is an additional problem of eddy current heating. The technique employed in this work was to install a small 1 Ω resistor, constructed from manganin wire, in parallel with the superconducting pickup coil and SQUID input coil.

This resistor was thermally grounded to the mixing chamber of the dilution refrigerator. Thus the Johnson noise temperature could be minimized and actually contributed a negligible amount of noise at 10 mK when compared to the intrinsic noise of the SQUID and associated electronics.[8] The value of the time constant was chosen carefully so as not to distort any signal within the full bandwidth of the flux-locked loop and was calculated to be 2.6×10^{-6} sec.

A block diagram of the measuring system used in this work is shown in Fig. 2. The SQUID measures the change in the average value of the z-component of the magnetization, either as the frequency of the rf field is swept through resonance or after the application of an rf pulse. For either the cw NMR measurements or the adiabatic fast passage measurements, the output of the flux-locked loop was fed to the vertical channel of an x-y recorder while the horizontal channel was driven by a voltage proportional to the frequency of the H_1 field. The output of the generator was fed through a large resistor to the H_1 coil. For pulsed work the generator was operated in the gated mode with the pulse width determined by a pulse generator. The output of the flux-locked loop was then recorded on either a strip-chart recorder or an oscilloscope. In low static fields it was expected that the effectiveness of the 1 Ω resistor in keeping the rf field away from the SQUID would be reduced. Consequently, a small nulling mutual inductor with a superconducting secondary was placed in series with the pickup-coil and input-coil to the SQUID. Its purpose was to cancel most of the rf current in the superconducting input circuit. The rf null electronics was to provide both the in-phase and out-of-phase signal fed to the nulling mutual inductor. However, even in 3.7 gauss this null scheme proved unnecessary and was not used.

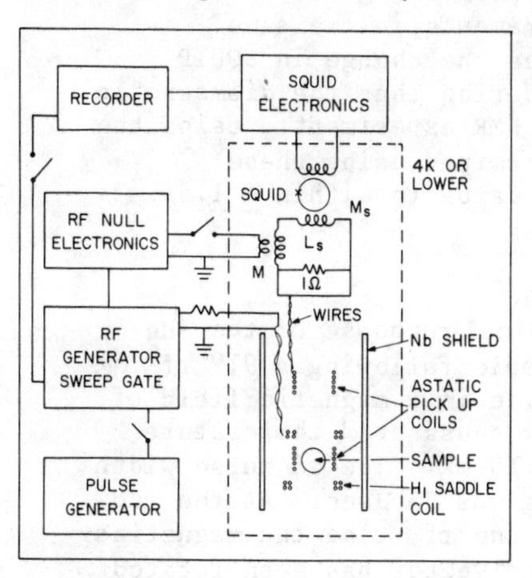

Fig. 2. Block diagram of the measuring system.

CALIBRATIONS

The absolute calibration of the SQUID magnetometer was performed in the temperature range 0.3 to 1.2 K using three different techniques. The first was to measure the change in flux coupled to the SQUID, in a field H_0, as the liquid ^3He sample slowly filled the region containing the detection coils. The change in flux $\Delta\phi$ coupled to the SQUID is equal to

$$\Delta\phi/\phi_o = \lambda(\chi_p + \chi_D) \quad , \qquad (1)$$

where χ_p and χ_D are the paramagnetic and diamagnetic susceptibilities of ^{3}He respectively, and λ is the calibration constant in the field H_o. Using the known value for χ_p[9] and from measurements of $\Delta\phi$ obtained at many different temperatures, it was determined that $\lambda = (6.38 \pm 0.15) \times 10^7$ and χ_D (V/N) = $-2.08 \pm 0.05 \times 10^{-6}$ cm^3/mole, where V/N is the molar volume of ^{3}He. A check on the value of λ was made by performing both adiabatic fast passage and 180° pulse NMR measurements in the same temperature range. For these two cases the change in SQUID output should be $\Delta\phi/\phi_o = 2\lambda\chi_p$, remembering that the diamagnetic susceptibility is not coupled to in a NMR experiment. Using the known value of χ_p, the value of λ determined using these two techniques agreed with the static value to within ± 1.5%.

PULSED NMR

Figure 3 shows a tracing of a typical response of the ^{3}He sample following a 91° rf pulse in a magnetic field of 309 gauss at a temperature of 15 mK. The rf pulse width was was 25 μsec. At the end of the rf pulse the magnetization vector has been rotated by 91°, but this initial change was faster than the flux-locked loop could respond and it momentarily lost lock. The new lock point is in general not simply related to the old one and cannot be used for the baseline determination.

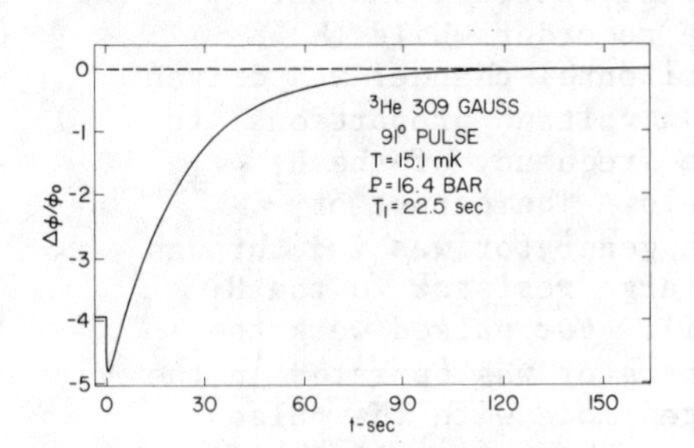

Fig. 3. Response of SQUID magnetometer following a 91° rf pulse.

However, shortly after the magnetization change has occurred, the SQUID can continually track the recovery of the magnetization back toward equilibrium. The baseline is determined by a straight line extrapolation of the end of the recovery process back to the beginning of the trace. The absolute value of the magnetization change can be determined by an extrapolation of the recovery curve back to the time of the end of the rf pulse.

The calibration of the current into the rf coil necessary to produce a known rotation of the magnetization vector is determined by measuring the change in magnetization obtained for a number of different currents. A typical calibration is shown in Fig. 4, where the extrapolated flux change immediately following a 24.5 μsec rf pulse is plotted as a function of the voltage applied to the superconducting rf coil. The impedance, z,

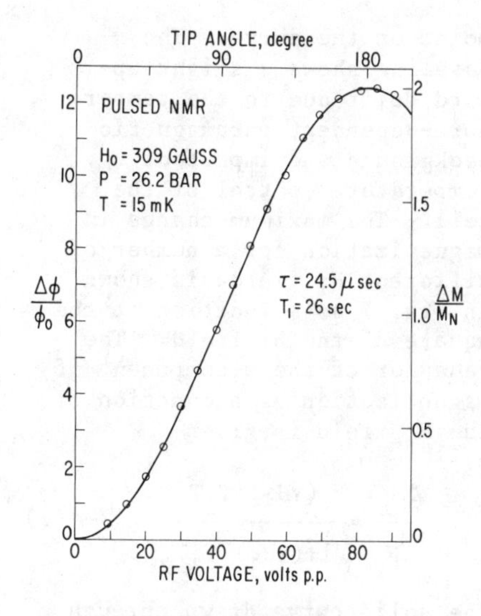

Fig. 4. Maximum change in flux coupled to the SQUID as a function of the applied rf voltage to the H_1 coil for a time $\tau = 24.5$ μsec.

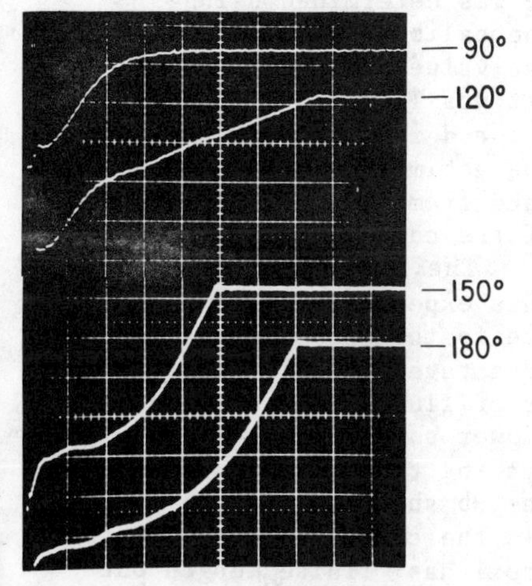

Fig. 5. Four examples of recovery of the magnetization of superfluid ^{3}He obtained at 2 mK in a 180 gauss field. The time scale for the 90° and 120° data is 0.5 msec/cm and 2.0 msec/cm for the 150° and 180° data.

of this coil was about 500 Ω in this field. The solid curve is a fit to the data using $M = M_N(1 - \cos\theta)$. The value of M_N was determined using the calibration constant λ. The relationship between the magnitude of the H_1 field and the rotation angle is $H_1\tau = \theta\pi/180°$, where τ is the pulse width. The maximum pulse width is determined by the inhomogeneity in H_o over the sample region and must be kept short enough so as to completely couple to all the sample spins in the detection region. An example of the real advantage of SQUID pulsed NMR is shown in Fig. 5 where four typical traces of the recovery of the magnetization in superfluid ^{3}He-B are shown[10] at a temperature of 2 mK for four different rotation angles. Relaxations that have exponential, linear, and square-root time dependences are observed and easily studied using a single rf pulse.

CW NMR

A typical cw NMR trace obtained in 309 gauss at 15 mK is shown in Fig. 6. This trace was obtained by sweeping the frequency of the H_1 field through resonance in the time shown. The two peak resonance curve is due to the gradient in magnetic field over the sample and will be discussed below. The direction of the sweep was from low frequencies to high frequencies. The high-frequency tail on the resonance curve is due to the fact that the sweep rate was slightly faster than the rate required to establish equilibrium conditions at each

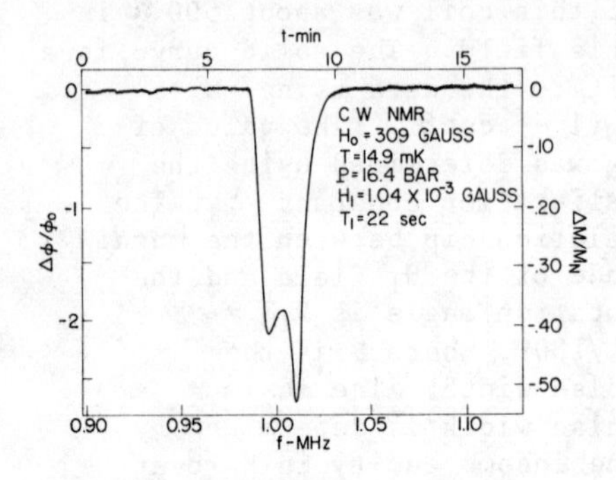

Fig. 6. Tracing of a typical cw NMR experiment.

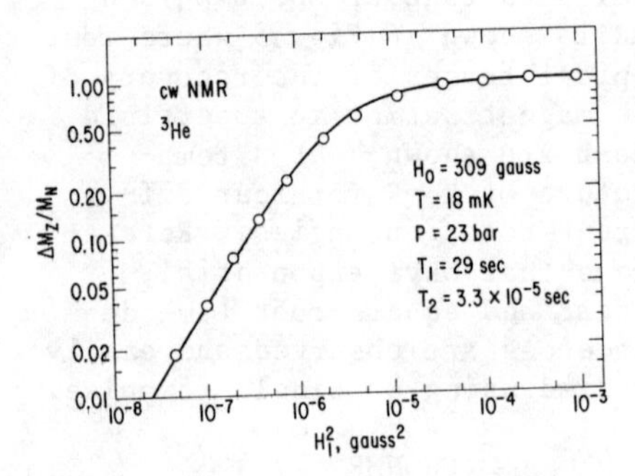

Fig. 7. Maximum change in magnetization occurring during a NMR experiment as a function of $H_1{}^2$.

point on the curve. The baseline shows a slight upward drift due to the temperature-dependent paramagnetic background and imperfect temperature control of the cell. The maximum change in magnetization for a number of different H_1 fields is shown in Fig. 7 as a function of the square of the H_1 field. The behavior of the z-component of magnetization as a function of the H_1 field is given by[11]

$$\frac{\Delta M_z}{M_N} = \frac{(\gamma H_1)^2 T_1 T_2}{1+(\gamma H_1)^2 T_1 T_2} \quad . \tag{2}$$

The solid curve drawn through the data is a fit that uses Eq. (2) with only T_2, the spin-spin relaxation time, as a free parameter. T_1 was measured using pulsed NMR and M_N was determined using the calibration constant. The value of T_2 used in the fit was $T_2 = 3.3 \times 10^{-5}$ sec and is in agreement with the estimate of T_2 one can make from the observed width of the curve displayed in Fig.6.

The linewidth observed in this experiment is completely due to the inhomogeneity in H_o over the sample and is the major disadvantage of SQUID NMR. The Nb shield traps an integral number of flux lines, and the field H_o at any cross section of the tower should be $H_o = d\phi/dA_{eff}$, where ϕ is the total flux threading the cylinder and A_{eff} is the cross-sectional area of the Nb shield minus the area of any superconducting material in the cross section considered. The superconducting pickup coil has finite length but is also sensitive to magnetization changes occurring outside the coil. It is expected that a two-peak resonance curve should be observed because of the nearly discontinuous change in H_o on going from the region just outside the coil to the region just inside the coil. The field change is calculated to be H = 0.013 H_o for the present experiment, while the measured peak separation is H = (0.016 $\pm$ 0.0005) H_o.

ADIABATIC FAST PASSAGE

Adiabatic fast passage is a cw technique used to invert the magnetization vector by applying a large H_1 field and quickly sweeping the frequency of the rf field through resonance. The conditions for an adiabatic fast passage in a SQUID NMR experiment are[3,7]

$$|\gamma H_1|^2 >> \frac{d\omega}{dt} >> \frac{\gamma \Delta H}{T_1} \quad , \qquad (3)$$

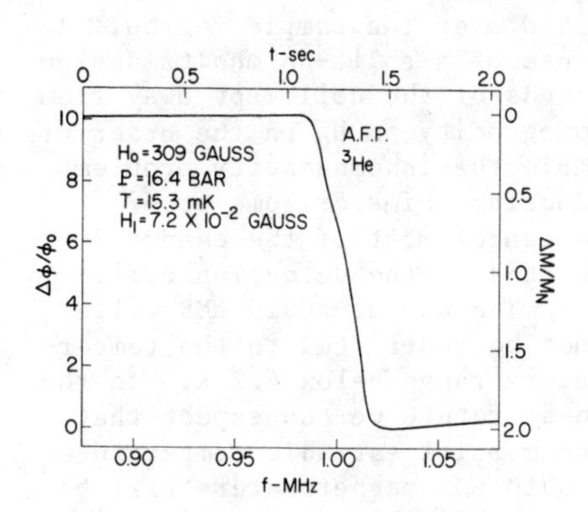

Fig. 8. Tracing of a typical adiabatic fast passage experiment obtained on He[3].

where ΔH is the inhomogeneity in H_o and T_1 is the spin lattice relaxation time. The first part of the inequality insures that the magnetization vector will be reversed adiabatically,[11] and the second part insures that the resonance curve is traversed in a time short compared to T_1. A typical adiabatic fast passage experiment on ^{3}He at 15 mK is shown in Fig. 8. Only the passage through resonance is shown and not the recovery of the magnetization back toward equilibrium. The experimental conditions under which this trace was obtained are

$$|\gamma H_1|^2 = 2.2 \times 10^6 >> \frac{d\omega}{dt} = 6 \times 10^5 >> \frac{\gamma \Delta H}{T_1} = 8 \times 10^3 \qquad (4)$$

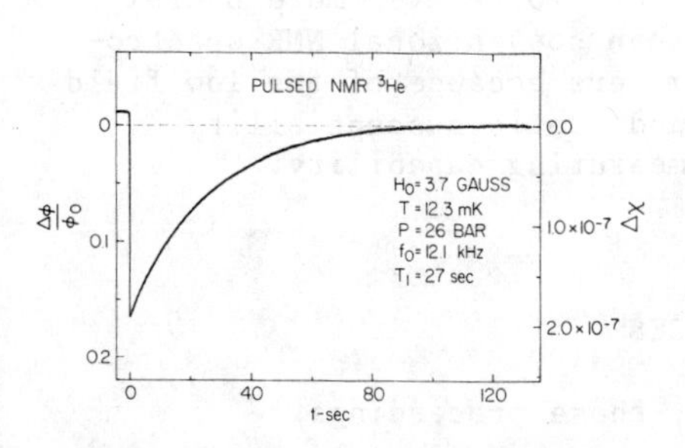

Fig. 9. Tracing of a pulsed NMR experiment in 3.7 gauss.

where the value of ΔH used was the full experimental linewidth. After the rf field has been swept through resonance, the spin lattice relaxation time as well as the value of the susceptibility the sample can be obtained.

FUTURE OF SQUID NMR

An example of the extreme sensitivity of the SQUID for the detection of nuclear magnetism is shown in Fig. 9. A tracing is shown of a pulsed NMR experiment performed in 3.7 gauss at a temperature of 12 mK following a nominal 180° pulse. The H_1 field was 0.154 gauss, and the pulse width was 1×10^{-3} sec. The change in susceptibility,

ΔX, is shown on the right-hand vertical scale. The signal-to-noise ratio for this trace was better than 100/1, and the bandwidth was 10 Hz. This trace demonstrates that zero field NMR is possible on some solids, using only the local field of the atoms for orientation. Although SQUID NMR is more sensitive than conventional NMR at low fields, in very large magnetic fields conventional NMR should in general be more sensitive because the detection sensitivity[7] is proportional to $\omega_o^{3/2}$ as compared to ω_o for SQUID NMR.

The major technological problem that must be solved is that of the inhomogeneity in the static field over the sample region. I believe that for a small sample, the use of a split-Helmholtz design for the SQUID pick up coil, with the ends of the coil kept away from the sample, can easily result in a homogeneity in H_o on the order of 1×10^{-4}. For the case of a long sample the inhomogeneity problem is more severe. The use of superconducting shims or some other suitable technique may be required to cancel most of the change in the effective area occurring near the ends of the detection coil.

The use of SQUID NMR will not be restricted to the temperature range below 4.2 K. In the near future we can expect that commercial variable temperature SQUID NMR magnetometers will be available that operate in the temperature range 1.2 K to 400 K. Figure 10 shows a schematic diagram of such a proposed variable temperature SQUID NMR magnetometer. If the frequency response of the system can be increased to 0.1 to 1 μ sec, this type of instrument may prove to be even more useful than conventional NMR spectrometers because of the low field and static susceptibility measuring capability.

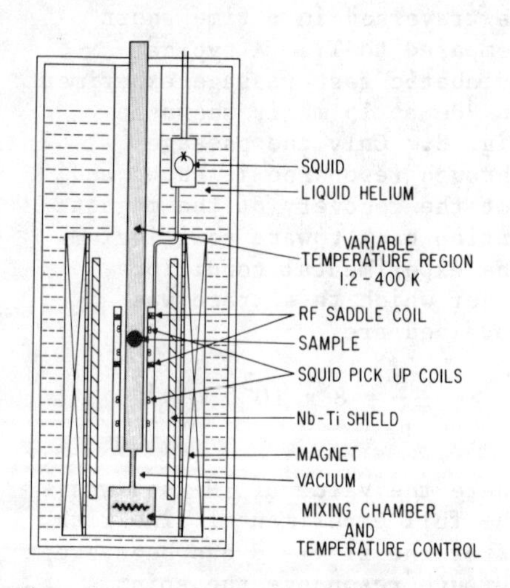

Fig. 10. Schematic drawing of a 1.2 K to 400 K variable temperature SQUID NMR magnetometer.

REFERENCES

1. R. P. Giffard and M. B. Ketchen, these proceedings.

2. A. H. Silver and J. E. Zimmerman, Appl. Phys. Lett 10, 142 (1967).

3. E. P. Day, Phys. Rev. Letters 29, 540 (1972).

4. D. J. Meredith, G. R. Pickett, and O. G. Symko, J. Low Temp. Phys. 13, 607 (1973).

5. R. A. Webb, Phys. Rev. Letters 38, 1151 (1977).

6. R. E. Sager, R. L. Kleinberg, P. A. Warkentin, and J. C. Wheatley, Phys. Rev. Letters 39, 1343 (1977).

7. R. A. Webb, Rev. Sci. Instr. 48, 1585 (1977).

8. R. P. Giffard, R. A. Webb and J. C. Wheatley, J. Low Temp. Phys. 6, 533 (1971).

9. H. P. Ramm, P. Pedroni, J. R. Thompson and H. Meyer, J. Low Temp. Phys. 2, 539 (1970).

10. R. A. Webb, to be published.

11. A. Abragam, "The Principles of Nuclear Magnetism" (Clarendon Press, 1961).

SQUIDS FOR MEASURING THE MAGNETIC PROPERTIES OF MATERIALS*

B. S. Deaver, Jr., Thomas J. Bucelot and James J. Finley
Physics Dept. University of Virginia, Charlottesville, VA 22901

ABSTRACT

SQUID magnetometers in conjunction with superconducting magnets and shields are now being used for susceptometers that provide unique combinations of sensitivity, speed, resolution, accuracy, stability and convenience. They are being integrated into systems with automated data acquisition and closed cycle helium refrigerators. These systems promise to become increasingly important for measurements of the magnetic properties of materials.

INTRODUCTION

The unique capabilities of superconducting susceptometers depend on combining the extremely high sensitivity and resolution of SQUID magnetometers with superconducting magnets and shields which are capable of providing the stable magnetic environment in which the SQUID characteristics can be exploited. For more than ten years it has been evident that devices combining these attributes would have great potential for measurements of the magnetic properties of materials. During the last year a susceptometer has been reported that has sensitivity, speed and resolution probably exceeding that of all other devices for magnetization measurements.

In this paper we briefly trace the development of superconducting susceptometers over the last ten years. We describe the principles and performance of current instruments and give some examples of measurements being made with them. Finally we describe some possibilities for the future.

BACKGROUND

The first superconducting susceptometers[1,2] were built just before the advent of the SQUID. They made use of modulated inductance detectors, devices that thermally switched a superconducting rod or cylinder between the superconducting and normal states to pump flux between two coils. It is interesting that the model[2] used to describe these devices is also quite useful for discussing the operation of SQUIDs since the Josephson junction can be regarded as a kinetic inductor that is varied cyclically in the SQUID.

Almost from its inception[3] the SQUID has been used for measurements on magnetic samples, the first measurement being made by Zimmerman and Mercereau[4] who observed quantized flux pinned in solid Nb wires. At the Symposium on the Physics of Superconducting Devices[5] held in Charlottesville in 1967 there were several precursors of the work to come with SQUIDs. Silver[6] summarized work on the new single junction device,[7] now the rf SQUID, Mercereau[8] described a thin film rf SQUID and Beasley and Webb[9] reported the use of a flux transformer to couple a sample to a dc SQUID and permit operation in a high magnetic field.

ISSN: 0094-243X/78/058/$1.50

There soon followed many applications including measurements of fluctuations in the diamagnetic susceptibility of superconductors,[10,11] of nuclear paramagnetism for low temperature thermometry,[12,13] and of fluxoid quantization in superconductors.[14]

The first relatively general-purpose susceptometer for measurements over a wide temperature range was reported by Hoenig et al.[15] and Cerdonio et al.[16] at the California Institute of Technology. Concurrently, similar instruments were being developed at Stanford by E. P. Day,[17] who reported the first observation of NMR with a SQUID magnetometer,[18] and at the University of Virginia by Cukauskas et al.[19] Measurements of NMR were also reported by Meredith et al.[20] and Cukauskas et al.[19] and the evolution of this technique to the present time is described by R. Webb in a paper at this conference.

The first commercial systems[21,22] were introduced in 1971 and were designed for measuring the remanent and induced magnetization of geological samples. About 30 of these superconducting rock magnetometers are now in service. They provide simultaneous measurement of up to three orthogonal components of the magnetization with typical magnetic moment sensitivity of 10^{-8} emu and have room temperature access ranging from 0.5 to 10 cm diameter. A review of these systems has been presented by Goree and Fuller.[23]

The first commercial general purpose susceptometers[24] were delivered in 1974. They used thin film SQUIDs[25] and provided continuous variation of the sample temperature from below 4K to room temperature by flowing temperature-controlled helium gas through the sample chamber. A second generation of susceptometers[26] used toroidal SQUIDs[25,27,28] which provided considerably improved slew rates and better stability.

Concurrently new instruments were being developed in research labs and used for a variety of measurements.[29-32] Very recently Philo and Fairbank[33] reported a susceptometer with resolution nearly a factor of 100 better than that of any other susceptometer. We discuss this system in more detail below.

Third generation commercial susceptometers are now available from two companies,[34] and they are beginning to offer performance that approaches that of the Philo and Fairbank instrument, although the requirement for operation with sample temperatures from helium temperatures to above room temperature imposes some difficult design problems. There are currently 9 to 10 commercial instruments in operation in various laboratories.

DESCRIPTION OF SUSCEPTOMETER

Conceptually a SQUID susceptometer consists of a superconducting pickup loop L_1 located in a uniform magnetic field and coupled to a SQUID (Fig. 1). When a sample is introduced into the loop, the resulting flux change is detected by the SQUID. Alternatively, instead of a single loop, two loops in a gradiometer configuration can be used, or the pickup loop can be coupled directly to the SQUID ring. Nearly all present susceptometers use rf SQUIDs operated using essentially the techniques (Fig. 2) originated by Silver and Zimmerman[7] that have been described in detail in earlier papers in this conference.

A diagram of a miniature susceptometer of the type shown at the

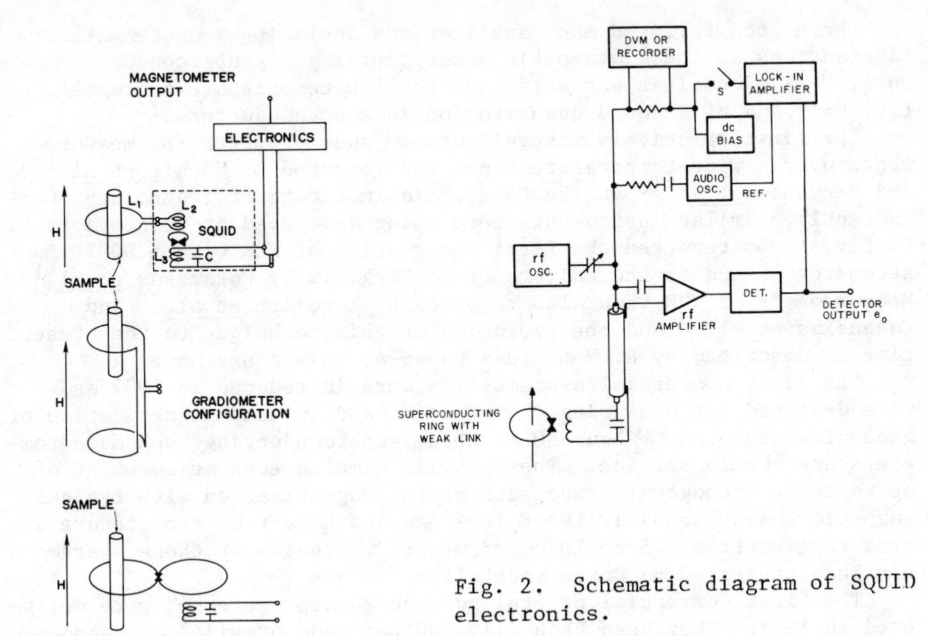

Fig. 1. Coupling Configurations a) Pickup loop L_1 and coil L_2 form a superconducting flux transformer, L_3 and C form r.f. tank circuit. b) Two pickup loops oppositely wound to form gradiometer. c) Sample directly coupled to SQUID.

Fig. 2. Schematic diagram of SQUID electronics.

bottom of Fig. 1 is shown in Fig. 3. It has excellent spatial resolution and is capable of measurements at the single flux quantum level. The sample can be moved in and out of the pickup loop, and its temperature can be varied and controlled. A superconducting shield surrounds the entire assembly. We have been using such a device for studying very small single crystals of layered superconductors.

A diagram of a general purpose superconducting susceptometer (Scχ) is shown in Fig. 4. Many of the details given in this section and the next are for a commercial instrument in use at the University of Virginia,[35] however they are representative of many of the instruments in current use. Two superconducting pickup coils in a Helmholtz configuration are joined by a superconducting stripline to a SQUID sensor. The sample is isolated from the cryogenic environment of the coils by a dewar, inside which the temperature is controlled by a flow of helium gas passed through a porous copper plug equipped

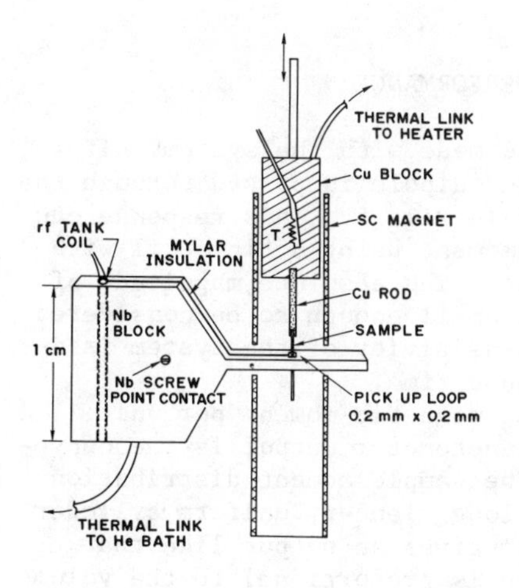

Fig. 3. Diagram of SQUID sensor and sample heater-thermometer unit. Entire assembly is enclosed in a vacuum can surrounded by a superconducting lead shield.

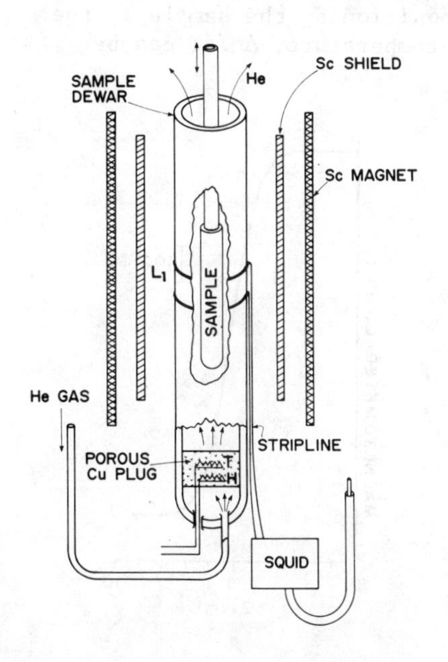

Fig. 4. Diagram of susceptometer. Cutaway of sample dewar shows location of sample in pickup loops (L_1). Thermometer (T) and heater (H) are shown in porous Cu plug.

with a heater and thermometer. The sample temperature can be varied from below 4K to well above room temperature.

Applied fields from 0-30 kG are provided by a superconducting solenoid operated in the persistent mode. The field is further stabilized by a superconducting shield, (a NbTi cylinder) cooled into the superconducting state after the field is persistent. This shield also provides isolation from external magnetic fields. The magnet, shield and pickup coils are mounted rigidly in a single unit to minimize vibration-induced noise.

The susceptometer is mounted inside a fiber-glass and aluminum helium dewar with a capacity of 30 liters of helium which permits operation for about a week without refilling. Sample access is through an air-lock at the top of the sample dewar which permits evacuation of the sample holder before introducing it into the dewar.

The sample is usually contained in an ultra high purity, thin-walled quartz tube up to 6 mm o.d. and mounted at the end of a glass or quartz rod which extends through an O-ring seal at the top of the sample dewar. The quartz rod also serves as a convenient light pipe for optically exciting the sample.

The top of this rod is connected to a miniature hydraulic cylinder which is used to vary the sample position smoothly and reproducibly. A potentiometer connected to the cylinder provides a voltage analog of the sample position.

TECHNIQUES AND PERFORMANCE

Several types of measurements are made with the system. If a very small sample approximating a point dipole is passed through the pickup coil, the response is as shown in Fig. 5. This response can be calibrated absolutely in magnetic moment using a tiny coil with precisely known dimensions and current. The absolute magnitude of the total magnetic moment of a sample small enough to be considered a dipole can then be measured. The sensitivity of the system is about 10^{-8} emu with a one second response time.

Also for long slender samples the magnetic moment per unit length can be determined since the magnetometer output is the convolution of this dipole response with the sample moment distribution. The simplest distributed sample is a long, slender, uniform cylinder which on being withdrawn from the coil gives an output like that shown in Fig. 6. The total deflection is proportional to the volume susceptibility of the sample. This type response can also be calibrated absolutely using a long solenoid of precisely known dimensions. The sensitivity for volume susceptibility changes is 3×10^{-11} emu/cm^3 for a 0.4 cm^3 sample in a 1 kG field.

One of the most valuable features of the susceptometer is its capability for sensing small changes ΔM even in the presence of a large steady magnetization. Thus by positioning the sample in the pickup coil and then incrementing the temperature, $\Delta M/\Delta T$ can be

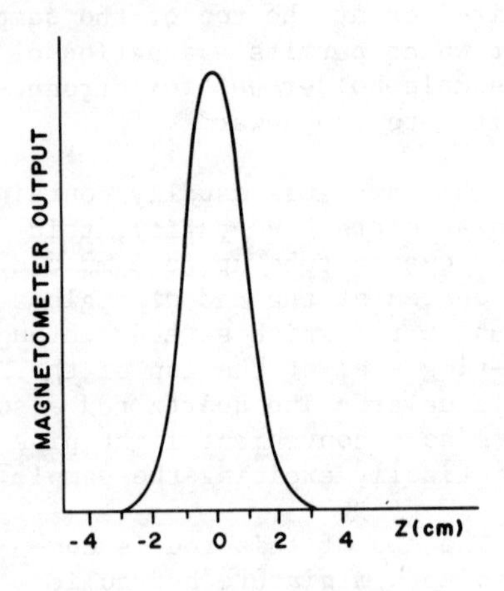

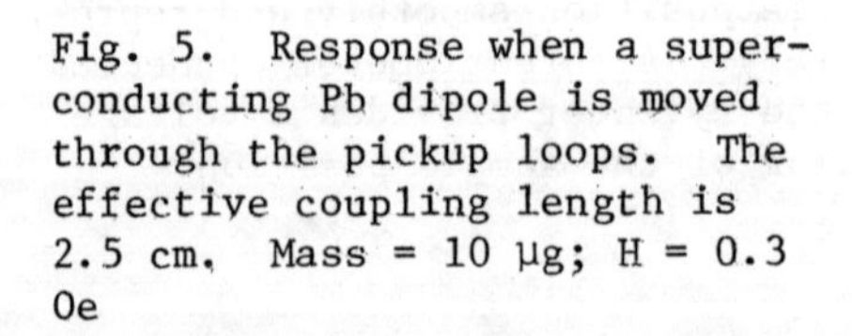

Fig. 5. Response when a superconducting Pb dipole is moved through the pickup loops. The effective coupling length is 2.5 cm. Mass = 10 μg; H = 0.3 Oe

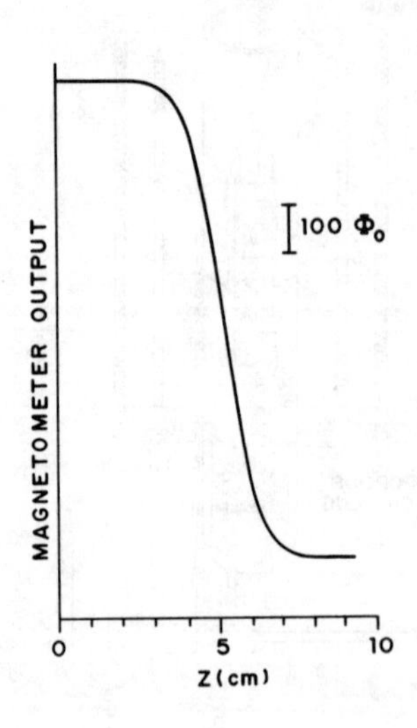

Fig. 6. Response when a fused quartz rod (d = 5 mm) is withdrawn from the pickup loops. H = 100.8 Oe

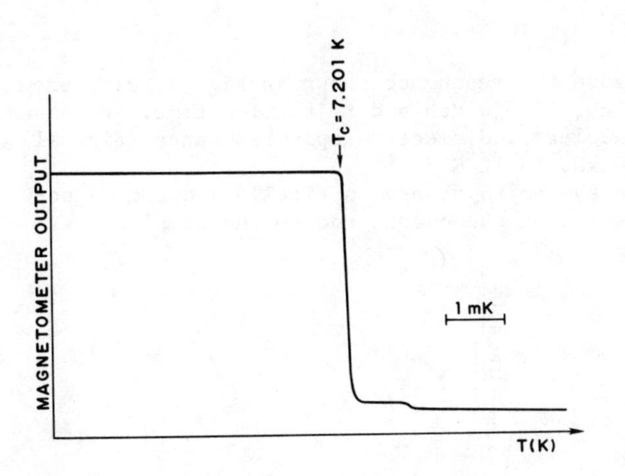

Fig. 7. Response when a Pb sample is kept stationary in the pickup loop while the temperature is swept through T_c. H = 0.3 Oe

measured. An example of this type measurement is shown in Fig. 7. This curve is actually a temperature calibration with a standard Pb sample using the superconducting transition in a very low field. The temperature can readily be maintained within better than ± 1 mK at cryogenic temperatures and the temperature can be reset to within about a millikelvin.

As already described by R. Webb, nuclear magnetic resonance is readily observed with a SQUID susceptometer. We use a coaxial sample holder of the type shown in Fig. 8 for resonance measurements. With the sample in the pickup loop, an rf sweeper is connected to the probe and the magnetometer output plotted versus frequency. We routinely use this probe to measure the magnetic field at the pickup loop using the proton resonance in Delrin. An example is shown in Fig. 9. A small amount of hydrogen

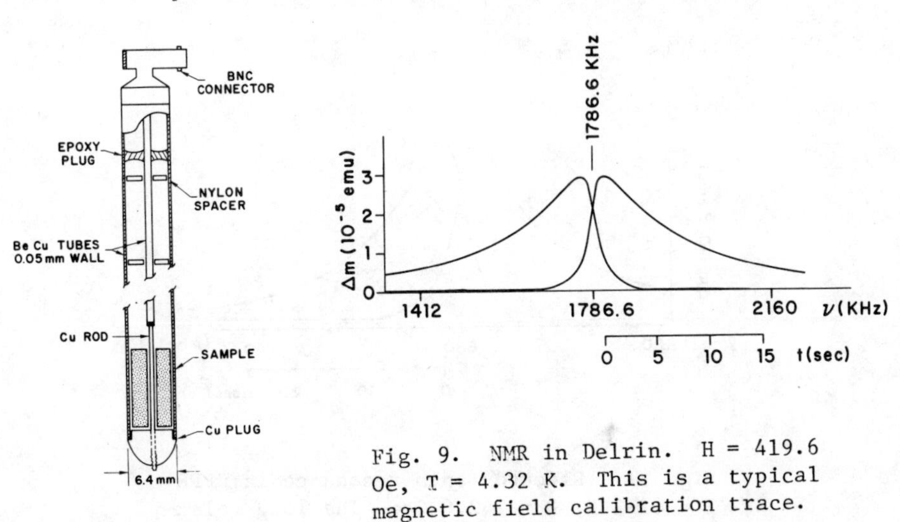

Fig. 9. NMR in Delrin. H = 419.6 Oe, T = 4.32 K. This is a typical magnetic field calibration trace.

Fig. 8. Magnetic Resonance probe.

condensed into the tube gave the resonance shown in Fig. 10 with the expected resonant frequency, line width and relaxation time. We have used the same probe for nuclear and electron spin resonance (Fig. 11) with frequencies from ~10 kHz to 12 GHz.

We have also used the system to observe optically induced magnetization by shining a laser down the quartz rod to the sample.

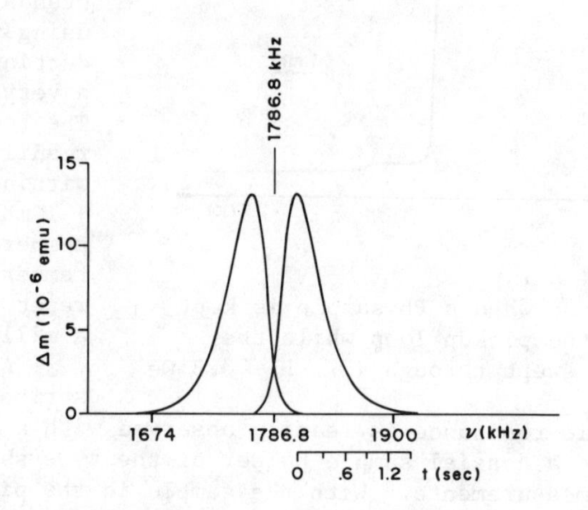

Fig. 10. NMR in solid hydrogen.
T = 4.34 K, H = 419.6 Oe

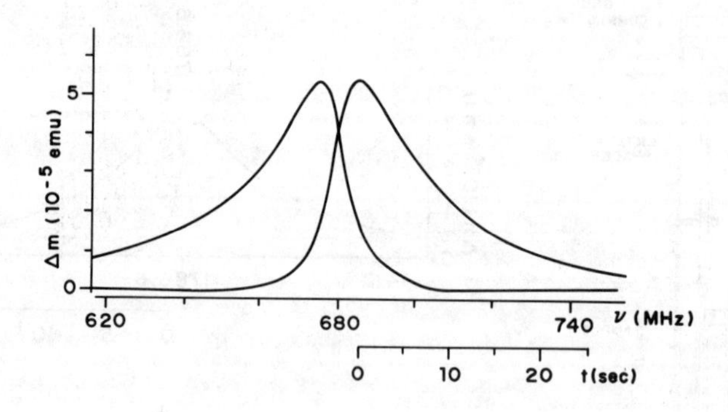

Fig. 11. Electron spin resonance in DPPH
T = 4.32 K, H = 242.6 Oe. The long relaxation time observed is a result of sample heating.

NOISE SOURCES AND LIMITATIONS

The intrinsic noise of the SQUID would permit, in principle, measurements of $\Delta\chi \approx 10^{-15}$ emu/cm^3. However, other factors limit the performance well before that limit is reached. The sources of noise have been thoroughly discussed[18,19,33] and include Johnson noise in conducting materials near the pickup coils, variation in the magnetization of paramagnetic construction materials due to temperature fluctuations (a susceptibility noise), and vibration of the pickup coils in the slightly inhomogeneous applied field. Because the susceptibility noise tends to dominate, making measurements at high fields does not improve the signal-to-noise ratio.

PHILO-FAIRBANK ScX

Several of the noise sources that limit previous systems have been enormously reduced by Philo and Fairbank.[33] Some details of their design are shown in Fig. 12. Two particularly important features are that the temperatures of the cold surfaces between the sample and pickup coils are highly regulated and the magnet winding density is adjusted to produce a very homogeneous field which reduces noise due to vibrations. With their improvements in design they achieve a sensitivity that is more than 100 times better than most other superconducting susceptometers.

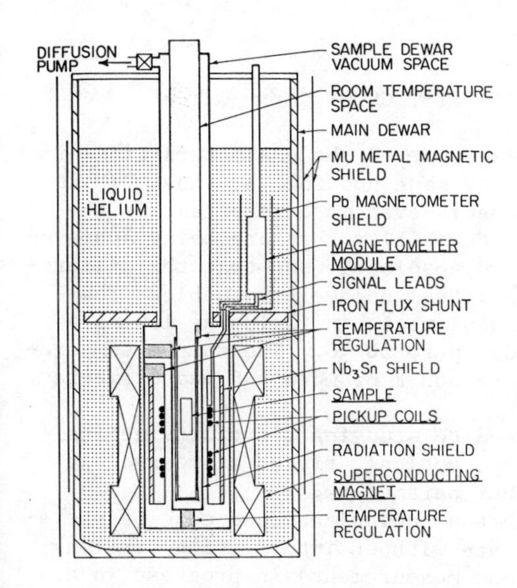

Fig. 12. Diagram of Philo-Fairbank susceptometer. (After Philo and Fairbank Ref. 33.)

With this system, by exploiting the exactly periodic response of the SQUID, they have been able to achieve one part per million resolution. An example of this precision is shown in Fig. 13 which shows data for the temperature dependence of the diamagnetic susceptibility of water.

Also with this system they have achieved a response time of ~300 μ sec. and have demonstrated the usefulness of the device for observing reaction kinetics. They measured the magnetization versus time of a hemoglobin solution containing CO as it returned to equilibrium after having CO removed from the hemoglobin with a short light pulse (Fig. 14).

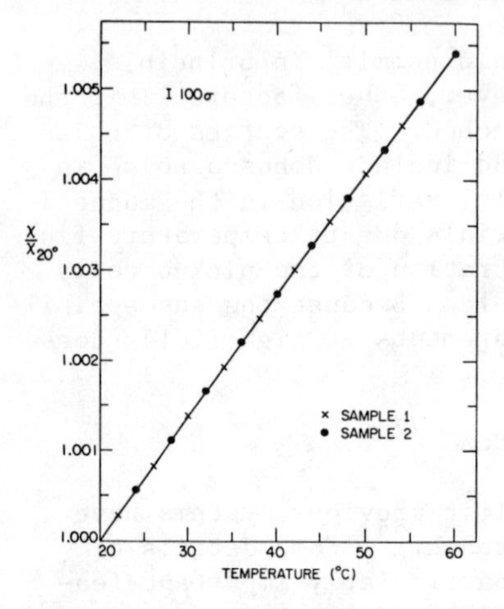

Fig. 13. Change in mass susceptibility of water with temperature. The solid line is a best fit third order polynomial. The width of the line is approximately 12 standard deviations. (After Philo and Fairbank, Ref. 33.)

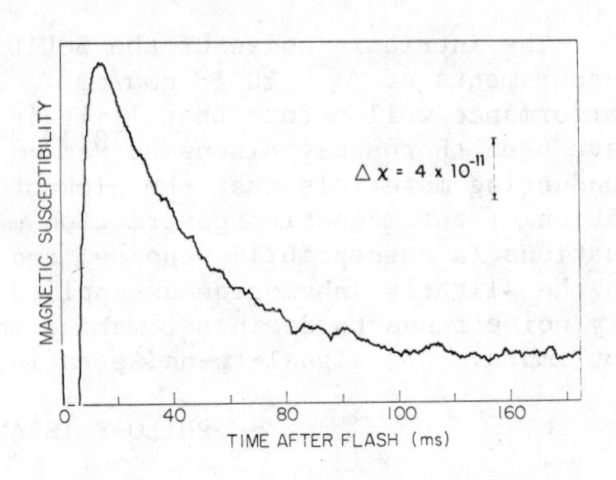

Fig. 14. Magnetic susceptibility changes after flash photolysis of HbCO. Total [Hb] = 38.5 μM; [CO] = 130 μM; 20° C; .1 M phosphate buffer pH 7.3. Average of 64 transients. (After Philo and Fairbank, Ref. 33.)

SUMMARY OF CHARACTERISTICS OF Scχ

In the table below we summarize some of the characteristics of Scχs. The extremely high sensitivity made possible by the SQUID is now beginning to be realized. Probably even more important is the high resolution and excellent reproducibility that are being achieved. The fact that they can be calibrated absolutely and that the calibration is stable for many months makes them uniquely valuable. They also provide the capability for measurements in fields as low as a few milligauss as well as in fields up to 50 kG. The sample temperature is readily varied from 1.5 K to 300 K or more and measurements are particularly fast and easy.

The instruments are well suited to studying reaction kinetics with reactions stimulated by mixing, optical, rf, or microwave excitation, temperature changes or other parameters.

Some practical advantages are the convenience and ease of operation and the fact that it can operate without interruption for long periods. A sample can be loaded and measurements in progress in a matter of minutes. Our susceptometer has been in continuous operation for more than six months.

Along with this recital of virtues we must also mention some limitations and disadvantages. Perhaps the most serious deficiency is that the applied field cannot be swept. Also after field changes there may be drifts over periods of a few minutes and sometimes even a few hours if the system has been at high fields. At high fields there are flux jumps and the superconducting shields are not so effective.

Anisotropy measurements are difficult because the field is parallel to the sample access so the field direction cannot be varied by simply rotating the sample.

Depending on the operating temperature, liquid helium consumption is 3-5 liters per day. And because of the unitized construction the device is difficult and expensive to repair.

Table I Characteristics of ScX

Sample Access:	cylinder 6 mm o.d. ℓ_{eff} = 2.5 cm
Magnetic Field:	0-30 kG measured to 0.01% with NMR probe
Temperature:	1.5 - 300 K
Range:	10^{-8} - 10^{-1} emu total moment
Sensitivity:	$\Delta m = 10^{-8}$ emu with 1 sec response $\Delta\chi = 3 \times 10^{-11}$ emu/cm^3 For 0.4 cm^3 sample, 1 kG. $\Delta\chi = 7 \times 10^{-13}$ Philo and Fairbank
Absolute calibration with coil with accuracy $\sim$ 1%	
Stability of Calibration $\pm$ 0.1% for 18 mo.	
Response Time:	10^{-3} sec
Resolution:	10^{-4} 10^{-6} Philo and Fairbank
Measures M_z, M_x, $\Delta M/\Delta T$, $M(\lambda)$, $M(t)$	
Very low field measurements	
Convenient and easy to use - continuous operation for more than 6 mo.	

SOME MEASUREMENTS BEING MADE WITH ScXs

From the beginning it was recognized that ScXs would make possible new types of measurements on biologically important materials. The early systems were used for measurements on hemerythrin[15] and carbonic anhydrase,[36] and to look for susceptibility changes during biomolecular phase changes.[17] More recently there have been measurements of the magnetic properties of oxyhemoglobin,[37] lysoxyme,[38] of the kinetics of hemoglobin-carbon monoxide reactions[39] and of the susceptibility of frozen aqueous solutions of biochemical compounds.[40]

ScXs are being used for the study of transition metal compounds and structure dependent magnetism,[41-44] for studies of new types of superconductors,[45] and for searching for new superconductors.[46]

FUTURE DEVELOPMENT AND POTENTIAL

Some work has already been reported and much more will be done on automating the operation and data acquisition of ScX systems. Some commercial rock magnetometers are equipped with online computers and Owen and Nadler[47] have described the automatic operation of a ScX for measurements on biological samples.

At the University of Virginia we are equipping our ScX with a DEC pdp11/03 system (Fig. 15) to control sample position, velocity, temperature, rf frequency, laser parameters and to record the longitudinal and transverse components of the magnetization, position, temperature and values of external parameters. A graphics terminal will be used to display individual scans and to compare them, and the computer will be used for signal averaging and data analysis.

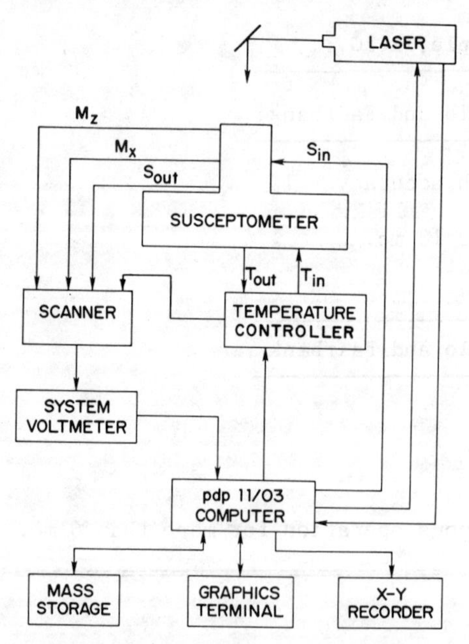

Fig. 15. Diagram of computer controlled ScX. S is sample position.

Also we are completing the adaption of a small helium refrigerator (CTI Model 1020 with a Joule-Thomson stage) to reliquify the boil off gas and return the liquid to the ScX dewar to permit essentially closed cycle operation.[48] A diagram of this sytem is shown in Fig. 16. The potential magnetic interference from the refrigerator has already been determined not to be a problem. However, the vibrations and pressure variations in the system have yet to be investigated.

Finally the system will be used with a pulsed Nitrogen laser and a tuneable dye laser to study optically induced magnetism and the magnetization of

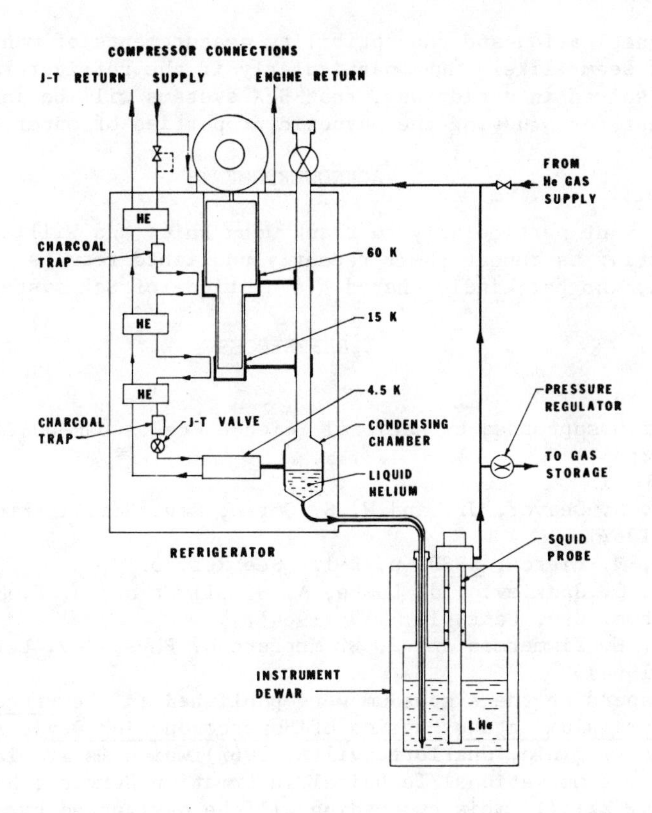

Fig. 16. Closed circuit, helium reliquefaction system.

excited states by measuring M(λ).

ScX systems are well suited to computer control and data acquisition. Deconvolution of the magnetometer output can be performed to extract the magnetic moment distribution. It seems probable that by recording M(t) and making use of the frequency spectrum of the various noise sources the signal-to-noise ratio can be improved.

It seems safe to assume that there will be rapid development of the use of ScX systems for observing time-dependent magnetization to study reaction kinetics and other transient magnetic phenomena. The response time can be decreased to 10^{-6} sec and with a microwave SQUID to perhaps 10^{-9} sec. However, for these very short times it will be necessary to consider whether direct observation of the magnetic pulses may be better than using the SQUID.

The extremely large dynamic range and high resolution made possible by flux counting techniques will almost certainly be exploited.

Also it should not be overlooked that even when the extreme sensitivity and speed are not required, ScX systems provide convenience, ease of operation, and accuracy that make them extraordinarily useful for magnetization and susceptibility measurements of many kinds.

It seems likely then, particularly if the refrigeration problem can be solved in a tidy way, that ScX systems will be increasingly important for studying the magnetic properties of materials.

ACKNOWLEDGMENTS

We want particularly to thank John Philo and William Fairbank for letting us repeat their recently published results, and George Candela, who has kindly shared his critique of ScX systems with us.

REFERENCES

*Research supported by National Science Grants DMR-00424-A03 and DMR-77-12268

1. B. S. Deaver, Jr. and W. S. Goree, Rev. Sci. Instrum. 38, 311 (1967).
2. J. M. Pierce, SPSD, p. B-1. See Ref. 5.
3. R. C. Jacklevic, J. Lambe, A. H. Silver and J. E. Mercereau, Phys. Rev. Lett. 12, 159 (1964).
4. J. E. Zimmerman and J. E. Mercereau, Phys. Rev. Lett. 13, 125 (1964).
5. Papers at the symposium were published in the Proceedings of the Symposium on the Physics of Superconducting Devices, (University of Virginia, Charlottesville, 1967) which is available as AD 661 848 from National Technical Information Service, Springfield, Va. 22151. This proceeding will be designated by SPSD in these references.
6. A. H. Silver, SPSD, p. F-1. See Ref. 5.
7. A. H. Silver and J. E. Zimmerman, Phys. Rev. 157, 317 (1967).
8. J. E. Mercereau, SPSD, p. U-1. See Ref. 5.
9. M. R. Beasley and W. W. Webb, SPSD, p. V-1. See Ref. 5.
10. J. P. Gollub, M. R. Beasley, R. S. Newbower, and M. Tinkham, Phys. Rev. Lett. 22, 1288 (1969).
11. J. E. Lukens, R. J. Warburton, and W. W. Webb, Phys. Rev. Lett. 25, 1180 (1970).
12. J. M. Goodkind and D. L. Stolfa, Rev. Sci. Instrum. 41, 799 (1970).
13. R. P. Giffard, R. A. Webb, and J. C. Wheatley, J. Low Temp. Phys. 6, 533 (1972).
14. W. L. Goodman, W. D. Willis, D. A. Vincent and B. S. Deaver, Jr., Phys, Rev. B-4, 1530 (1971).
15. H. E. Hoenig, R. H. Wang, G. R. Rossman, and J. E. Mercereau, Proceedings of the Applied Superconductivity Conference, Annapolis, Maryland. 1972. IEEE PUB. No. 72-CH0682-5-TABSC, p. 570 (1974).
16. M. Cerdonio, R. H. Wang, G. R. Rossman, and J. E. Mercereau, Proc. Low Temp. Phys. Conf. LT13 (1972), edited by E. Timmerhaus (Plenum, New York, 1974) p. 525.
17. E. P. Day, Ibid. p.550.

18. E. P. Day, Phys. Rev. Lett. 29, 540 (1972).
19. E. J. Cukauskas, D. A. Vincent, and B. S. Deaver, Jr., Rev. Sci. Instrum. 45, 1 (1974).
20. D. J. Meredith, G. R. Pickett, and O. G. Symko, J. Low Temp. Phys. 13, 607 (1973).
21. These systems were developed by W. S. Goree, W. L. Goodman, V. W. Hesterman and L. H. Rorden at Develco, Inc., Mountain View, California.
22. W. S. Goree, Proceedings of the Applied Superconductivity Conference, Annapolis, Maryland, 1972. IEEE PUB. No. 72-CH0682-5-TABSC, p. 640 (1974).
23. W. S. Goree and M. Fuller, Reviews of Geophysics and Space Physics 14, 591 (1976).
24. These systems were developed by W. S. Goree, W. L. Goodman and V. W. Hesterman at Superconducting Technology, Inc., Mountain View, California.
25. W. L. Goodman, V. W. Hesterman, L. H. Rorden and W. S. Goree, Proc. of the IEEE 61, 20 (1973).
26. These systems were developed by D. A. Vincent, W. L. Goodman, and W. S. Goree at Superconducting Technology, Inc., Mountain View, California.
27. D. A. Vincent, Ph.D. Dissertation, University of Virginia, 1971.
28. R. E. Sarwinski, Advances in Instrumentation, Vol. 31, paper No. 616, ISA-76 Annual Conference (Instrument Society of America, Pittsburgh, 1976).
29. M. Cerdonio and C. Messana, IEEE Trans. Mag. MAG-11, 728 (1975).
30. F. R. Fickett and D. B. Sullivan, J. Phys. F4, 900 (1974).
31. M. Cerdonio, C. Cosmelli, G. L. Romani, C. Messana, and C. Gramaccioni, Rev. Sci. Instrum. 47, 1 (1976).
32. K. Gramm, L. Lundgren and O. Beckman, Physica Scripta 13, 93 (1976).
33. J. S. Philo and W. M. Fairbank, Rev. Sci. Instrum. 48, 1529 (1977).
34. S.H.E. Corp., San Diego, California. Superconducting Technology, Division of United Scientific Corp., Mountain View, California.
35. The University of Virginia susceptometer was manufactured by Superconducting Technology, Inc. and is operated under the joint direction of B. Deaver of the Physics Department and E. Sinn of the Chemistry Department.
36. P. H. Haffner and J. E. Coleman, J. Biol. Chem. 248 (no. 19), 6630 (1973).
37. M. Cerdonio, A. Congiu-Castellano, F. Mogno, B. Pispisa, G. L. Romani and S. Vitale, Proc. Natl. Acad. Sci., USA, 398 (1977).
38. G. Careri, L. DeAngelis, E. Gratton and C. Messana, Phys. Lett. 60A, 490 (1977).
39. J. S. Philo, Proc. Natl. Acad. Sci., USA, 74 2620 (1977).
40. C. S. Owen, E. K. Jaffe and D. F. Wilson, Rev. Sci. Instr. 48 1541 (1977).
41. E. J. Cukauskas, B. S. Deaver, Jr., and E. Sinn, J. Chem. Soc. Chem. Comm., 698 (1974).
42. E. J. Cukauskas, B. S. Deaver, Jr., and E. Sinn, J. Chem. Phys.

67, 1257 (1977).

43. E. J. Cukauskas, B. S. Deaver, Jr., and E. Sinn, Inorg. Nucl. Chem. Letters 13, 283 (1977).
44. W. E. Hatfield, H. W. Richardson and J. R. Wasson, Inorg. Nucl. Chem. Letters 13, 137 (1977).
45. R. Spal, C. K. Chiang, A. Denenstein, A. J. Heeger, N. D. Miro, and A. G. MacDiarmid, Phys. Rev. Lett. 39, 650 (1977).
46. Work on alkali tungsten bronzes with emphasis on H_xWO_3 is being done at the University of Virginia by C. Friedberg, J. Ruvalds, I. Lefkowitz, J. Snare, T. J. Bucelot and B. S. Deaver, Jr.
47. C. S. Owen and W. Nadler, Rev. Sci. Instrum. 48, 1537 (1977).
48. The refrigerator is one manufactured a number of years ago by CTI for the Navy and made available to us by the Office of Naval Research. The adaption of this unit to the ScX system is being done by D. A. Vincent of Vincent Processes, Tampa, Florida.

APPLICATION OF SQUIDs TO MEASUREMENTS IN FUNDAMENTAL PHYSICS

B. Cabrera
Physics Department, Stanford University, Stanford, CA 94305

ABSTRACT

Several experiments applying SQUID instrumentation to the study of diverse fields in fundamental physics are reviewed. Included from gravitational physics research are the relativity gyroscope experiment, an equivalence principle experiment and work on second generation gravitational wave detectors. Experiments studying the properties of elementary particles at rest include the He^3 nuclear gyroscope, fractional electric charge detector and a magnetic monopole charge detector, where new preliminary measurements on the magnetic charge of the fractional electric charge candidates are presented. Measurements of 2e/h and h/m_e are included as significant contributions to the determination of the fundamental physical constants. In addition a brief description of ultra-low magnetic field shields used in four of these experiments is given.

INTRODUCTION

Since the development of the first SQUIDs in the 1960's an ever-increasing number of applications have been developed outside of low temperature physics research. In this paper several experiments in gravitation, elementary particles and the determination of the fundamental physical constants are reviewed as examples of diverse fundamental physics research which make use of SQUID instrumentation. Applications in fields other than physics are presented in the succeeding papers of the SQUID Symposium.

ULTRA-LOW MAGNETIC FIELDS

Over the past decade a technique has been developed using superconducting shields for obtaining magnetic field regions below 10^{-7} gauss. Ultra-low fields are needed for several of these experiments to shield sensitive SQUID pick-up loops from noise sources such as trapped flux jumping and mechanical motions in larger ambient fields. Briefly, cylindrical shields are constructed from lead foil, pleated, folded flat and slowly cooled through their transition inside of a vacuum-jacketed glass tube; then mechanically expanded with a plunger (Fig. 1). The lowest fields are produced by repeatedly cooling a shield inside an outer previously cooled one (3 or 4 times). The field profile of an eight inch diameter forty inch long shield (Fig. 2) used for the relativity gyroscope experiment[2] was measured with a two-inch diameter superconducting flip coil coupled to a torodial rf-SQUID magnetometer. The flip coil structure is made entirely of quartz tubing with thin teflon bushings and provides a background magnetic field below 10^{-8} gauss, allowing absolute magnetic field measurements in the 10^{-9} gauss range. Using this magnetometer, shields with no trapped flux will be attempted within the next few years.

ISSN: 0094-243X/78/073/$1.50

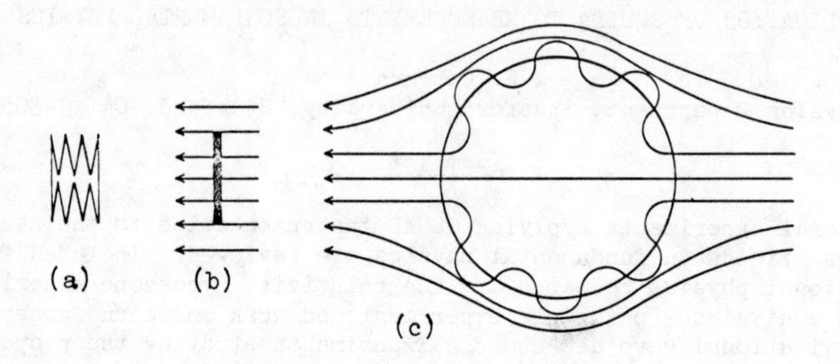

Fig. 1 Cross-sectional View of Pleating Pattern (a), Field Trapped at Transition (b) and Reduced Central Field After Expansion (c).

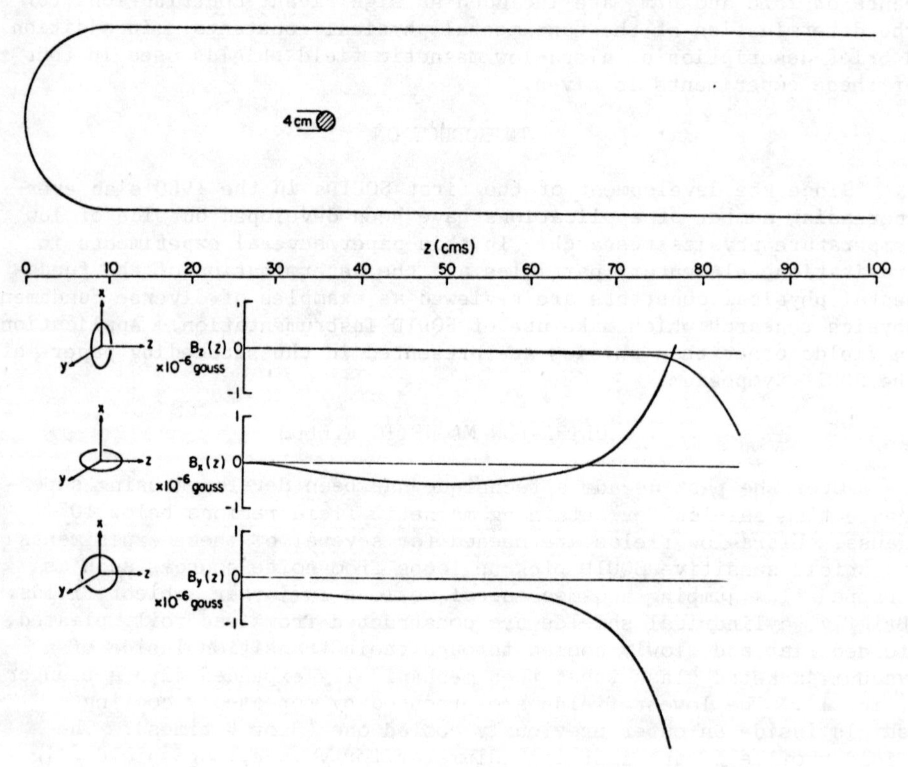

Fig. 2 Magnetic Field Components Along the Axis of an Eight-Inch Ultra-Low Field Shield (Gyroscope Rotor Position Shown)

GRAVITATIONAL PHYSICS

The Relativity Gyroscope Experiment. A cryogenic gyroscope and reference telescope satellite package (Fig. 3) is being developed to measure both the geodetic precession and the smaller mass-current precession to an uncertainty of 1 milliarc-sec/year (10^{-16}rad/sec).[3,4,5] In a 500 km polar orbit the magnitude of the effects is 7 arc sec and .05 arc sec respectively integrated over one year. Each gyroscope

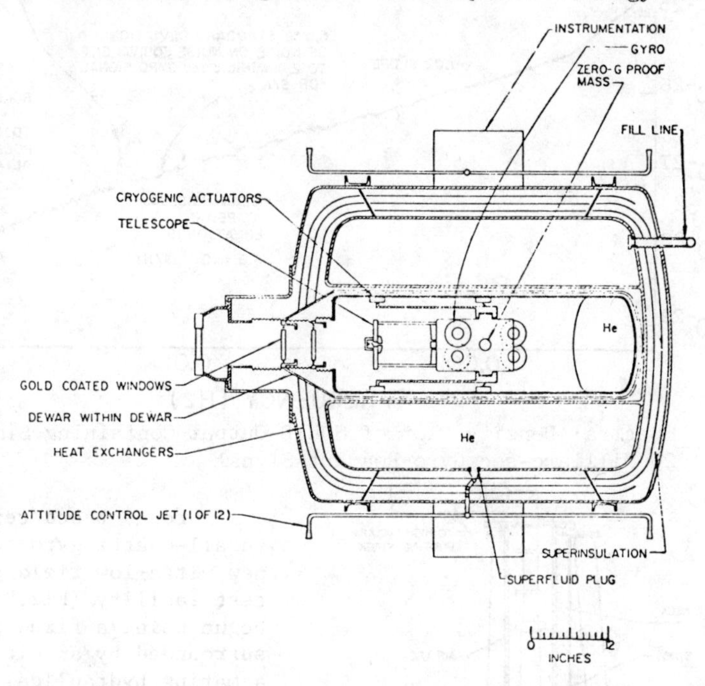

Fig. 3 Gyro Relativity Satellite: Cross-Section

consists of a"perfect"quartz sphere (1.5 in. dia.) coated uniformly with 5000 Å of niobium and electrostatically suspended in a close fitting quartz housing while spinning at 200 Hz. The spin axis orientation relative to its case and thus to the telescope is determined by measuring the London moment vector using three mutually orthogonal pick-up loops coupled to SQUID magnetometers.

The trapped flux in the rotor must be less than 10^{-7} gauss as seen in the pick-up loops in order not to saturate the dynamic range of the SQUID electronics. The low field environment is provided by an ultra-low magnetic field shield as described above, which also shields the pick-up loops from ambient field changes. The signals seen by the magnetometers are modulated by the roll of the satellite about the line of sight to the star at a frequency of 0.002 Hz. In order to investigate the integration properties of low frequency signals in the 1/f noise of a commercial SQUID system, a 140 hr data set was taken on a system using a calibration signal at 0.017 Hz corresponding to a 20 milliarc-sec signal from a gyro. The signal is clearly visible (Fig. 4), integration times agree well with white noise theory, and in fact for a S/N = 1, a resolution of 1 milliarc sec is obtained in 60 hr. More information is provided in a contributed paper on this subject.

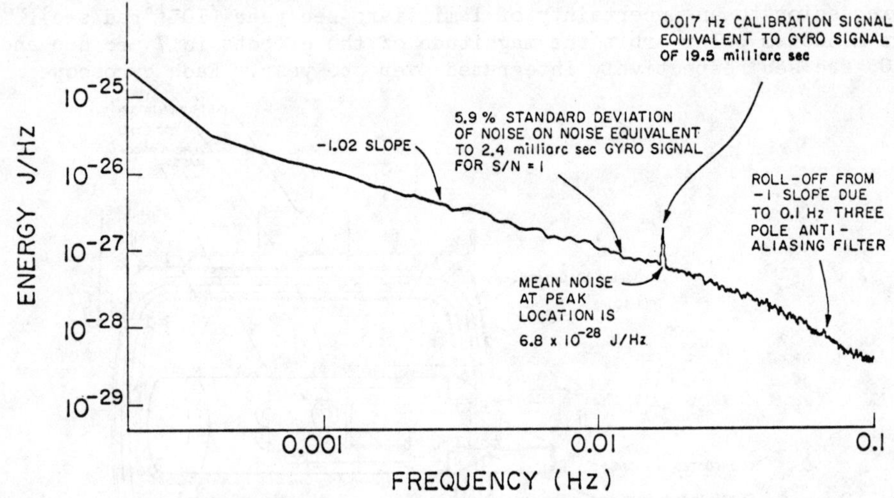

Fig. 4 Spectral Density Plot of SQUID Output Containing Simulated 20 Milliarc-sec Gyro Readout Signal

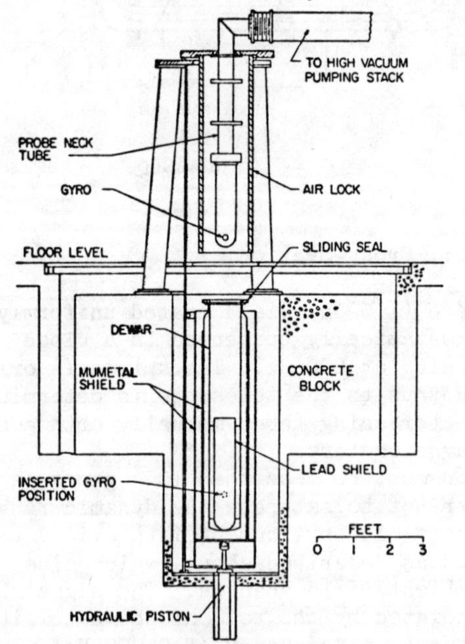

Ground based testing of an all-quartz gyroscope in a new ultra-low field shielded test facility (Fig. 5) has begun using a fixed probe surrounded by an airlock and a mating hydraulically activated movable dewar containing the superconducting shield. Gyro operations can be continuously monitored from room temperature to 4K. The performance of the London moment readout is expected to approach the 1 arc sec resolution limit estimated for a 1 g environment. The complete system is scheduled to fly during the first half of the next decade.

Fig. 5 Schematic Cross-Section of the New Apparatus with Dewar in Lowered Position

The Equivalence Principle Experiment. Another experiment being developed for future orbit is an improved test of the equivalence principle.[6] The Eötvös ratio

$$\eta = 2\left[\left(\frac{M}{m}\right)_A - \left(\frac{M}{m}\right)_B\right] / \left[\left(\frac{M}{m}\right)_A + \left(\frac{M}{m}\right)_B\right] \qquad (1)$$

where m and M are the inertial and passive gravitational masses of the materials A and B, has been shown to be less than 3×10^{-11}. An apparatus at Stanford aimed at working out the design of a flight experiment and for providing a ground-based check of equivalence to 1 part in 10^{13} with the sun as source has been built and is being tested (Fig. 6).

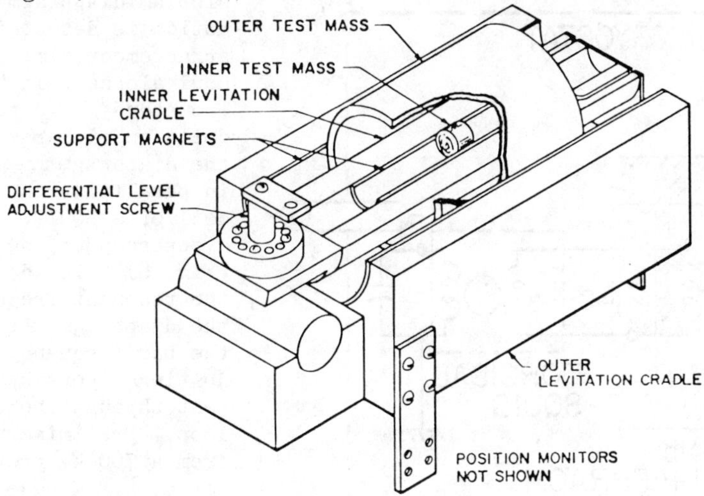

Fig. 6 The Equivalence Principle Accelerometer

The differential accelerometer consists of an outer hollow cylindrical mass made of niobium-coated aluminum and an inner concentric cylinder of pure niobium. Both are levitated in semi-cylindrical superconducting magnetic levitation cradles. This geometry greatly reduces gravity gradient effects. Small trimming coils perpendicular to the main windings are used to apply correction currents which redistribute the magnetic field within the trough and so tilt the support plane. (Fig. 7) If the mass moves from its equilibrium, the current I_2 amplified by a SQUID becomes non-zero and is used to drive a centering control loop using the tilt coil. In the space version where η could be measured to one part in 10^{17}

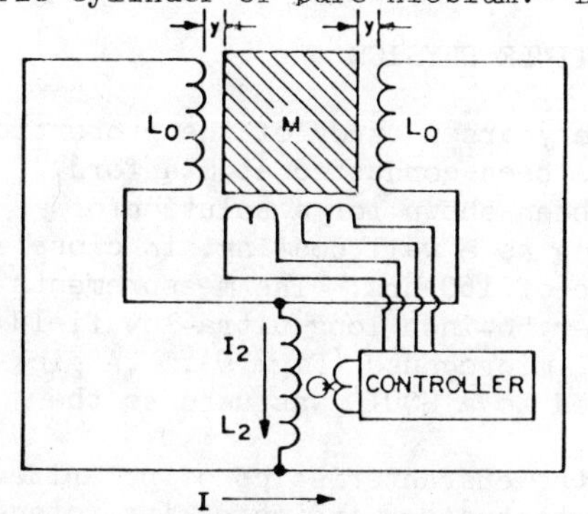

Fig. 7 Position Detector for Equivalence Principle Experiment

the tilt coils are not needed, the restoring force from the end coil being sufficient.

Gravitational Wave Detectors. Second generation resonant gravitational wave detectors are now being developed with sensitivities 10^3 to 10^4 higher than their room temperature predecessors.[7] Several of these liquid helium cooled systems used SQUIDs as low noise impedance matched transducer amplifiers. In Rome the currents from a piezoelectric transducer mounted on a 5-ton aluminum bar will be detected using a toroidal impedance matching transformer and a SQUID as a voltage sensor.[8]

The accelerometer used on a 5-ton aluminum bar being set up at Stanford in a coordinated program with Rome and LSU consists of a niobium diaphragm whose motion is detected with an arrangement similar to the equivalence principle accelerometer.[9] Two pancake coils on either side of the diaphragm are connected in parallel with the input coil of a SQUID, forming a superconducting loop (Fig. 8). In addition the resonant frequency of the diaphragm is tuned to the bar frequency by adjusting a persistent current through the L_1-L_2 loop. Preliminary results from a 700 kg prototype bar at 4.2 K have demonstrated an effective noise temp. of 0.39 K using this accelerometer.

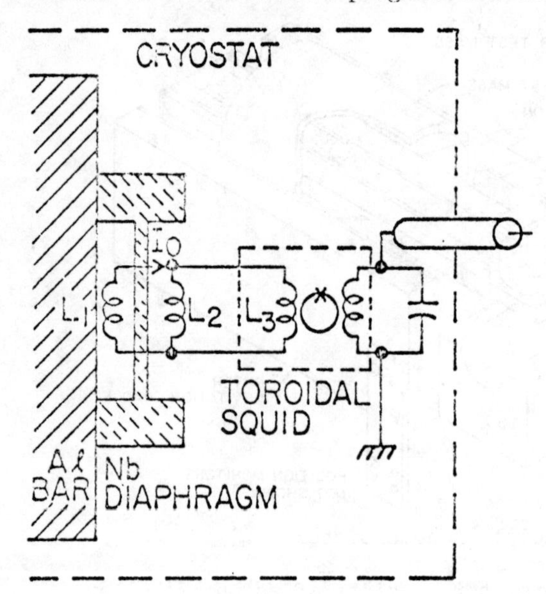

Fig. 8 Schematic of Superconducting Accelerometer

ELEMENTARY PARTICLE PHYSICS

He^3 Nuclear Gyroscope. For some years a study of the properties of polarized He^3 samples at 4.2 K has been conducted at Stanford.[10] A relaxation time T_1 of 140 hrs has been shown for a solution of 0.07% He^3 in liquid He^4 using hydrogen as a wall coating, in close agreement with a theoretical estimate of 160 hrs. The measurements were carried out in an 8-inch diameter 56-inch long ultra-low field shield in an ambient field of several microgauss (Fig. 9). A pick-up loop around the sample bulb coupled to a SQUID was used as the readout.

Recently a general analysis of the characteristics of polarized He^3 based on current low-temperature technology indicates its potential for a competitive nuclear navigational gyroscope. A test probe for investigation of this potential has been designed and consists of

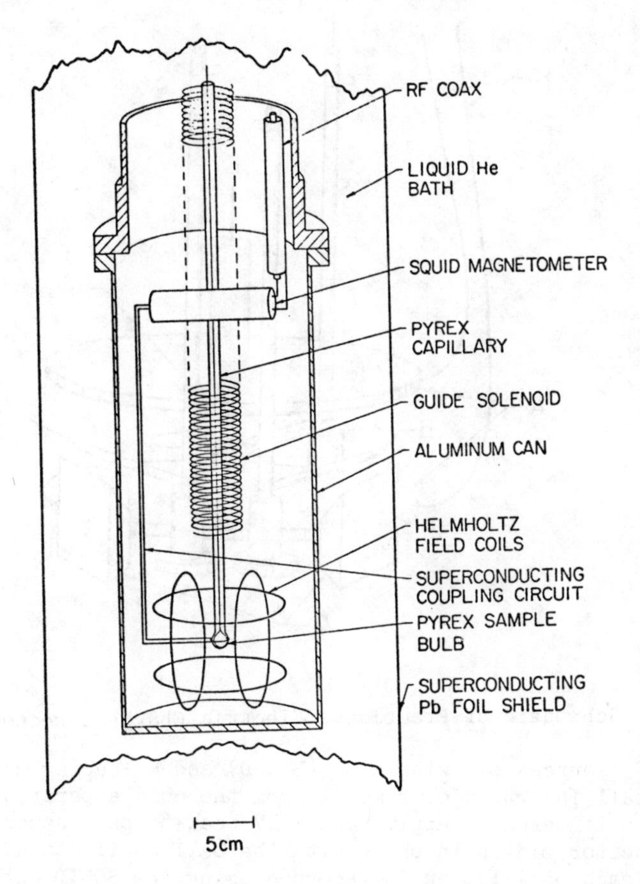

Fig. 9 Relaxation Measurements of Polarized He^3 in He^4 at 4 K.

a 1.5" diameter "perfectly" spherical He^3 sample cell in a homogeneous field with a pick-up loop coupled to a SQUID magnetometer. An outer ultra-low field shield will provide an ambient field of 5×10^{-8} gauss or less. If the phase of the precession signal is compared with that of a stable clock, the instrument becomes a uniaxial gyroscope. Alternatively if a large electric field is applied across the sample, a sensitive search for an electric dipole precession can be made, which may someday be competitive with the neutron measurements. A non-zero result would be a direct violation of time-reversal invariance.

Fractional Electric Charge Detector. Some evidence for the existence of fractional electric charge on matter has been reported recently by LaRue, Fairbank and Hebard.[11] Their present apparatus consists of a magnetically levitated niobium ball of mass $\sim 9 \times 10^{-5}$ g oscillating vertically at a frequency of ~ 0.8 Hz (Fig. 10G). The ball is located between two capacitor plates with a 1 cm separation (A) and its motion is detected using a SQUID magnetometer coupled to a pick-up loop (E). The charge on the ball can be changed using

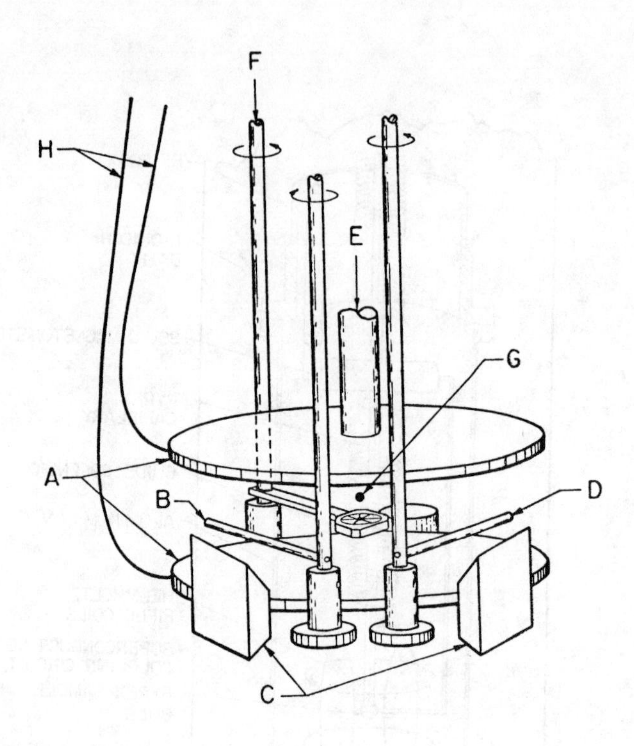

Fig. 10 Schematic of Fractional Electric Charge Detector.

β_+ and β_- sources on swing arms (B & D), and a loading mechanism (F) allows ball insertion or removal from the cold apparatus.

Data is taken by applying a 2 kV peak-to-peak square wave to the capacitor plates in phase with the ball oscillation. The amplitude of each oscillation is recorded using the SQUID output (Fig. 11), and every 50 cycles the phase of the square wave is revised. The charge is determined by measuring the difference between the time rate of change before and after reversal, since it is independent of the damping and proportional to the change (Fig. 12). A typical amplitude after 50 oscillations for a net charge of one or two e is 10μm. The calibration is performed by observing 1 e charge changes. In addition a study of spurious effects is performed for each ball before final detection of the residual charge. Two of the balls were found to have a non-zero residual charge of $(+0.337 \pm 0.009)$ e and (-0.331 ± 0.070) e. Work is continuing on both improving the apparatus and increasing the number of balls measured.

Magnetic Monopole Charge Detector. The possibility of the existence of free fractional electric charges has allowed a direct preliminary test of a suggestion that the theoretical quarks may also be magnetically charged. Assuming the elementary electric charge to be that of an electron, the elementary magnetic charge g_o must be $g_o = \hbar c/2e$ or a total flux of $4\pi g_o = hc/e = 2\phi_0$. If both electric and magnetic charge are allowed on the same particle, the elementary charge must be $g_d = 2g_o$, and if in addition the elementary

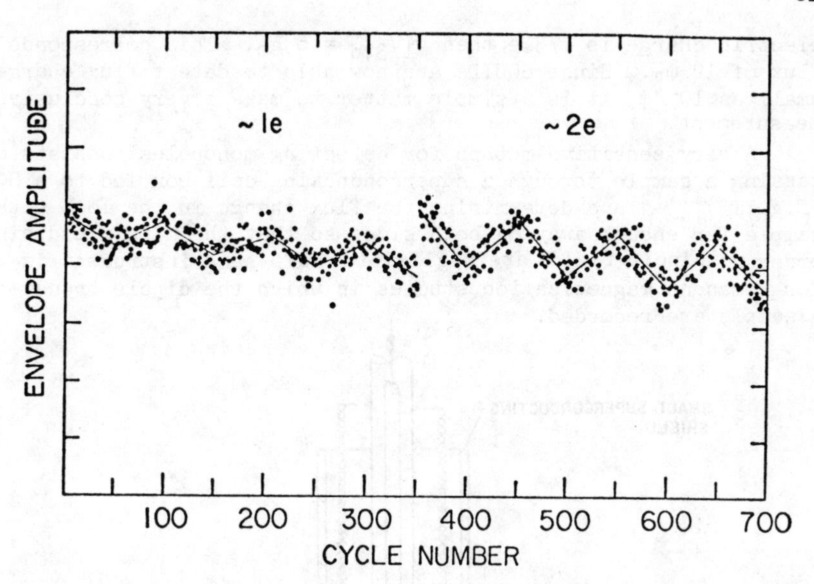

Fig. 11 Amplitude of Increasing and Decreasing Oscillations Showing a Charge Change.

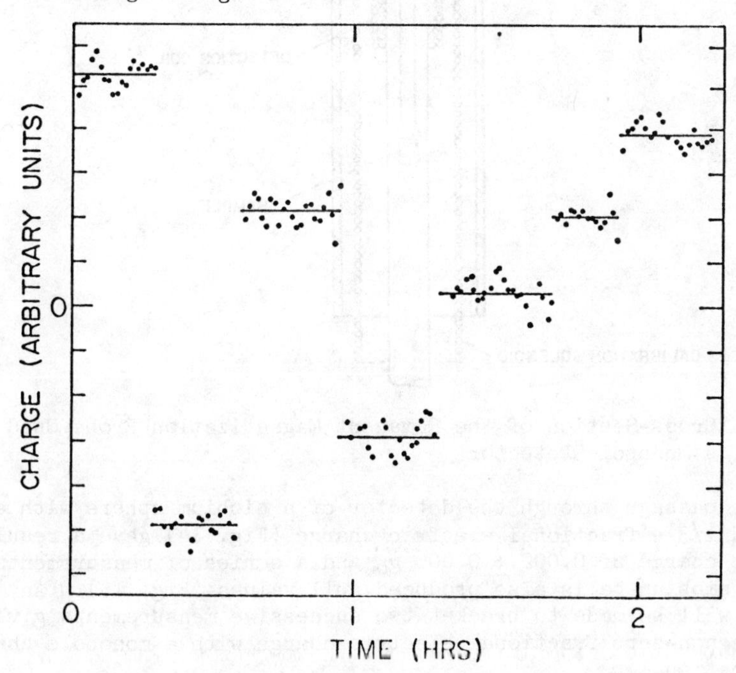

Fig. 12 Electric Charge Computed Once Per Vertex in Fig. 11.

electric charge is 1/3 e, then $g_{1/3,d} = 6\ g_o$. This corresponds to a flux of 12 ϕ_0. Since SQUIDs are now able to detect flux charges as small as $10^{-4}\phi_0$, it is a simple matter to make a very conclusive measurement.

A very sensitive method for detecting monopoles consists of passing a sample through a superconducting coil coupled to a SQUID (Fig. 13)[11,12] and determining the flux change in the loop with the sample far enough away on both sides so that the dipole and higher order multipole fields are negligible. This same instrument is used for remanent magnetization studies in which the dipole transients of a sample are recorded.

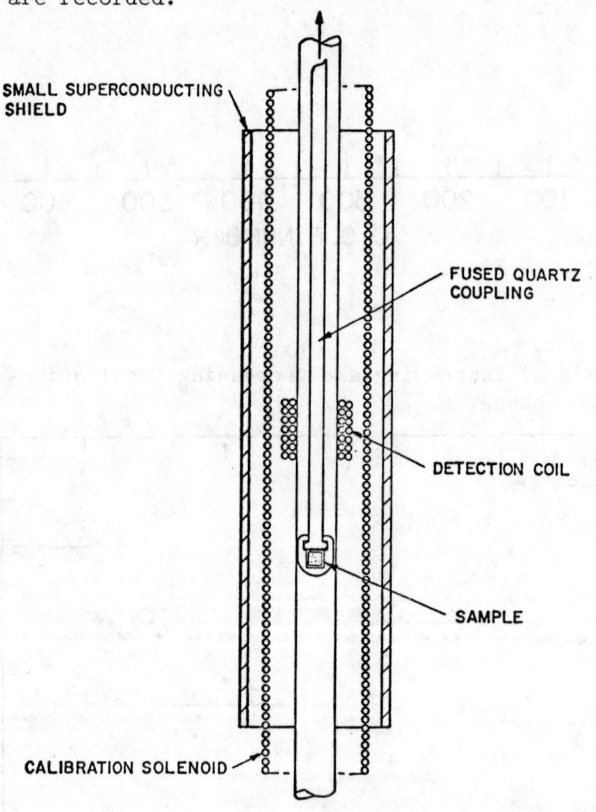

Fig. 13 Cross-Section of the Remanent Magnetization Probe Used as a Monopole Detector

The passage through the detector of a niobium sphere with a measured 1/3 e fractional electric charge (Fig. 14) gave a resultant monopole charge of $0.002 \pm 0.006\ g_o$ and a series of measurements on 7 other niobium balls also produced null values (Fig. 15). An attempt will be made to bracket two successive measurements giving the same non-zero fractional electric charge with a monopole charge determination.

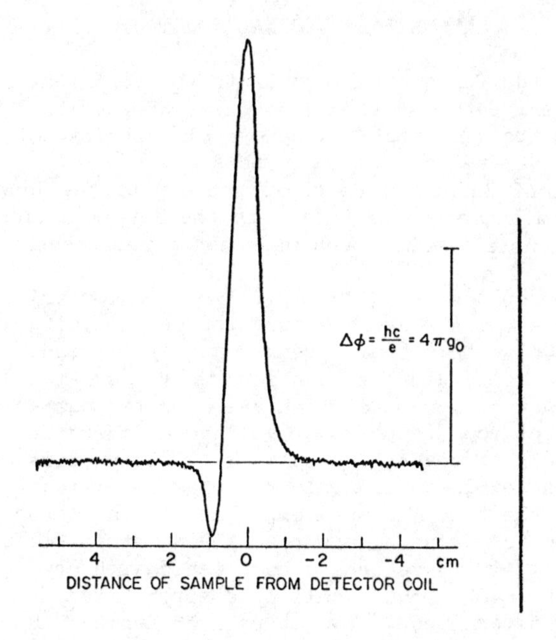

Fig. 14 Passage of Niobium Sample Through Detector

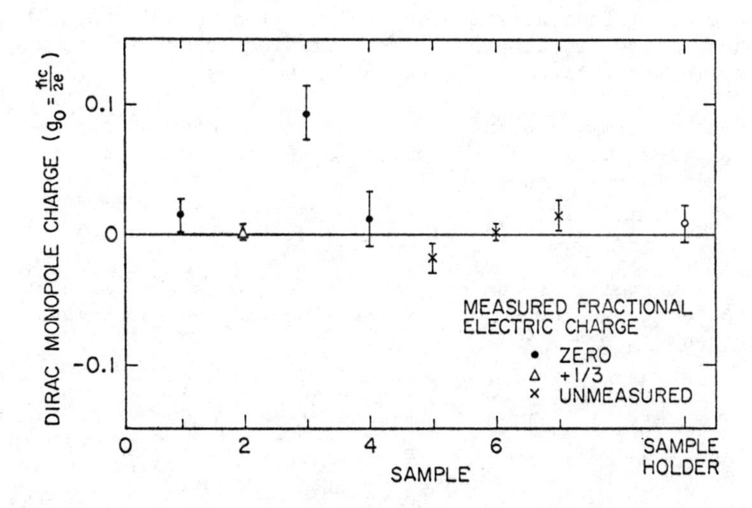

Fig. 15 Data from a Preliminary Monopole Run

FUNDAMENTAL PHYSICAL CONSTANTS

Determining the fundamental constants to even greater accuracy by utilizing experiments in different fields of physics is an extremely powerful tool for testing the overall consistency and validity of the basic theoretical models in physics.

Measurement of 2e/h. Since the discovery of the Josephson effects a very important contribution to the determination of the fine structure constant α has been made and more recently refined, using a SQUID.[13,14]

Using the Josephson relationship $2eV = h\nu$ a comparison can be made between the as maintained volt and the output voltage from a junction irradiated with a known frequency of radiation. Since the highest available voltage from junctions is $\sim$10 mV, at least a factor of 100 step up must be accurately achieved for the comparison. The first generation of room temperature voltage dividers obtained a determination of 2e/h of several parts in 10^6. Using this value a new value for α was determined which cleared up a significant discrepancy existing between the then accepted experimental value for α and the theoretically predicted one. More recently determinations of 2e/h to several parts in 10^8 have been accomplished with cryogenic voltage dividers employing a SQUID as a null detector. First a resistance ratio $R:N^2R$ is determined in two steps R:NR and $NR:N^2R$ utilizing a SQUID as a voltage comparitor. And then a smaller current is passed through N^2R in series with R. The current is adjusted until a galvinometer in series with a standard cell shows no deflection, and then the frequency of the radiation on the junction is adjusted until a null is found in the SQUID output. The limiting error is now due to the heating in the resistors during the higher current resistance calibration.

Measurement of h/m_e. Another measurement of fundamental constants using superconductivity and SQUIDs is the determination of h/m_e. For a thick superconducting ring

$$\int \vec{p} \cdot d\vec{\ell} = \int (2m\vec{v} + \frac{2e}{c}\vec{A}) \cdot d\vec{\ell} = nh \tag{2}$$

where the line integral is taken around the ring, $\vec{p}$ is the canonical momentum and $\vec{A}$ the vector potential. If the ring rotates about its axis at a frequency ω:

$$4m\omega S + \frac{2e}{c}\iint \vec{B} \cdot d\vec{S} = nh \tag{3}$$

and by detecting flux nulls for each n

$$h/m_e = 4S\,\Delta\omega \tag{4}$$

is measured. Using a weak link directly deposited on a 2 mm quartz rod, h/m has been determined to one part in 10^4.[15]

Recent advances in roundness measurement and absolute diameter determination make it possible to determine the area to better than one part in 10^6. In addition null detection can be improved to the necessary level using ultra-low field superconducting shields. A determination of h/m to say 4 parts in 10^7 would provide a determination of α to 2 parts in 10^7 through the relation

$$\alpha^2 = \frac{2R_\infty}{c} \frac{h}{m_e} \qquad (5)$$

and an additional data point for a least squares adjustment of the fundamental constants. In addition an interesting relativistic mass correction should be seen at 10 ppm.

CONCLUSION

Many additional SQUID applications to the study of fundamental physics have been reluctantly omitted in bringing the paper to a manageable length. In the future, with further reductions in the noise of SQUID instruments, existing measurements will be further improved and many new applications will undoubtedly develop.

REFERENCES

1. B. Cabrera, Ph.D. Thesis, Stanford University, 1975.
2. B. Cabrera and F. J. van Kann, "Ultra-low Magnetic Field Apparatus for Continuing Development of a Cryogenic Gyroscope," Acta Astronautica (to be published).
3. "The Gyroscope Experiment I: General Description and Analysis of Gyroscope Performance," C. W. F. Everitt, Experimental Gravitation, B. Bertotti, Ed., pp. 331, Academic Press, 1973.
4. "The Gyroscope Experiment II: Development of the London Moment Gyroscope and of Cryogenic Technology for Space," J. A. Lipa, C. W. F. Everitt and W. M. Fairbank, Experimental Gravitation. B. Bertotti, Ed., pp. 361-380, Academic Press, 1974.
5. J. A. Lipa, "Status of the Gyro Relativity Experiment," Proceedings of the International School of General Relativistic Effects in Physics and Astrophysics: Experiments and Theory (3rd course) 1977, p. 129.
6. C. W. F. Everitt, "General Relativity and Precision Experiments," See Ref. (5) p. 104.
7. R. P. Giffard, S. P. Boughn, W. M. Fairbank, M. S. McAshan, H. J. Paik and R. C. Taber, "Gravitational Wave Detectors of Increased Sensitivity," presented at the International School of Physics "Enrico Fermi," 1976, Varenna, Italy. In Proceedings of the International School of Physics "Enrico Fermi," R. Ruffini, ed. (in preparation).
8. E. Amaldi, "The Gravitational Wave Experiment at the University of Rome," See Ref. (5) p. 174.
9. S. P. Boughn, W. M. Fairbank, R. P. Giffard, J. N. Hollenhorst, M. S. McAshan, H. J. Paik and R. C. Taber PRL 38, 454-457 (1977).
10. M. Taber, Ph.D. Thesis, Stanford University (in preparation).
11. G. S. LaRue, W. M. Fairbank and A. F. Hebard, Phys. Rev. Lett. 38, 1011 (1977).
    See Ref. (1) p. 168.
12. B. Cabrera, "Generating Ultra-Low Magnetic Field Regions and Their Use with a Sensitive Magnetic Charge Detector," Proceedings

of the 14th International Conference of Low Temperature Physics, Vol. IV, pp. 270, Matti Krusius and Matti Vuorio, Eds., North Holland/American Elsevier, 1975.

13. R. F. Dziuba, B. F. Field and T. F. Finnegan, IEEE Trans. on Instr. and Meas. I.M. 23, 264 (1974).
14. V. Kose, IEEE Trans. on Instr. and Meas. IM25, 483 (1976).
15. W. H. Parker and M. B. Simmonds, NBS Special Publication #343, 243 (1970).

SQUIDS AND MAGNETOTELLURICS WITH A REMOTE REFERENCE

John Clarke, Thomas D. Gamble, and Wolfgang M. Goubau
Department of Physics, University of California,
and Materials and Molecular Research and Earth Sciences Divisions,
Lawrence Berkeley Laboratory, Berkeley, California 94720

ABSTRACT

Two 3-axis SQUID magnetometers were used to perform simultaneous magnetotelluric measurements at two sites separated by 4.8 km near Hollister, California. The data obtained at each site were analyzed by using a standard least-squares method, and also using the two magnetic field measurements at the other site as a lock-in reference. The apparent resistivities obtained from the conventional analysis were nearly always biased downward by noise picked up in the magnetic channels, at some frequencies by as much as two orders of magnitude. By comparison, the same data analyzed with the remote reference technique showed no such bias.

INTRODUCTION

Magnetotellurics[1] (MT) is a technique for estimating the surface impedance of the earth. The method involves the simultaneous measurement of naturally-occurring magnetic and electric field fluctuations at the earth's surface. The usual frequency range of interest is 10^{-4} to 10^{3} Hz. Below about 1 Hz, most of these fluctuations at ground level are generated by the currents in the ionosphere and magnetosphere that are produced by solar activity. The spectral density in this range is typically f^{-3}, where f is the frequency. At higher frequencies, the magnetic fluctuations are predominantly generated by electrical thunderstorms around the world. Because of the high conductivity of the earth and the high contrast in conductivity between the earth and the atmosphere, these electric and magnetic fluctuations are related (to a good approximation) in the same way as if they were generated by normally incident plane waves. Most of the energy is reflected, and an evanescent wave penetrates into the ground with a characteristic decay length given by the skin depth, δ, at the frequency of interest. The direction of the evanescent waves is normal to the surface. If the ground is homogeneous and isotropic, the electric field is perpendicular to the magnetic field, but in general it is not. By simultaneously measuring the magnetic and electric fields one can obtain estimates of the impedance tensor of the earth as a function of frequency. Thus, with suitable modeling, one can estimate the conductivity of the ground to depths of tens of kilometers. MT is a widely-used tool in surveying for petroleum and geothermal resources.

One has to measure the horizontal orthogonal components of electric and magnetic field fluctuations, $E_x(t)$, $E_y(t)$, $H_x(t)$, and $H_y(t)$.

ISSN: 0094-243X/78/087/$1.50

In addition, one usually measures the vertical magnetic field component, $H_z(t)$, from which further valuable information can be obtained. The electric field fluctuations are usually measured by means of porous ceramic pots, each containing a saturated solution of copper sulfate in water in which is suspended a copper electrode. The pots are buried just below the surface of the ground at a separation of (say) 500 m, and the electrodes are connected to the input of a differential preamplifier. The voltages typically vary from 10 to 100 $\mu V\ Hz^{-1/2}$ per km around 1 Hz, and their measurement is relatively straightforward. The magnetic field fluctuations have been measured conventionally with induction coils. A typical coil consists of 30,000 turns of copper wire wound on a laminated molypermalloy core 2 m long and 100 mm in diameter. The coil is connected to a differential chopper-stabilized amplifier, specially designed for a high-inductance source. The optimum sensitivity at around 1 Hz is typically 1 to 10 $pT\,Hz^{-1/2}$. The magnetic fluctuations to be measured vary widely, but are often 1 $pT\,Hz^{-1/2}$ or less at higher frequencies. At lower frequencies, the noise increases rapidly with decreasing frequency.

In this paper, we describe an MT survey that we made near Hollister, California using two 3-axis SQUID magnetometers.[2] These magnetometers offer substantial improvements in sensitivity compared with induction coils. In addition, we propose and describe a test of a new technique that uses a remote magnetic reference to lock-in detect the MT signals. This method has resulted in a substantial improvement in the accuracy of the estimates of the impedance tensor.

THEORY OF MAGNETOTELLURICS

In MT, estimates are obtained for the impedance tensor $\underline{\underline{Z}}$ from the equations

$$E_x(\omega) = Z_{xx}(\omega)\, H_x(\omega) + Z_{xy}(\omega)\, H_y(\omega), \tag{1}$$

$$\text{and } E_y(\omega) = Z_{yx}(\omega)\, H_x(\omega) + Z_{yy}(\omega)\, H_y(\omega). \tag{2}$$

$E_x(\omega)$, $E_y(\omega)$, $H_x(\omega)$, and $H_y(\omega)$ are the Fourier transforms of the fluctuating magnetic and electric fields. In the past, estimates of the impedance tensor elements from Eqs. (1) and (2) have used a least-squares analysis that minimizes the mean square error caused by the noises in two of the four fields.[3] This analysis produces an expression for each impedance element that contains an autopower, for example, $|H_x(\omega)|^2$. Thus, if there is noise present in some of the components of E(t) and H(t), the measured autopowers will always be increased by the noise power.[3,4] The influence of the bias on the estimates of the impedance tensor can be very substantial, as we shall see below. Schemes to avoid this problem have been suggested,[5] but none has been wholly successful. In particular, these schemes have failed when the noises in two or more of the channels are correlated, as is often the case.[6]

Our new technique involves the simultaneous measurement of

$E_x(t)$, $E_y(t)$, $H_x(t)$, and $H_y(t)$, and of the magnetic field components, $H_{xr}(t)$ and $H_{yr}(t)$, at a site remote from the MT station. One measures many data segments and Fourier transforms each segment. We multiply Eqs. (1) and (2) in turn by $H^*_{xr}(\omega)$ and $H^*_{yr}(\omega)$ and average the various crosspowers over a narrow band of frequencies and over all data segments to obtain

$$\overline{E_x H^*_{xr}} = Z_{xx} \overline{H_x H^*_{xr}} + Z_{xy} \overline{H_y H^*_{xr}}, \tag{3}$$

$$\overline{E_x H^*_{yr}} = Z_{xx} \overline{H_x H^*_{yr}} + Z_{xy} \overline{H_y H^*_{yr}}, \tag{4}$$

$$\overline{E_y H^*_{xr}} = Z_{yx} \overline{H_x H^*_{xr}} + Z_{yy} \overline{H_y H^*_{xr}}, \tag{5}$$

and

$$\overline{E_y H^*_{yr}} = Z_{yx} \overline{H_x H^*_{yr}} + Z_{yy} \overline{H_y H^*_{yr}}. \tag{6}$$

Equations (3) to (6) can be solved for the impedance elements, for example,

$$Z_{xy} = \frac{\overline{E_x H^*_{xr}}\;\overline{H_x H^*_{yr}} - \overline{E_x H^*_{yr}}\;\overline{H_x H^*_{xr}}}{\overline{H_y H^*_{xr}}\;\overline{H_x H^*_{yr}} - \overline{H_y H^*_{yr}}\;\overline{H_x H^*_{xr}}}. \tag{7}$$

These impedance elements involve only crosspowers of the MT and reference fields, and cannot be biased by noise unless the noise in one or both of the reference fields is correlated with the noise in one or more of the MT fields. Since all of the terms in Eqs. (1) and (2) are multiplied by the same reference fields, the estimates of the impedance tensor do not depend on the amplitude of the reference fields or phase shifts between the reference fields and MT signals. The remote magnetic fields are used to lock-in detect the four MT fields.

MEASUREMENTS AND DATA PROCESSING

To test the remote-reference technique we set up two complete MT stations 4.8 km apart on La Gloria Road near Hollister, California (Fig. 1). The four MT signals and $H_z(t)$ at Lower La Gloria were transmitted by FM telemetry to Upper La Gloria, where they were recorded on a digital tape recorder, together with the five signals at Upper La Gloria. We used three dc SQUIDs[7] as a 3-axis magnetometer at Lower La Gloria. The SQUIDs were mounted in a 5-liter fiberglass dewar that had a liquid helium hold-time of about 6 days. The electronics, which were battery operated, were mounted on top of the cryostat, and a metal can was placed around the cryostat as a low-pass filter with a 3 dB roll-off frequency of 55 Hz. The sensitivity was approximately 10^{-2} pT $Hz^{-1/2}$ at frequencies above 10^{-2} Hz. The whole assembly was surrounded by a wooden wind- and sun-shield.

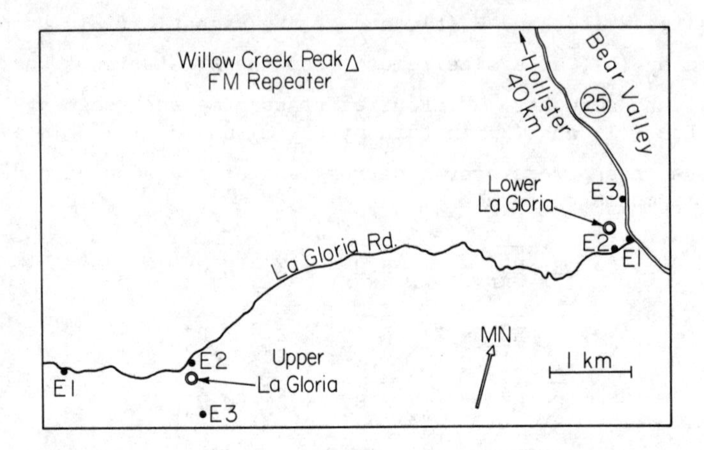

Fig. 1. Magnetotelluric measurement sites in Bear Valley, California. ◎ magnetometer; • electrode.

We used a 3-axis rf SQUID magnetometer[8] at Upper La Gloria. The magnetometer at each site was used as the reference for the MT signals at the other site.

We recorded data continuously for 40 h (except for brief interruptions to change frequency bands and batteries) in the four overlapping bands 0.02 to 1 s, 0.33 to 5 s, 3 to 100 s, and 30 to 100 s. Data were rejected that were obviously useless on a visual inspection, for example, because of the temporary failure of equipment or magnetic interference from passing vehicles. For processing, we grouped the data from each band into segments of equal time, subtracted the mean value and any linear trend, multiplied the ends of each segment by a cosine taper, and computed the fast Fourier transform. The crosspowers were calculated by averaging crossproducts of the various fields over Fourier harmonics in a window of Q = 3 and over all data segments.

The impedance tensor was calculated at each station, and corrected for the non-orthogonality of the electrode arrays (Fig. 1). At each frequency, we rotated the coordinate axes of the tensor to maximize $|Z_{xy}|^2 + |Z_{yx}|^2$, thereby aligning one of the axes along the horizontal direction in which the resistivity was constant, if such a direction existed. We computed the <u>apparent</u> resistivities, $\rho_{xy}(\omega)$ and $\rho_{yx}(\omega)$ defined by

$$\rho_{xy} = 0.2\ T\ |Z_{xy}|^2, \tag{8}$$

$$\text{and } \rho_{yx} = 0.2\ T\ |Z_{yx}|^2, \tag{9}$$

where ρ_{xy} and ρ_{yx} are in Ωm, T is the period in seconds, and Z_{xy} and Z_{yx} are in (mV/km) nT^{-1}. If the ground is homogeneous and isotropic, one can show that $\rho_{xy} = \rho_{yx} = \rho_o$, the true resistivity of the ground.[1]

For comparison, from the same data we also calculated the impedance tensor using a least-squares method in which the local fields $H_x^*(\omega)$ and $H_y^*(\omega)$ replace $H_{xr}^*(\omega)$ and $H_{yr}^*(\omega)$ in Eqs. (3) to (6). The apparent resistivities from this estimate of the impedance tensor are biased downwards to lower values by the presence of noise in the magnetic channels. As an example, consider Z_{xy} in the simple case where the resistivity of the ground varies only with depth, so that $Z_{xx} = 0$. Then, from Eq. (4) with H_{yr}^* replaced with H_y^*, we find

$$Z_{xy} = \frac{\overline{E_x H_y^*}}{\overline{|H_y|^2}} . \qquad (10)$$

Z_{xy} is obviously reduced by noise in H_y.

RESULTS

The apparent resistivities for the remote reference and standard analyses at Upper and Lower La Gloria are shown as functions of period in Figs. 2 and 3, and 4 and 5, respectively. The remote reference apparent resistivities are reproduced as dashed lines in Figs. 3 and 5. All data that survived the initial screening have been included.

It is evident that the apparent resistivities for the remote reference analysis vary much more smoothly with period than those for the least-squares analysis. Furthermore, the apparent resistivities obtained from the least-squares method are consistently lower than those for the remote reference method, as one would expect when there is noise in the magnetic fields. At periods near 10 s at Lower La Gloria (Fig. 5), the bias in ρ_{xy} obtained from the least-squares analysis was more than two orders of magnitude. The corresponding value of ρ_{xy} obtained from the remote reference technique showed no sign of bias.

Where the 0.02 to 1 s and 0.33 to 5 s bands overlapped, the apparent resistivities computed with the remote reference method agreed to within 1.8%. By comparing apparent resistivities obtained from data collected at different times, we estimated the standard deviation at periods less than 3 s to be 1.3%.

SUMMARY AND DISCUSSION

We have successfully obtained MT data using SQUID magnetometers in this and several other field tests. We have not experienced any difficulty with the SQUID magnetometers; in fact they have proved as reliable and as easy to operate as any of the other electronic equipment that we have used in our MT experiments. We have found the new remote reference technique to be significantly better than

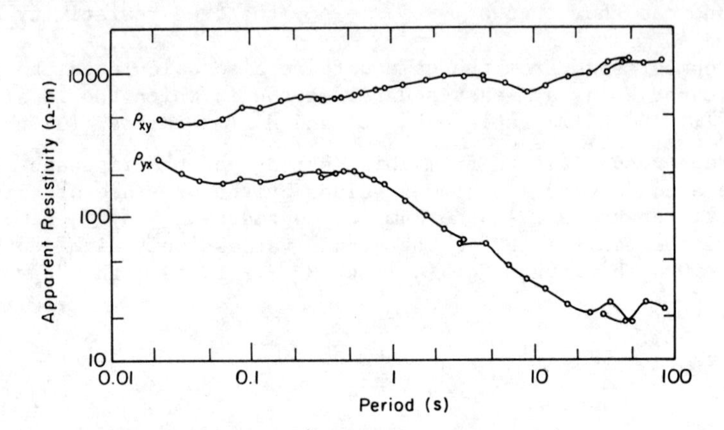

Fig. 2. Remote reference method apparent resistivities vs. period, Upper La Gloria.

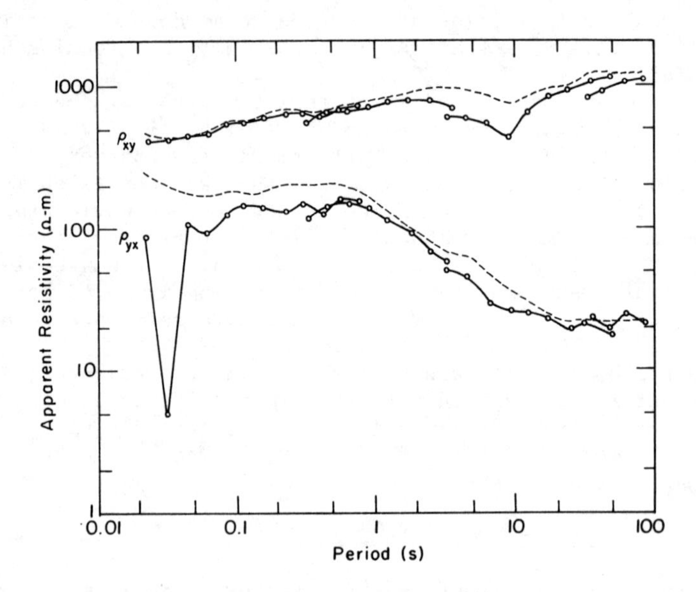

Fig. 3. Standard method apparent resistivities vs. period, Upper La Gloria. --- remote reference results.

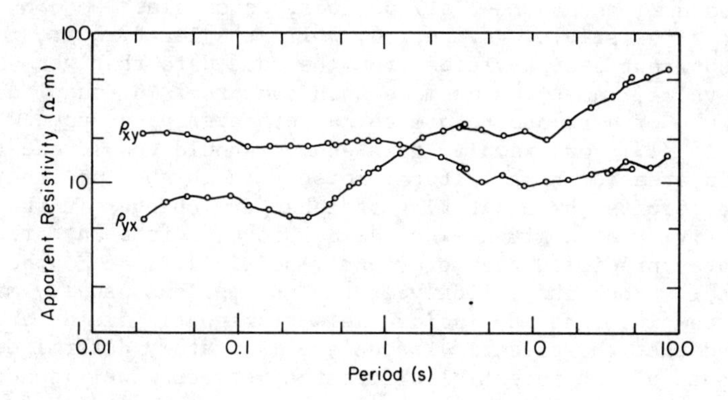

Fig. 4. Remote reference method apparent resistivities vs. period, Lower La Gloria.

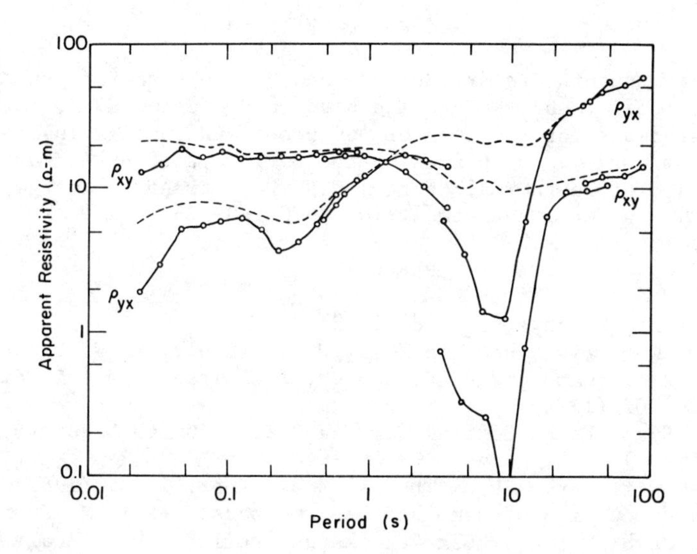

Fig. 5. Standard method apparent resistivities vs. period, Lower La Gloria. --- remote reference results.

the standard MT method. At all periods, we obtained smooth curves of apparent resistivity vs. period, while the least-squares method yielded apparent resistivities from the same data that were always lower in value, sometimes by more than two orders of magnitude.

The use of a remote reference magnetometer opens up new possibilities in MT. For example, this scheme should enable one to perform MT in areas of high cultural noise. The higher accuracy now available permits the monitoring of long-term changes in the apparent resistivity at a given site (as may occur before an earthquake) to a greater precision than previously possible. One may be able to critically test the validity of the assumptions usually made in MT, for example, that the incident electromagnetic fields are plane waves, and that the electric fields are adequately determined by measurements of the potential difference between widely-spaced electrodes. Another intriguing question concerns the nature of the magnetic and electrical noises that may be the signature of some process below the ground. Once accurate apparent resistivities have been obtained, one may investigate these noises by using the measured magnetic fields to predict the electric fields and vice versa, and comparing the predicted fields with the measured fields.

Finally, it should be noted that this technique enables one to make use of almost all of the data collected. Thus, in addition to being more accurate, the apparent resistivities can be measured in a significantly shorter time, thereby reducing survey costs appreciably.

ACKNOWLEDGMENTS

We are grateful to Mr. Melendy and Mr. De Rosa for granting us access to their land. We are indebted to Professor H. F. Morrison and his students for the loan of equipment and for invaluable assistance. This work was supported by the Divisions of Basic Energy Sciences and of Geothermal Energy, U.S. Department of Energy, and by the U.S.G.S. under grant number 14-08-0001-G-328.

REFERENCES

1. K. Vozoff, Geophys. 37, 98 (1972).
2. SQUIDs were first used in MT by N. V. Frederick, W. D. Stanley, J. E. Zimmerman, and R. J. Dinger, IEEE Trans. Geosci. Electron. GE-12, 102 (1974).
3. W. E. Sims, F. X. Bostick, Jr., and H. W. Smith, Geophys. 36, 938 (1971).
4. C. M. Swift, Jr., Ph.D. Thesis, Mass. Inst. Tech. (1967).
5. For a discussion of some of these methods, see W. M. Goubau, T. D. Gamble, and J. Clarke, "Magnetotelluric Data Analysis: Removal of Bias", to be published in Geophys.
6. T. D. Gamble, W. M. Goubau, and J. Clarke, unpublished.
7. J. Clarke, W. M. Goubau, and M. B. Ketchen, J. Low Temp. Phys. 25, 99 (1976).
8. Rented from S.H.E. Corporation.

MEASUREMENTS OF MAGNETIC GRADIENTS FROM OCEAN WAVES

Walter Podney and Ronald Sager
Physical Dynamics, Inc., La Jolla, CA 92038

ABSTRACT

The unprecedented sensitivity of superconductive magnetic gradiometers affords means of measuring fluctuating gradients of magnetic fields from eddy currents generated by ocean waves moving seawater across the earth's magnetic field. Our aim is to tell wave spectra from measurements of fluctuating magnetic gradients over oceans. As a first step in development of requisite experimental and analytical techniques, we report measurements of fluctuating magnetic gradients from surface waves passing an oceanographic research tower located about one mile off Mission Beach near San Diego, California, in 18m of water.

The measurements give experimentally-determined frequency response functions for gradients that substantially confirm response functions calculated using Ampere's law. Before making measurements over an open ocean from aircraft, however, we require, first of all, means of continually maintaining magnitudes of gradiometer imbalance less than 10^{-6} m^{-1}. Development of methods and techniques for suppressing noise from aircraft platforms at frequencies of ocean waves as well as as radio frequencies, and for constructing cryostats suited to aircraft, are presently of secondary importance.

INTRODUCTION

Ocean waves moving seawater across the earth's magnetic field drive, along their crests, eddy currents that generate weak magnetic fields above the oceans. The fields fluctuate in concert with wave motions, so they tell motions of seawater below the surface. Our aim is to read wave motions from measurements of fluctuating magnetic gradients over oceans. Because magnetic fluctuations at sea also come from large scale electric currents in the ionosphere, we measure spatial gradients of magnetic fields to distinguish fluctuations coming from ocean waves. The unprecedented sensitivity of superconductive quantum interference devices (SQUIDS) provides heretofore unavailable means of measuring the small magnetic gradients from ocean waves.

Although so-called alkali vapor magnetometers have been used on fixed, floating, and airborne platforms to measure fluctuations of magnetic fields at sea[1-5], sensitivities of about $0.1(\mathrm{pT/m})/\sqrt{\mathrm{Hz}}$ required to measure magnetic gradients from ocean waves have been attained with superconductive gradiometers only recently[6,7]. We are now just beginning to develop experimental and analytical techniques affording the capability of telling motions of seawater from measurements of fluctuating magnetic gradients made over oceans

ISSN: 0094-243X/78/095/$1.50

from airborne platforms. As the first step, we report here first measurements of magnetic gradients from surface waves. We use a superconductive magnetic gradiometer[7] fixed 7 m above the surface to measure transverse gradients from swells passing the oceanographic research tower operated by the Naval Ocean Systems Center. The tower is located about one mile off Mission Beach near San Diego, California, in 18 m of water.

The measurements show that frequency response of the gradiometer follows the response expected using Ampere's law and superposition of wave motions to describe gradients from ocean waves. Moreover, techniques developed to suppress noise from magnetization currents in the steel structure of the tower suggest means of suppressing noise from magnetization and eddy currents in airframes. In what follows, we first describe the expected response to gradients from surface waves, then compare measured response with expectations, and conclude by outlining requirements, as presently perceived, for an instrument suited to measuring gradients from an airborne platform at sea.

GRADIENTS EXPECTED FROM SURFACE WAVES

We represent gradients of a magnetic field, $\vec{b}(\vec{R},t)$, at position $\vec{R}$, and time t by a matrix G having elements

$$g_{ij}(\vec{R},t) = \hat{x}_i \cdot \vec{\nabla}[\hat{x}_j \cdot \vec{b}(\vec{R},t)] \tag{1}$$

in an orthogonal basis $\{\hat{x}_i\}$. An element g_{ij} represents the gradient in a direction $\hat{x}_i$ of the component of magnetic field in a direction $\hat{x}_j$. Gradients in another basis, $\{\hat{y}_i\}$, say, are determined from gradients in the basis $\{\hat{x}_i\}$ by the similarity transformation

$$G_y = \tilde{R} G_x R \, , \tag{2}$$

where the matrix G_x represents gradients in the basis $\{\hat{x}_i\}$, G_y represents gradients in the basis $\{\hat{y}_i\}$, the orthogonal matrix R represents a rotation from the basis $\{\hat{x}_i\}$ to the basis $\{\hat{y}_i\}$, and tilde marks a transposed matrix.

Because the divergence of a magnetic field vanishes, the sum of diagonal elements of a gradient matrix vanishes. Furthermore, the identity

$$g_{ij} - g_{ji} = (\hat{x}_i \times \hat{x}_j)\cdot(\vec{\nabla} \times \vec{b}) \tag{3}$$

tells us that matrices representing gradients of an irrotational magnetic field are symmetric; $\tilde{G} = G$. Displacement currents are negligibly small at frequencies of seawater oscillations, so magnetic fields from ocean waves are irrotational above the surface. Matrices representing magnetic gradients above an ocean, then, are symmetric and have a vanishing trace.

We express the magnetic field generated above an ocean by waves in terms of a Fourier integral of the field generated by a horizontally progressing wave, $\vec{\beta}(\vec{k},\omega,z)$, having frequency ω and wave vector $\vec{k}$; namely

$$\vec{b}(\vec{R},t) = \iint d\omega\, d\vec{k}\, \vec{\beta}(\vec{k},\omega,z)\, e^{i(\omega t-\vec{k}\cdot\vec{r})} , \qquad (4a)$$

where $\vec{r}$ is a horizontal coordinate vector, $\hat{z}$ is a unit vector directed vertically downward, and $\vec{R} = \vec{r} + z\hat{z}$. The vertical coordinate, z, is positive below the surface and negative above the surface. We reckon heading of a horizontally progressing wave with respect to a magnetic meridian plane as shown in Figure 1. In an earlier

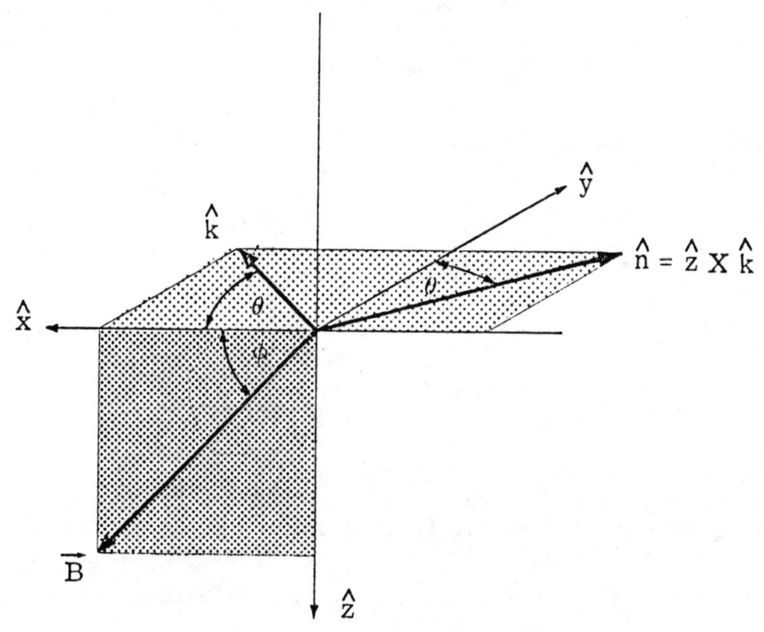

Fig. 1. Heading of a horizontally-progressing wave. The horizontal x axis is directed northward; the y axis, eastward; and the earth's magnetic field vector lies in the vertical x-z plane and is inclined at an angle ϕ to the x axis.

paper[8], we demonstrated that the expression

$$\vec{\beta}(\vec{k},\omega,z) = i\omega A(\vec{k},\omega) \left(\frac{\mu_0\sigma}{2k}\right) [(\vec{B}\cdot\hat{\upsilon}) + f(kD)(\vec{B}\cdot\hat{\upsilon}^*)] e^{kz} \hat{\upsilon}^* , \quad (4b)$$

where

$$f(x) = \frac{xe^{-x}}{\sinh x} , \quad (4c)$$

gives the magnetic field produced above the surface ($z < 0$) by a surface wave progressing horizontally on an ocean of depth D and having an amplitude $A(\vec{k},\omega)$. Here, $\vec{B}$ is the earth's steady magnetic field; σ, the electrical conductivity of seawater (~4 mhos/m); $\mu_0 = 4\pi \times 10^{-7}$ H/m, and $\hat{\upsilon}^*$ is the complex conjugate of the circularly polarized unit vector defined by the relation

$$\hat{\upsilon} = \frac{1}{\sqrt{2}} (\hat{z} + i\hat{k}) \quad . \quad (4d)$$

From equations 1 and 4a, we find that the expression

$$g_{ij}(\vec{R},t) = \iint d\omega d\vec{k} \, (\hat{x}_i\cdot\hat{\upsilon}^*)(\hat{x}_j\cdot\hat{\upsilon}^*) g(\vec{k},\omega,z) \, A(\vec{k},\omega) \, e^{i(\omega t-\vec{k}\cdot\vec{r})} , \quad (4e)$$

where

$$g(\vec{k},\omega,z) = \left(\frac{\mu_0\sigma}{\sqrt{2}}\right) i\omega \, [(\vec{B}\cdot\hat{\upsilon}) + f(kD)(\vec{B}\cdot\hat{\upsilon}^*)] \, e^{kz} , \quad (4f)$$

gives elements of a matrix representing gradients of magnetic fields generated above an ocean by surface waves.

To determine spectra of fluctuating gradients above an ocean, we suppose that wave fields are statistically homogeneous and stationary and specify a covariance, $C_{pq}^{\ell m}(z,\tau)$, between gradients g_{pq} and $g_{\ell m}$ as the averaged product

$$C_{pq}^{\ell m}(z,\tau) = \iint d\vec{r} \, dt \, g_{pq}(\vec{R},t) \, g_{\ell m}(\vec{R},t+\tau) \quad . \quad (5a)$$

Its spectrum, $P_{pq}^{\ell m}(z,\omega)$, is the Fourier integral

$$P_{pq}^{\ell m}(z,\omega) = \frac{1}{2\pi} \int d\tau \; C_{pq}^{\ell m}(z,\tau) \; e^{-i\omega\tau} \; . \tag{5b}$$

By using Equation 4e, we then find that the expression

$$P_{pq}^{\ell m}(z,\omega) = \int d\vec{k} \; \chi_{pq}^{\ell m}(\hat{k}) \; |g(\vec{k},\omega,z)|^2 < |A(\vec{k},\omega)|^2 > \tag{5c}$$

gives spectra of fluctuating gradients above an ocean. Here,

$$\chi_{pq}^{\ell m}(\hat{k}) = (\hat{p}\cdot\upsilon)(\hat{q}\cdot\hat{\upsilon})(\hat{\ell}\cdot\hat{\upsilon}^*)(\hat{m}\cdot\hat{\upsilon}^*) \tag{5d}$$

and $<|A(\hat{k},\omega)|^2>$ is a spectrum of surface wave amplitudes. In keeping with a linear description of surface waves, we use the dispersion relation

$$\omega^2 = gk \tanh kD \quad , \text{ with } g = 9.8 \text{ m/(sec)}^2 \; , \tag{6a}$$

to express Equation 5c in terms of a wave number spectrum, $\Lambda(k,\theta)$, as

$$P_{pq}^{\ell m}(z,\omega) = \frac{k}{C_g} \int_0^{2\pi} d\theta \; \chi_{pq}^{\ell m}(\theta) |g(k,\theta,z)|^2 \Lambda(k,\theta) \; , \tag{6b}$$

where θ is the angular heading of a wave reckoned eastward of magnetic north, $k(\omega)$ is wave number at frequency ω, and C_g is the group speed; namely, $C_g = d\omega/dk$. Equation 6b tells us spectra of gradients are proportional to a weighted integral of a directional wave spectrum over wave heading.

To compare spectra of gradients expected from surface waves with measured spectra, we simultaneously measure spectra of wave amplitude and gradients and empirically determine a frequency response function that relates fluctuating gradients to wave motions. We measure a single transverse gradient, so $\ell = p$ and $m = q$. Moreover, wave spectra at the tower, for our purposes, are dominated by swell running largely eastward toward Mission Beach, so the relation

$$P(z,\omega) \cong |H(k,\bar{\theta},z)|^2 W(\omega) \tag{7a}$$

approximately describes spectra of fluctuating gradients expected from swell passing the tower at an average heading $\bar{\theta}$. Here the relation

$$W(\omega) = \frac{k}{C_g} \int_0^{2\pi} d\theta \; \Lambda(k,\theta) \tag{7b}$$

gives the frequency spectrum of wave amplitudes, and

$$H(k,\overline{\theta},z) = (\hat{p}\cdot\hat{\upsilon})(\hat{q}\cdot\hat{\upsilon})g(k,\overline{\theta},z) \ , \tag{7c}$$

gives the expected frequency response function.

MEASURED GRADIENTS FROM SURFACE WAVES

We use a superconductive magnetic gradiometer[7] comprising a single axis formed by two coplanar pickup loops to measure gradients from ocean waves passing the tower. The cryostat fits in a gimbal mount fixed to a nonmagnetic cantilever truss that supports it about 7m above the surface and 20m from the south face of the tower. The two axes of rotation and one axis of tilt of the gimbal allow every orientation of the pickup loops within a cone fixed by the tilt limits (± 45°). Three pressure transducers in a triangular array measure wave heights and give an estimate of wave direction.

We use the gimbal to orient pickup loops in order to suppress noise from magnetization currents in the steel structure of the tower. As we demonstrate elsewhere[9], gradients of irrotational magnetic fields at a point are equal to gradients of a dipole positioned on a sphere of unit radius about the point. Moment of the dipole and four angles, giving its location and orientation, specify gradients of an irrotational magnetic field at a point. We represent fluctuating gradients from the tower, then, as coming from small fluctuations in moment, orientation, and location of an equivalent dipole giving gradients of the tower at the position of the gradiometer. Orienting pickup loops so that an equivalent dipole representing tower gradients lies in the plane of the loops and is located along a vector perpendicular to the line joining centers of the loops suppresses noise from apparent fluctuations in moment, orientation, and location of the dipole. Fluctuations result both from small motions of the gradiometer in the steady gradient field of the tower and from ionospherically-induced fluctuations of magnetization currents in the tower.

With pickup loops oriented to suppress noise from the tower, we simultaneously record data once a second from three pressure transducers and the gradiometer. Figure 2 compares a spectrum of fluctuating gradients with a spectrum of wave amplitudes determined from a time series record of 2048 seconds. The dashed curve gives the spectrum of wave amplitudes and corresponds to the right hand coordinate scale marked in units of m^2/Hz. The solid curve gives the corresponding spectrum of gradient fluctuations in units of V^2/Hz, as marked on the left-hand coordinate scale. The gradiometer calibration constant is 4.5 V/(nT/m). Values of kD are marked along the upper coordinate scale.

As is evident, features of the spectrum of gradient fluctuations replicate features of the spectrum of wave amplitudes. The sharp drop at the upper end of the frequency scale comes from a lowpass

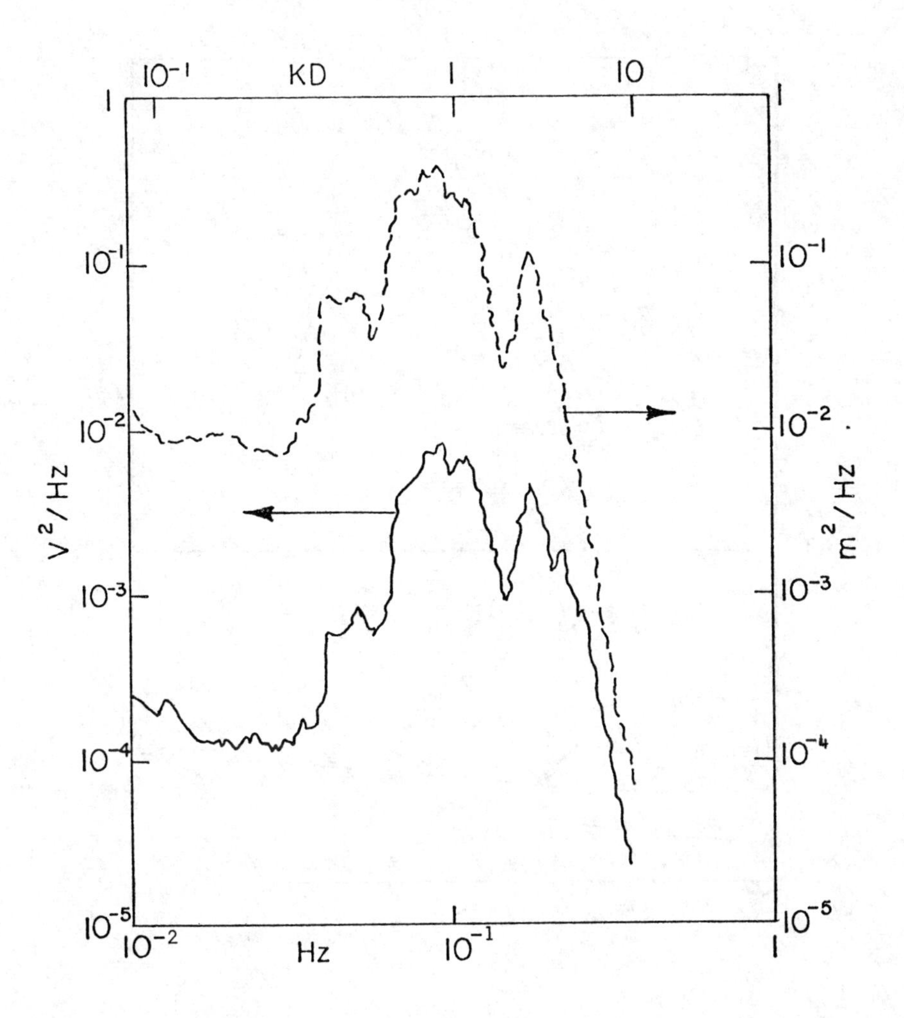

Fig. 2. Spectral densities of wave amplitudes (dashed curve) in units of m^2/Hz and gradient fluctuations (solid curve) in units of V^2/Hz.

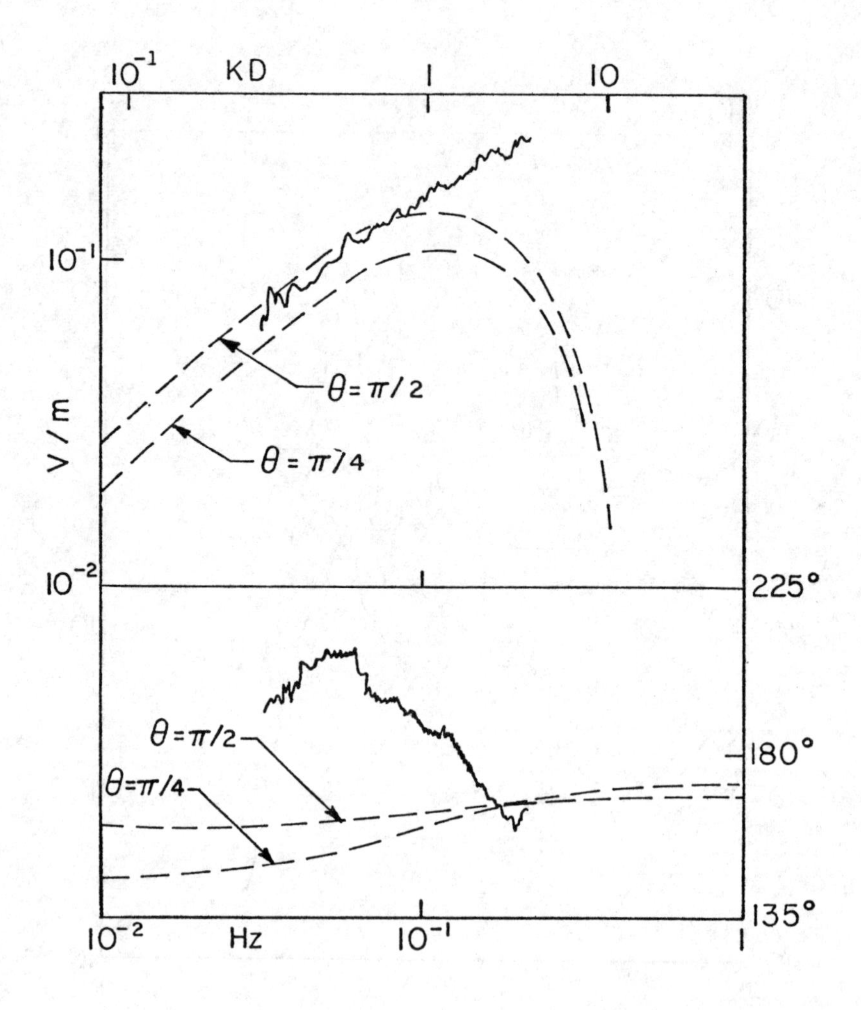

Fig. 3. Comparison of measured (solid curves) and expected (dashed curves) gain and phase of the gradiometer frequency response.

filter having a cutoff frequency of 0.3 Hz to preclude aliasing. The dominant swell component is at about 0.1 Hz with sidebands at 0.045 Hz and 0.178 Hz. The plateau at the lower end of the frequency scale lies at frequencies below the swell, which are poorly resolved for a record length of 2048 seconds when averaged over 32 frequencies.

The ratio of the two spectra determines gain of the frequency response, as approximated by Equation 7a. We empirically determine gain and phase of the frequency response from a cross spectrum between wave amplitudes and gradient fluctuations[10]. Figure 3 compares gain and phase of the frequency response determined from time series having spectra shown in Figure 2 with gain and phase expected for average headings of 90° and 45° east of true north. Dashed curves delineate expected values, and solid curves, measured values. Measured gain agrees well with expected values over the frequency band of the swell, but it does not show the expected decrease at the upper end of the frequency scale. Measured phase, however, differs noticeably from expected values.

We have not resolved differences between expected and measured frequency response functions yet. Phase lags can result from the pressure transducers or from the electronic low pass filter. Absence of the expected decrease in gain indicates presence of an unresolved contribution at the upper frequencies, which likely comes from motion of the gradiometer. We are presently analyzing time series for a number of other days to determine consistency and variability of estimates of frequency response. Nonetheless, the data do establish, for our purposes here, that a superconductive magnetic gradiometer responds to gradients of magnetic fields from surface waves and that Ampere's Law together with superposition of waves describe amplitude of its response per unit wave amplitude.

CONCLUSION

Measurement of fluctuating gradients using a superconductive gradiometer fixed above the surface on a stable platform in shallow water is the first step in developing experimental and analytical techniques affording means of telling motions of seawater from measurements of fluctuating magnetic gradients made over an open ocean from an airborne platform. First measurements presented here show that our preconceptions of dominant effects are fundamentally correct and so provide a basis for setting requirements for an instrument suited to making measurements from an airborne platform at sea. Platform motion not only introduces additional noise but also aliases frequencies of magnetic fluctuations by shifting them in accordance with the Doppler effect.

Measurements from aircraft require, first of all, means of continually maintaining imbalance of a gradiometer less than 10^{-6} m^{-1} in order to suppress noise from motion through the earth's magnetic field. Although present methods of balancing by mechanical adjustment of small superconducting discs are adequate for measurements from fixed platforms, they are not stable enough for measurements from moving platforms. Electronic means of attaining and maintaining an imbalance less than 10^{-6} m^{-1} must be developed.

Second, means of suppressing noise at frequencies of ocean waves from magnetization and eddy currents in an airframe are required. To avoid limitations on cryostats and electronics imposed by towing an instrument far enough from the airframe to avoid its magnetic field, we suggest developing techniques to suppress noise for an instrument fixed to the airframe. For example, orienting pickup loops can suppress noise from an airframe, in the same manner as from the tower structure, to the extent that fluctuations in moment, orientation, and location of an equivalent dipole represent fluctuating gradients from currents in an airframe.

Third, directionality of gradients must be used to develop, in effect, a "sidelooking" gradiometer that preferentially sees gradients from waves progressing perpendicular to the flight path. Aliasing resulting from the Doppler effect is then minimal. At least two gradiometer axes are required to provide enough collimation. For example, coplanar pickup loops preferentially respond to gradients from waves heading normal to the plane of the loops, and coaxial pickup loops, to gradients from waves heading along their axis. Consequently, subtracting the response of two coplanar pickup loops from two coaxial pickup loops, with each pair having a common axis joining their centers, and cross-correlating the difference with response from the coaxial loops, give the spectrum of waves progressing preferentially along the common axis.

Finally, cryostats for airborne instruments pose unique problems to be considered as an integral part of overall instrument development. Development of cryostats for inboard instruments is less exacting than for outboard instruments.

ACKNOWLEDGEMENT

We greatly appreciate unstinting support provided by the Ocean Measurements Group of the Naval Ocean Systems Center, under the direction of Dale Good, and by the staff of the La Posta Astrogeophysical Observatory, directed by Ilan Rothmuller.

REFERENCES

1. K. C. Maclure, R. A. Hafer, and J. T. Weaver, Nature, 204, 1290 (1964).
2. A. N. Kozlov, G. A. Fonarev, and L. A. Shumov, Geomagn. Aeron., Engl. Transl., 11, 633 (1971).
3. A. G. Kravstov, and G. V. Sokolov, Geomagn. Aeron., Engl. Transl. 11, 783 (1971).
4. Ye. M. Groskaya, R. G. Skrymnikov, and G. V. Sokolov, Geomagn. Aeron., Engl. Transl., 12, 131 (1972).
5. R. Baker and P. W. Graefe, 5th Symposium on Remote Sensing of

Environment, U. S. Office of Nav. Res., Univ. of Mich., Ann Arbor, 1968.

6. W. M. Wynn, C. P. Frahm, P. J. Carroll, R. H. Clark, J. Wilhoner, and M. J. Wynn, IEEE Trans. Magn., MAG-11, 701 (1975).
7. G. H. Gillespie, W. N. Podney, and J. L. Buxton, Jour. Appl. Phys., 48, 354 (1977).
8. W. N. Podney, Jour. Geophys. Res., 80, 2977 (1975).
9. W. N. Podney, and G. H. Gillespie, RADC-TR-77-100, 12 (1977).
10. J. S. Bendat and A. G. Piersol, Random Data: Analysis and Measurement Procedures (Wiley-Interscience, N. Y., 1971), p. 196.

BIOMEDICAL APPLICATIONS OF SQUIDs

S.J. Williamson, D. Brenner, and L. Kaufman
Neuromagnetism Laboratory*
Departments of Physics and Psychology
New York University, New York, N.Y. 10003

ABSTRACT

SQUID magnetic field detectors are currently used in nearly a dozen laboratories in the United States and Europe to study biomagnetic fields produced by various organs of the human body. Some of the more promising findings of clinical and fundamental interest are briefly reviewed. Emphasis is placed on which technological improvements in the performance of SQUID systems may prove beneficial in future biomedical applications.

INTRODUCTION

The demonstrations that SQUID systems can detect weak magnetic fields from the human body in rural[1] and urban[2] environments without recourse to magnetic shielding has made biomagnetic techniques more generally available for clinical and research purposes. The results to date clearly show that magnetic techniques offer advantages in several applications over conventional measurements of voltages between points on the skin, and indeed provide information that may not be attainable by skin electrodes. Simply put, a magnetic field detector is most sensitive to the strongest electrical currents nearby, and the relatively strong currents within the active organ of interest compared with the weak currents which have propagated to the dermis may make the latter inconsequential despite their closer proximity to the probe. Thus in many cases magnetic fields provide a direct indication of internal activity, whereas skin voltages are indirect. Biomagnetic fields derive from different spatial averages over the primary currents within the active organ and the accompanying secondary currents within the surrounding conducting tissue than do skin voltages. Therefore, the two differently emphasize contributions from various regions. The magnetic measurement technique itself, when incorporating a SQUID, provides several advantages in biomedical studies such as wideband sensitivity, including response down to dc. And not to be overlooked is the capability for continuous spatial mapping of field patterns by simply moving the probe over the body. Fine resolution, limited by the size of the pickup coil, is proving useful in studies of the brain where significant variation in field over distances of 1 cm have been observed.[3]

*Supported by the Office of Naval Research (00014-76-C-0568).

ISSN: 0094-243X/78/106/$1.50

Nearly a dozen laboratories now have programs in biomagnetism. The topics include studies of the relatively strong remanent moment of magnetic contaminants in the lung (~ 100 picotesla); the somewhat weaker fields associated with the susceptibility of portions of the body responding to the earth's steady field (~ 10 pT); time-varying fields generated by the flow of current in and around the heart (~ 50 pT); the weaker field near the head caused by spontaneous activity in the brain such as the alpha rhythm (~ 1 pT); and the weakest of all biomagnetic fields--those caused by brain activity evoked by sensory stimuli (~ 0.1 pT). Here we can only touch on some of the highlights of the most recent advances in these areas. The reader is referred to recent reviews[4,5] for a more complete discussion of the relevant background.

SENSITIVITY

Measurements of weak biomagnetic fields in the presence of varying background fields some four to seven orders of magnitude greater are made possible by the technique of spatial discrimination. In practice, the SQUID senses the field of interest by means of a superconducting flux transporter whose primary (or "detection" coil) has a geometry that renders it insensitive to fairly uniform fields generated by distant sources. In rural settings a detection coil wound in the form of a gradiometer may suffice.[1] But with the higher noise level in an urban environment, fluctuating field gradients must also be avoided, and a second-derivative gradiometer (two coaxial gradiometers in opposition) may be essential for suitable noise suppression. High sensitivity to the field from a source placed near the end coil (or "pickup" coil) of the detection coil can nevertheless be retained, provided that the field strength diminishes so rapidly with distance that it has negligible coupling with the other coils comprising the detection coil.

Two examples of noise spectra from second-derivative gradiometer systems operating on the ninth story of a steel frame

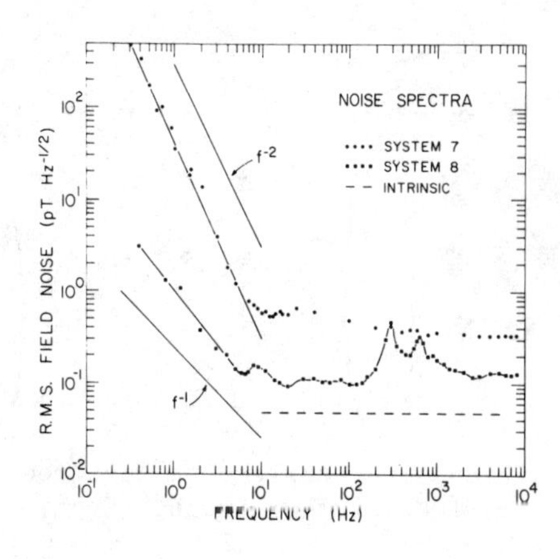

Fig. 1. Noise spectra for two unshielded SQUID systems. The intrinsic noise for System 7 was 0.3 pT $Hz^{-\frac{1}{2}}$ and for System 8, 0.04 pT $Hz^{-\frac{1}{2}}$. For the latter, cancellation techniques were used to eliminate noise fields from the power lines at 60 Hz and 180 Hz. The diameter of the pickup coil in both systems was 2.3 cm.

building in Manhattan are shown in Fig. 1. Above ~ 10 Hz the noise is reasonably flat except for components at odd harmonics of the power line frequency. Below ~ 6 Hz the noise for our most recent detector (System 8) varies approximately as f^{-1}, whereas for the previous detector (System 7) it displayed a stronger frequency dependence. The latter effect is believed due to excessive gradient noise which arose because remote coils in the flux transporter were used to trim the balance of the detection coil; whereas System 8 has small superconducting tabs placed near the detection coil itself. In comparison with the noise exhibited by System 8 in our laboratory even better performance can be obtained by operating the system in a rural location. The high frequency sensitivity is then limited by the intrinsic noise of the SQUID system, with the increase in noise at low frequency being observed only below a few Hertz. SQUID gradiometers operating in well-designed magnetically-shielded[6] or eddy current-shielded[7] enclosures will exhibit the intrinsic SQUID noise to even lower frequencies.

REMANENT FIELDS

Magnetic contaminants in the lung after being magnetized by an externally applied field can be detected by sensing the remanent field outside the chest.[5,8] This finding has led to ongoing studies on humans and animals of deposition patterns and clearance rates, using benign ferrimagnetic Fe_3O_4 as a magnetic tracer. A pilot study on nine normal subjects is being conducted by D. Cohen at M.I.T. and a group under J. Brain at the Harvard School of Public Health. Moderately small amounts of magnetite yield a remanent field that easily can be detected by conventional fluxgate magnetometers having

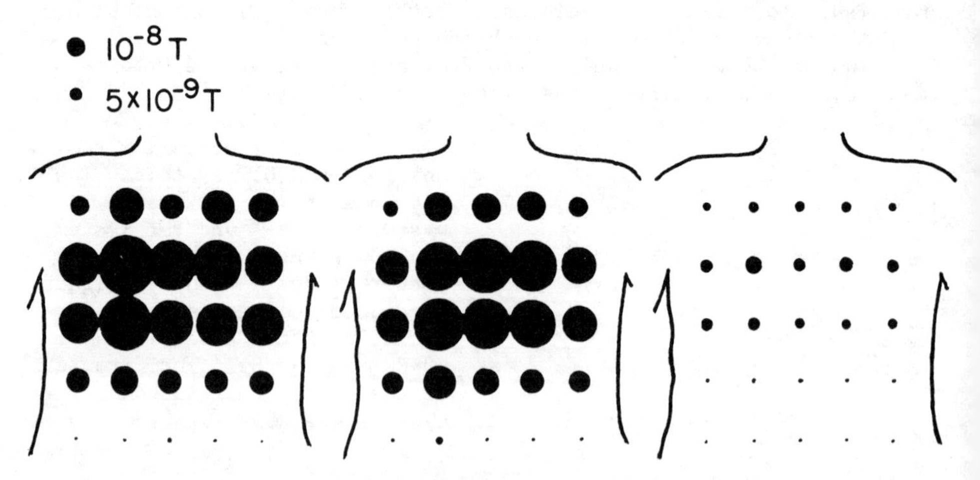

Fig. 2. Remanent field at various positions over the chests of three arc welders is indicated by the relative areas of the circles (From reference 9).

a noise level of ~ 30 pT $Hz^{-\frac{1}{2}}$. Such techniques are now employed by M. Lippmann and collaborators at the Institute of Environmental Medicine at New York University to investigate long-term clearance characteristics in the lungs of donkeys, as indicated by both magnetic and conventional radioactive techniques.

SQUID systems have been used by a group at the Helsinki University of Technology to map the magnetic loading in the lungs of arc welders (Fig. 2).[9] A positive correlation was found between the strength of the remanent field and indications of abnormality in chest radiographs even when respiratory function was found normal. An extension of magnetic surveying to an assessment of dust retention in the lungs of coal miners is underway by S. Robinson at the Hahnemann Medical College in Philadelphia. Here the high sensitivity of the SQUID is essential because the concentration of magnetic contaminants in coal is low. One question yet to be answered is whether long-term exposure in a coal mine can increase the remanent field significantly above the level for the non-exposed control population.

SUSCEPTIBILITY

A weak diamagnetic field from the human body responding to the earth's field can easily be detected by a SQUID system when a person moves close to the pickup coil. The change in field at the coil is typically ~ 30 pT. Increased sensitivity can be attained by applying a more intense dc field as was done at Stanford University by J.P. Wikswo, Jr., J.E. Opfer and W.M. Fairbank for monitoring changes in the volume of blood in the heart.[10] An especially interesting observation was recently reported by a group at Case Western Reserve University who found with an unshielded SQUID system in a rural location that the normally diamagnetic human liver may become paramagnetic from the accumulation of excess iron in patients suffering from hemochromatosis, a disease that may prove fatal.[11] The possibility of monitoring the amount of iron in the form of ferritin and hemosiderin opens interesting prospects for more widespread clinical applications. But reduction in the detected fluctuations of background fields at low frequencies must be achieved before this could be done without shielding in a hospital setting.

CARDIOMAGNETISM

The heart's magnetic field, which was first observed in 1963 by G. Baule and R. McFee,[12] continues to attract the greatest attention. Studies have been performed or are underway at M.I.T., Helsinki, T.R.W. Systems, Stanford and Vanderbilt Universities, Dalhousie University in Halifax, Twente University of Technology in Holland, Case Western Reserve University, and the Scientific and Medical University at Grenoble. The cardiac field near the chest is sufficiently strong (~ 50 pT) so that even in a hospital it can be detected with a Faraday induction coil fitted with a high permeability core. Clinically useful records can be obtained by signal averaging about 30 beats using the electrocardiogram as a reference trigger,

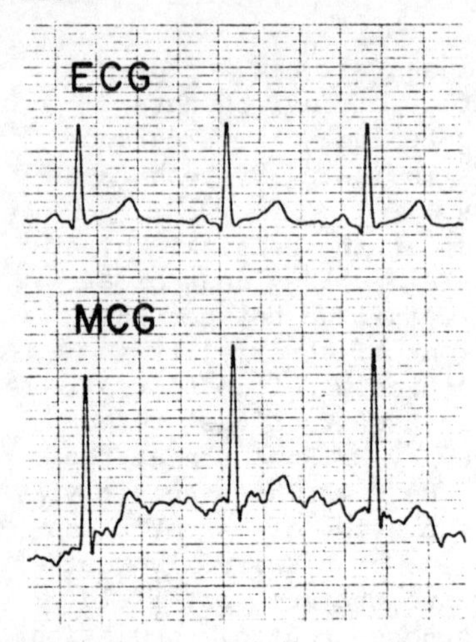

Fig. 3. Electrocardiogram and magnetocardiogram of a normal patient as recorded in an unshielded environment. The MCG at this position over the chest has a P-wave with inverted polarity preceding the large QRS peak. (Provided by R. Walter-Peters of the Twente University of Technology).

as is done by B. Denis and D. Matelin at Grenoble.[13] A second-derivative SQUID system operated by the group under R. Walter-Peters at Twente in an unshielded building only 50 m from a busy highway has produced the data illustrated in Fig. 3. The noise at low frequencies can be greatly reduced by filtering and signal averaging. Other examples of results from Stanford are given in the paper by J.P. Wikswo, Jr. and collaborators which appears in this proceedings.

In studies with cardiograms it is desirable to minimize contributions from secondary currents in tissue surrounding the heart so as to increase relative sensitivity to the primary currents within the heart itself. J.P. Wikswo, Jr., et al.[14] provide evidence that the magnetocardiogram when recorded directly over the heart has substantially reduced sensitivity to secondary currents than when measured elsewhere over the chest or in comparison with the electrocardiogram. An effective format for summarizing the MCG taken over the heart has been described by W. Barry et al.[15] Data are obtained by placing the SQUID detector above the chest of a supine patient with the instrument axis directed vertically through the center of the ventricular chambers. With the detection coils tilted at an angle of $54^{o}44'$ to the instrument axis, successive measurements at 120° intervals of rotation about the axis yield data along orthogonal directions from which the orientation and magnitude of the magnetic heart vector (MHV) can be deduced. The MHV is the magnetic dipole positioned at the center of the heart that would account for the observed signals from the SQUID detector. Fig. 4 compares features of the MHV throughout the cardiac cycle with the analogous electric heart vector (EHV) deduced from skin voltages measured with the Frank arrangement of 12 electrodes. The patient characterized in Fig. 4 suffers from a bundle branch block that causes the left ventricle to be activated later than the right, thus displaying separated R peaks in the QRS complex of the MHV cycle. As illustrated, the MHV may differently emphasize aberrant behavior when compared with the EHV and consequently may prove a useful supplement in clinical and research studies.

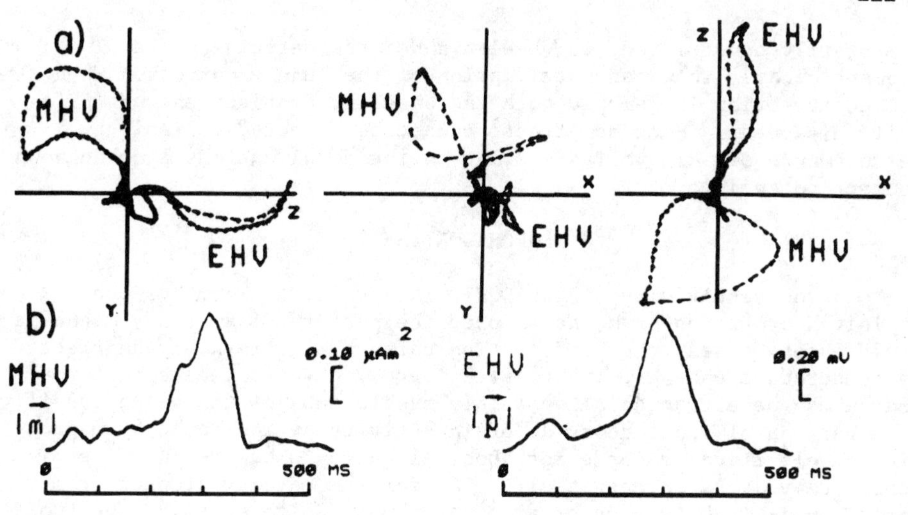

Fig. 4. (a) Loci traced by the heads of the magnetic and electric heart vectors throughout one cardiac cycle, projected onto three orthogonal planes. The x-axis is directed toward the patient's left, y toward the feet, and z toward the back. (b) Variation of the magnitudes of the MHV and EHV throughout one cycle (From reference 15).

Compared with measuring skin voltages, a unique advantage of the SQUID is its capability for dc measurements. An example described by D. Cohen and L.A. Kaufman[16] is given in Fig. 5. An anomaly in the electrocardiogram known as the "ST shift" was commonly believed to represent the effect of a new current due to cardiac injury which appears only during the ST portion of the cycle. Experiments on dogs demonstrated that this need not be the case; to the contrary, the ST shift may result from an <u>interruption</u> of the injury current, which at other times is steady. The field from this dc current of injury can be seen in Fig. 5 as a baseline displacement when the dog, with coronary artery

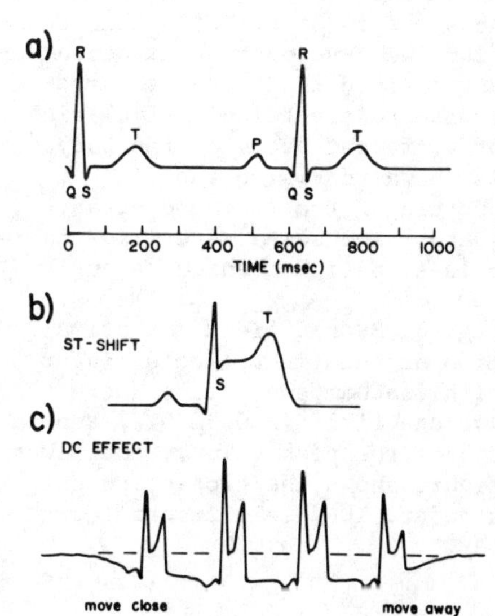

Fig. 5. (a) Normal ECG; (b) an example of an ST shift in the ECG; (c) MCG as a dog with constricted coronary artery is wheeled toward and then away from the SQUID detector (Adapted from reference 16).

temporarily constricted, is wheeled under the detector. The ST segment clearly shows an interruption of the injury current. When the constriction is removed both the baseline displacement and ST shift disappear. Such dc effects cannot be detected reliably by measurements of skin voltages owing to the complications from unknown contact potentials.

NEUROMAGNETISM

The dc sensitivity of the SQUID in a shielded room has been exploited by D. Cohen who has mapped the pattern of a steady magnetic field observed near the head.[17] The pattern is remarkably unaffected by temperature changes, light, sleep, etc., which suggests that the source may be extracranial, possibly muscle and not the brain itself. A comparison of spontaneous ac brain activity as indicated by magnetic and electrical records for abnormal patients has recently been reported by J. R. Hughes et al.[18] As for the earlier studies of normal subjects by Hughes et al.[19] and by M. Reite et al.[20], the two records display different features, with alpha rhythms occasionally but not always being correlated. The clinical implications of these differences have not yet been assessed.

A major investigation of how the brain responds to various sensory stimuli is underway by our group at New York University. Unshielded measurements when complemented by signal averaging have allowed us to detect responses to repetitive stimuli as weak as 40 femtotesla with a signal-to-noise ratio of 2. With this technique we have observed a neuromagnetic field that appears in response to electrical stimulation of different areas of the body.[3] As shown in Fig. 6 the field pattern of the averaged steady-state response to repetitive current pulses applied transcutaneously

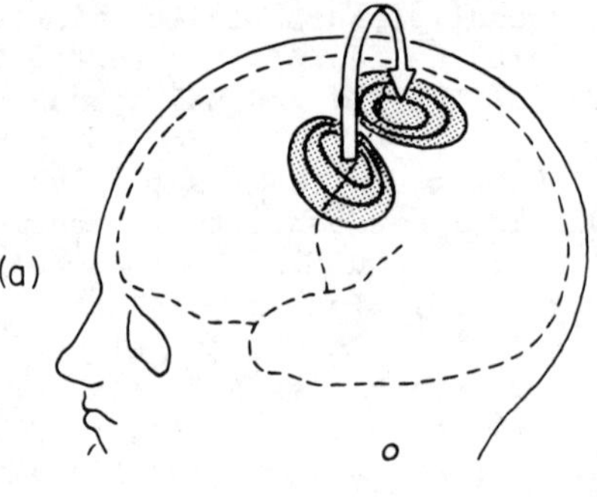

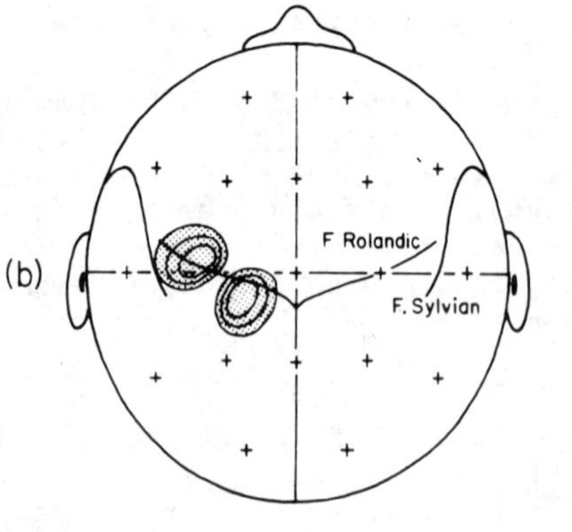

Fig. 6. Evoked field for stimulation of the right little finger, with isochamps denoting where the average field is 0.9, 0.7, and 0.5 of the peak value. The lower figure shows the projection onto the standard 10-20 electrocardiogram map.

across the right little finger, appears on the contralateral side of the head. One half-cycle after the time illustrated the field direction is reversed. The center of the pattern can be located with an accuracy of 1 cm, but the spatial resolution of the details of the pattern is limited by the diameter of the pickup coil which is 2.3 cm. When the thumb is stimulated instead, the pattern shifts downward by 2 cm. These positions on the scalp coincide with the expected locations of activity in the underlying somatosensory cortex of the brain, as established on other patients during brain surgery by W. Penfield [21] and others. The peak field intensity occurs about 70 ms following a stimulation pulse, and this latency is interpreted as the sensory response time. Despite increasing background noise at low frequencies (Fig. 1), responses to stimulation at repetition rates as low as 0.8 Hz could be detected after a 1 min average, thus indicating that this technique is nearly able to monitor transient responses.

In earlier studies of the visually-evoked response we established that the evoked neuromagnetic field is localized near the visual cortex, at the back of the head.[2] This contrasts with evoked scalp potentials, which are more widespread. The latency for the appearance of the peak field was found to depend on features of the pattern used as a stimulus.[4] A particularly simple and useful pattern is formed by a sinusoidal variation of luminance horizontally across the screen of an oscilloscope, with the parallel vertical bars of the grating formed in this way being periodically shifted back and forth so that dark regions are periodically replaced by light. The contrast of any position on the screen with respect to adjacent areas is thereby periodically reversed. The neuromagnetic latency measured from the time of contrast reversal is found to vary with the spatial frequency of this pattern or the number of bars that are encompassed by 1 degree of visual angle.[4,22] It does not depend on the number of bars *per se*. This finding supports the modern notion in psychology that at least for simple patterns the relevant information to which the brain responds is the Fourier transform of the scene.

The neuromagnetic latency increases from approximately 80 ms for low spatial frequencies of less than 1 cycle per degree to more than 200 ms for high spatial frequencies over 10 c/deg. Subjects' reaction times to the appearance of similar grating patterns in simple button-press experiments vary in the same way but are $\sim$ 115 ms longer. This suggests that reaction time can be divided into a sensory response time indicated by the neuromagnetic latency and a motor response time. A similar value for the motor response time is deduced from studies with electrical stimulation of a finger. Thus the peak magnetic field occurs at a significant moment in the brain's response to sensory stimuli.

More recent work in our laboratory bears directly on a modern theory for large-scale organization of the human visual system. The notion is that at least two specialized channels exist which interpret features of the scene at the retinal level and separately

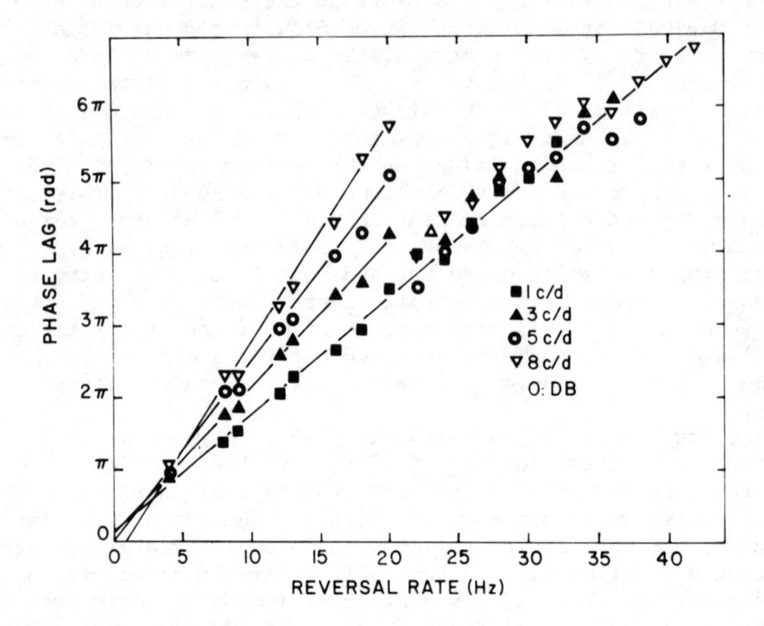

Fig. 7. Phase lag of the visually-evoked field relative to the stimulus cycle for grating patterns of various spatial frequencies.

transmit information back to the visual cortex for further processing. One called the "transient" channel is preferentially sensitive to aspects in the visual field involving rapid movement (or high temporal frequencies) and broad variations in luminance (low spatial frequencies). The second or "sustaining" channel is preferentially sensitive to low temporal frequencies and high spatial frequencies. Signals in the transient channel are sent by a faster route to the visual cortex. This could conceivably have survival value since such channels respond rapidly to moving and therefore possibly threatening objects. By this model the pattern dependence of the latencies that we obtain for low temporal frequencies of contrast reversal should disappear for high temporal frequencies, endowing all responses with the same short latency regardless of spatial frequency.

Fig. 7 shows data for the phase lag relative to the contrast reversal cycle for the neuromagnetic field emerging perpendicular to the scalp on the midline close to the base of the skull at the back of the head.[23] A linear variation of phase with increasing reversal rate can be characterized by a latency. The latencies deduced from the slopes of the fan-like trends at low reversal rates range from 80 ms for 1 c/deg to 140 ms for 8 c/deg for this observer. The 1 c/deg trend continues in linear fashion to the highest reversal rate where an averaged response can be detected, suggesting that response to this low spatial frequency is governed by the transient channel for all reversal rates. But for the 3 c/deg pattern there is a departure in the phase trend at ~ 22 Hz, above which the phases closely follow the trend for 1 c/deg, thus indicating that at high reversal rates they have identical, short latencies. Similar effects are seen for 5 and 8 c/deg: the trends above ~ 22 Hz generally follow the same slope as that for 1 c/deg (Notice that there is always ambiguity in the modulus of 2π chosen to interpret the measured phase if discontinuous changes occur; this ambiguity however has no effect on the slope of linear trends on which our interpretation is based). Other subjects display similar crossover behavior at ~ 20 Hz between domains of low and high temporal frequency. These findings constitute important physiological evidence that the organization of the human visual system includes specialized channels that are dependent on spatial and temporal frequencies of the scene. This evidence is consistent with inferences from many psychophysical experiments.

SUMMARY

Increasing interest in biomagnetic studies is due in part to the successful use of SQUID systems operating in an unshielded environment. Many opportunities are available for exploratory work; and ongoing research of the lungs, liver, heart, and brain is producing results of clinical and fundamental interest. One technical challenge is that of reducing the detection of comparatively strong background fields at low frequency, either by improving spatial discrimination or by using adaptive filtering techniques. The second need is to reduce the sensitivity of SQUIDs to rf interference, since the inclusion of an rf shield within the dewar, or outside, may compromise the balance of a gradiometer detection coil. The eddy currents induced by low frequency components of the background field contribute to the noise. Finally, with effective shielding such as the enclosures at M.I.T.[6] and the National Bureau of Standards at Boulder[7], background fields are below the equivalent intrinsic noise of the SQUID system, which remains the limiting factor. Even in an urban environment, an unshielded system with a balanced second-derivative gradiometer exhibits noise above 10 Hz which may be only a factor of 2 greater than the intrinsic SQUID noise. Thus we have reached the point where technological improvements to reduce SQUID and electronic noise would be beneficial in biomedical applications.

We wish to thank D. Crum, R. Walter-Peters, and J.P. Wikswo, Jr., for informative discussions.

REFERENCES

1. M. Saarinen, P.J. Karp, T.E. Katila, and P. Siltanen, Cardiovascular Res. 8, 820 (1974).
2. D. Brenner, S.J. Williamson, and L. Kaufman, Science 190, 480 (1975).
3. D. Brenner, J. Lipton, L. Kaufman, and S.J. Williamson, Science 199, 81 (1978).

4. S.J. Williamson, L. Kaufman, and D. Brenner, "Biomagnetism," in Superconductor Applications: SQUIDS and Machines, B.B. Schwartz and S. Foner, Eds. (Plenum, N.Y., 1977), pp. 355-402.
5. D. Cohen, IEEE Trans. Magn. MAG-11, 694 (1975).
6. D. Cohen, Rev. de Phys. Appliquee 5, 53 (1970).
7. J.E. Zimmerman, J. Appl. Phys. 48, 702 (1977).
8. D. Cohen, Science 180, 745 (1973).
9. P.-L. Kalliomäki, P.J. Karp, T.E. Katila, P. Mäkipää, P. Saar, and A. Tossavainen, Scand. J. Work. Environ. and Health 4, 232 (1976).
10. J.P. Wikswo, Jr., J.E. Opfer, and W.M. Fairbank, A.I.P. Conference Proc. 18, 1335 (1974).
11. J.W. Harris, D.E. Farrell, M.J. Messer, J. Tripp, G.M. Brittenham, E.H. Danish, and W.A. Muir, Clinical Res. 26, 3 (1978).
12. G.M. Baule and R. McFee, Am. Heart J. 66, 95 (1963) and J. Appl. Phys. 36, 2066 (1965).
13. B. Denis, D. Matelin, C. Favier, M. Tanche and P. Martin-Noel, Arch. Mal. Coeur 69, 299 (1976).
14. J.P. Wikswo, Jr., J.C. Griffin, M.C. Leifer, W.M. Barry, and D.C. Harrison, Clinical Res. 26 (1978).
15. W.H. Barry, W.M. Fairbank, D.C. Harrison, K.L. Lehrman, J.A.V. Malmivuo, and J.P. Wikswo, Jr., Science 198, 1159 (1977).
16. D. Cohen and L.A. Kaufman, Circulation Res. 36, 414 (1975).
17. D. Cohen, Francis Bitter National Magnet Laboratory Annual Report, October, 1977, p.129.
18. J.R. Hughes, J. Cohen, C.I. Mayman, M.L. Scholl, and D.E. Hendrix, J. Neurol. 594, 1 (1977).
19. J.R. Hughes, D.E. Hendrix, J. Cohen, F.H. Duffy, C.I. Mayman, M. L. Scholl, and B.N. Cuffin, Electroenceph. Clin. Neurophysiol. 40, 261 (1976).
20. M. Reite, J.E. Zimmerman, J. Edrich, and J. Zimmerman, Electroenceph. Clin. Neurophysiol. 40, 59 (1976).
21. W. Penfield and T. Rasmussen, The Cerebral Cortex of Man (Macmillan, N.Y., 1950), p. 21.
22. S.J. Williamson, L. Kaufman, and D. Brenner, Vision Res., in press.
23. L. Kaufman, D. Brenner and S.J. Williamson, to be published.

THE USE OF SQUIDs IN LOW-FREQUENCY COMMUNICATION SYSTEMS

M. Nisenoff
Naval Research Laboratory
Washington, DC 20375

ABSTRACT

Antennas employing SQUID amplifiers can be used as relatively compact, highly sensitive receiving elements in communication systems operating in the frequency range from the ULF band (below 30 Hz) up through the HF band (3 to 30 MHz). Antenna elements operating in this frequency region tend to be small compared to the wavelength of interest and thus are poorly matched to the radiation field. Furthermore, their response is often limited by the input noise of the amplifier connected across it. However, a combination of small antenna elements and SQUID amplifiers, which can have noise temperatures lower than 10^{-3}K, could result in receiving antennas with superior noise characteristics compared to conventional ones. A submarine towed ELF (30 to 300 Hz) antenna using SQUID devices, for which preliminary design and development work has been performed at NRL, will be described in some detail. Several other potential applications for SQUIDs in low-frequency communications systems will be outlined. Some performance limitations of current SQUID systems are identified, and recommendations made on how to improve the performance of future SQUID systems.

INTRODUCTION

Electromagnetic radiation with frequencies at the lower end of the spectrum, that is, for frequencies below about 30 MHz, can propagate around the earth and, therefore, are used for worldwide communications, for ship-to-shore communications and for navigational systems. In addition, since the skin depth of electromagnetic radiation in sea water increases with decreasing frequency, the Navy uses low frequencies to communicate with its submerged submarines.

Although these wavelengths do propagate readily, the detection of radiation at low frequencies poses some serious problems. Recalling that the free space wavelength λ is given by

$$\lambda = 3 \times 10^8/f$$

(where λ is in meters and f is in Hz), most antennas for low-frequency applications tend to be short compared to the wavelength of interest. Since the radiation resistance of small antennas varies as $(a/\lambda)^n$, where n is 4 for circular loops and 2 for linear dipoles, these antennas couple very inefficiently to the radiation field. In addition, amplifiers in this region, especially at frequencies below 100 Hz, exhibit 1/f noise which further degrades the overall performance of low-frequency antenna systems.

SQUID amplifiers have been constructed with effective noise temperatures of 10^{-3} K or less.(1) When such an amplifier is connected to a loop antenna fabricated from superconducting wire, the combination of an inefficient loop antenna and a very low noise amplifier can result in an extremely sensitive antenna, one that can be orders of magnitude more sensitive than conventional antennas of comparable size.

Atmospheric noise at the low end of the electromagnetic spectrum is shown in Figure 1.(2) Note that

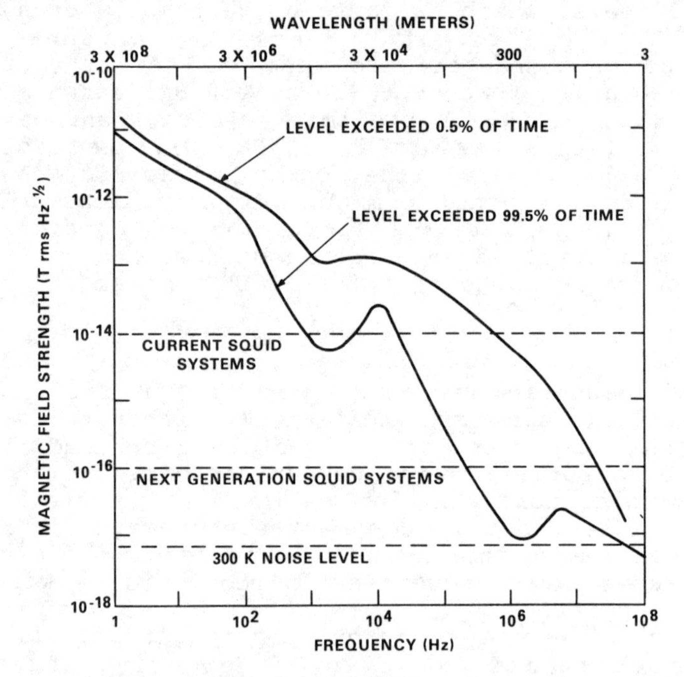

Fig. 1. Atmospheric noise as function of frequency. The horizontal dashed lines represent levels of performance of several SQUID magnetometer systems.

the noise increases approximately inversely with frequency from the HF (3 to 30 MHz) band down. Most antennas designed for operation in this frequency region are built so that their instrument noise level is comparable to or slightly less than atmospheric background noise. This level of performance is adequate for most applications, especially those in which the antenna is located at the surface of the earth. However, SQUID antennas (that is, SQUID magnetometers) have been built with instrument noise well below background noise. The upper dashed curve in Figure 1 represents the level of performance typical of current SQUID systems.

The next generation of SQUID magnetometers (antennas) will have instrument noise levels appproaching 10^{-16} T rms $Hz^{-1/2}$. The lower dashed line in the figure represents the performance of a SQUID magnetometer with an effective noise temperature of 300K. Such a system would have an instrument noise lower than atmospheric noise for all frequencies below about 100 MHz. Near 1 Hz, this system would have an instrument noise level five orders of magnitude below background noise!!!!

There are a number of low-frequency communications applications where an antenna with instrument noise several orders of magnitude below background noise can be used to great advantage. These include:

(1) detection of low frequency signals at appreciable depths below the surface of the ocean,
(2) the use of SQUID antennas in arrays and super-gain arrays where the detection pattern can be modified and lobes and nulls steered to desired directions.

These applications will be discussed in the remainder of this communication.

SUBMARINE-TOWED RECEIVING ANTENNAS

One of the critical problems facing the Navy today is the need to communicate with submarines on station.[3] Since sea water is a conducting medium, high frequency signals are strongly attenuated, and appreciable penetration does not occur except at signal frequencies below 1 kHz. This is illustrated in Figure 2, where the magnitude of geomagnetic background noise at various depths below the surface of the ocean is given as a function of frequency. The operational requirements on data rates for a given system will impose a minimum acceptable signal-to-noise ratio for satisfactory reception of information. Both the signal of interest and the background noise in a frequency band containing the signal will attenuate at the same rate as they penetrate through sea water. The greatest depth at which useful information can be received will be that depth at which the attenuated background noise becomes comparable to or smaller than the instrument noise of the antenna. By inspection of the plots in Figure 2, it can be seen that

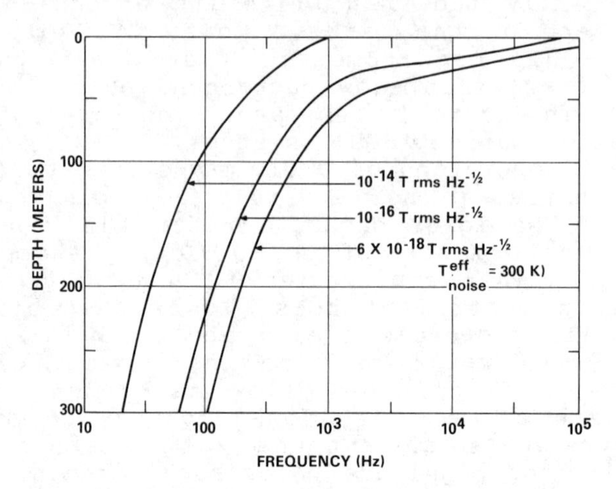

Fig. 2. Atmospheric noise at various depths below sea level as function of frequency.

instrument noise levels of 10^{-14} T rms $Hz^{1/2}$ will be useable for submerged antennas at signal frequencies below 100 Hz. These data show that submerged antennas operating in the tens of kHz region must have instrument noise levels of the order of 10^{-16} to 10^{-18} T rms $Hz^{-1/2}$ to receive information at depths of even several tens of meters. Conventional antennas deployed from submarines with these instrument noise levels tend to be very long and cumbersome and have undesirable directional receiving patterns. However, SQUID antennas appear to have instrument noise levels adequate to receive signals in these frequency ranges at useful depths with omni-directional receiving patterns.

MOBILE SQUID SYSTEMS

The exceptionally low noise response of SQUID antennas (magnetometers) frequently given in the literature was obtained with a stationary antenna. However, the response of a SQUID antenna or any totally superconducting circuit depends on the magnetic flux threading the circuit. Magnetic flux ϕ is given by

$$\phi = \vec{B} \cdot \vec{A} \tag{1}$$

where $\vec{B}$ is the ambient magnetic induction and $\vec{A}$ is the cross-sectional area of the circuit. Thus any change in the orientation of a superconductive circuit with respect to the earth's magnetic field will result in a change in the magnetic flux linking the circuit.

Consider the case of a superconductive loop located in the magnetic induction of the earth $\vec{B}_e$ ($B_e = 10^{-4}$ T), and assume that $\vec{B}_e$ is in the plane of the circuit. If the orientation of the loop with respect to $\vec{B}_e$ changes by an angle θ, then the motion-induced change in its projection on the normal to the plane of the circuit is given by

$$\Delta B_{MOTION} = B_e \sin\theta \sim B_e\theta \quad \{\theta << 1\} \tag{2}$$

If the motion-induced output signal is to be less than instrument noise of the SQUID magnetometer, which will be assumed to be 10^{-14} T rms $Hz^{-1/2}$, the platform must be maintained stable to better than

$$\theta \sim \frac{\Delta B_{Motion}}{B_e} \leq \frac{\Delta B_{Noise}}{B_e} \leq 10^{-10} \text{ rads} \quad . \tag{3}$$

Platform stability of this order is totally out of the question.

If the magnetic induction B_e is perpendicular to the plane of the circuit, then the platform stability requirements are not quite as severe. Consider the signal that would result if the normal to the circuit is along the earth's field and then is tilted by an angle θ. In this case

$$\Delta B_{motion} = B_e - B_e \cos\theta \sim B_e \theta^2/2 . \tag{4}$$

Again, if the motion noise is to be comparable to the instrument noise of the device, then

$$\frac{\theta^2}{2} \simeq \frac{\Delta B_{Motion}}{B_e} \leq \frac{\Delta B_{Noise}}{B_e} \simeq 10^{-10} \text{ rads} \tag{5}$$

or

$$\theta \leq 10^{-5} \text{ rads} .$$

In this case, the requirements on platform stability are reduced but still beyond present technology. One can, with a great deal of effort, maintain a sensor along the earth's field to within about 10^{-3} radians, but 10^{-5} radians is definitely beyond state-of-the -art. These considerations appear to indicate that it is not meaningful to consider the use of a single SQUID system with an instrument noise level of 10^{-14} T rms $Hz^{-1/2}$ or less on a mobile platform. However, if multiple SQUID sensor arrays can be used, there may be some way to configure the sensors so that the antenna system could operate with a detection threshold comparable to the instrument noise of the individual sensors.

TRIAXIAL MAGNETOMETER CONCEPT

Consider an array of three ideally orthogonal component (vector) magnetic field sensors located in the earth's magnetic field, and assume that the platform supporting the array is stable to some angle θ.[3] Motion-induced signals in the output of the three individual channels will be given either by $B_e \theta$ or $B_e \theta^2/2$, depending on the orientation of the given sensor with respect to the earth's field. However, if the output of the three orthogonal channels were squared and then added together, a rotationally invariant quantity would be formed, viz,

$$B_e^2 = B_x^2 + B_y^2 + B_z^2 \quad . \tag{6}$$

This relationship would be maintained as long as the individual channels were linear and were able to track changes in the respective component fields.

If SQUID magnetometers are to be used in such a system, there are complications due to the fact that the SQUID does not measure absolute magnetic field but only changes in field. The component of the ambient magnetic field along the axis of a SQUID magnetometer can be represented as

$$B_i = G_i(A_i + M_i(t)) \tag{7}$$

where G_i = gain of the sensor

$$= \frac{\text{Change in Output Signal}}{\text{Change in Incident Magnetic Field}}$$

$M_i(t)$ = Output Signal From Sensor

$G_i A_i$ = Offset Magnetic Field, that is, magnetic field when $M_i(t) = 0$.

When this form for the response of a magnetic field sensor is inserted into Eq. (6), the expression for the rotationally invariant quantity takes the form

$$\frac{1}{G_1^2} B_e^2 = \sum_{i=1}^{3} g_i^2 \, (A_i + M_i \, (t) \,)^2 \quad . \tag{8}$$

There are five parameters that must be determined in order to produce an invariant output signal

$$A_1, A_2, A_3, \; g_2 = G_2/G_1, \; g_3 = G_3/G_1, \; .$$

The parameter g_1 $(g_1 = G_1/G_1)$ is, by definition, unity. If the magnitude of B_e is of interest, the quantity G_1 must also be determined.

Non-Orthogonal Sensors

If the three sensors are not precisely orthogonal, then an expression of the form shown in Equ. 8 will not be rotationally invariant. In this case, a coordinate transformation must be performed to obtain an equivalent set of readings in an orthogonal frame[4]. For example, it can be shown with quite general considerations that, if the three directions in the "skew" frame are not coplanar, then a minimum of three angles must be specified in transforming from the skew frame to an equivalent orthogonal frame. The detailed arguments in regard to carrying out these transformations are treated in detail elsewhere.[4] It is sufficient to state here that for the situation of a triaxial sensor system where the sensors are not mutually orthogonal, there are three additional parameters, that is, the three non-orthogonality angles, that must be determined in addition to the three offset field values and the two gain stability parameters in order to form a sum-of-squares quantity that will be rotationally invariant.

Determination of Unknown Parameters

There are at least two techniques that can be used to determine the parameters in the sum-of-squares formalism. If the sensors are not precisely orthogonal, then eight simultaneous sets of output readings must be taken and inserted into eight equations of the form of Equ. 8, and these equations are then resolved for the eight parameters, viz, the three offset fields, the two relative gain settings and the three angles describing the non-orthogonality of the sensors. Another version of this approach is to take more than eight sets of readings and to solve for the parameters using least-squares techniques.[5]

In the second approach an adaptive technique is used.[4] If the parameters A_i's , g_i's, and Ω_{ij}'s are not precisely known, and the sum-of-squares quantity is formed, then the resulting expression will vary as the sensor arrays translate and oscillate in the earth's field. Using suitable computational procedures, it is possible to "hunt" for a set of parameters which will reduce the excursions of the sum-of-squares expression to below some acceptable level, that is, the desired instrument noise level of the overall system.

These two approaches have been explored in some detail for three-axis SQUID systems.[4,5] The choice of procedure to be used for a given application must be made for each case, taking into consideration the operational constraints of the system.

NRL ELF SQUID System

Preliminary feasibility studies of a submarine-towed ELF receiving antenna using a three-axis SQUID system have been performed at the Naval Research Laboratory.[3,4] A suitable three-axis SQUID system[6] and a 100-day hold time horizontal helium dewar[7] were procured, and adaptive processing algorithms were developed. Simulations of the motion-induced signals expected for the mobile antenna were carried out by placing the SQUID system on a "shake table" consisting of a set of inflated automobile inner tubes. Unfortunately, the SANGUINE/SEAFARER program for which this antenna was developed has been drastically curtailed and thus the final phase of the SQUID antenna program, packaging the antenna and dewar in a towed body and performing actual sea tests, has been postponed for the time being.

ARRAYS AND SUPER-DIRECTIVE ARRAYS

Field patterns of a single-element antenna are usually fairly broad and thus, in some cases, multi-element antenna systems have been used to obtain more directive patterns.[8] This is illustrated in Figure 3[9] where the familiar cosine pattern from a single-element array is shown along with the patterns for several multi-element arrays. In the case of a three-equally-spaced-element array during the summing process, outputs from the end elements were multiplied by a different gain factor than that used for the center element. The result is a beam pattern narrower than that of the single element (curve a) or for a similar three-element array in which the three outputs are summed with identical gain settings. This improvement in directivity, however, is achieved at the loss of sensitivity, about 5 dB, compared to the single-element array. In curve c, the pattern for a five equally-spaced-element array is shown in which the output of the center element was multiplied by one gain setting, the inboard elements by another, and the outboard by yet another gain setting. For the particular set of gain settings chosen, the pattern is still peaked in the same direction but has a much narrower angular pattern and, furthermore, its peak response has been reduced by about

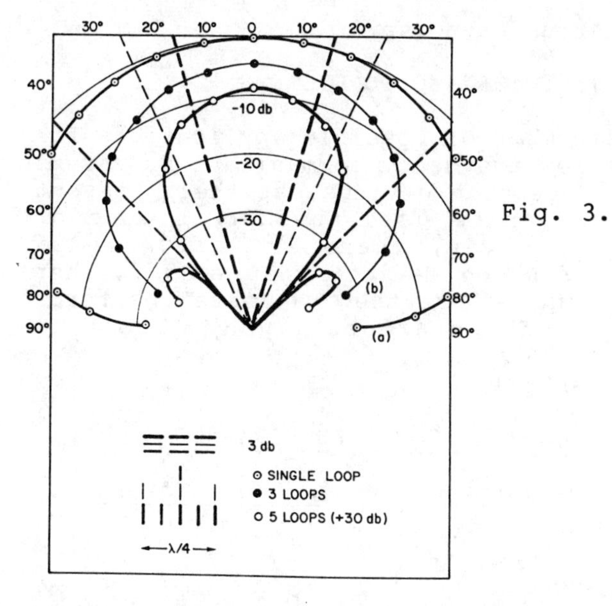

Fig. 3. Calculated antenna patterns for several different array configurations. See text for details of assumptions made in each case.

40 dB compared to peak response of the single-element antenna.

In Figure 3, the peak in the response curves for the multiple-element arrays coincided with that of the single-element antenna. By properly selecting the phase and gain factors applied to the outputs of the several elements before combining, the peak (and, correspondingly, the nulls) can be steered in any desired direction. Thus, the peak of the response curve can be directed toward an incoming signal of interest, while it might also be desired to steer a null in the pattern toward a strong interfering signal in order to discriminate against this signal in preference of the desired signal.

The feasibility of these arrays is based on the assumption that the individual elements are sufficiently small that the mutual coupling between elements is not significant. SQUID magnetometers used as low-frequency antennas tend to be noticeably smaller than antennas with comparable sensitivities made with competing technologies Furthermore, the elements of super-directive arrays must have matched responses and exceedingly low values of internal instrument noise in order to compensate for the overall loss in response of the multiple-element arrays compared to the simple antenna. SQUID technology appears to be able to meet all the requirements placed on the elements of super-directive arrays especially at the low frequency end of the spectral region of interest discussed here. However, even in the HF band, SQUID antenna elements may have applications in arrays due to their small size and low instrument noise.

IMPROVEMENTS IN FUTURE SQUID SYSTEMS

In this paper, a number of possible applications for SQUIDs in the area of low-frequency communication systems have been outlined. Josephson devices, at their present level of development, can be used to address some of these opportunities. However, there are a number of features of current Josephson devices that need further improvement or refinement before the "ultimate" performance can be realized. These are outlined in Table I.

TABLE I

FACTORS LIMITING CURRENT SQUID PERFORMANCE

Parameter	Current Performance	Limiting Factors
Energy Sensitivity		
RF SQUID	30×10^{-30} J/Hz	Pump Frequency Input noise of preamplifier
DC SQUID	7×10^{-30} J/Hz	Johnson noise in shunting resistors
Analog Dynamic Range	140 dB (i.e., 10^7)	Voltage swing from 1μ V to 10 volts
Slewing Rate	~10^6 ϕ_o/sec	SQUID Noise Modulation Frequency
Signal Bandwidth	"DC" to ~ 100 kHz	SQUID Noise Modulation Frequency

There are three areas where future work should be directed in order to improve overall SQUID performance:

(1) Lower SQUID Noise. Lowering SQUID noise would result in an improvement in energy sensitivity, slewing rate and signal bandwidth. In the case of RF SQUIDs this can be done by increasing the pump frequency, while in the case of the DC SQUID, a decrease in device noise should result from decreasing the tunnel junction capacitance (by reducing its area), thus permitting an increase in the value of the shunting resistor needed to eliminate hysteresis. Research activities in both these areas are currently underway.

(2) _Lower Input Noise of Preamplifier_. In many cases, state-of-the-art room temperature circuits are in use. In principle, improvement could be realized by cooling the preamplifier or placing the preamplifier directly in the helium bath along with the SQUID itself. However, this would greatly increase the heat leak into the helium bath and complicate the refrigeration requirements. Another possible means of improving the noise characteristics of the detection electronics is to use very expensive equipment such as masers and paramps in SQUID systems operating in the microwave frequency region. However, the complicating of the refrigeration system or the addition of expensive components may not be considered cost-effective even though overall system performance may be improved.

(3) _Develop Hybrid Flux Counting/Analog Output Electronics_. A unique feature of SQUID magnetometer systems is their very great dynamic range and the spectacular loss of information when the dynamic range is exceeded. The current SQUID technology provides analog outputs with dynamic ranges of 140 dB (factor 10^7) referred to instrument noise. To digitalize an analog signal of this dynamic range would require 24 bit analog-to-digital (A/D) convertors. The best A/D convertors presently available are 16-bit devices. Furthermore, when the dynamic range is exceeded, the feedback system "unlocks" and resets at some new arbitrary "offset" operating point.

However, both of these problem areas could be circumvented if the digital nature of the SQUID response were fully utilized. Shortly after the invention of the SQUID, Forgacs and Warnick[10] built a digital/analog system in which the SQUID was allowed to run free with the number of quanta counted and an interpolation scheme employed to determine changes that were fractions of a quantum. Such a system would have an unlimited dynamic range, subject to slewing rate restraints, and would provide an output in digital format which would be directly available for processing.

CONCLUSIONS

In this paper the uses of SQUID systems in several low-frequency communications systems have been outlined. Possible areas of interest include receiving antennas for communications with submerged submarines and multiple sensor arrays where the outputs can be suitably processed to shape and steer the overall antenna pattern. In addition, some of the performance limitations of current generation SQUID systems are identified and recommendations made on how to improve the performance of future generation systems based on the Josephson Effect.

ACKNOWLEDGEMENTS

The development at the Naval Research Laboratory of the SQUID ELF antenna for submarine deployment was a joint effort involving R. J. Dinger, J. R. Davis, J. Goldstein, S. A. Wolf and the author. Several discussions on the processing of the signals from Triaxial SQUID systems with Drs. Robert Dinger, Charles Sinex and John Claassen are acknowledged with appreciation. The author would also like to thank Dr. R. A. Hein for critically reading this manuscript.

REFERENCES

1. R. P. Giffard, R. A. Webb and J. C. Wheatley, J. Low Temp. Phys., 6 53 (1972).

2. CCIR Report 322, International Radio Consultative Committee, Xth Plenary Assembly, Geneva, 1963, (International Telecommunications Union, Geneva, 1963).

3. S. A. Wolf, J. R. Davis and M. Nisenoff, IEEE Trans. of Communications, COM-22 549 (1974).

4. J. R. Davis and M. Nisenoff; SQUID: Superconducting Quantum Interference Devices and Applications, Ed. by H. D. Hahlbohm and H. Lubbig, (Walter de Gruyter, Berlin, 1977), pp. 439-484. (See cited references for details).

5. C. H. Sinex, "A Signal Processing Method for Superconducting Total Field Magnetometers", Report POR 3168, (Aug. 1977) (Applied Physics Laboratory, Laurel MD)(Unpublished).

6. Procured from the SHE Corporation, San Diego, CA.

7. Procured from Superconducting Technology, Inc., Mountain View, CA.

8. See for example, "Fields and Waves in Communications Electronics", S. Ramo, J. R. Whinnery and T. VanDuzer, (John Wiley & Son, New York, 1965, 2nd Ed.) p. 690-692.

9. N. K. Welker and F. D. Bedard, "Proceedings of the 1973 Workshop on Naval Applications of Superconductivity", Ed. by M. Nisenoff, NRL Report 7822 (30 Oct. 1974) (Naval Research Laboratory, Washington, D. C. 20375), p. 48.

10. R. L. Forgacs and A. Warnick, Rev. Sci. Instr. 38 214 (1967).

COMMERCIAL SUPERCONDUCTING SQUID SYSTEMS

William Goree and John Philo
Superconducting Technology Division United Scientific Corporation
Mountain View, California 94043

INTRODUCTION

The first commercial entry and manufacture of SQUID-based electronic instruments began in late 1969, with the formation of the Superconducting Instrument Division of Develco Inc. and S.H.E. Corporation. A third company, Superconducting Technology, was formed early in 1973. Develco Inc. is no longer in the SQUID instrument business, leaving only Superconducting Technology and S.H.E. in the United States. To the best of our knowledge, there are three other companies manufacturing SQUID-based systems. These are Canadian Thin Films, Canada; Instruments for Technology, Finland, and Cryogenic Consultants in England. During these past eight years numerous papers have been written describing the technology and the instruments manufactured by these companies.[1,2,3]

The initial market for SQUID's and SQUID-based instruments was divided between research labs interested in studying the SQUID itself and/or incorporating the SQUID in their own research apparatus, and very special one-of-a-kind instruments, for example, those used in geophysical studies to measure magnetic samples. Most of the instruments built during these first few years of commercialization were special systems built with close cooperation and discussion with the customers who were generally low-temperature physicists. It was seldom that more than one of a given type of instrument was sold in a year. Recently the market has matured, so that now two or more of the same type of instrument may be sold in a single year - still not mass production.

We estimate that the total annual sales of SQUID systems in the United States has grown from about $200,000 in 1970 to about 1.5 million dollars for 1978. Most of this growth occurred during the first three to four years (1970-73) and has been relatively stable during the past two to three years. The market character has changed appreciably, though, in recent years, and now most of the customers buying superconducting instrument systems are likely to be geophysicists, chemists and other scientists interested only in the unique measurement capability that the SQUID system offers them.

In the context of this conference we have collected together our ideas (opinions) on the market areas with the most potential impact on science, medicine, and high technology industry. We have also considered the advances in SQUID's and other components of SQUID systems that may accelerate many of these applications, or that might be required to enter new market areas. Further, we have reviewed several unique properties of the SQUID systems that have not been fully capitalized or utilized with respect to commercial applications.

ISSN: 0094-243X/78/130/$1.50

POTENTIAL MARKETS

We feel that SQUID systems will have their largest impact in the areas of analytical instruments, magnetic gradiometers for medical measurements, geophysical exploration systems, and instrumentation for Department of Defense requirements. The largest market for analytical instruments appears to be in the use of SQUID's to measure the magnetic susceptibility of small samples. The presently available SQUID detectors and associated electronics are perfectly adequate to provide extremely large improvements in measurement capability for DC and low-frequency susceptibility measurement. SQUID's also offer unique advantages in very high frequency or rapid time response susceptibility measurements. In this case, some advances in SQUID technology would be required to give time resolutions of better than one millisecond. The largest market for susceptometers appears to be in chemistry and biochemistry. The engineering problems associated with building a useful susceptibility system are many, but as the result of the large research and development effort in this area,[4] these problems are now well in hand.

The second market that could grow very rapidly is medical applications of SQUID systems. Numerous research laboratories are investigating measurement of the magnetic field from the human heart and brain using superconducting first-derivative and second-derivative gradiometers. These efforts are primarily directed by physicists. These measurements could have profound medical diagnostic capabilities.[5] The technology required to manufacture instruments suitable for these medical applications is well in hand, and several standard superconducting instruments have been delivered within the last three years for medical applications. The recent demonstration that second-derivative instruments can successfully be used in typical urban laboratory environments <u>without</u> shielded rooms will have important implications for growth in this area. Widespread use, however, probably must await both more basic research to firmly establish the unique clinical capabilities of these instruments and more widespread market acceptance by the medical profession.

The geophysical market for SQUID systems is growing slowly, although more geophysical survey companies are converting their magnetic field measurement instruments to SQUID's. The major use at present is for multi-axis superconducting magnetometers to be used in stationary magnetotelluric studies.[6] The requirements for these systems are straightforward, and standard SQUID magnetometers and rugged liquid helium dewars in off-the-shelf configurations meet these requirements with ease. The use of superconducting gradient measurements for airborne magnetic surveys has not yet been applied to geophysical surveying. The basic technology is well-developed and could be easily transferred to geophysical surveying. The major impediment here seems to be a reluctance of survey companies to procure unique measurement capabilities until their customers, primarily oil and mineral producing companies, voice a strong requirement for these new measurements. Therefore, it appears that the technological advances needed to utilize SQUID's for geophysical airborne surveying lie in the area of analysis techniques which could show that multi-axis gradient measurements provide more useful data for the end customer.

A peripheral market that is of interest to many areas of technology is the use of the A.C. Josephson effect to produce a high-precision voltage standard.[7] This is not a SQUID system in the same context used for the instruments described earlier in this paper, although it is a superconducting tunneling device. The U.S. standard volt has been determined by a Josephson effect device for about five years, and several other countries, e.g. Australia, England and Germany are also using a Josephson effect standard. A portable Josephson effect voltage standard is now commercially available and six units are being produced for delivery in the fall of 1978. The instrument is unique in that it utilizes the A.C. Josephson effect to determine the volt with an absolute accuracy of about 1 part in 10^7. The device uses a precision quartz crystal oscillator whose frequency is converted to a stable voltage by the Josephson Junction. The device is unique in that its absolute calibration can be confirmed at anytime by the user without sending the device back to the National Bureau of Standards for re-calibration as is required with the chemical cells.

CLOSED-CYCLE REFRIGERATORS

The complementary technology which may have the biggest potential to expand the market for SQUID systems is that of small closed-cycle refrigerators.[8,9] The expense and time involved in liquid helium transfers, as well as difficulties in obtaining liquid helium in certain areas of the world, constitute substantial impediments to market growth and acceptance.

Many commercially-available refrigerators are suitable for cooling systems to the 12^oK range and these units are suitable for some applications with high Tc SQUID's. Falco[10] and Beasley[11] have discussed the "state of the art" of high temperature SQUID's, and commercial devices operating in the 12K and higher range could be available within a few years.

A number of systems have already been fitted with refrigerators. While the existing refrigerators can, in some cases, reduce helium use by nearly an order of magnitude, further improvements in this technology are certainly needed.

Overall, it would be difficult to overestimate the impact of the development of a reliable, low-cost, non-magnetic, vibration-free 4K refrigerator on the SQUID systems market.

OTHER POTENTIAL APPLICATIONS FOR SQUID's

Several fairly unique properties of SQUID's have not yet been extensively utilized. In particular, the SQUID and associated electronics constitute an amplifier with extraordinary linearity and dynamic range. Linearized SQUID's have been shown to have dynamic range (ratio of largest to smallest signals measurable without changing scale factors) of 10^7 (140 db), and a linearity of 1 in 10^7. (These properties are limited by the room-temperature electronics). The use of flux-counting techniques can extend the linearity to at least as high as 1 in 10^8 and the dynamic range to at least 10^8 (160 db). In principle, with flux counting, it should be possible to extend the dynamic range to well above 10^8, but the problems of flux flow and hysteresis associated with such large currents flowing in the superconducting signal paths have, to our knowledge, not been studied.

This ability of the SQUID to resolve very small changes in a large signal, or to measure very small signals at one frequency in the presence of very large signals at nearby frequencies, offers unique measurement capabilities. In addition, because the output of the SQUID is linked to a fundamental physical constant, the flux quantum, the SQUID may be used for measurements requiring extremely high precision and long-term calibration accuracy. Indeed, with flux-counting techniques it is possible for a SQUID system to measure low-frequency signals with a resolution and accuracy approaching 1 part in 10^8 (27 binary bits!). These capabilities have, to a certain extent, been exploited in magnetic susceptibility systems,[12] but may prove useful in virtually any measurement which can be converted to a voltage, current, or magnetic field.

CONCLUSION

In conclusion, we feel that the SQUID-based instrument's market is well-developed with a commercially-available line of instruments covering measurements from geophysical surveying to medical diagnostics. The major limitation to increased use of these instruments is not technological but rather lack of awareness by potential users, need for product evolution in appearance and ease of use, and the requirements for liquid helium.

REFERENCES

1. M.B. Simmonds, "Superconductor Application: SQUIDs and Machines", in N.A.T.O. Advance Study Series, Series B: Physics, Ed. by B. Schwartz and S. Foner,(Plenum Press, N.Y.- 1977) p.403.
2. W.S. Goree and M. Fuller, Rev. of Geophysics and Space Physics, 14 no. 4, Nov. 1976, pp. 591-608.
3. R.E. Sarwinski, Advances in Instrumentation, 31 no. 2, paper 616, I.S.A., 1976. Annual Conference, Institute Soc. Amer., Pittsburgh, Pa. 1976.
4. B.S. Deaver, Jr., T.J. Bucelot and J.J. Finley, - to be published in proceedings of this conference.
5. S. Williamson, Biomedical Applications of SQUID's, ibid.
6. W.M. Goubau, J. Clarke and T. Gamble, ibid.
7. D.B. Sullivan, L.B. Holdeman and R.J. Soulen, The Future of Superconducting Instruments in Metrology, ibid.
8. J.E. Zimmerman, Comments on Small Cryocoolers, ibid.
9. W.A. Little, Design and Construction of Microminiature Cryogenic Refrigerators, ibid.
10. C.M. Falco, Properties of High Temperature SQUID's, ibid.
11. M.R. Beasley, Improved Materials for Superconducting Electronic Devices - Problems, Progress and Prospects, ibid.
12. J.S. Philo and W.M. Fairbank, Rev. Sci. Instrum. 48, 1529 (1977).

SQUID SIGNAL STRENGTH VERSUS FREQUENCY - SOME SURPRISES

Robert A. Peters
Catholic University of America, Washington, D.C., 20064

Stuart A. Wolf
Naval Research Laboratory, Washington D.C., 20375

Frederic J. Rachford
University of Maryland, College Park, MD, 20745

I. ABSTRACT

Measurements of SQUID signal strength as a function of frequency, obtained with a single probe having constant rf coupling, are reported here. Signal strength is found to be proportional to frequency from 20 MHz to 400 MHz. Between 400 MHz and 1500 MHz, signal strength remains approximately constant. This saturation coincides with an increase in the rf critical current of the SQUID, suggesting that an enhancement of the energy gap of the weak link may be associated with a degradation of the Josephson junction response.

II. INTRODUCTION

Traditionally, rf SQUIDS have been biased between 2 and 30 MHz. In the past few years, the trend has been to raise the biasing frequency with the expectation that the SQUID signal will increase proportionally.[1-4] It has previously been shown that the signal does increase as the frequency is raised; however, the actual frequency dependence of the SQUID signal has never been reported using the same probe and with constant rf coupling. Therefore it was decided to measure this frequency dependence using broadband electronics.

Here, this response is reported using a niobium variable thickness bridge in a SQUID configuration. The SQUID used has nearly "ideal"[5] dissipative mode characteristics at 20 MHz. It is found that the signal strength increases proportionately with frequency from 20 MHz to 400 MHz and then unexpectedly saturates from 400 MHz to 1500 MHz. The high-frequency saturation appears to be correlated with an enhancement in the critical current observed for this sample in the same frequency range. Such enhancement in the critical current indicates a strengthening of the coupling in the weak link that may give rise to a degradation of the SQUID properties.

III. EXPERIMENT

The bulk of the data were obtained with a broadband, nonresonant, mutual inductance rf bridge. The output of this bridge is amplified by an Advantek rf amplifier and then detected by a Tektronix 7L12 Spectrum Analyzer. A Tektronix TR501 Tracking Generator whose frequency is synchronized to the spectrum analyzer

ISSN: 0094-243X/78/135/$1.50

provided the rf drive. This system is used in the "zero frequency scan" (fixed frequency) mode, with the frequency changed after each measurement. A low frequency, magnetic field modulation signal is coupled into the same loop which rf biases the SQUID. The detected output of the spectrum analyzer goes to a lock-in detector synchronized with the modulation signal. The lock-in output or signal amplitude is graphed as a function of rf drive current with the drive current measured with a H.P. 8405A Vector Voltmeter (with a useful upper frequency range of 1500 MHz). In this way the signal amplitude is obtained for different frequencies. While the system was carefully designed to minimize standing waves, some standing waves are unavoidable and contribute to the scatter in the results.

By recording the direct output of the spectrum analyzer as a function of the rf drive current to the bridge (without low frequency modulation), the critical current is obtained. Critical current is defined by the onset of the typical "staircase" I-V pattern. By dividing this onset value by the difference in current between steps, the critical current is obtained directly in units of Φ_0/L_s, where Φ_0 is the flux quantum and L_s the SQUID inductance. This ratio is independent of instrumentation gain and effects of standing waves.

In addition to the broadband measurements, data on critical currents were obtained using conventional 20 MHz resonant tank circuit electronics and a 9.2 GHz spectrometer employing a resonant cavity.

The SQUID used in this work is a thin film niobium ring, 240 A thick with a granular weak link 1 micrometer long by 20 micrometers wide and thinned to approximately 35 A by anodization. This sample, designated N67E in reference 6, has a step height to step rise parameter[5] of 0.18 as determined by the 20 MHz system which is consistent with the noise being intrinsic Kurkijarvi[7] thermal flux noise. The T_c obtained from extrapolating 9.2 GHz critical current data is 3.85 K. Current phase and direct flux noise measurements on this sample were carried out by Prof. Buhrman at Cornell with results indicating that a sinusoidal current phase relationship, and intrinsic flux noise was observed above 3 K.

IV. RESULTS

Signal strength and critical current versus frequency data are shown in Figure 1. Data are for three temperatures, all above 3 K where the SQUID response is ideal. Signal strength in this region is independent of temperature (and critical current). This is expected[3] since, even for the smallest value of critical current, $2\pi i_c L_s/\Phi_0 > 1$. From 20 MHz to approximately 400 MHz the signal strength increases linearly with frequency as illustrated by the line of slope one on the log-log plot.

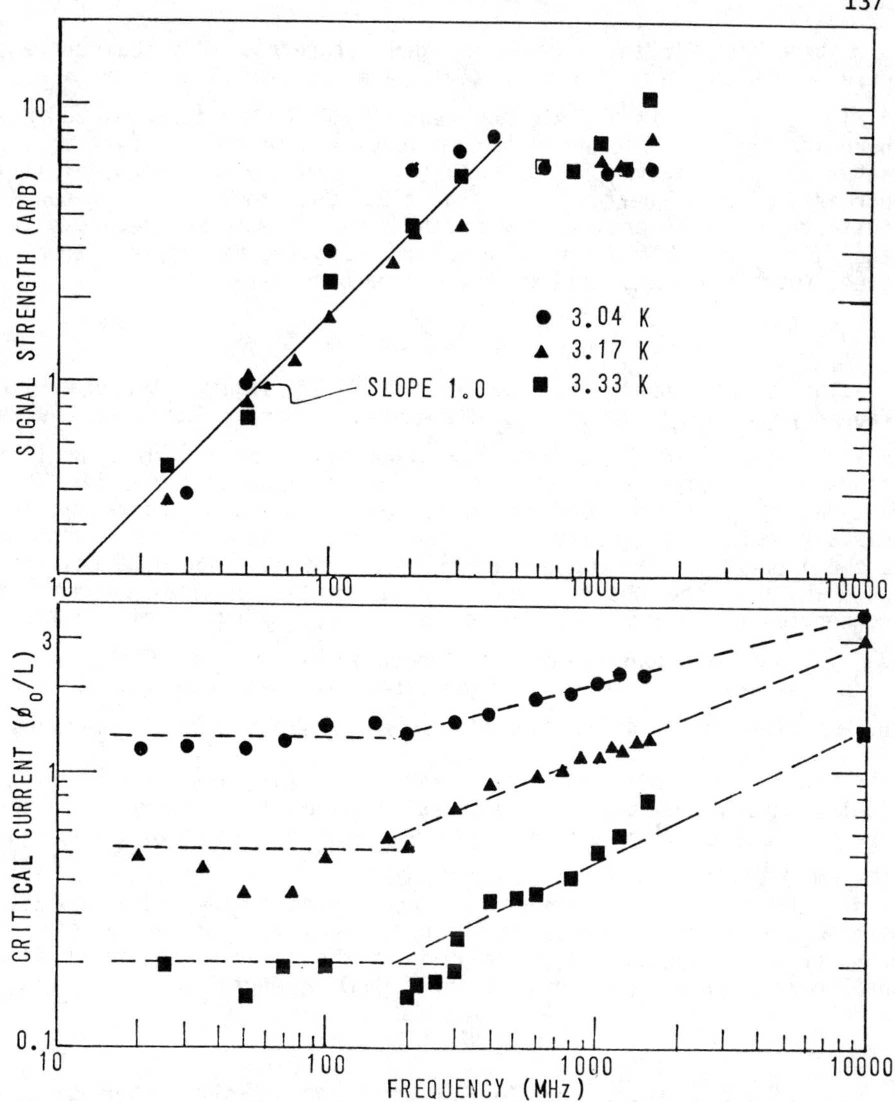

Figure 1. This figure depicts the logarithm of signal strength (upper portion) and the logarithm of critical current (lower portion) versus the logarithm of frequency for a granular niobium SQUID. A slope of one is shown on the upper plot indicating a linear increase of signal strength with frequency. The signal strength departs from the linear regime at frequencies where an enhancement of critical current is observed. The 20 MHz and 9200 MHz critical current data were obtained using resonant systems; the other data were obtained using a broadband mutual inductance system. Temperatures at which the measurements were made are indicated in the upper plot.

Above 400 MHz the signal strength saturates. Critical current versus frequency, in units of Φ_0/L, are shown in the lower half of Fig. 1. There is a frequency region from 20 MHz to about 200 MHz where the critical current is constant to within the accuracy of the measurements. Above 200 MHz the critical current monotonically increases with frequency. The data at 9.2 GHz, taken with a resonant cavity, are quite consistent with the trend of the broad-band data. Data at 20 MHz taken using conventional resonant circuit electronics agrees (within the scatter) with the broadband data.

V. DISCUSSION

The proportional increase of the SQUID signal strength with frequency observed between 20 MHz and 400 MHz is consistent with various models of SQUID response.[3] However, these models neglect relaxation effects associated with the superconducting state that must enter as the period of the rf biasing current approaches the relevant relaxation times. In the region where these phenomena become important, various non-equilibrium effects should be observed and, in fact, the observed enhancement of the critical current at frequencies above about 200 MHz corresponding to a time of about 8×10^{-9} seconds is consistent with measurements[7] on a similar SQUID which have determined that the time associated with the relaxation of the quasiparticle diffusion distance is about 5×10^{-9} seconds.

Enhancement of critical currents has also been observed in aluminum microbridges and has been associated with an increase in the superconducting energy gap by high frequency excitation of a non-equilibrium distribution of quasiparticles.[8,9] Such "strengthening"[10] of the superconducting properties of the bridges might be associated with a breakdown of the sinusoidal current phase relation of the granular weak link and may give rise to the observed departure from the linear frequency response of the signal strength.

VI. CONCLUSION

Broadband measurements have beeen made of the frequency response of the signal strength and the critical current of a granular niobium SQUID. As expected from various SQUID models, the signal strength increases linearly with frequency from 20 to 400 MHz. Above 400 MHz the signal strength saturates concurrent with an enhancement in the critical current. This enhanced critical current implies a strengthening of the superconducting properties of the weak link and possibly explains the degradation of the pure Josephson-type coupling associated with the "ideal" response at lower frequencies.

VI. ACKNOWLEDGEMENTS

We acknowledge the support of the Office of Naval Research, the Research Corporation, and the National Science Foundation. We would like to thank Dr. Martin Nisenoff and Prof. Jack Leibowitz for their initial advice and encouragement in this project and Mr. James Kennedy for sample preparation.

VIII. REFERENCES

1. J. E. Mercereau, Rev. Phys. Appl., 5, 13 (1970).
2. M. Nisenoff, Rev. Phys. Appl. 5, 21 (1970).
3. L. D. Jackel and R. A. Buhrman, J. Low Temp. Phys., 19, 201 (1975).
4. R. A. Buhrman and L. D. Jackel, IEEE MAG-13, 879 (1977); D. Pascal and M. Sauzade, J. Appl. Phys., 45, 3085 (1974).
5. J. Kurkijarvii, J. Appl. Phys., 44, 3729 (1973).
6. F. J. Rachford, S. A. Wolf, J. K. Hirvonen, J. Kennedy and M. Nisenoff, IEEE MAG-13, 875 (1977).
7. R. A. Peters, F. J. Rachford and S. A. Wolf, Phys. Rev. Lett., 40, 810 (1978).
8. T. M. Klapwijk and J. E. Mooij, Physica 81B, 132 (1976).
9. T. Kommers and J. Clarke, Phys. Rev. Lett., 38, 1091 (1977).
10. J. J. Chang and D. J. Scalapino, Phys. Rev. B15, 2651 (1977).

ANALYTIC STUDY ON THE INFLUENCE OF THE WEAK LINK PARAMETERS ON THE RESONANT CIRCUIT OF THE NONHYSTERETIC RF SQUID ($\omega \leq 10^8 s^{-1}$)

H. Lübbig

Physikalisch-Technische Bundesanstalt, Institut Berlin,
Berlin (West), Germany

ABSTRACT

An analytic expression for the ac impedance of a one-junction SQUID is derived for the nonhysteretic mode of operation. This impedance involves completely the nonlinear influence of the weak links' current phase relationship on the driving resonant circuit.

INTRODUCTION

The description of the influence of the nonlinearity of the current-phase-relation (CPR) on the conventional (normal-conducting) components of the circuit is one of the basic problems of an analysis of electromagnetic circuits which include structures with Josephson-effect-like behavior [1-4].

In the region of low-frequency excitation, i.e. $\omega \ll \omega_\Delta$, $\omega_\Delta = 4\Delta/\hbar$, where Δ denotes the energy gap, some different weakly superconducting structures can be described satisfactorily by means of the resistively shunted junction model (RSJM)

$$I_s(\varphi,\dot{\varphi}) = \left(1 + \tau_\alpha \frac{d}{dt}\right) I_c \sin\varphi + (\Phi_0/2\pi R_q)\,\dot{\varphi}\,. \quad (1)$$

Here the nonlinearity of the current I_s flowing through the weak link is characterized by the critical current I_c, the resistance R_q of the quasiparticle current, and the coefficient α of the $\cos\varphi$ -term, $\tau_\alpha = \alpha\,(\Phi_0/2\pi R_q I_c)$.

Based on the RSJM, in this paper the complete reaction of an inductively coupled one-junction SQUID on the exciting signal is studied analytically.

ISSN: 0094-243X/78/140/$1.50

DYNAMICS OF THE SQUID CIRCUIT

If the weak link is included in a superconducting ring, and coupled inductively to an external flux $\Phi_a = M\, I(t)$, as shown in Fig. 1a, the dynamic behavior of flux Φ enclosed in the SQUID ring is governed by the flux balance equation

$$\phi(t) = \int_0^t \frac{dt'}{\tau_q} e^{-\frac{t-t'}{\tau_q}} \left[\phi_a(t') - \left(1 + \tau_\alpha \frac{d}{dt'}\right) 2\pi\beta \sin\phi(t') \right] . \quad (2)$$

Here $\phi_a, \phi = 2\pi \Phi / \Phi_0$ denote the normalized fluxes, and $\beta = L_s I_c / \Phi_0$, $\tau_q = L_s / R_q$ are the normalized critical flux and the time constant of the SQUID circuit resp. Under practical conditions one has 10^{-12} s $\leq \tau_q , |\tau_\alpha| \leq 10^{-10}$ s, and, consequently, in the low-frequency regime $\omega \leq 10^8\, s^{-1}$ equ.(2) can be replaced by its stationary limit

$$\phi = \phi_a - 2\pi\beta \sin\phi . \quad (3)$$

If the magnetization is assumed to be reversible, i.e. $2\pi\beta < 1$, equ.(3) can be solved analytically, leading to the following diffraction-pattern-like expression

$$\phi = \phi_a \left(1 + \sum_{n=1}^{\infty} 2(-1)^n J_n(2\pi n\beta) \frac{\sin n\phi_a}{n\phi_a} \right) . \quad (4)$$

The SQUID parameter $2\pi\beta$ influences the pattern by means of the coefficients $J_n(2\pi n\beta)$; J_n denotes the ordinary Bessel Function of the first kind.

In order to solve the complete flux balance (2), the adiabatic relationship (4) can be applied to treat the nonlinear mixing term, $\sin\phi$, in equ.(2). By this way the internal flux ϕ can be calculated for arbitrary external excitations ϕ_a by means of a single quadrature.

AC IMPEDANCE OF THE SQUID CIRCUIT

To study the reaction of the nonlinear elements of the SQUID circuit on the driving resonator, the ac impedance $Z = V(t)/I(t)$, arising in the inductive path, (Fig.1a), is introduced for the case of a monofrequent excitation $I(t) = |I| \exp i\omega t = \Phi_a/M$. Applying the solution of equ (2), as discussed above, the interdependence between the SQUID circuit and the resonant circuit is represented by a generalized Schlömilch series[5]

$$Z(|\phi_a|) = Z_o + \sum_{n=1}^{\infty} 2 Z_n \cdot J_1(n|\phi_a|)/n|\phi_a| . \quad (5)$$

The term

$$Z_o = R\left[1 + iQ\frac{\omega}{\omega_o}\left(1 - K^2(1 - \mathcal{R}_q)\right)\right]$$

describes the impedance in the normal-conductive state of the SQUID, while the influence of the parameters $\omega\tau_q$, $\omega\tau_\alpha$, and $2\pi\beta$ of the superconductive state are involved completely in the coefficients

$$Z_n = i\frac{\omega}{\omega_o}K^2 Q R \mathcal{R}_q \mathcal{R}_\alpha \, 2(-1)^n J_n(2\pi n\beta)\cos n\phi_{dc} \quad (6)$$

$$= -\frac{1}{2\pi}\int_0^1 (1-u^2)^{-3/2} du^2 \int_{-\pi}^{\pi} Z(uv)\cos nv \, dv \quad (7)$$

Here

$$\mathcal{R}_q = \left[1 + (\omega\tau_q)^2\right]^{-1/2} \exp(-i \operatorname{arctg} \omega\tau_q) ,$$

$$\mathcal{R}_\alpha = \left[1 + (\omega\tau_\alpha)^2\right]^{1/2} \exp(i \operatorname{arctg} \omega\tau_\alpha)$$

describe the influence of the quasiparticle- and the $\cos\varphi$-term resp. Furthermore, K denotes the coupling coefficient, and $\omega_o/2\pi$ the resonant frequency of the high-Q tank circuit. In addition, ϕ_{dc} denotes the normalized dc-flux component enclosed in the SQUID.

Two properties of this relationship shall be pointed out:

1) If the constitution of the weakly superconductive structure is known, i.e. if the temperature dependences of τ_q, τ_α and β are known, the coefficients Z_n, equ.(6), are determined, and their influence on the resonant circuit can be described by introducing the "SQUID-circuit-equivalent" impedance Z, equ.(5), as shown in Fig.1b. The action of the nonlinear SQUID elements on the impedance $Z_{res}=\hat{V}/\hat{I}_{RF}$ of the resonant circuit for a given RFdrive current amplitude,

$$\frac{1}{Z_{res}} = \frac{1}{Z} + i\frac{\omega}{\omega_0}\frac{1}{QR},$$

can be calculated from the equation

$$\frac{\hat{I}_{RF}}{|I|} = 1 + i\frac{\omega}{\omega_0}\frac{1}{QR}Z(|I|).$$

This equation has to be solved numerically. Under special conditions analytic solutions are possible.

2) As a consequence of the fact that Z has the form of a Schlömilch series, the coefficients Z_n, equ.(6), are represented by a twofold quadrature of Z, equ.(7). This indicates that the RF SQUID can be applied for a weak-link diagnostic, if the ac impedance Z can be determined as a function of $|I|$ under special conditions.

CONCLUSION

The interdependence between the one-junction SQUID and the driving resonant circuit is shown to be represented by a Schlömilch series. This result shows the coefficient Z_n, n = 1,2,..., equ.(6), to be the only invariant parameters which describe the interaction between both circuits independent of ac currents.

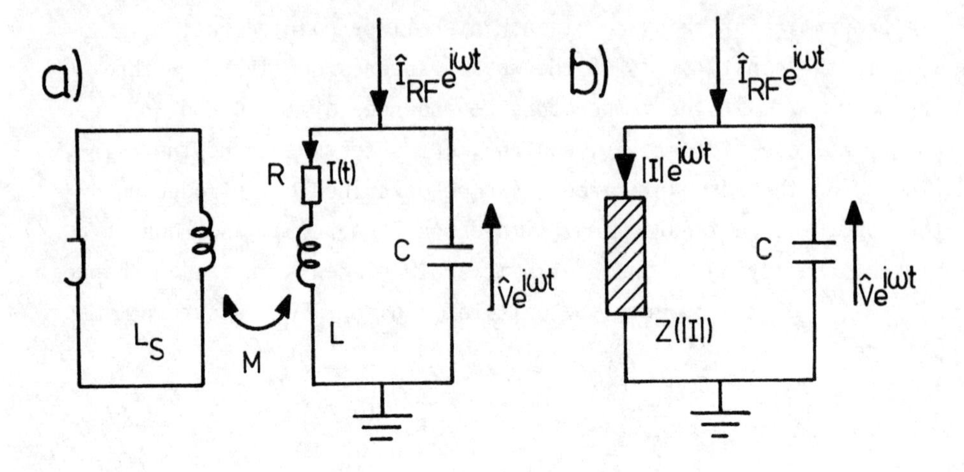

Fig.1 a) Model circuit for the RF biased SQUID; the symbol J denotes the CPR and M the mutual inductance coefficient;

b) high-Q tank circuit including the "SQUID circuit equivalent" ac impedance Z.

REFERENCES

1. P.K.Hansma, Phys.Rev. B 12, 1702 (1975).
2. M.Nisenoff, S. Wolf, Phys.Rev.B 12, 1712 (1975).
3. R.Rifkin, D.A. Vincent, B.S. Deaver, P.K. Hansma, J. Appl.Phys. 47, 2645 (1976).
4. S.N.Erné, H.-D.Hahlbohm, H.Lübbig. J.Appl.Phys. 47, 5440 (1976).
5. G.N. Watson, A Treatise on the Theory of Bessel Functions, Cambridge, 1958.

OPTIMIZATION OF SQUID DIFFERENTIAL MAGNETOMETERS

John P. Wikswo, Jr.
Vanderbilt University, Nashville, Tn. 37235

ABSTRACT

This paper outlines the design criteria for SQUID differential magnetometers and presents several measures of instrument sensitivity. Various magnetometer configurations are compared.

INTRODUCTION

The application of Superconducting Quantum Interference Devices (SQUIDs) to the measurement of biomagnetic fields has occurred because of their sensitivity, their stability, and the measurement flexibility afforded by using a pick-up coil that is part of a persistent-current, superconducting flux transformer. Simple SQUID magnetometers are sensitive to magnetic noise and to rotation in the geomagnetic field. This limitation can be partially overcome by using gradiometers or first-order differential magnetometers[1-3] which in theory are insensitive to uniform magnetic fields. The performance of these instruments is generally limited by noise fields with substantial gradients or by imbalance in the pick-up coil which makes the instrument sensitive to a uniform noise field. Second-order differential SQUID magnetometers, which are insensitive to both uniform fields and fields with a constant gradient, can measure to $10^{-13}\,T\sqrt{hz}$ in a typical, unshielded laboratory environment[4,5]. Since the performance of these magnetometers is limited by the intrinsic noise of the SQUID, it is important to optimize the coupling between the magnetic field source, the flux transformer, and the SQUID. This paper addresses this problem, with particular emphasis on the design of instruments for biomagnetic measurements.

DESIGN CRITERIA

Several factors affect the design of the magnetometer pick-up coil, including the desired sensitivity of the system, the physical size and location of the magnetic field source relative to the magnetometer, and the need to match the pick-up coil inductance to that of the signal coil within the SQUID. Biomagnetic signals are as small as 10^{-13}T with a measuring bandwidth of 1 Khz, the source volume ranges from less than 1 cm^3 to 0.1 m^3, and it may be desirable to have the pick-up coil within a few millimeters of the source. Generally, it is more useful to measure the biomagnetic field rather than its gradient, so the baseline of the differential magnetometer should be somewhat longer than the characteristic dimension of the source. If the source is placed near one end of the differential pick-up coil, the desired field will couple primarily to one of the loops, termed the face loop, while the more distant noise sources will couple to all loops, resulting in a common mode rejection of up to 10^7.

By using a Taylor's series expansion to describe the field change between loops, it can be shown that the following conditions

ISSN: 0094-243X/78/145/$1.50

on the number of turns N and area A of each loop must be met for proper differential performance

1st ORDER: $N_1A_1 = N_2A_2$ Rejects Uniform Fields (1)

2nd ORDER: $N_1A_1 + N_3A_3 = N_2A_2$ Rejects Uniform Fields (2)

$N_1A_1a = N_2A_2b$ Rejects Uniform Gradients (3)

where a is the distance between loops 1 and 3 and b is that between 2 and 3.

While early differential magnetometers[1-5] met these conditions by using coils of equal area, Zimmerman[6] has shown that if the field source couples primarily to one of the loops of the pick-up coil, the coupled energy is divided among all of the loops and the signal coil. Thus the sensitivities of symmetric (equal coil areas) 1st and 2nd order magnetometers to the field of a nearby object are a factor of $\sqrt{2}$ and 2, respectively, lower than the corresponding 0th order (single loop) magnetometer. However, in the asymmetric configuration, the loops have different areas so that the inductance of the face loop can be made much larger than that of the other loops. Thus the energy coupled into the pick-up coil is divided almost equally between the face loop and the signal coil in the SQUID, so that asymmetric differential magnetometers can have essentially the same sensitivity as a 0th order one if the source is close to the face loop. While it is possible to build SQUIDs with a wide range of signal coil inductances, this analysis assumes that a SQUID sensor with a 2 μH signal coil[7] will be used. If the pick-up coil has approximately a 2 μH inductance, the pick-up and signal coils will be properly matched and the SQUID will have a current sensitivity I_{min} of 8.0×10^{-12} amps/$\sqrt{hz}$.

SENSITIVITY DEFINITIONS

Because of the large number of pick-up coil parameters that can be varied, optimization of the pick-up coil design for biomagnetic measurements requires a realistic measure of magnetometer sensitivity. The simplest way to compute sensitivity to a source near the face coil is to assume that the source couples <u>only</u> to the face coil. In this case, the minimum detectable field at the face coil is related to I_{min} by the expression

$$B_{min} = I_{min} L/(N_1A_1) \tag{4}$$

where L is the sum of the pick-up and signal coil inductances. The simplicity of Eq. (4) is offset by its failure to account for the field at the distant loops. This is of particular importance in biomagnetic instruments, since decreasing the baseline of the difference measurement will improve the signal-to-noise ratio only until the baseline is somewhat shorter than the distance to the source; then the instrument acts as a true gradiometer for both the distant noise sources and the nearby signal source.

Another way to compute sensitivity is to determine the minimum detectable magnetic dipole at a given location on the coil axis. Since the dipole distance, measured with respect to the face coil, could be comparable to the face coil diameter, it is necessary to ac-

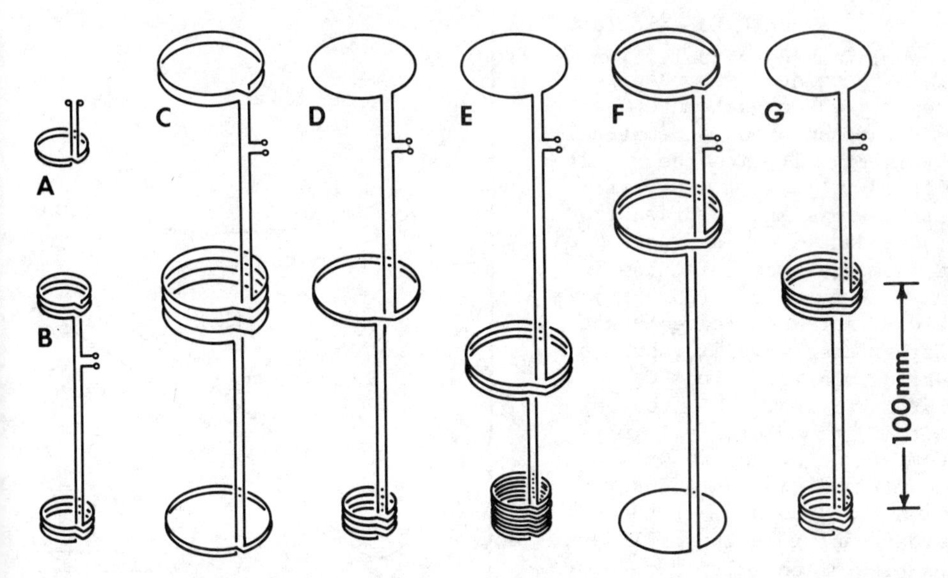

Fig. 1. Pick-up loops for a simple magnetometer, and for 1st and 2nd order differential magnetometers.

count for the variation of the dipole field over the loop. Rather than determine the net flux Φ by integration of the field over each loop, it is more convenient to use the reciprocity theorem to show that

$$\Phi = (\vec{B}_{pc} \cdot \vec{m})/I_{pc} \tag{5}$$

where $\vec{B}_{pc}$ is the magnetic field produced at the dipole by passing a current I_{pc} through the pick-up coil. The flux linking a loop of radius r_ℓ at a distance z from a dipole m_z on the loop axis is thus

$$\Phi = \mu_o m_z r_\ell^2/2(z^2 + r_\ell^2)^{3/2} \quad . \tag{6}$$

The minimum detectable magnetic dipole can then be determined by

TABLE I. B_{min} and the minimum detectable magnetic dipole m_z at (z-z') meters from the face loop of the coils in Fig. 1.

Config. & Order	L_{TOT} μH	B_{min} $T/\sqrt{hz}$	m(0.01) $Am^2/\sqrt{hz}$	m(0.1) $Am^2/\sqrt{hz}$	m(1.0) $Am^2/\sqrt{hz}$	m(10.0) $Am^2/\sqrt{hz}$
A-0	2.3	1.8×10^{-14}	3.9×10^{-13}	9.3×10^{-11}	0.9×10^{-7}	8.9×10^{-5}
B-1	2.9	1.5×10^{-14}	3.8×10^{-13}	9.4×10^{-11}	3.1×10^{-7}	2.6×10^{-3}
C-2	4.4	0.9×10^{-14}	9.3×10^{-13}	6.3×10^{-11}	5.7×10^{-7}	3.4×10^{-2}
D-2	3.6	1.4×10^{-14}	3.4×10^{-13}	9.5×10^{-11}	9.2×10^{-7}	6.1×10^{-2}
E-2	5.1	1.0×10^{-14}	3.4×10^{-13}	8.6×10^{-11}	9.5×10^{-7}	6.5×10^{-2}
F-2	3.9	1.5×10^{-14}	1.6×10^{-12}	1.0×10^{-10}	7.9×10^{-7}	5.0×10^{-2}
G-2	4.1	1.6×10^{-14}	4.4×10^{-13}	1.1×10^{-10}	1.0×10^{-6}	7.0×10^{-2}

using I_{min} and L to calculate the minimum detectable flux. This technique was used to compare the magnetometer configurations shown schematically in Fig. 1, with the results listed in Table I. Several conclusions can be drawn from these data. For a dipole 0.01 m from the face coil, the smaller face coil (25.4 mm) is always more sensitive than a larger one (50.8 mm), but comparing configurations C through E shows that the opposite is true for z = 10 m. Comparison of the m(.01) data confirms Zimmerman's observation that the asymmetric 2nd order configuration will have as high a sensitivity to near sources as a simple (0th order) magnetometer. The m(1.0) data shows that the 2nd order systems have one tenth the sensitivity to a dipole at 1.0 m that a 0th order one has. At 10 m, the 2nd order systems have only one-thousandth the sensitivity of the 0th order one, demonstrating the efficiency of 2nd order systems in rejecting noise from distant sources. Configuration F, which has its middle coil farther from the face loop than the other 2nd order coils, has the lowest near-field sensitivity because most of the pick-up coil inductance in configuration F is provided by the two upper loops rather than the face loop. This could be remedied by flipping the pick-up coil over, making it similar to configuration E. Configurations D, E, and G all have comparable near-field sensitivities. However, G has three different-sized coils and is more difficult to fabricate and balance. E requires twice the number of face coil turns as D but has no significant difference in sensitivity. It is also interesting to note that B_{min} computed using Eq. (5) has little correlation with the dipole sensitivities.

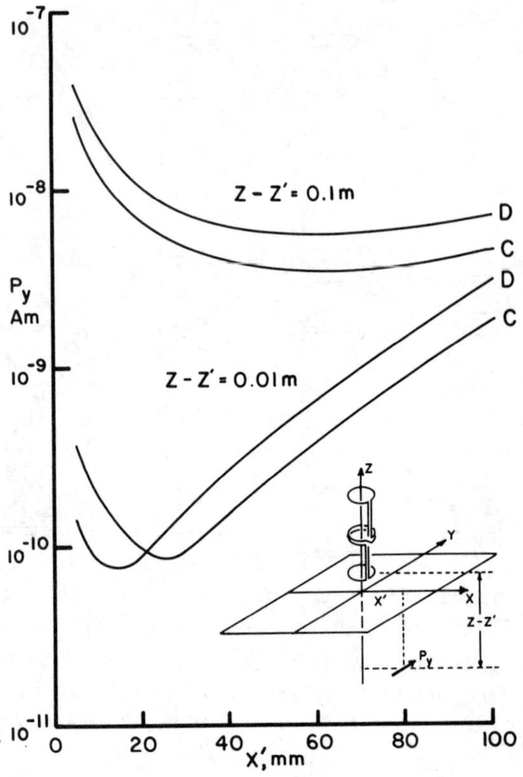

Fig. 2. The minimum detectable current dipole p_y for coil configurations C and D in Fig. 1.

While magnetic noise sources may have fields similar to those of a magnetic dipole, some biomagnetic fields are produced by electric currents in conducting media and are not necessarily magnetically dipolar. The corresponding measure of sensitivity is the ability to detect a dipolar electric current source $\vec{p}$ located at a point $\vec{r}'$ in a conducting half-space. If the air-conductor interface is the x-y plane, B_z at a point $\vec{r}$ in the air is given by[8]

$$B_z(\vec{r}') = \frac{\mu_o}{4\pi} \frac{p_x(y-y') - p_y(x-x')}{[(x-x')^2 + (y-y')^2 + (z-z')^2]^{3/2}} \quad . \tag{7}$$

This is equivalent to the magnetic vector potential

$$A_x = \frac{\mu_o p_x}{4\pi|\vec{r}-\vec{r}'|} \qquad A_y = \frac{\mu_o p_y}{4\pi|\vec{r}-\vec{r}'|} \quad \text{with } B_z = \frac{\partial A_y}{\partial x} - \frac{\partial A_x}{\partial y} \quad . \tag{8}$$

Since $\vec{p}$ is a current source, the conductivity does not enter into the expressions. The vector potential can be integrated around each loop of the pick-up coil to obtain the flux linking that loop. If the loop axis corresponds to the z axis and the dipole lies beneath the positive x axis as shown in Fig. 2, the integral reduces to

$$\Phi = \frac{\mu_o p_y}{\pi k} \sqrt{r_\ell/x'} \left[(1 - \frac{k^2}{2})K(k) - E(k)\right] \tag{9}$$

where r_ℓ is the loop radius, K(k) and E(k) are the complete elliptic integrals of the 1st and 2nd kind, and k is given by

$$k^2 = 4x'r_\ell / \left[(x+r_\ell)^2 + (z-z')^2\right] \quad . \tag{10}$$

This result can also be obtained using the reciprocity theorem[9].

Equation (9) was used to compute the minimum detectable electric current dipole p_y as a function of x' for configurations C and D. The results, plotted in Fig. 2, show that dipoles on the order of 10^{-10} $Am/\sqrt{hz}$ can be detected 1 cm from the face loop, and that the asymmetric design (D) has a slightly higher sensitivity and better spatial resolution for dipoles near the face loop (z - z' = 0.01 m) than does the symmetric one. The opposite is true for z - z' = 1.0 m.

CONCLUSION

It is possible to use m_z and p_y to compare various pick-up coil configurations and to determine the loop parameters that provide the optimum sensitivity and spatial resolution for a particular measurement. An interactive computer program, available from the author, has been developed to perform the required inductance and sensitivity calculations.

ACKNOWLEDGEMENT

The author is indebted to Drs. Duane Crum and Mike Simmonds of the SHE Corporation for their suggestions. This work has been funded in part by a Vanderbilt/NIH Biomedical Support Grant.

REFERENCES

1. J. E. Zimmerman and N. V. Frederick, Appl. Phys. Lett. 19, 16 (1971).
2. W. Goodman, et al., Proc. IEEE 61, 20 (1973).
3. J. P. Wikswo and W. M. Fairbank, IEEE Trans. Mag-13, 354 (1977).
4. J. E. Opfer, et al., IEEE Trans. Mag-10, 536 (1974).
5. E. Brenner, et al., IEEE Trans. Mag-13, 365 (1977).
6. J. E. Zimmerman, J. Appl. Phys. 48, 702 (1977).
7. SHE Model 330X Low Noise SQUID.
8. B. N. Cuffin and D. Cohen, IEEE Trans. BME-24, 372 (1977).
9. J. Malmivuo, Acta Polytech. Scand., El. 39 (1976).

APPLICATION OF A MICROWAVE BIASED RF-SQUID-SYSTEM TO INVESTIGATE MAGNETIC AFTER-EFFECTS

H. Rogalla, C. Heiden
Institut für Angewandte Physik der Justus Liebig-Universität Gießen, Heinrich Buff-Ring 16, D-6300 Gießen, Germany

ABSTRACT

For the investigation of magnetic after-effects in superconductors low-drift high-resolution magnetization measurements are required. An rf-biased (9 GHz) closed-loop SQUID-system was developed for this purpose having a slew rate of $2\cdot10^{+7}$ $\emptyset_0$/s and a dynamic range of 1400 $\emptyset_0$. The upper limit for a possible drift referred to the input of the used flux transformer is 3 μT/h. This low drift and fast response allows registrate magnetization curves over time periods of the order of 10^5 s even under the presence of spontaneous magnetization changes (flux jumps). This system was used in combination with a microprocessor controlled-precision generator for the magnetic field to study the behavior of periodically swept minor hysteresis loops of superconducting vanadium samples. In general the observed loops are not stationary but exhibit a more or less pronounced creep effect that can be approximated by a logarithmic dependence on the number of minor hysteresis cycle. For sufficient low rates of change of the magnetic field (dH/dt < 20 A/cms) a dependence of the creep effect on the total measuring time could not be observed. Contradictory to a solely thermal activated flux creep is also the fact that flux density gradients inside the investigated specimens tend to increase with the number of swept cycle.

THE MEASURING SYSTEM

Magnetization measurements using electronic integration usually suffer from a more or less noticeable drift in particular due to thermal emf's. Vibrating sample magnetometers in principle allow high-resolution low-drift measurements; however, vibrations and a periodically varying interaction between sample and measuring system provide a handicap, for instance, for the investigation of magnetic after effects. Using a SQUID-system has the advantage that there is no need for electronic integration and the interaction between flux transformer and specimen is small and can be kept constant. Due to spontaneous magnetization changes like flux-jumps, a high slew-rate of the SQUID-system however is required. First attempts[1] with a commercially-available SQUID-system failed because of

ISSN: 0094-243X/78/150/$1.50

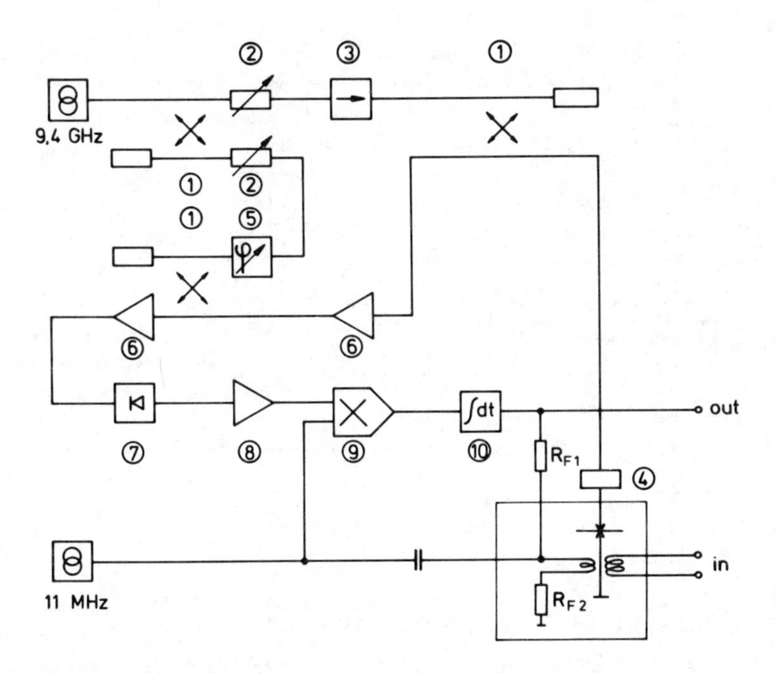

Fig. 1. Schematic diagram of microwave biased rf SQUID electronics. 1) directional coupler, 2) attenuator, 3) isolator, 4) impedance transformer, 5) phase shifter, 6) tunnel diode amplifier, 7) Schottky diode, 8) video amplifier, 9) phase detector, 10) integrator

the insufficient slew rate. Therefore a faster rf-SQUID-system [2] was developed (Fig. 1). It operates at a bias frequency of 9.4 GHz and a modulation frequency of 11 MHz. A fast feedback-system allows slew rates up to $2\cdot10^{7}$ ϕ_o/s in a dynamic-range of $\pm$ 700 ϕ_o. The equivalent input noise is white between 10^{-3} Hz and 10^{5} Hz at a value of $3\cdot10^{-4}$ $\phi_o/\sqrt{Hz}$. When operated at a restricted dynamic range the equivalent input noise was observed to be as small as $6\cdot10^{-6}$ $\phi_o/\sqrt{Hz}$.

An adjustable differential flux transformer system (Fig. 2) is used to measure the magnetization or the induction of a specimen under test. It consists of three flux transformers a,b, and c. By proper adjustment of the input loop of transformer c, a differential flux proportional to the induction or magnetization of the specimen is coupled to the SQUID, resulting in a sensitivity of about 6 mT/V and a resolution of ca 1 μT. Drift of the system is less than 3 μT/h.

The magnetic field is supplied by a superconducting solenoid and a precision current generator providing a

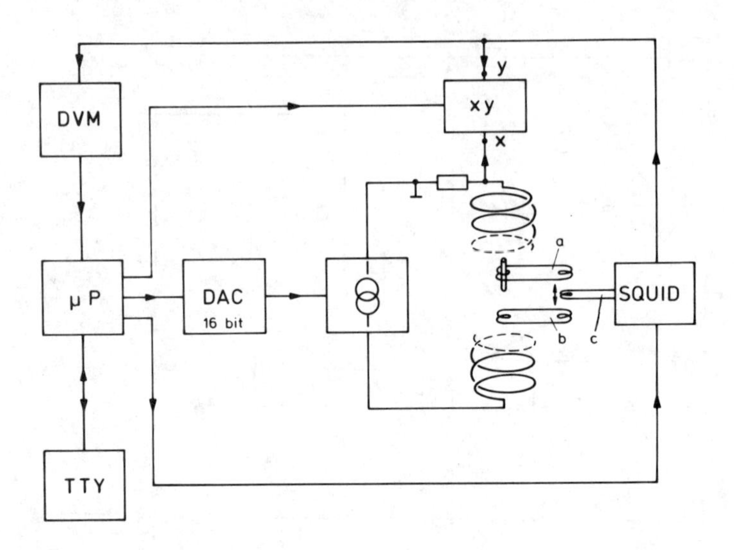

Fig. 2. Schematic diagram of measuring system

field range of 0 ... 1424 A/cm with an accuracy of $2\cdot10^{-5}$.

Well defined field sweeping programs can be run by means of a microprocessor, which via a 16-bit digital-to-analog-converter (DAC) provides the input signal of the current generator and controls also the operation of the SQUID-system. The output voltage of the SQUID-system is measured by a digital volmeter (DVM) and printed and punched at the console (TTY). Further evaluation of the data is done by a PDP 11-computer. With this system magnetization measurements with negligible drift can be performed over measuring times of the order of 10^5 s, that are at present limited only by helium boil-off.

MAGNETIC AFTER-EFFECT

In the simple critical state model it is assumed that the local critical current density of a type-II-superconductor is a single valued function of flux density and temperature. Periodically swept minor hysteresis loops therefore should be stationary. This was checked with the measuring system described above.

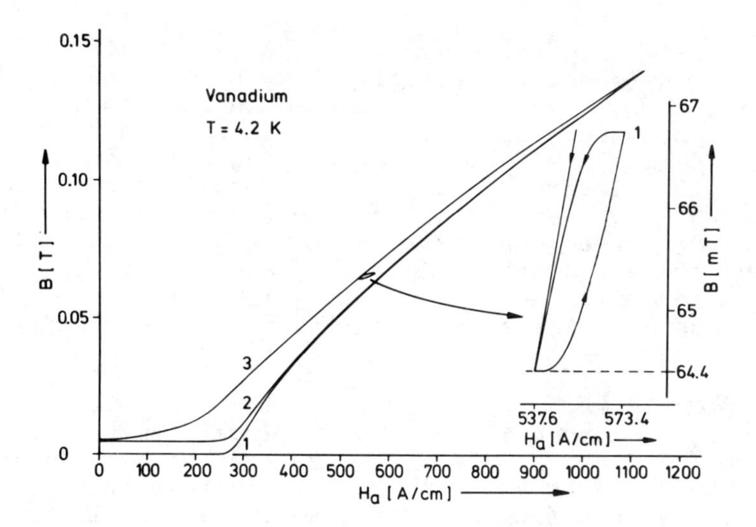

Fig. 3. Initial and secondary magnetization curve and minor loop of a circular vanadium cylinder of 28 mm length and 0.9 mm diameter with applied field parallel to the cylinder axis. The specimen of 99.96 % purity was degassed in UHV of 10^{-9} torr for 60 min. at 1420° C.

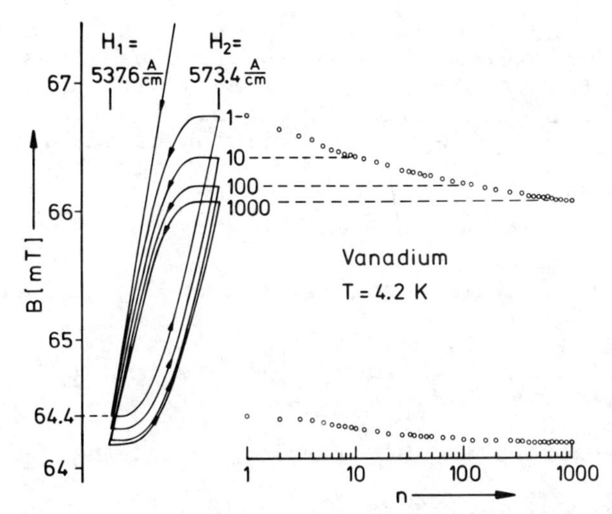

Fig. 4. Creep effect of minor magnetization loops of the vanadium sample and logarithmic dependence on the number of swept cycles.

For this purpose, the same minor hysteresis loop as in Fig. 3 was swept periodically many times, the 1., 10., 100., and 1000., cycle being plotted in Fig. 4. There is a rather pronounced creep effect[3] Both endpoints of the loop creep into the interior of the magnetization curve, however with different creep rates, resulting in a change of shape of the hysteresis loop. To first approximation, the observed creep depends logarithmically on the number of swept cycle and within measuring accuracy is independent of the measuring time, as long as field sweep rates do not exceed 20 A/cms (no significant sample heating due to flux flow!).

Obviously, we are not dealing with a purely thermal-activated flux creep, which shows a logarithmic time dependence [4] as for example in the relaxation experiments using cylindrical tubes [4,5]. This is further supported by the fact that the flux density gradient inside the specimen on the whole seems to increase with increasing cycle number. To show this, derivatives of the minor hysteresis loops were taken which provide a measure for the flux density inside the specimen [6,7]. The after-effect described so far can be observed also at other dc-bias fields and field-sweeping amplitudes.

Altogether, it appears that the continued cycling leads to a more effective pinning of flux lines in the superconductor. A similar effect, only with domain walls instead of flux lines, may play a role in ferromagnetic materials, where creeping and tilting of minor hysteresis loops has been observed already many years ago [8].

REFERENCES

1. H. Rogalla, Diploma thesis, University of Münster (1972)
2. H. Rogalla and C. Heiden, Appl. Phys. 14, 161 (1977)
3. H. Rogalla and C. Heiden, Phys. Letters 63A, 63 (1977)
4. Y.B. Kim, C.F. Hempstead, A.R. Strnad, Phys. Rev. 131 6, 2486 (1963)
5. G. Antesberger and H. Ullmaier, Phil. Mag 29, 1101 (1974)
6. A.M. Campbell, J. Phys. C2, 1492 (1969)
7. R.W. Rollins, H. Küpfer, W. Gey, J. Appl. Phys. 45, 5392 (1974)
8. L. Néel, C.R., Acad. Sci. 244, 2668 (1957)
   J.A. Ewing, Phil. Trans. 176, 569 (1885)

DEVELOPMENT OF A LONDON MOMENT READOUT FOR A SUPERCONDUCTING GYROSCOPE*

J. T. Anderson and R. R. Clappier
Stanford University, Stanford, CA 94305

ABSTRACT

A gyroscope for verifying certain predictions of General Relativity is under development at Stanford. A method of measuring the gyroscope orientation based on the use of the London moment and a SQUID magnetometer is described. It is estimated that the required sensitivity of 1 milliarc-sec can be achieved in a bandwidth of 2.5×10^{-6} Hz.

INTRODUCTION

In 1961, Schiff[1] predicted two small precessional effects of an orbiting gyroscope attributable to general relativity. The larger of these, the geodetic effect, is about 7 arc-sec/yr, and is due to the orbital motion of the gyroscope about the earth. The other, the Schiff motional effect, is about 50 milliarc-sec/yr, and is due only to the rotation of the earth. The gyroscope drift rate required even to detect the motional effect is orders of magnitude beyond conventional state-of-the-art gyroscopes.

A program to develop a gyroscope and associated systems to test general relativity[2] has been underway at Stanford since 1961, and additionally, at the National Aeronautics and Space Administration's Marshall Center since 1967. Only those details of the experiment related to the readout of the gyroscope orientation will be considered here.

PRINCIPLE OF THE LONDON MOMENT READOUT

The gyroscope is a 1.5" diameter niobium-coated fused quartz sphere, round and homogeneous in density to about 3 parts in 10^7. Support and centering of the gyroscope in its housing are maintained by a combination capacitive bridge position sensing system and electrostatic suspension system. To eliminate mechanical instabilities and take advantage of the unique properties of superconductors, the gyroscope will be cooled to below 4K. The gyroscope is initially spun up by a gas-spin up system, but thereafter operates in a high vacuum.

The readout of a conventional electrostatically-supported gyroscope uses a surface pattern or a small reflecting flat, either of which has a known orientation with respect to a previously established principle axis of inertia. The low drift rate required at the relativity gyroscope precludes any readout system that specifically requires unequal inertia axes. Additionally, neither method

*Work supported by NASA Contract NAS8-32355

ISSN: 0094-243X/78/155/$1.50

can achieve the 1 milliarc-sec sensitivity needed to measure the motional effect to within a few percent.

The solution proposed by Everitt[3] is that the magnetic moment of a spinning superconductor, the London moment,[4] be used. For a spherical superconductor, the field is uniform and has a magnitude

$$B_L = 7.14 \times 10^{-11} \text{ f Tesla} \tag{1}$$

where f is the rotation rate in Hz. The niobium coating on the gyroscope provides the necessary superconductor. The London moment is exactly aligned with the spin axis so that its orientation is also that of the gyroscope.

The elements of the London moment readout are shown in Fig. 1.[5,6] The gyroscope is initially spun up with its spin axis lying in the plane of the readout loop and pointed at a reference star with respect to which the gyroscope precessions are measured. The experimental geometry is chosen such that in the absence of gyroscope precessions, the spin axis remains in the plane of the loop. Precession

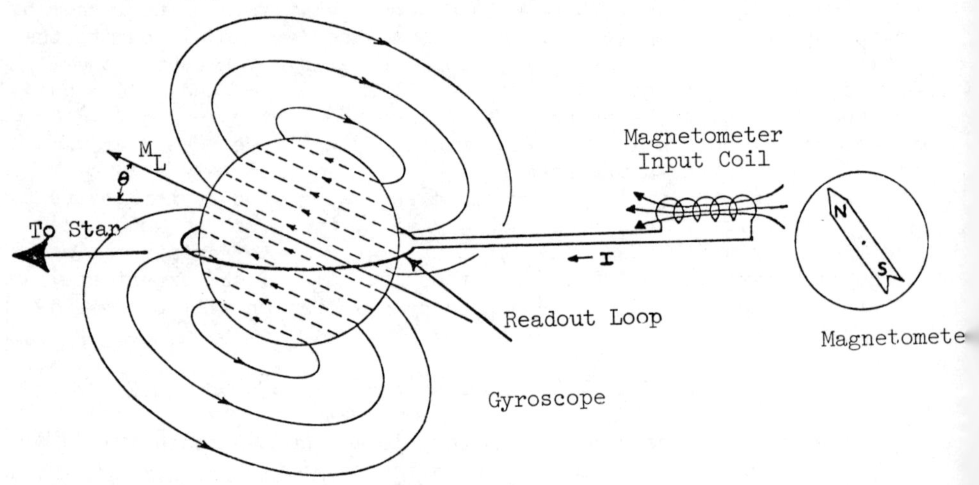

Fig. 1 Basic London Moment Readout

of the gyroscope tips the London moment out of the plane, and magnetic flux will couple the readout loop. The readout loop and associated circuitry form a superconducting path, so that the flux threading the loop induces a current in the readout circuit. A magnetic field generated by that current is then measured as the indicator of θ. Two orthogonal coils can be operated in the null configuration; a third coil receives the full London moment.

READOUT SENSITIVITY

The calculation of the expected sensitivity of the London moment readout is straightforward. Referring to Fig. 1 and Eq. 1, the total flux threading the readout loop for a loop radius $r_R \sim r_G$, where r_G is the gyroscope radius, is

$$\phi_T = 7.14 \times 10^{-11} f\pi r_G^2 \sin\theta \,. \qquad (2)$$

The current flowing in the readout circuit, i_R, is

$$i_R = \frac{\phi_T}{L_R + L_M} \qquad (3)$$

where L_R and L_M are the readout loop and SQUID input coil inductances respectively. Finally, the energy appearing in the input coil of the SQUID, E_i, is

$$E_i = \frac{1}{2} L\, i_R^2 \,. \qquad (4)$$

The maximum energy transfer occurs when $L_R = L_M$. For our application, $f = 200$ Hz, $L_R \approx 0.3$ μH, $r_G = 1.9 \times 10^{-2}$ m so that for a 1 milliarc-sec angular displacement of the gyro, $E_i = 2.6 \times 10^{-33}$ J.

Shown in Fig. 2 are simplified power spectra for several SQUIDs. Attempts to simply integrate the magnetometer output down to the 10^{-33} J level will fail because the bandwidths required place the noise to be integrated into the 1/f region of the magnetometers.

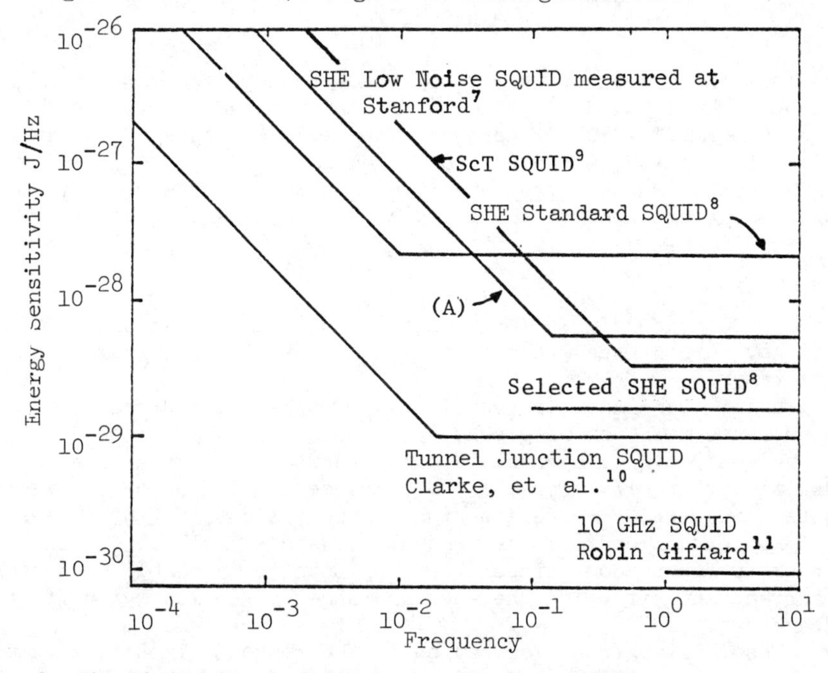

Fig. 2 Simplified Power Spectra for Several SQUIDs

To avoid the integration of a dc signal, the satellite will rotate about the line of sight to the reference star. The effect of the roll can be seen from Fig. 1. As the readout loop rotates with the satellite, M_L, which is now assumed to have precessed out of the loop, points alternately into and out of the loop. Note, however, that a misalignment of the readout loop with the roll axis will not modulate the signal, and thus will not produce a readout error.

The projected roll is in the 1-10 minutes range. Using, for example, a period of 1 min and curve (A) of Fig. 2, the required bandwidth, ΔB, to observe 1 milliarc-sec is

$$\Delta B = \frac{2.5 \times 10^{-33} \text{ J}}{5 \times 10^{-28} \text{ J/Hz}} = 5 \times 10^{-6} \text{ Hz.} \tag{5}$$

In this case, the integration is achieved by narrowbanding around 0.0167 Hz rather than by integrating deeper into the 1/f regime.[12]

READOUT SYSTEM TORQUES ON GYROSCOPE

To insure the required low drift rate of the gyroscope, any potential source of torque on the gyroscope must be considered.[13] Pertinent to the readout are the interactions of currents in the readout circuit with the London moment. The drift rate Ω_i of the gyroscope in response to a torque Γ_i perpendicular to the spin axis is

$$\Omega_i = \frac{\Gamma_i}{I\omega_s} \tag{6}$$

where I is the moment of inertia of the gyroscope, and ω_s its angular velocity. The total drift allowed from all sources, Ω_{max}, is 1 milli-arc-sec/yr or 1.5×10^{-16} rad/sec. The maximum torque, Γ_{max}, is 1.7 $\times 10^{-18}$ nt-m, taking $I = 9.0 \times 10^{-6}$ kg-m^2.

The torque on the gyroscope from a current in the readout loop is that on a current loop in the field of a dipole, and is given by

$$\Gamma_\ell = \frac{3}{4} i\pi r_G^2 B_L \cos\theta. \tag{7}$$

For the worst case of $\Gamma_\ell = \Gamma_{max}$ and $\theta = 0$, the current, $i_{max} = 10^{-7}$ amps. Any current less than this will not, by itself, produce a torque in excess of Γ_{max}.

A brief discussion of the effects of selected sources of readout loop currents will now be given.

The most obvious current in the readout loop is that due to the London moment itself. For $|\theta| < 800$ arc-sec, a factor of 16 greater than that required for the experiment, the readout current is less than i_{max}. Additionally, the readout for space will use current feedback to the readout circuit (rather than to the SQUID), reducing the current and extending the angular range by the loop gain of the feedback system.

Johnson noise from dissipation in the readout circuit, such as from the damping cylinder, is less than i_{max} for an equivalent resistance in parallel with the readout ring greater than 2μΩ (the damping cylinder is ∿600 μΩ). Further, Johnson noise is a fluctuating perturbation, and drift from it will average to zero.

If the gyroscope contains trapped magnetic field, readout circuit losses acting on the ac current generated by the rotating magnetic field will reduce the component of angular velocity lying in the plane of the readout loop. By extension of the arguments already given, it can be shown that for a trapped field less than 3×10^{-10} T, which we have achieved, the minimum parallel equivalent resistance must be greater than 5 mΩ. When feedback is applied, the minimum tolerable resistance is reduced by the loop gain.

LABORATORY IMPLEMENTATION OF THE LONDON MOMENT READOUT

The form of the London moment readout as currently in use in the laboratory is shown in Fig. 3. A SQUID magnetometer detects the readout current. To isolate the SQUID from electrical interference generated by the gyroscope suspension system, a "damping cylinder" is in-

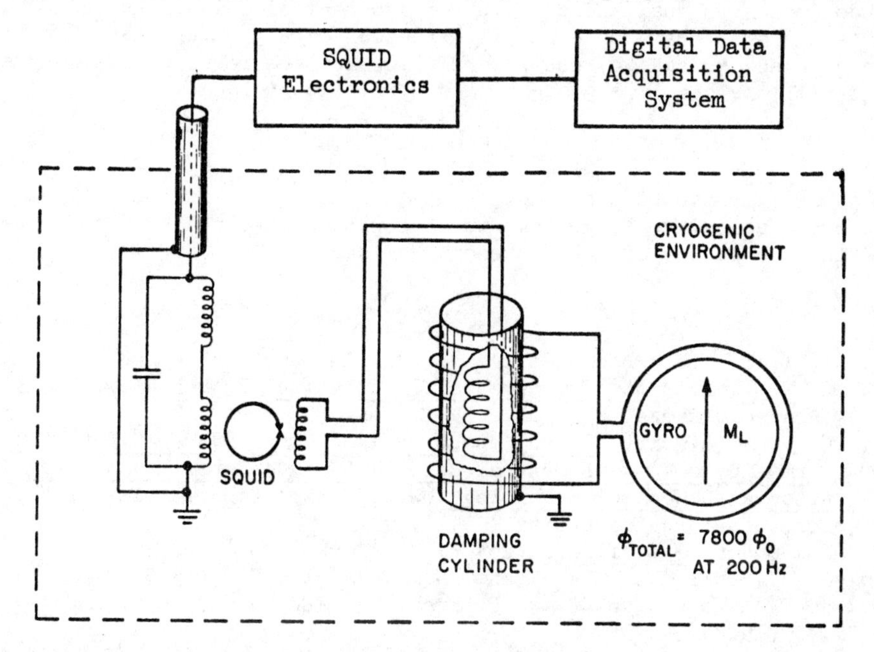

Fig. 3 Laboratory Form of the London Moment Readout

terposed between the readout loop and the SQUID. The damping cylinder is a combination low pass filter and electrostatic shield. Without it, straight capacitive coupling of the 1 kV rms 20 kHz support voltage easily saturates the SQUID rf electronics. Further, the suspension currents flowing through the gyroscope generate milligauss level fields that would otherwise be superimposed on the London moment signal. The 1 MHz position sensing signal contributes to the problem at a lower level.

The damping cylinder is made of 5 mils electroplated copper on a 5 mil wall, 0.20" diameter copper cylinder. Grounding the cylinder provides capacitive isolation. For magnetic signals having periods short compared to the L/R time constant of the cylinder, it shields the internal coil. Low frequency signals are passed. The damping cylinder described has a bandwidth ∿ 100 Hz and provides 40 db attenuation at 20 kHz. The resultant 20 kHz signal appearing at the SQUID is about $\phi_0/4$ p.p. In orbit, where the relativity experiment must actually be done, little suspension support will be needed, suspension signals will be small, and the damping cylinder will not likely be needed.

Limited tests of the readout have been performed in the laboratory. Although we have observed the London moment[14], technical difficulties in areas other than readout have prevented our using it for readout purposes. The equivalent readout drift rate in the laboratory is approximately 1 arc-sec/min. Resolution to 1 milliarc-sec cannot be achieved in the laboratory because the experimental apparatus has no provision for rotation, and the damping cylinder attenuates the London moment signal and contributes substantial noise. Further, the gyroscope drift rate on earth is expected to be greater than 10 milliarc-sec/sec, so that gyroscope drift in the time required to integrate to 1 milliarc-sec is expected to be more than one degree.

ACKNOWLEDGMENTS

An experiment of this complexity has many contributors. The authors gratefully acknowledge the stepping stones placed before them and the assistance of their present colleagues.

REFERENCES

1. L. I. Schiff, Proc. Nat. Acad. Sci. Am. 46, 1333 (1960).
2. C. W. F. Everitt, Experimental Gravitation, B. Bertotti, ed. (Academic Press, New York, 1973), pp. 331-358.
3. C. W. F. Everitt, Final Report on NASA Grant 05-020-019 to Perform a Gyro Test of General Relativity in a Satellite and Develop Associated Control Technology (W.W. Hansen Laboratories of Physics, Stanford University, 1977), p. 3.
4. F. London, Superfluids (Dover, New York, 1964), p. 83.
5. J. B. Hendricks, IEEE Trans. Magn. MAG-11, 712 (1975).
6. J. T. Anderson and C. W. F. Everitt, IEEE Trans. Magn. MAG-13, 377 (1977).
7. B. Cabrera and J. T. Anderson, "Signal Detection in 1/f Noise of SQUID Magnetometers," a paper contributed to Conference on Future Trends in Superconductive Electronics, Charlottesville, VA, 1978. Only a portion of the data justifying the curve is given.
8. SHE Corporation, 4174 Sorrento Valley Blvd, San Diego, CA 92121.
9. Superconducting Technology Division, United Scientific Corp., 1400 Stierlin Rd., Mountain View, CA., from Ref. (10).
10. John Clarke, SQUID Superconducting Quantum Interference Devices and Their Applications, ed. H. D. Hahlbolm and H. Lübbig (Walter de Gruyter, New York, 1977), pp. 213.
11. Robin Giffard, Stanford University, private communication.
12. See Reference (7).
13. See Ref. (2), pp. 344-358, for more on gyroscope drift.
14. J. A. Lipa, J. R. Nikirk, J. T. Anderson and R. R. Clappier, Proceedings of the 14th International Conference on Low Temperature Physics, Vol. 4, ed. Matti Krusius and Matti Vuorio, (North Holland/American Elsevier, New York, 1975), pp. 250.

SIGNAL DETECTION IN 1/f NOISE OF SQUID MAGNETOMETERS*

B. Cabrera and J. T. Anderson
Physics Department, Stanford University, Stanford, CA 94305

ABSTRACT

We have shown that the variance on the SQUID power spectrum in the 1/f low frequency region is well behaved, i.e., any small frequency band may be treated as white noise in standard power spectrum estimation theory. Specifically we have taken a calibration signal at 0.017 Hz with an equivalent energy referred to the SQUID input coil of 1×10^{-30} J and digitally recorded and analyzed a record of 140 hr duration obtaining an optimum S/N better than 400. The results are in good agreement with theory. In addition we have seen no deviation from the 1/f dependence of the noise energy spectrum down to frequencies below 10^{-5} Hz. A commercially available SQUID and electronics system were used.

INTRODUCTION

The technological feasibility of the component systems has been demonstrated for a high accuracy cryogenic gyroscope being developed to perform a satellite experiment to test general relativity.[1] The gyroscope rotor is a 1.5 inch diameter "perfect" quartz sphere uniformly coated with 5000 Å of superconducting niobium. The London moment along the 200 Hz spin axis is read out using three mutually orthogonal pick-up loops coupled to SQUID magnetometers. The roll of the satellite about the line of sight to a reference star modulates the high sensitivity readout data, thus providing a low frequency ac signal in the millihertz range with an information bandwidth of 10^{-6} hertz. Details of the gyro readout system are presented in an accompanying paper.[2] Here we present data taken to demonstrate the feasibility of using SQUIDs to detect gyro motions of one milliarc second corresponding to an energy resolution of 3×10^{-33} J.

DATA ACQUISITION

In order to investigate recovery of a low-frequency signal from noise in a SQUID output to the desired sensitivity, a digital system was used for obtaining integration times of days. An S.H.E. TSQX SQUID sensor and model 330 X electronics were used for the testing. The SQUID was placed in an ambient magnetic field of 3×10^{-7} gauss with a superconducting short across the input leads (Fig. 1). The magnetometer signal was filtered using the internal two pole 0.1 Hz filter of the SQUID electronics, amplified x1000, and again filtered by an additional pole at 0.1 Hz; simultaneously a strip chart recording of the raw data was made. The data was stored on magnetic tape at a sample interval of 5 seconds. In addition, a calibration signal

*Work supported by NASA Contract NAS8-32355.

SSN: 0094-243X/78/161/$1.50

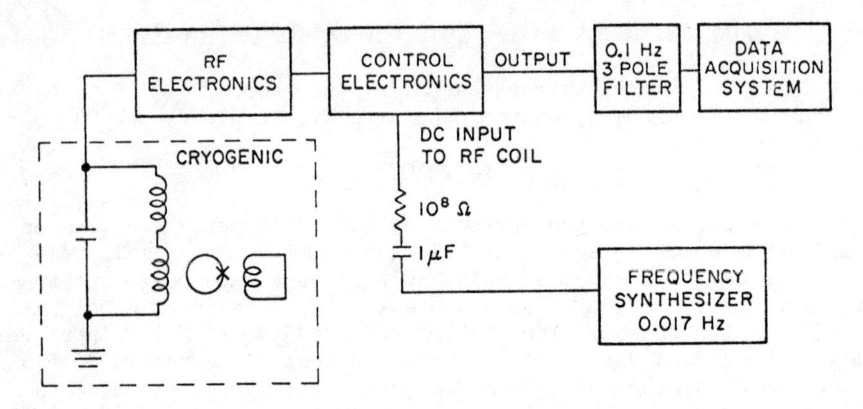

Fig. 1 Block Diagram of System used to Measure SQUID noise

was injected into the SQUID rf coil at a frequency of 0.017 Hz with an amplitude of 1 x 10^{-5} ϕ_0 at the SQUID (Fig. 2). The clocks on the data acquisition system and the frequency synthesizer used for calibration were in agreement to one part in 10^6. In this mode a 140 hour data set was recorded. The signal detection analysis presented here was made from this data set.

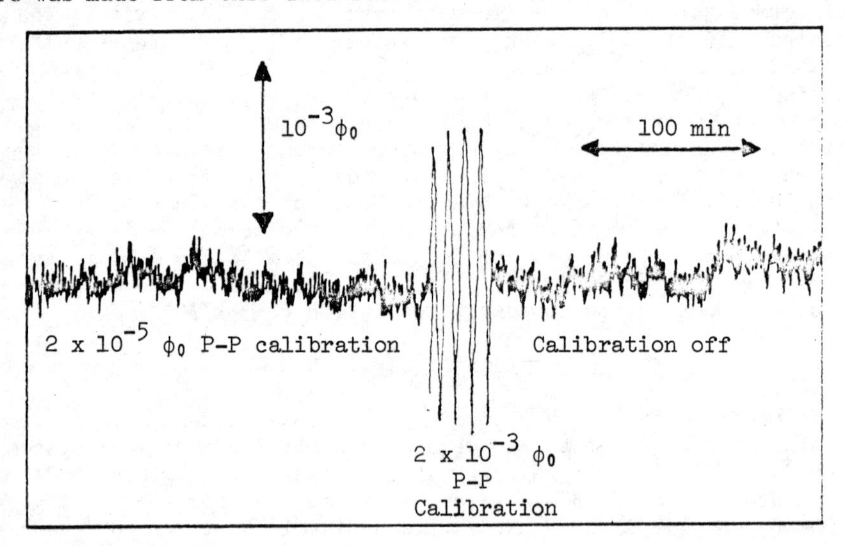

Fig. 2 SQUID Output through 0.1 Hz Filter with and without 0.017 Hz Calibration Signals

RESULTS

Preliminary analysis of the data was done using 1024 point fast Fourier transforms on 198 data segments to form a 512 point autocorrelation function of the entire data set. The final 512 frequency/energy spectrum, folded with a Hamming window, was then found from the autocorrelation function.[3] The Nyquist frequency was 0.1 Hz, thus each frequency interval, and the lowest non-zero frequency is 2 x 10^{-4} Hz. On a log-log plot (Fig. 3), the data forms a straight line of slope -1 indicating a characteristic 1/f spectrum. The roll off towards the higher frequency end is due to the three-pole anti-aliasing filter; the calibration signal is also clearly visible. A

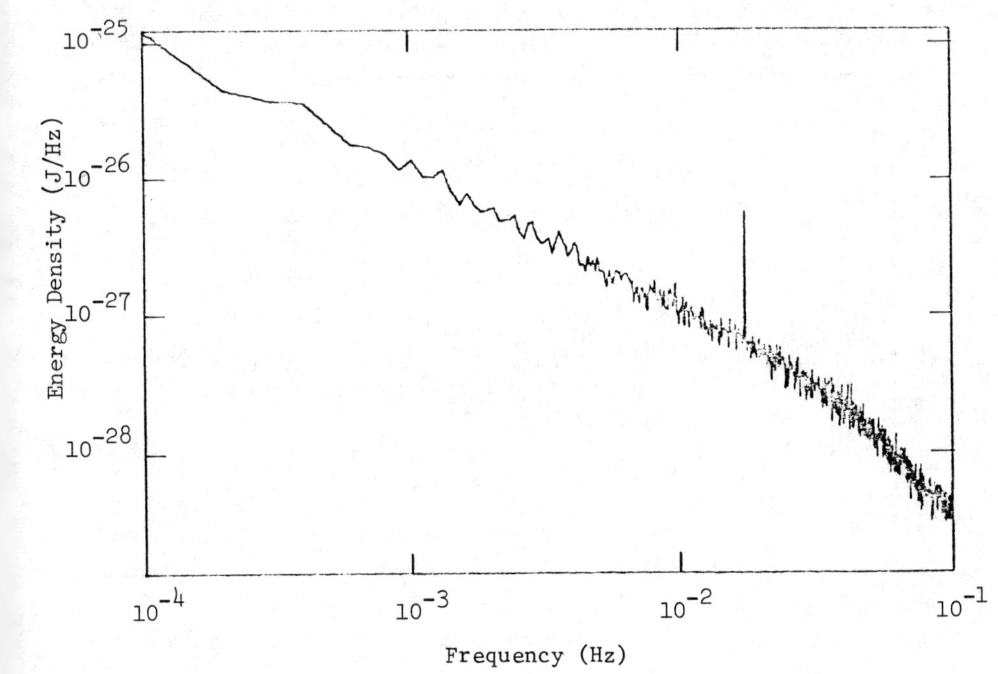

Fig. 3 SQUID Noise Spectrum in 1/f Region with Calibration Signal Clearly Visible.

one hundred point fit along the 1/f line provides an experimentally determined slope of -1.02 and a standard deviation ratio for the noise on the noise energy spectrum of 5.9%. The calculated standard deviation ratio σ_R = 6.3% obtained from a white noise model is given by:[4]

$$\sigma_R = \sqrt{\frac{2M}{N}\left(\alpha^2 + \frac{\beta^2}{2}\right)} \text{ for } M >> 1 \tag{1}$$

where 2M - 1 is the window width, N is the data length and α = 0.54 and β = 0.46 for the Hamming window. From the measured mean noise at the calibration peak $S(\omega_0) = 6.8 \times 10^{-28}$ J/Hz, a signal-to-noise ratio of 1 corresponds to an energy resolution δE

$$\delta E = (\sigma_R/\alpha)\ S(\omega_0)\ \ \Delta f = 1.55 \times 10^{-32} \text{ J} \tag{2}$$

Optimum signal-to-noise from this data can be achieved by performing a single FFT on the entire data set. Then $\Delta f = 2 \times 10^{-6}$ Hz and for signals with this bandwidth or less

$$\delta E = \bar{S}(\omega_0)\ \Delta f = 1.34 \times 10^{-33}\ \mathrm{J} \quad . \tag{3}$$

Recently we have demonstrated this resolution by expanding a small region around $\omega_0 = 0.017$ Hz with resolution $\Delta f = 2 \times 10^{-6}$ Hz (Fig. 4). Although the analysis is not yet complete the expected energy resolution increase is qualitatively correct. In addition

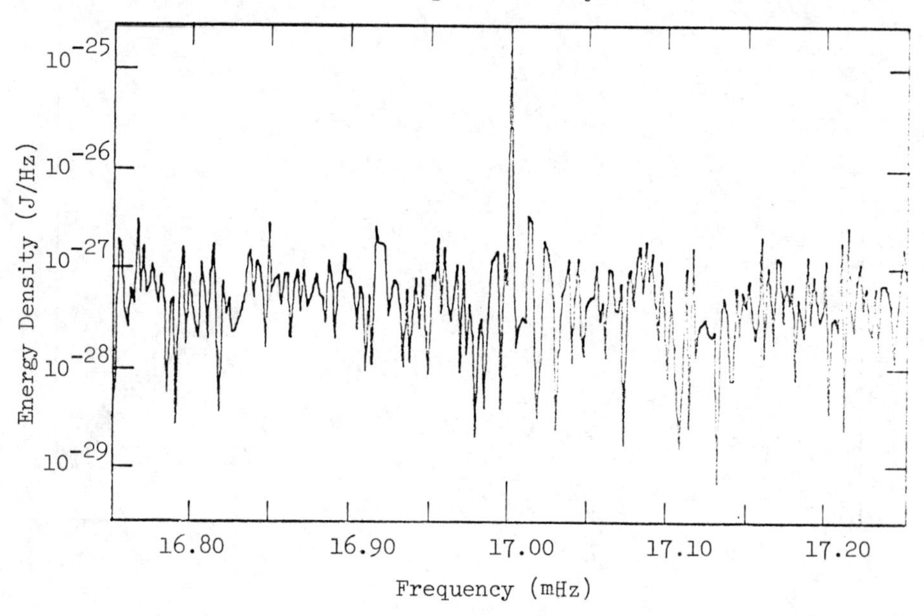

Fig. 4 Increased Signal Resolution by Narrowing of Frequency Bin Width

we have investigated the 1/f behavior down to a frequency below 10^{-5} Hz and found no deviation (Fig. 5). We are in the process of extending the plot to 2×10^{-6}, the lowest frequency in the data set.

CONCLUSIONS

Narrow band detection of signals within the 1/f region of SQUID magnetometers is straightforward and yields the results expected from white noise theory. For low-frequency detection we have shown that the use of digital systems allows the expected improvement for integration times up to 5×10^5 sec. No deviation from the 1/f energy spectrum dependence is observed for frequencies down to 1×10^{-5} Hz.

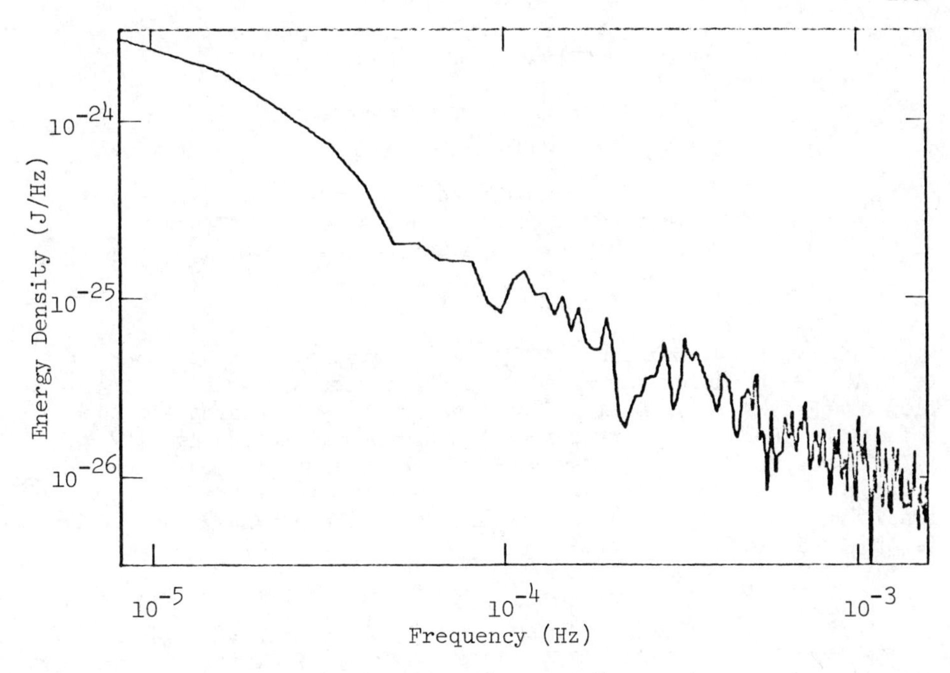

Fig. 5 Continuation of 1/f Spectrum to 10^{-5} Hz.

ACKNOWLEDGMENT

The authors wish to thank J. Hollenhorst for setting up the digital data acquisition system used in this experiment, computing some of the power spectra, and for general discussions of power spectrum analysis. We also wish to thank R. Giffard for many helpful discussions.

REFERENCES

1. J. A. Lipa, "Status of the Gyro Relativity Experiment," p. 129 Proceedings of the International School of General Relativistic Effects in Physics and Astrophysics: Experiments and Theory (3rd Course), Nov. 1977.
2. J. T. Anderson and R. R. Clappier, "Development of a London Moment Readout for a Superconducting Gyroscope."
3. Similar results are obtained by J. Clarke, et al, e.g., J. of Low Temp. Phys. 25 Nos. 1/2, 1976 p. 99.
4. A. V. Oppenheim and R. W. Schafer, Digital Signal Processing Prentice-Hall (1975) p. 573, also includes much information on FFT software.

SUPERCONDUCTING GRAVITY GRADIOMETERS

H. J. Paik and E. R. Mapoles
Department of Physics

K. Y. Wang
Department of Aeronautics and Astronautics
Stanford University, Stanford, Ca. 94305

ABSTRACT

Two types of superconducting gravity gradiometers have been developed. Both employ superconducting test masses suspended by niobium diaphragms whose motions modulate persistent currents in superconducting loops. A current signal proportional to the gravity gradient is detected by a 19 MHz SQUID. The balancing of common acceleration has been verified experimentally for both systems. A gradient sensitivity better than 2×10^{-10} s^{-2} $Hz^{-\frac{1}{2}}$ with a baseline of 10 cm is expected of these gradiometers.

INTRODUCTION

A very sensitive superconducting accelerometer has been developed for the Stanford gravitational wave detector.[1] A pair of these accelerometers can be used in a subtraction mode to detect gravity gradients. Two superconducting gravity gradiometers using different principles of subtracting the common mode signal have been assembled and tested. Both systems employ unique properties of superconductors to obtain high transducer coupling, increased dynamic range and sensitivity. In this paper, we discuss the basic features of design and report experimental results which verify the essential properties of the two gradiometers.

CURRENT-DIFFERENCING GRAVITY GRADIOMETER

The first version of gradiometer that we have developed, which we call "Superconducting Current-differencing Gravity Gradiometer (SCGG)", is simply a rigid assembly of two identical niobium-diaphragm accelerometers of the type used for gravitational wave experiments.[1] The details of the basic accelerometer have been published elsewhere.[2] In the gradiometer, the signal currents from the two accelerometers are fed into the same SQUID with opposite polarity in order to accomplish signal differencing.

A schematic of SCGG is shown in Fig. 1. Two test masses, m_1 and m_2, represent two circular niobium diaphragms with fundamental frequencies of 850 Hz. The four pickup coils, L_{11}, L_{12}, L_{21}, L_{22}, were wound on G-10 glass epoxy coil forms with 0.05 mm Nb-Ti wire in the shape of a pancake. In assembly these coils are held rigidly facing the surfaces of the two diaphragms with approximately 0.1 mm clearance.

ISSN: 0094-243X/78/166/$1.50

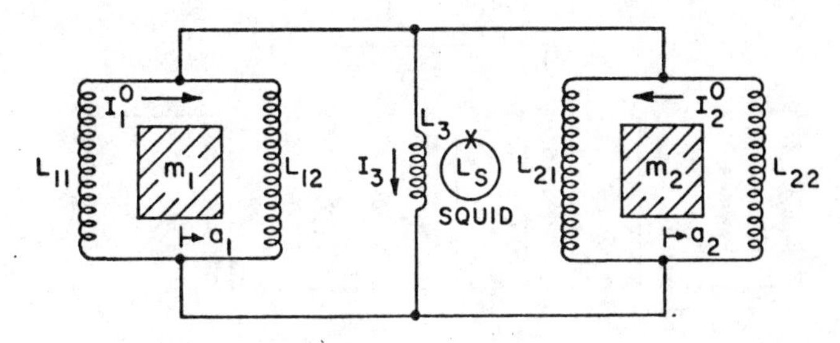

Fig. 1. Schematic diagram of SCGG.

When persistent currents I_1^o and I_2^o are stored with opposite relative sense in the two pickup loops, the resulting current through the SQUID input coil L_3 due to accelerations of test masses, $a_1(t)$ and $a_2(t)$, can be shown to be

$$I_3 = \frac{2}{1+\gamma}\,\frac{1}{\eta}\left[\left(\frac{I_1^o}{\omega_1^2 d_1^o} - \frac{I_2^o}{\omega_2^2 d_2^o}\right)\bar{a} - \left(\frac{I_1^o}{\omega_1^2 d_1^o} + \frac{I_2^o}{\omega_2^2 d_2^o}\right)\Delta a\right], \qquad [1]$$

where $\bar{a} \equiv (a_1+a_2)/2$ is the "common mode" acceleration, $\Delta a \equiv a_2-a_1$ is the differential acceleration, $\gamma \equiv L_3(L_{11}^{-1}+L_{12}^{-1}+L_{21}^{-1}+L_{22}^{-1})$, d_1^o and d_2^o are average pickup coil spacings in the two accelerometers, and η is a geometrical factor of the order of unity. We have assumed that signal frequencies are low compared to the resonant frequencies of the test masses, ω_1 and ω_2, and ignored terms of the order of $(a_1^o/\omega_1^2 d_1^o)^3$. From Eq. [1], it is clear that one can "balance" the common-mode by choosing a current ratio $I_2^o/I_1^o = \omega_2^2 d_2^o/\omega_1^2 d_1^o$.

The ability to accurately match the scale factors by adjusting the amount of stored currents after assembly of the system is a very convenient feature of superconducting gravity gradiometers. Another unique feature is the possibility of electronically differencing the electrical signals before detection. This signal differencing before amplification is an important property which can enhance the dynamic range of the device.

DISPLACEMENT-DIFFERENCING GRAVITY GRADIOMETER

The second version of the superconducting gradiometer, called "Superconducting Displacement-differencing Gravity Gradiometer (SDGG)", is one in which a magnetic property of the superconductor is used to detect the differential displacement of the two masses directly. Figure 2 is a schematic of SDGG. Each test mass is a circular niobium cylinder suspended from a solid ring by means of an annular niobium diaphragm. A flat superconducting pickup coil L_1 is mounted on one of the test masses facing the other and is connected to the input coil of a 19 MHz SQUID through a stepdown transformer. When a persistent current is stored in the pickup loop, the inductance of L_1 is modulated by the relative displacement of the test masses, sending to the SQUID a signal current directly proportional to the differential acceleration. Two separate superconducting loops are located at the

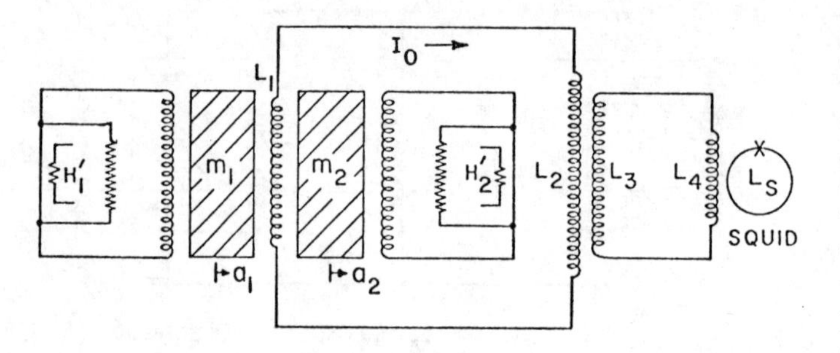

Fig. 2. Schematic diagram of SDGG.

outer surfaces of the test masses in order to tune the frequencies of the test masses. The low resistance shunts are inserted for the purpose of fine control of persistent currents. When the heat switches, H_1' and H_2' are turned on, a current decay time of about 600 s is obtained.

With a persistent current I_o flowing through the pickup loop, the signal current through L_2 becomes

$$I - I_o = \frac{1}{1+\gamma}\frac{I_o}{2d_o}\left[\left(\frac{1}{\omega_1^2} - \frac{1}{\omega_2^2}\right)\bar{a} - \left(\frac{1}{\omega_1^2} + \frac{1}{\omega_2^2}\right)\Delta a\right] , \qquad [2]$$

where d_o is the average spacing between the pickup coil and the test masses and $\gamma \equiv L_2/L_1$. The "balancing" condition is now simply $\omega_1=\omega_2$. This frequency matching is obtained by adjusting currents in the tuning loops. This second system can be made more compact and has the advantage of employing a single pickup coil which simplifies the gradiometer both mechanically and electronically. However, when a longer baseline is needed for better gradient sensitivity, the differential-current device with four pickup coils will be suitable.

As one can see from Eqs. [1] and [2], the gradient sensitivity improves as one reduces the resonant frequencies of the test masses. In SDGG, we were able to obtain frequencies near 60 Hz by mass-loading the center of the niobium diaphragms. The SQUID noise in this device corresponds to a gradient sensitivity of 10^{-9} s^{-2} $Hz^{-\frac{1}{2}}$ with a baseline of 2 cm. Because of the high mechanical Q of the niobium diaphragms ($Q = 2 \times 10^5$), the Brownian motion noise of the test masses is small and is found to be one order lower than this value.

TEST RESULTS

Both gradiometers have been assembled and tested. The mechanical frequencies of the test masses in SDGG were 59.4 Hz and 80.6 Hz. It took a persistent current of 3.0 A to tune the lower frequency test mass to 80.6 Hz. According to Eq. [2], a complete common-mode balancing should be obtained in this state. This was confirmed in the following experiment. A 20 Hz common-mode acceleration of constant amplitude was provided by driving the entire cryostat with a solenoid at 10 Hz. The resulting SQUID output was measured with a two-phase lock-in amplifier as tuning current I_1 was varied. Figure 3 shows the amplitude and phase of the SQUID output. The signal am-

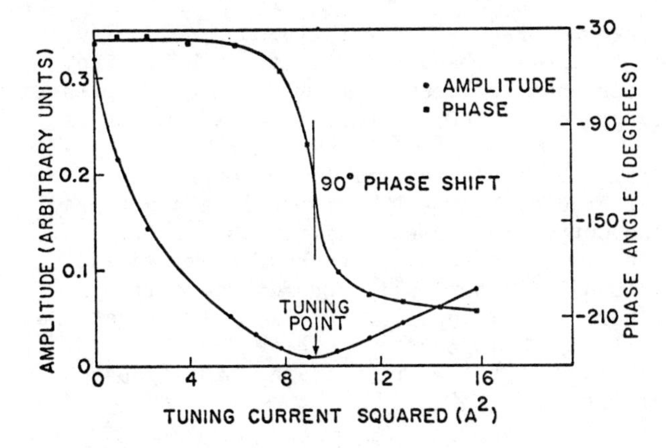

Fig. 3. The amplitude and phase response of SDGG at 20 Hz as a function of tuning current.

plitude does reach a minimum at $I_1 = 3.0$ A with an accompanying phase shift of 180° across the tuning point. The small residual amplitude at the tuning point and the finite width of phase change arise from a coupling to a third mode, the lowest dipole mode of one diaphragm. This coupling is due to a departure from cylindrical symmetry in the fabrication of loaded diaphragms. This proves that the device does become a true gradiometer when the fundamental modes of the test masses are tuned to each other. It is important to realize that a scale factor matching was obtained electronically when the mechanical matching of the diaphragms was quite poor as can be seen from their initial frequency difference.

A similar balancing characteristic was observed with SCGG. In this system the ratio of the two persistent currents was adjusted until the SQUID output reached a minimum. Figure 4 consists of the frequency responses of SCGG in a balanced and off-balanced state measured with a spectrum analyzer. The common-mode acceleration was produced by driving a piezoelectric bimorph with a constant-voltage reference signal from the spectrum analyzer. A bandwidth of 10 Hz was used for the frequency sweep. In order to simulate a slightly off-balanced state, we first stored I_1^o=0.1A and I_2^o=0A in the two pickup loops. Curve (a) is the result plotted on a semi-logarithmic scale. The same experiment was repeated after balancing with I_1^o=1.0A and I_2^o=1.35A, and the result is shown in curve (b). A low frequency peak (≲ 50 Hz) arising from resonances of the suspension system appears in both curves. Arrows represent positions where the magnetometer begins to unlock due to increased

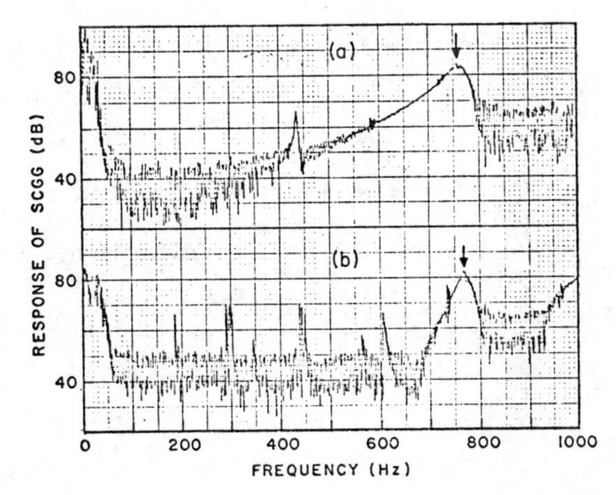

Fig. 4. Frequency responses of SCGG.

response near the diaphragm resonances. The background noise level around 40 db corresponds to the wideband noise of the SQUID. From Fig. 4, one can clearly see that the common-mode signal is suppressed over a wide frequency range, dc to 700 Hz, when the gradiometer is balanced.

The long-term stability of scale factors is an important requirement for a sensitive gradiometer. In a cryogenic gravity gradiometer, a dc drift in the output is caused by irreversible flux motion in current-carrying superconductors. We have examined the output of SDGG and have found an essentially logarithmic drift as predicted theoretically for type-II superconductors.[3] The fractional drift rate reached a level of 4×10^{-11} s^{-1} at the end of 2 hours. This drift rate should go down at t^{-1} where t is the time elapsed since the storage of current.

CONCLUSIONS

Essential features of the two superconducting gravity gradiometers have been verified. When a sufficiently accurate balancing of common-mode acceleration is obtained, these devices should have a gradient sensitivity of 2×10^{-10} s^{-2} $Hz^{-\frac{1}{2}}$ with a baseline of 10 cm. As a lower frequency suspension of test masses becomes available, the sensitivity will improve further. The instruments will be calibrated with a known gravity gradient from a rotating dumbbell.

With enhanced sensitivity, these devices could be used to map the gravitational field of earth anomalies for inertial navigation systems, and to carry out sensitive experiments on gravitation in the presence of uniform interfering fields from distant objects. We are preparing a laboratory experiment to test for a possible departure from the inverse square law of gravitation using these instruments. More recently, a new type of gravitational wave detector, which is essentially a gigantic version of our SCGG, was discussed.[4]

We gratefully acknowledge valuable suggestions and encouragement from Professors W. M. Fairbank, D. B. DeBra, and R. P. Giffard.

REFERENCES

1. S. P. Boughn, W. M. Fairbank, M. S. McAshan, J. E. Opfer, H. J. Paik, and R. C. Taber, Proceedings of the 13th International Conference on Low Temperature Physics (Plenum, New York, 1974) Vol. 4, p. 559.
2. H. J. Paik, J. Appl. Phys. 47, 1168 (1976).
3. P. W. Anderson, Phys. Rev. Lett. 9, 309 (1962).
4. R. V. Wagoner, C. M. Will, and H. J. Paik, submitted to Phys. Rev. Lett.

THE FUTURE OF SUPERCONDUCTING INSTRUMENTS IN METROLOGY *

D.B. Sullivan
Electromagnetic Technology Division
National Bureau of Standards, Boulder, Colorado 80303

L.B. Holdeman
Electricity Division
National Bureau of Standards, Washington, D.C. 20234

R.J. Soulen, Jr.
Heat Division
National Bureau of Standards, Washington, D.C. 20234

ABSTRACT

Superconductivity has played a major role in the development of useful new methods for standards and measurements, particularly for dc and high-frequency instruments and for thermometry. In this paper we speculate on future activities in this field including the development of commercial instruments, the improvement of devices for standards laboratories and the adaptation of some of these systems to small refrigerators.

INTRODUCTION

Over the past decade significant effort has been spent in standards laboratories throughout the world on the application of superconductivity to measurement problems, and as a consequence, superconducting devices have become important tools in the metrology laboratory. Devices of particular note include: voltage standards, noise thermometers, current comparators, null detectors, attenuation calibrators, rf power meters, frequency multipliers and highly stable oscillators. For the most part these have been rather specialized, one-of-a-kind instruments, suitable for operation by highly trained personnel in the primary or secondary standards laboratory. Future development of these devices will probably concentrate on simplification and improvement in convenience, although extension of the range of measurements will not be ignored. A good bit of this development is inexorably tied to technology advances discussed by other speakers at this conference, and one purpose of this paper is to attempt to identify the opportunities afforded by these advances.

The developments to date have been well reviewed[1,2] and no effort will be made in this paper to provide a complete background. Although the title of this paper encompasses the whole spectrum of applications in metrology, superconducting cavity-stabilized oscillators will not be discussed since that subject is of sufficient importance to warrant separate coverage.[3] In looking toward the future one is more often concerned about the evolution of ideas which are currently in voque,

---

and that is the thrust of this paper, but we might also expect to see entirely new concepts which could shape future trends in yet unforeseen ways. To some degree the future will certainly see an expansion of commercial developments and we would be remiss if we did not discuss this subject.

JOSEPHSON VOLTAGE STANDARDS

Because of its sensitivity and broad range of applications, the SQUID has gained wide recognition as a measurement instrument. In the standards laboratory, however, the Josephson voltage standard has had the most notable impact. The ac Josephson effect is used for voltage maintenance in the national standards laboratories of all the major industrial nations, and "second-generation" instruments are being developed and tested. Standards laboratories in many of the technologically less-developed countries are establishing Josephson voltage maintenance. A commercial Josephson voltage standard is now being marketed.[4] Junction arrays, applications of integrated circuit techniques to Josephson devices, and high-T_c devices in closed-cycle cryocoolers portend yet another generation of voltage sources and standards. In this section we shall discuss voltage standards applications of the ac Josephson effect, beginning with a brief historical review.

In his original paper on superconductive tunneling, Josephson predicted that the ac supercurrent which flows when a potential difference V is maintained across the electrodes of a junction would be frequency-modulated in the presence of electromagnetic radiation of frequency ν, and that a dc supercurrent could flow when V and ν satisfied the relation $V = n(h/2e)\nu$. This prediction was experimentally confirmed shortly thereafter. In the ensuing years the ac Josephson effect was extensively studied, both experimentally and theoretically, and the voltage-frequency relation is believed to be free of corrections to better than a part in 10^8. The relation was used to measure 2e/h at the University of Pennsylvania in a series of increasingly precise experiments that culminated in a sub-pmm determination.[5] Subsequent high-precision measurements of 2e/h carried out in the early 1970's at several national laboratories revealed that, in general, the as-maintained national units of emf were changing with time (the U.S. legal volt, for example, was drifting at the rate of approximately -0.4 μV per year). These precise measurements of 2e/h demonstrated the feasibility of using the ac Josephson effect to maintain a time-invariant voltage standard at a precision of a few parts in 10^8. On July 1, 1972, the National Bureau of Standards officially converted to voltage maintenance via the ac Josephson effect, and stabilized the U.S. volt by adopting the value of 2e/h for that date, $2e/h = 483593.420\ \mathrm{GHz}/V_{NBS}$. However, most other national laboratories have since adopted the valued $2e/h = 483594.0\ \mathrm{GHz}/V_{BI69}$, where V_{BI69} is the January 1, 1969, value of the volt that is maintained at the Bureau International des Poids et Mesures (BIPM). This value for 2e/h was derived by extrapolating experimental drift rates back to the year in which most countries adjusted their voltage units to agree with the BIPM volt, which had been adjusted to agree with the best estimate of the absolute (SI) volt. Although voltage units can be maintained with a precision of a few parts in 10^8 by adopting a value for 2e/h and using the ac Josephson effect, it should perhaps be emphasized that accuracy in terms of the absolute volt is determined by the accuracy to which 2e/h is known in SI units. The absolute volt is realized in terms of the absolute ampere and the absolute ohm. Whereas the absolute ohm has been determined to parts in 10^8 through the calculable capacitor, the experimental uncertainty of the absolute ampere is a few ppm.[6] Improved ampere determinations (possibly through applications of superconductivity[7]) could result in a change in the adopted values for 2e/h.

The first generation Josephson voltage standards incorporate a point-contact or tunnel junction biased on a microwave-induced step at one to ten millivolts. The junction voltage is compared to the emf of a Weston-type cadmium-sulfate standard cell using a room temperature resistor network.[5] The second-generation Josephson voltage standards also employ junctions operating on a microwave-induced step at a few mV, but the room temperature resistor network for comparing junction voltages to standard cell voltages is replaced by cryogenic resistive dividers and superconducting switches. Several different approaches to divider calibration are used. For example, the cryogenic resistive divider can be calibrated to parts in 10^9 in situ using a cryogenic current comparator.[8,9] These "all-cryogenic" Josephson voltage standards may result in voltage maintenance to a few parts in 10^9.

A prototype for a commercial Josephson-effect voltage-standard instrument has been developed at NBS,[4] and, like the instruments described above, is designed for standard cell calibration. With a suitable atomic frequency standard, a secondary standards laboratory can calibrate standard cells independently of NBS; when traceability to NBS is required, the frequency standards broadcast by NBS over radio stations WWV can be utilized. A single tunnel junction is incorporated into a microwave integrated circuit in this instrument, and produces steps at several millivolts ($\sim$ 5 mV) using rf input from a fixed-frequency low-power solid-state source. Cost, reliability, and ease-of-use considerations necessitated some trade-off with precision; even so, the commercial Josephson voltage standard still has an overall precision (2σ) of 0.4 ppm.

What's in the future? One promising prospect is the coherent array, which could produce higher voltages with less power dissipation per junction. Another is the high-T_c tunnel junction which, operating in a closed-cycle cryocooler, could eliminate liquid helium as a requisite for Josephson voltage standards (more on this later). An interesting "inverse ac Josephson-effect" voltage standard was recently suggested by Levinson, Chiao, Feldman, and Tucker.[10] For an hysteretic tunnel junction, the lower-order microwave-induced steps, like the zero-voltage supercurrent, can cross the voltage axis (this point was noted in several early experimental papers but applications were not recognized). Thus, for small n, quantized voltages can be induced across the junction with no dc current through the junction! The suggestion is that irradiating a number of such junctions in series might produce usable voltages without the need of a stable current supply for junction bias. Note that in this situation every junction is either on an induced step or the zero-voltage step; no possibility exists for intermediate states.

Under appropriate conditions, when an array of N nearly identical junctions is irradiated with microwaves of frequency ν, steps can be induced in the I-V characteristic of the array at integral multiples of $N(h/2e)\nu$, i.e., at voltages $V_n = nN(h/2e)\nu$.[11] With appropriate leads to tap sub-arrays of N_k junctions, and judicious variation of the parameters n, N_k, and ν, it might be possible to produce step-locked voltage sources that are tunable over a large voltage range. For voltage standards applications, the higher voltages produced by arrays would ease the constraints on the rest of the instrument. There are advantages even for small N. For a single junction, the heat flux from the combined rf and dc power necessary for microwave-induced steps at high voltages can exceed the critical flux for transition from nucleate to film boiling in liquid helium. The temperature of a heated surface in the film boiling mode is above T_c for elemental superconductors in a 4.2 K helium bath. For steps at $V_n = nN(h/2e)\nu$, the microwave power required to induce the nth step is considerably less than that required to produce a step at the same voltage in a single junction, and the dc power dissipated per junction is reduced by N^2. Because high-T_c superconducting compounds have low thermal conductivities, and because heat would be extracted along the films or through the substrate in a closed-cycle refrigerator, arrays may prove essential for operation without liquid helium if nonresonant junctions are used. Indeed, with this in mind, the frequency of the commercial instrument[4] was chosen so that the junction will operate on the 240th step, and the microwave integrated circuit incorporating the junction is a simple plug-in unit. Since 240 = 2 x 3 x 4 x 5 x 2, the rest of the instrument is immediately adaptable to arrays of N = 2,3,4,5,6,8, . . ., 240.

HIGH-FREQUENCY MEASUREMENTS

In this section we discuss radio frequency measurements of current, power, and attenuation and, to a lesser extent, frequency synthesis into the infrared portion of the spectrum. The application of SQUID's to rf measurements was first reported by Kamper and Simmonds[12] and has recently been reviewed.[13] A quick reminder of the method follows.

In the concept the amplitude of an rf signal is measured in terms of the fundamental period, ϕ_o, of response of the SQUID just as length is measured in terms of a wavelength. The SQUID, which is biased at some suitably high frequency, is operated in an overdamped mode so that the nearly sinusoidal output is proportional to $\cos(2\pi\phi/\phi_o)$. A lower frequency signal, $I = I_m \sin(\omega_m t)$ is coupled to the SQUID through a fixed coil. The ratio ϕ/ϕ_o can be replaced by I/I_o where I_o is the current required to put one flux quantum into the SQUID and the output is then proportional to

$$\cos\left(\frac{2\pi I_m}{I_o} \sin(\omega_m t)\right) \quad . \qquad (1)$$

This is the familiar function which, when time-averaged, gives the Bessel function of zero order, $J_o(2\pi I_m/I_o)$. Since the Bessel function is accurately tabulated, the amplitude I_m can be determined in units of I_o. Within limits set by the coil parameters and line impedance, a dc determination of I_o calibrates the system as an absolute rf ammeter, and a knowledge of terminating impedance yields absolute power. Attenuation is obtained from ratios of rf currents, a measurement which is especially easy to handle at zero crossings of J_o because these are readily located by null methods.

Using this concept, instruments have been developed which offer 0.002 dB uncertainty in attenuation and 0.1 dB uncertainty in absolute power over the frequency range from 0 to 100 MHz. Power sensitivity is as low as 10^{-16} W with the 0.1 dB uncertainty. Several systems have been developed for measurements to 1 GHz although the sensitivity has been somewhat less. Perhaps the most attractive feature of these instruments is their broad bandwidth, since conventional systems are most often built for a single frequency. In addition, the precision for attenuation measurements is derived from the highly linear response of the SQUID rather than from some very precise machine work, as is usually the case for conventional systems. For these reasons and because the cost of such instruments is competitive (at least for the best standards laboratory work), there is significant interest in adapting the devices to routine calibration work. There may well be an opportunity for the development of commercial instruments since the concept may free one from the need for periodic calibrations referred to a national laboratory.

To date, only point contact SQUID's have been used in this application. The development of a thin-film SQUID with a tunnel-type junction would represent a major improvement. In order to achieve appropriate damping, many of the point contacts have been flame oxidized. This yields shunt resistances of the order of 1 Ω for a critical current of the order of 10 μA, but the temperature dependence of i_c for such contacts is large. Such temperature dependence and the concurrent stability problems could virtually be eliminated through use of a properly shunted tunnel junction. The use of thin-film techniques, particularly lithographic fabrication methods, would facilitate the production of many well-matched SQUID's. This could offer a real advantage for multiple sensor applications. A clear example of such an application is the six-port measurement system.[14] With four amplitude sensors this system allows one to completely characterize an rf signal. Obviously point contact SQUID's could be used, but the problems would be greatly diminished with thin-film sensors.

Another device worthy of investigation and development for rf measurements is the double-junction SQUID.[15] Clearly this type of SQUID will respond to rf signals, at least qualitatively, in the same manner as the rf-biased SQUID. The question to be answered is whether the periodic response can be made to be as purely sinusoidal, a prime requisite for minimal error. Published data suggests that the interference pattern becomes more sinusoidal as the average bias voltage is increased although the signal amplitude drops. This presumably results from overdamped operation much as is found with the rf-biased SQUID.

The double-junction SQUID would offer significant advantage because it is self-biased by its own internal oscillations. That is, no external oscillation or detection equipment is required. The effective bias frequency is proportional to bias voltage. This means, for instance, that the frequency for measurements could be extended to higher ranges without the need for expensive and cumbersome microwave oscillators (the bias frequency must be considerably higher than the frequency of the signal to be measured). The reduction in signal processing components would offer further advantage for multiple-sensor applications as mentioned above.

A thin-film version of the double-junction SQUID would probably be most desirable. This would simplify junction matching and would be most amenable to reproduction of identical SQUID's. One interesting possibility would be the development of integrated rf circuits using strip-line technology. For instance it might be possible to build a six-port measurement system with four SQUID sensors on a monolithic chip. The use of planar geometries would obviously require careful development of new coupling and shielding schemes.

There is one last ac application, at very high frequencies, which should be mentioned. The Josephson junction has such a strongly nonlinear characteristic that it is considered to be the most efficient harmonic generator known today. As such it has great potential for single-step frequency synthesis from the microwave region into the infrared. The development of simpler methods for frequency measurements of infrared lasers could offer the path to a new time standard or perhaps even to a unified standard of length and time.[16] To date this work has been done with very finely-pointed contacts which are not only difficult to fabricate but are also short-lived. What is needed is a reliable diode of very small dimensions ($<< 1\ \mu m$), a device which could be constructed using electron-beam lithographic methods. As e-beam techniques become more widespread we should expect to see development work in this direction. Frequency synthesis at a very high harmonic order requires an extremely stable oscillator at the low frequency end. Stein and Turneaure address this need in another paper in these proceedings.[3]

THERMOMETRY

Thermal noise voltages, as can be seen by the defining equation,

$$\langle V^2 \rangle = \int 4RkT d\nu \qquad (2)$$

are so small at low temperatures that amplifiers must be used before electronic averaging to obtain a value for temperature. Unfortunately, room-temperature amplifiers contribute a significant amount of their noise to the measured output. It is only recently, by using SQUID amplifiers with their very low-noise characteristics, that noise temperatures as low as a few millikelvins above absolute zero have been measured by this technique. In fact, two distinct methods for cryogenic SQUID noise thermometry have evolved.

The first, developed by Webb, Giffard and Wheatley,[17] consists of observing the current fluctuations generated by a resistor which is inductively coupled by a coil to a SQUID magnetometer. The output voltage of the SQUID amplifier circuit is proportional to the noise voltage; the constant of proportionality must be determined experimentally by a calibration process in order to measure absolute temperature. The noise contributed by the SQUID amplifier was found to be less than 0.1 mK, making measurements of temperature as low as 0.002 K possible. Webb, et al.,[17] used one noise thermometer to measure temperature between 0.005 K and 4.2 K with a statistical imprecision of ± 1% and an inaccuracy (i.e., uncertainty in the proportionality constant) of ± 3%. A second noise thermometer was used to test the accuracy of two Curie-law magnetic thermometers (cerium magnesium nitrate and cerium "dipicolinate") from 0.002 K to 0.020 K. The averaging time required to obtain the 1% statistical imprecision in these two series of experiments was 2000 s. This type of SQUID noise thermometer is now available commercially.

On the basis of the aforementioned experiments, one should expect to measure temperatures from a few mK to perhaps 10 K with a statistical imprecision of 1% using 2000 s averaging times and a very small device temperature of less than 0.1 mK. Even further reduction of the SQUID noise using either higher rf bias signals or "dc" SQUID's should mean that the same imprecision would be obtained with averaging times of 20 - 40 s, but with a somewhat higher device temperature (0.15 mK). At least one laboratory intends to attempt to define a temperature scale below 3 mK based solely on noise thermometry in order to more accurately define the phase diagram of He^3.

In all of these situations, however, the absolute accuracy is expected to be limited to ± 1% owing to the difficulty in the determination of the circuit values controlling the proportionality constant between the SQUID output signal and the noise (source) signal. If one cryogenic fixed point is accepted as given, however, the gain could be calibrated in situ with greater accuracy. In sum, this type of SQUID noise thermometer represents a very significant advance in noise thermometry in that the amplifier noise (i.e., device temperature) is extremely low; however, it has not contributed significantly to the removal of the inaccuracy inherent in the determination of the proportionality constant.

A second form (although first historically) of SQUID noise thermometer eliminates the need for calibration and appears capable of performing truly absolute temperature measurements. This scheme, originally conceived by Kamper[18] and subsequently developed by Kamper, et al.,[19] consists of connecting the noise-generating resistor directly across a point-contact Josephson junction. By the voltage-to-frequency-conversion feature of Josephson junctions, the noise voltage is converted into a time-varying frequency. Analysis of this SQUID-resistor circuit, often called an R-SQUID, shows that the variance of those frequency counts is related to absolute temperature by a coefficient only involving the magnitude of the resistance R and fundamental constants.[20] Thus, once R is determined (it is easily done in situ to ∿ 100 ppm accuracy), the relation between the variance and T is governed entirely by fundamental constants, not by circuit parameters;

hence the thermometer is absolute. The price paid for this feature is a greater device temperature (typically 1 mK) and longer averaging times -- 20,000 s for a 1% imprecision. This thermometer has been used by Soulen at NBS to define a noise temperature scale from 0.01 K to 0.5 K with a statistical imprecision of ± 0.5% at the lowest temperature and decreasing to ± 0.2% at the highest temperature. This scale compares well with another absolute scale developed by Marshak[21] based on the anisotropy of γ-rays emitted from Co^{60}, and with a scale obtained by Utton using nuclear magnetic resonance in copper.[22] The overall agreement is on the order of 0.5%. Future plans at NBS for noise thermometry call for using a higher rf bias (10 GHz) to decrease the averaging time and device noise by at least a factor of ten.[23] The likelihood that a commercial version of the R-SQUID noise thermometer will be produced is small because it is intrinsically slower than the other SQUID noise thermometer mentioned above; it is probable that one or more national standards laboratories (possibly Germany, England) will develop temperature scales based on this principle.

We see some other developments coming from the R-SQUID. Recently, Soulen and Giffard[24] studied the impedance of an R-SQUID as a function of the dc and rf currents injected into the circuit and obtained extremely good agreement with the resistively-shunted junction model of the Josephson junction. These experiments indicate that this model may be tested with greater accuracy than hitherto thought possible using R-SQUID circuits, thereby yielding important information about the properties of Josephson junctions. More precise experimental tests of this model for an R-SQUID are underway.

Experiments conducted by Zimmerman[25] on a more complicated R-SQUID circuit yielded some very intriguing results which may point the way to great simplification of the readout schemes for R-SQUID's. This circuit consisted of two R-SQUID's (i.e., two pairs of resistor-Josephson junction circuits) interconnected by a third resistor. When each R-SQUID was biased separately, an unexpectedly strong beat frequency was observed. The device could well have operated to parametrically up-convert the difference frequency controlled by the voltage across the third resistor. If this interpretation is correct, it has important ramifications for the amplification of low-frequency signals by R-SQUID's in that the need for high-frequency amplifiers is eliminated and instead is replaced by two dc power supplies and two resistive SQUID's! In a second paper Harding and Zimmerman[26] describe a double-junction SQUID (see ref. 15) which has two resistive sections. Again, this device seems to provide an amplification of the Josephson oscillation produced by sending a bias current through either or both of the resistive sections. We think that a revival and more careful study of these phenomena could lead to quite useful results for noise thermometry and signal amplification in general. If indeed these notions are correct, then effort should be directed toward thin-film development of the device. The advantages of thin-film technology for the double-junction SQUID mentioned in the section on rf measurements would apply here as well.

A final area of possible study for R-SQUID's is noise spectroscopy. By this we mean the identification of the presence of various power-law noise spectra in a "dc" signal. This is easily done with the noise thermometer used at NBS which is programmed to observe power spectra other than Johnson noise. The detection scheme depends on the fact that successive frequency counts which are taken can be analyzed by an on-line computer to calculate several variances,

$$\sigma_\alpha^2 = \frac{\Sigma(\nu_i - \nu_{i+\alpha})^2}{2N} \quad . \tag{2}$$

Here N is the total number of frequency counts taken, ν_i is a given frequency count and $\nu_{i+\alpha}$ is a frequency count taken α gate times later. If the power spectrum generating the frequency fluctuations is "white," all the σ_α^2 will be equal; whereas the presence of other noise spectra (e.g., flicker (ν^{-1}), ν^{-2} etc.) will cause σ_α^2 to depend in well-characterized ways[27] on the gate time, τ, and N. Therefore, if a dc power supply were used to bias the R-SQUID, the current noise generated by the instrument could be characterized.

Another application of superconductivity for metrology has been developed at NBS. It involves the dissemination, intercomparison, and preservation of cryogenic temperature scales by means of superconductive fixed points. Schooley, Soulen and Evans[28] showed that, for many materials, individual samples of a given element made from the same batch exhibit transitions within ± 1 mK of each other. It has thus been possible to fabricate units, distributed by the NBS Office of Standard Reference Materials, which provide five fixed temperatures from 0.5 K to 7.2 K for calibration purposes. The small physical size and solid-state nature of these units have the additional advantage that in situ calibrations can be carried out using rather simple electronic circuitry. Several units have already found their way into industrial, university, and government research laboratories; others have participated in an international comparison of national standards laboratory temperature scales at the NML in Australia.[29] They appear to have served well in these applications and will very likely be adopted as the appropriate fixed points for the cryogenic extension of the next international practical temperature scale.

An additional five superconductors have been developed for use below 0.5 K which extend from 0.015 K to 0.208 K. They will soon be available from the Office of Standard Reference Materials. Concerning higher-T_c superconductors, samples of Nb, V_3Ga and Nb_3Sn are being examined for use from 9 K to 18 K. The goal of the program is to offer from ten to thirteen superconductive fixed-points which span three decades of temperature.

DISCUSSION

Perhaps the first point to stress is that for most of these applications there is a need for improved operational convenience and simplicity. No single factor could affect the use of superconductivity in metrology any more than the development of inexpensive and reliable cryocoolers. Zimmerman[30] discusses such devices in his paper at this conference. Commercial interests are just beginning to consider superconducting devices for the metrology market, but more extensive commercial efforts would undoubtedly be pursued if the need for liquid helium transfers could be greatly reduced or eliminated. The ultimate solution is direct refrigeration of the device, but acceptable levels of helium handling could be achieved through use of evaporative cryostats with refrigerator-cooled radiation shields. Until this happens we expect to see only limited specialized applications at the primary or secondary standards laboratories.

An essential point to note is that the mating of a cryocooler and, for instance, a volt standard would be less demanding than would be a cryocooler and a magnetometer, where self-generated noise poses a severe constraint. This self-generated noise results both from the refrigerator's magnetic interference and from vibration of the SQUID in the earth's field. The devices for metrology are, in general, shielded from outside fields and are thus much less susceptible to magnetic and vibrational noise problems. Another important point is that one need not push the limits of noise performance to have an impact. That is, one can tolerate higher temperatures (high-T_c superconductors) without too much concern about added thermal noise. As simple cryocoolers capable of 8 K to 9 K operation become available, we envision further stimulation of high-T_c tunnel junction work (Beasley[31] describes junctions with at least one electrode composed of high-T_c material).

Lithographic fabrication of superconducting instruments will ultimately affect metrology applications. When this happens the problems will shift from junction and SQUID fabrication problems to those of system packaging. The costs of such systems would undoubtedly fall with increased sales volume, but that factor is difficult to assess. One great advantage of the Josephson volt standard, the noise thermometer, and the SQUID rf measurement system is the relative freedom from need for routine calibration by higher-level standards laboratories. This particular virtue could conceivably prompt the use of such instruments at lower-level standards laboratories and thus influence the size of the market.

In conclusion, we should stress that superconductivity has already made rather important inroads in metrology, but significant opportunities still exist. In this paper we have concentrated on a few familiar devices to illustrate these opportunities. Other superconducting instruments such as current comparators, null detectors and A/D converter will certainly be developed and improved to expand the role of superconductivity in metrology.

REFERENCES

1. R.A. Kamper, in Superconductor Applications: SQUID's and Machines, eds. B.B. Schwartz and S. Foner, Plenum Press, New York (1977). Chap. 5.
2. R.P. Hudson, H. Marshak, R.J. Soulen, Jr., D.B. Utton, J. Low Temp. Phys. 20, 1 (1975).
3. S.R. Stein and J.P. Turneaure, these proceedings.
4. L.B. Holdeman, B.F. Field, J. Toots and C.C. Chang, these proceedings.
5. T.F. Finnegan, A. Denenstein and D.N. Langenberg, Phys. Rev. 177, 1487 (1969).

6. B.N. Taylor and E.R. Cohen, in Atomic Masses and Fundamental Constants 5, eds. J.H. Sanders and A.H. Wapstra, Plenum Press, New York (1976),p 663.
7. D.B. Sullivan and N.V. Frederick, IEEE Trans. Mag. MAG-13, 396 (1977).
8. I.K. Harvey, Rev. Sci. Instrum. 43, 1626 (1972).
9. D.B. Sullivan and R.F. Dziuba, IEEE Trans. Instrum. Meas. IM-23, 256 (1974).
10. M.T. Levinson, R.Y. Chiau, M.J. Feldman and B.A. Tucker, Appl. Phys. Lett. 31, 776 (1977).
11. D.W. Palmer and J.E. Mercereau, Phys. Lett. 61A, 135 (1977).
12. R.A. Kamper and M.B. Simmonds, Appl. Phys. Lett. 20, 270 (1972).
13. D.B. Sullivan, R.T. Adair and N.V. Frederick. Proc. IEEE, April 1978, to be published.
14. G.F. Engen, IEEE Trans. Micro. Theory Tech. MTT-25, 1075 (1977).
15. J. Clarke, W.M. Goubau and M.B. Ketchen, J. Low Tem. Phys. 25, 99 (1976).
16. D.G. McDonald, A.S. Risley and J.D. Cupp, Low Temp. Phys. LT-13, K.D. Timmerhaus, W.J. O'Sullivan and E.F. Hammel, eds., Plenum Press, New York (1974), p 542.
17. R.A. Webb, R.P. Giffard and J.C. Wheatley, J. Low Temp. Phys. 13, 383 (1973).
18. R.A. Kamper in Proc. Symposium on the Physics of Superconducting Devices, Univ. of Virginia, Chalottesville, Virginia (1967). [Doc. No. AD661848, NTIS, Springfield, Virginia] p M-1.
19. R.A. Kamper, J.D. Siegwarth, R. Radebaugh and J.E. Zimmerman, Proc. IEEE 59, 1368 (1971); also R.A. Kamper and J.E. Zimmerman, J. Appl. Phys. 42, 132 (1971).
20. D.J. Scalapino, In Proc. Symposium on the Physics of Superconducting Devices (1967), see ref. 18, p G-17.
21. H. Marshak and R.J. Soulen, Jr., Low Temp. Phys. LT-14, eds. M. Krusius and M. Vuorio, North-Holland, Amsterdam, (1975), p 60.
22. D.B. Utton, National Bureau of Standards, Private Communication.
23. For an analysis of this problem, see R.J. Soulen, Jr. and T.F. Finnegan, Rev. Phys. Appl. 9, 305 (1974).
24. R.J. Soulen, Jr. and R.P. Giffard, to be published.
25. J.E. Zimmerman, J. Appl. Phys. 41, 1589 (1970).
26. J.T. Harding and J.E. Zimmerman, J. Appl. Phys. 41, 1581 (1970).
27. J.A. Barnes, A.R. Chi, L.S. Cutler, D.J. Healey, D.B. Leeson, T.E. McGunigal, J.A. Mullen, W.L. Smith, R. Aydnor, R.F.C. Vessot, G.M.R. Winkler, NBS Tech. Note 394 (1970), U.S. Gov't. Printing Office, Wash. D.C.
28. J.F. Schooley, R.J. Soulen, Jr. and G.A. Evans, Jr., NBS Special Publication 260-44 (1972).
29. L.M. Besley and W.R.G. Kemp, Metrologia, to be published.
30. J.E. Zimmerman, these proceedings.
31. M.R. Beasley, these proceedings.

PROTOTYPE FOR A COMMERCIAL JOSEPHSON-EFFECT VOLTAGE STANDARD*†

L. B. Holdeman, B. F. Field, J. Toots, and C. C. Chang§
Electricity Division, Institute for Basic Standards,
National Bureau of Standards, Washington, D.C. 20234

ABSTRACT

This paper describes a prototype commercial voltage standard in which a Josephson junction is used as a precise frequency-to-voltage converter. The ratio of the induced voltage to the applied frequency is an integral multiple of the magnetic flux quantum $\Phi_o = h/2e$, where h is Planck's constant and e is the elemental charge. A constant-voltage step is induced in the current-voltage characteristic of a Josephson tunnel junction by irradiating the junction with microwaves of known frequency. A specially designed potentiometer is used to scale the junction voltage up to 1.017+ V for calibrating Weston-type cadmium-sulfate standard cells. Although cost, simplicity, and ease-of-use considerations necessitated some tradeoff with precision, the instrument nevertheless has an overall accuracy (2σ) relative to the NBS volt of 0.4 ppm or better.

INTRODUCTION

The ac supercurrent that flows when a potential difference V is maintained across the electrodes of a Josephson junction can be frequency-modulated with electromagnetic radiation of frequency ν to produce constant-voltage steps in the current-voltage (I-V) characteristic of the junction at voltages which satisfy the relationship $V = n(h/2e)\nu$, where n is an integer. Since July 1, 1972, the National Bureau of Standards (NBS) has maintained the U.S. legal volt using the ac Josephson effect and the assigned value $2e/h = 483593.420$ GHz/V_{NBS}, but outside NBS the workhorse voltage standard is still the Weston-type cadmium-sulfate standard cell. Dissemination of the U.S. volt through the calibration services currently provided by NBS requires the transportation of standard cells to and from the NBS volt facility at Gaithersburg, Maryland. This is in principle unnecessary, since a voltage can be produced directly in terms of V_{NBS} by irradiating a Josephson junction with microwaves of known frequency and using the assigned value of 2e/h. In actual practice, however, rather sophisticated instrumentation is required in order to calibrate a one-volt working standard at the 1-ppm level from junction voltages. Figure 1 shows an instrument designed to be used by a secondary standards laboratory to calibrate standard cells. The instrument became fully

---

*Contribution of NBS. Not subject to copyright.

†Support for instrument development was provided in part by the Calibration Coordination Group of DoD. Additional partial support for junction research was provided by NASA under GO# H-27908B.

§Permanent address: Department of Electronics, The Chinese University of Hong Kong, Shatin, Hong Kong.

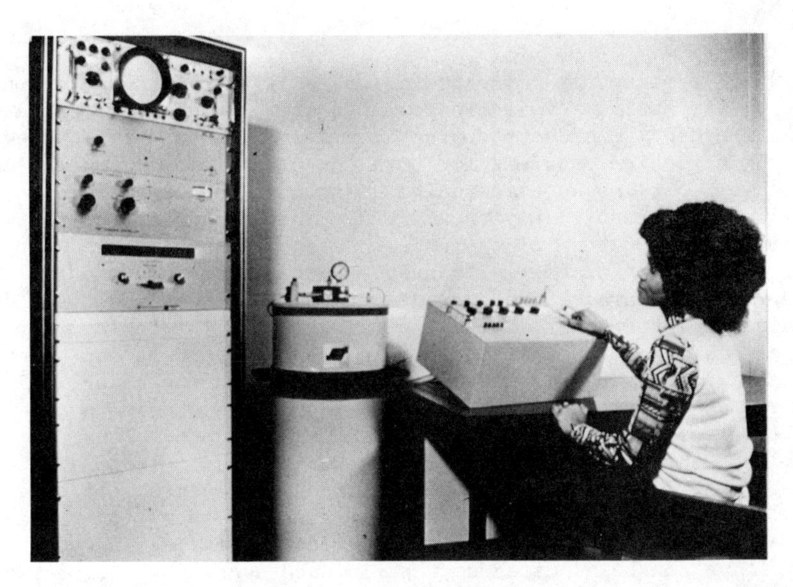

Fig. 1. Photograph of the voltage standard. The equipment shown consists of the potentiometer (right), a liquid helium dewar containing the Josephson junction, and a rack containing an oscilloscope, the microwave source, the junction bias supply, and a galvanometer.

operational in May 1977, and has been independently evaluated for the Department of Defense by a primary standards laboratory outside NBS. A version of this instrument is now commercially available.[1] Although somewhat less precise than the voltage standard in use at NBS, the commercial instrument is simpler, much less expensive to build, easier to use, and its overall precision of 0.4 ppm is generally more than adequate. As can be seen in Fig. 1, the instrument has three major components: the potentiometer, the Josephson junction and dewar system, and the junction current supply and the microwave source. These components are described briefly in the following sections, and in greater detail elsewhere.[2-4]

MICROWAVE SOURCE AND DC BIAS SUPPLY

The microwave source is a relatively inexpensive, commercially-available source consisting of a phase-locked transistor oscillator and a frequency multiplier. The source phase-locks only at certain discrete frequencies, with a capture range about each frequency, and hence for this application operates as a single-frequency source. The high-stability frequency reference for the source is provided by a temperature-controlled 10 MHz crystal oscillator which has an initial long-term stability of 0.015 ppm/week. Thus the frequency reference must be checked against a suitable frequency standard (e.g., WWV) and adjusted if necessary, since the instrument is direct-reading only at the design frequency. The output of the source is approximately 50 mW near 9 Ghz. However, a variable attenuator, a dc block (to eliminate ground loops),and an isolator (to absorb reflected power) are permanently connected in series with the source output, reducing the maximum available power to about 30 mW. Incorporating a single-frequency microwave source as a system component resulted in non-trivial design problems for both the potentiometer and the Josephson device, but cost considerations permitted no alternative.

The dc current supply provides a stable dc current to bias the junction at the center of the appropriate constant-voltage step. Since observation of the I-V characteristic is needed to establish the correct microwave power and dc current settings, the bias supply contains amplifiers with voltage offsets to permit display of the I-V characteristic on an oscilloscope. A small ac current can be added to the dc current to display the step pattern during setup, and the dc level is indicated by a bright dot in the trace using a low-voltage z-axis intensity control on the system oscilloscope. Thus the dc level can be easily set to the center of the proper step. Using the potentiometer described below, the correct step can easily be determined with a standard cell whose emf is known to 0.1%, or with a good digital voltmeter (0.1% accuracy, 1 mV resolution). A low-drift offset voltage can be adjusted so that in subsequent use the correct step automatically appears in the center of the oscilloscope screen.

POTENTIOMETER

The design of the potentiometer system is based on the method of series and parallel connection of n nearly identical resistors to provide step-up from the junction voltage to standard cell voltages: The ratio of the series-connected resistance to parallel-connected resistance is n^2, and the error introduced by resistor mismatch is of second order. The potentiometer consists of a constant current source, a series/parallel (Hamon) resistor network of fourteen 100 Ω resistors permanently connected together with 4-terminal (tetrahedral) connections, and a 10 kΩ Kelvin-Varley divider.[2,3] The voltage step-up is accomplished as follows: Using a photocell galvanometer as a null detector, the current source in the potentiometer is adjusted so that the potential drop across the parallel-connected network is equal to the junction voltage (5.188776 mV). The potential drop across the series-connected network is then 196 times the junction voltage, or 1.01700 V. Since the microwave source has fixed frequency, the usual method of adjusting the frequency so that the junction voltage equals the standard cell voltage divided by n^2 cannot be used. Instead, a shunted Kelvin-Varley divider in series with the Hamon network provides a small additional potential drop with which to match standard cell emf's, with the photocell galvanometer again used as a null detector. The divider dials are direct-reading in volts from 1.0170 V to 1.0199 V (spanning the emf range of both saturated and unsaturated cells) with a resolution of 0.1 μV. Because of the limited range of this adjustment, only moderate accuracy is required for the divider. The design of the potentiometer system thus makes the instrument very simple to use: A standard cell whose emf is to be measured is connected to the potentiometer, the divider is adjusted for null, and the cell emf is read from the dials. The Josephson junction is needed only to determine the initial potentiometer current setting and for periodic readjustment of the current to compensate for drift; i.e., for less than five minutes per hour of operation. The constant current source must be unaffected by the load changes this mode of operation requires, and must have good temporal stability and a low noise output. Changes in the current due to changing load resistances are less than 0.04 ppm for either position of the Hamon network and any setting of the Kelvin-Varley divider. The current source exhibits drift of about 0.2 ppm/hour, and noise from the source is approximately 0.02 ppm rms in a 10 Hz bandwidth.

JOSEPHSON JUNCTION AND DEWAR SYSTEM

The helium dewar is a superinsulated 30-liter metal container with three vapor-cooled shields. For ease of operation, the helium bath is not pumped, and the Josephson junction operates at 4.2 K. Helium boil-off is about 2 liters/day for a hold time of about two weeks. Hold times of these dewars can be greatly extended by using a closed-cycle cryocooler to refrigerate the radiation shields.[5] The all-metal construction of the dewar provides electromagnetic shielding for the junction. Additional shielding is provided by a cylindrical composite shield (superconducting Pb inside mu-metal) mounted around the bottom of the dewar probe. The leads connecting the junction to the potentiometer pass through ferrite beads at the top of the cryostat. Inside the 4.2 K shield are a superconducting shorting switch which is used when thermal emf's in the potentiometer leads are nulled, and a microwave integrated circuit (MIC) incorporating a Josephson tunnel junction.

The geometry of the MIC is shown in Fig. 2. The circuit is produced by evaporating Nb, Pb, or Pb-alloy films through stencil masks onto 25 mm x 25 mm x 0.8 mm glass substrates. The shaded and unshaded portions of the circuit are evaporated separately, and an oxidation process is carried out between evaporations so that a tunnel junction is formed where the two films overlap. The circuit is composed of short sections of microstrip transmission line, which are combined to form resonators, low-pass filters, coupling elements, and filter terminations and contact pads for attaching external dc leads. A ground plane beneath the substrate completes the transmission-line circuitry, and the device is mounted in an enclosed MIC package. Microwave power is brought into the cryogenic environment via a small-diameter stainless-steel 50 Ω coaxial cable and into the MIC package through a commercial adaptor to a short section of 50 Ω microstrip line. This section of microstrip line is end-coupled to a "half-wavelength" microstrip resonator, which in turn is coupled to a second resonator through the junction impedance and a section of high-impedance microstrip line. The resonators build up the rf amplitude to the level needed to induce higher-order 2e/h steps, and the "double-resonator" configuration results in excellent coupling of the microwave radiation to the junction. The thin-film filter in the dc leads is functionally a band-stop filter with 30 dB of attenuation between 8-11 GHz. The as-fabricated frequencies of these devices are about a percent higher than the fixed frequency of the microwave source, and each MIC is individually tuned to the correct operating frequency.

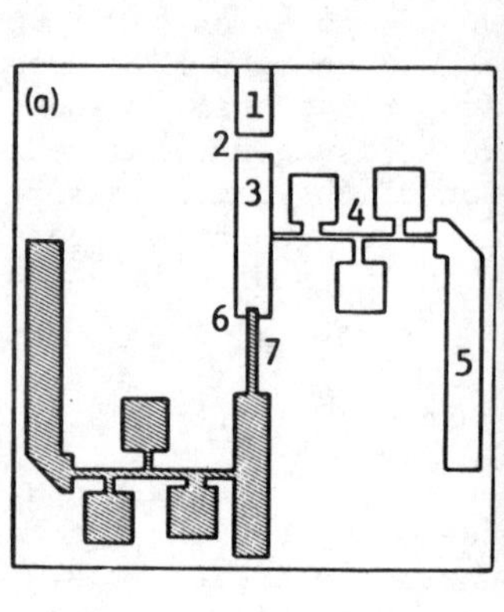

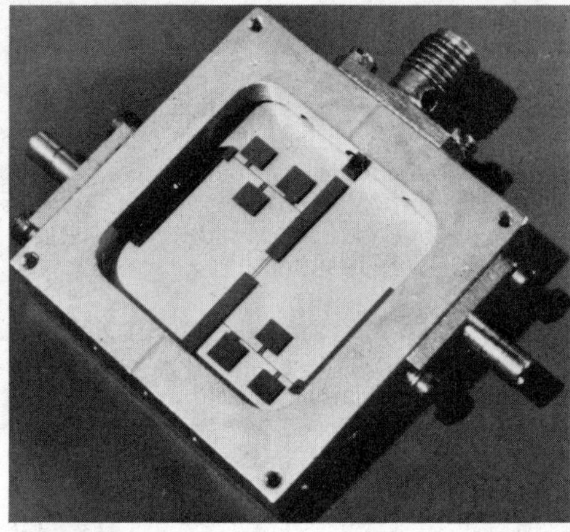

Figure 2. Josephson junction-microwave integrated circuit developed for the commercial voltage standard: (1) Section of 50 Ω microstrip transmission line for contact to coax-to-microstrip launcher, end-coupled to one of two microstrip resonators; (2) Gap between launch pad and resonator determines the loaded Q of the coupled resonators; (3) Half-wavelength microstrip resonator; (4) DC lead incorporating an elliptic-function low-pass filter; (5) Section of 50 Ω microstrip transmission line terminating the low-pass filter and forming a contact pad for attaching external dc leads; (6) Josephson tunnel junction formed at the overlap of the two superconducting films; (7) Section of high-impedance microstrip line coupling the two microstrip resonators.

The resonance frequency is a function of the relative dielectric constant ε_r of the substrate material and can be shifted slightly with a thin shim of different ε_r between the substrate and ground plane. This can be precisely accomplished by anodizing the ground plane.

In conclusion, a voltage standard instrument has been constructed. Operation of this prototype suggested a few minor changes (e.g., 15 resistors in the Hamon network, a frequency change to 9.1075751 GHz) which will be incorporated into the commercial version of the system.

REFERENCES

1. The Josephson-effect voltage standard is being marketed by Superconducting Technology, Inc., of Mountain View, Calif. This information is included for completeness and should not be construed as an endorsement or recommendation by the National Bureau of Standards. Technology developed by NBS is in the public domain.
2. B. F. Field and V. W. Hesterman, in Fifth Cal Poly Measurement Science Conference Proceedings (Dec. 5-6, 1975, California Polytechnic State University, San Luis Obispo, Calif.), Session 75-1.
3. B. F. Field and V. W. Hesterman, IEEE Trans. Instrum. Meas. IM-25, 509 (1976).
4. C. C. Chang, L. B. Holdeman, and J. Toots, to be published.
5. W. S. Goree, in Conference on Applications of Closed-Cycle Cryocoolers to Small Superconducting Devices (proceedings to be pub.)

HIGHER STANDARD VOLTAGE GENERATED BY MULTIPLE JOSEPHSON JUNCTIONS

Masao Koyanagi, Tadashi Endo, and Akira Nakamura
Electrotechnical Laboartory, 5-4-1, Mukodai-machi, Tanashi, Tokyo 188, Japan

ABSTRACT

Preliminary experiments have been performed to obtain a higher standard volt with ten tunnel junctions connected in series. A bias microwave source is used. The microwave power at each junction has to be controlled as closely as possible. Parameters and conditions for the microwave coupling are discussed. The voltage obtained is 55.5 mV at present, but no serious obstacles can be seen for obtaining voltages higher than 100 mV.

INTRODUCTION

The constant voltage produced by the ac Josephson effect has been used successfully as a reference voltage in many national standard laboratories.[1] This new technology has improved the accuracy of the standard voltage more than one order of magnitude, and the uncertainty has been reduced to below 0.1 ppm. The value of the uncertainty is determined principally by the induced voltage and the sensitivity of null detector.

If the generated voltage is increased, several improvements would result, in addition to reducing the uncertainty. One important point is the realization of the potentiometer with Josephson junctions. The unknown voltage may be determined by comparing with one of the constant voltage steps with adjustable frequency of the incident microwave. Here, the Josephson voltage is not considered as a constant reference voltage but a variable reference voltage.

At the present state, the generated voltage is not higher than about 10 mV.[2] A single junction cannot produce higher voltages because of the decrease in the supercurrent at higher voltage levels[3] and heating effects due to the higher level of microwave power.[4]

A possibility was proposed by Palmer and Mercereau[5] for the higher voltages generated by multiple superconducting bridges connected in series with separations of the orfer of 1 μm. Coherent supercurrent through whole bridges would produce the voltage steps of height N(h/2e)f where N is the number of bridges in the system, and f is frequency of the incident microwaves. Another suggestion was made by Levinsen et al.[6] based on the inverse ac Josephson effect. However, many efforts seem to be required before these proposals will be realized as usable voltage reference.

In this paper, preliminary experiments are reported on the higher voltage generation with multiple Josephson junctions which are presently used in the voltage standard.[7] As a first step, the aim is to generate 100 mV with 10 to 20 tunnel junctions connected in series. Direct current through each junction is adjustable separately, but only one microwave source is allowed for feeding all junc-

ISSN: 0094-243X/78/187/$1.50

tions. Microwave power in each junction should be controlled as closely as possible. Microwave characteristics of the junctions are the most crucial aspect in achieving the present aim.

MICROWAVE CHARACTERISTICS

Behavior of the constant voltage step produced by the ac Josephson effect depends strongly on the microwave characteristics of the junction and its environment. The stepwidth δI has a microwave power dependence somewhat like

$$\delta I \propto | J_n(2eV_{rf} / hf) | \tag{1}$$

when the junction size is smaller than the wavelength of the microwaves. Here, $J_n(x)$ means the n-th order Bessel function with argument x, and V_{rf} is the microwave voltage at the junction. From the property of the Bessel function, δI reaches its maximum when the argument x is nearly equal to n for large n. Thus, the n-th step has the maximum width when

$$V_{rf} = n\ (h/2e)\ f = V_o^{max}\ . \tag{2}$$

Here, V_o^{max} is just the dc voltage of the n-th step.

In the creation of voltage standards, the size of the tunnel junction is determined in such a way that it behaves as a resonant cavity. Eq(1) should be modified for the space dependence of V_{rf}, but the condition in eq(2) is still valid. This relation is obtained in the experiment as shown in Fig. 1. V_o^{max} is proportional to the square root of the incident power P.

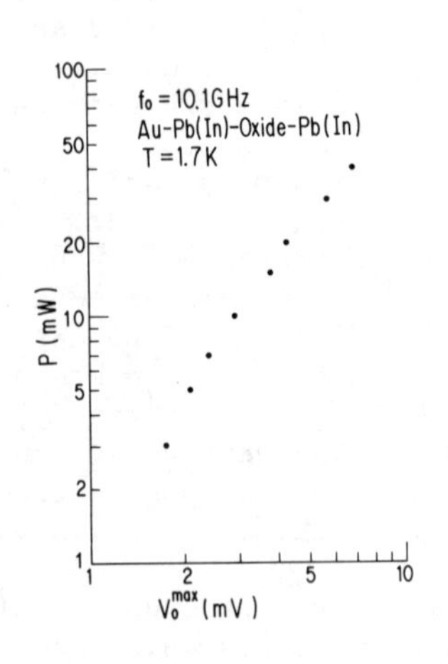

Fig.1. Incident microwave power vs dc voltage of the maximum width step.

The next problem is how the microwave voltage V_{rf} will be derived from the incident power. In other words, how the normal resistance R_N will play a role in the power vs step voltage relations. Incident power P required to obtain a step voltage V_o^{max} with maximum width is plotted against R_N of a junction in Fig. 2. Various values of R_N are obtained in a lead-indium alloy junction after recycling between liquid helium and room temperature. Normal resistance increases rapidly when the tunnel junction is left in open air, but it goes back towards the original value if the junction is stored in a desiccator as shown in Fig. 3. Thus, without any change in the microwave circuit, the results in Fig. 2 can be obtained.

Solid lines in Fig. 2 express the empirical relation

$$P = A\,(V_o^{max})^2 / R_N^2 , \tag{3}$$

where A is taken as 17.3 Ω by fitting to the experimental value drawn with the open circle. From eqs (2), (3), V_{rf} is found to be proportional to R_N for fixed P. This fact suggests that the microwaves couple to the junction through the current. In fact, the junction is a part of one conductor of the stripline which is completed by another conductor with an insulator (Si), and is terminated by a matched load of 50 Ω as shown in Fig. 4. Rf current is mainly determined by the line impedance, and the induced voltage V_{rf} across the junction is approximately proportional to the normal resistance R_N. Thus, for larger R_N, generation of a certain voltage with maximum step width requires a smaller incident power. There must be, however, an upper limit to R_N because of the considerable decrease in the step widths.

The length of the tunnel junction determines the resonant frequency, which should be nearly the same in every junction connected in series. If the Q-value is assumed to be 100, the fluctuation should be kept below 1/100

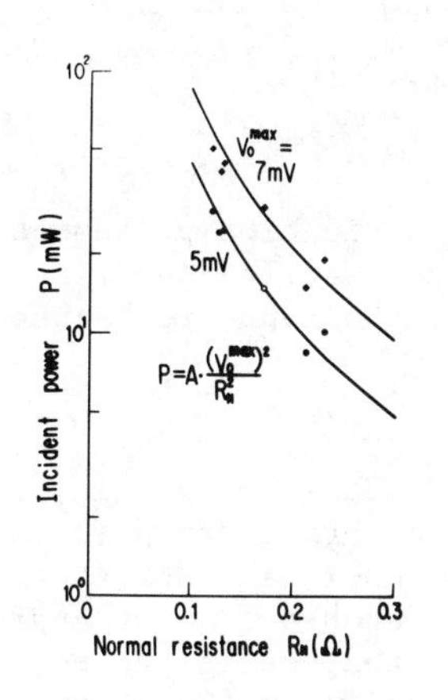

Fig.2. Microwave power P required to obtain certain V_o^{max} depends on the normal resistance of the junction.

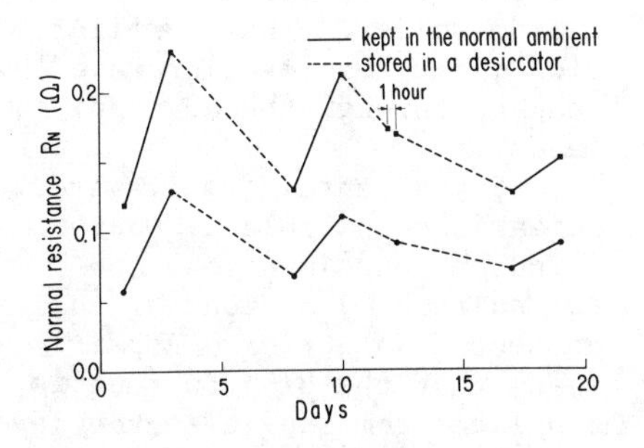

Fig.3. Variation of R_N observed in Pb(In)-oxide-Pb(In) junctions after recycling between liquid helium and room temperature.

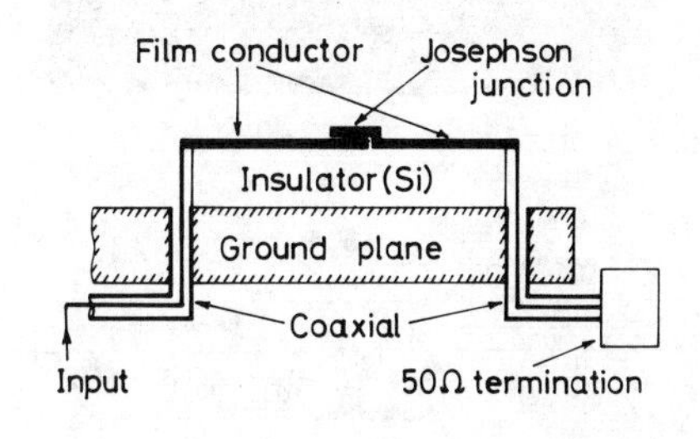

Fig.4. Stripline configuration with one Josephson junction.

of the length, i.e. 0.9 mm. This means that the fluctuations in sizes should be below several micrometers. Photolithography techniques can achieve this precision easily, but pure lead is not adequate because of its sensitivity to moisture, which is unavoidable in processing photoresist. Lead alloy with 10 percent indium is resistant against moisture and water. Thin gold films on the surface of the substrate are effective in improving the recycling characteristics. Fabrication process of the junction is shown in Fig. 5.

1 Si
2 AZ1350J
3 deposition
Au 10 nm
Pb(In) 150 nm
Oxide 8 nm
4 lift off
5 AZ1350J
6 deposition
Pb(In) 300 nm
7 lift off

Fig. 5. Fabrication process of Pb(In)-oxide-Pb(In) junction by photolithography techniques.

PRELIMINARY EXPERIMENT

Multiple junctions are prepared by photolithography on silicon substrate with a pattern as shown in Fig. 6. The broad strip forms one conductor of the stripline. Another conductor lies under the substrate. The stripline is terminated with a matched load of approximately 50 Ω. The tunnel junction is formed by the overlapped portion of the narrow strip with the broad strip. The oxide is formed by thermal oxidation in 10 kPa of oxygen at room temperature for 13 hours.

Ten junctions out of the 33 are used in the experiment. Others are all punctured in order to make a contact with wires from dc supplies. The wires are connected perpendicularly to the surface of the strip with indium solder. This configuration minimizes the microwave leakage through the dc supply leads.

Experimental results are summarized in Table 1, where 6 independent junctions and two pairs of junctions are all connected in series. A pair means that the bias current is supplied from single source, and the generated voltages and the measured resistances are just the summation for two junctions. Normal resistance of junctions R_N are within 10 to 20 percent of the mean value except for the

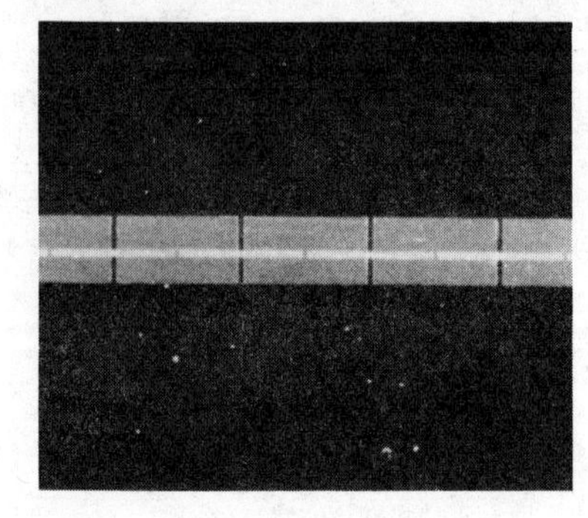

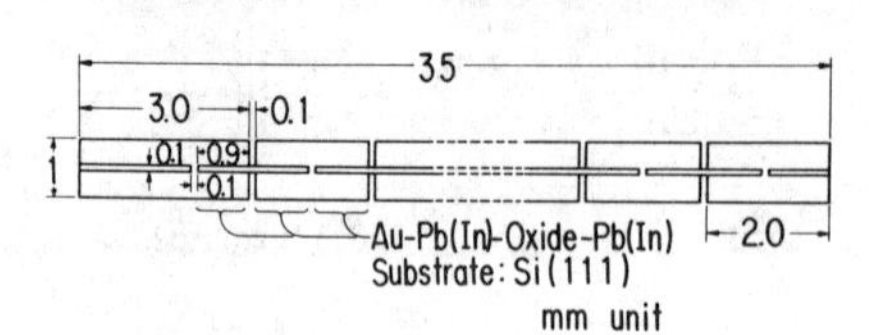

Fig.6. Pattern and sizes of the multiple Josephson junctions used in the experiment. Photo shows a part of them.

Junction number	Normal resistance	Max. width step voltage	
1	0.0625 (Ω)	4.0 (mV)	
2 + 2'	0.26	12.6	pair
3	0.13	7.8	
4 + 4'	0.24	12.2	pair
5	0.11	4.2	
6	0.12	7.4	
7	0.094	4.3	
8	0.082	3.0	
Total		55.5	

$f_0 = 9.22$ GHz P = 100 mW

Table 1. Experimental data for the multiple Josephson junctions connected in series.

two junctions at the far ends of the stripline. Voltages of maximum step width V_0^{max} follow roughly the rule expressed by eq (3). Only junction No. 5 gives the relatively smaller value of 4.2 mV rather than that expected from R_N=0.11 Ω. The reason is a slight deviation of the resonant frequency.

Total voltage produced is 55.5 mV in this experiment with single microwave source which has a frequency of 9.22 GHz and a power of 100 mW. Step widths are approximately 100 μA for single junctions, but 80 μA and 40 μA for pairs 4+4' and 2+2' respectively. This width for pairs coresponds to the overlapped portion of two steps. As a result, these steps are usable as voltage standards when the bias current in each junction is adjusted separately. With the present state of technology a total voltage of 100 mV will be obtained as a standard voltage in the near future. However, improvements of junction fabrication techniques will be required in the control of characteristics for further developments.

The authors thank Drs. H. Hayakawa and S. Kosaka for valuable suggestions on junction fabrication techniques.

REFERENCES

1. B. W. Petley: Atomic Masses and Fundamental Constants 5 (Plenum Press, New York - London, 1976) ed. J. H. Sanders and A. H. Wapstra, p. 450.
2. T. F. Finnegan, A. Denenstein, and D. N. Langenberg: Phys. Rev. B4, 1487 (1971).
3. I. O. Kulik and I. K. Yanson: The Josephson Effect in Superconducting Tunneling Structures (Israel Program for Scientific Translations, Jerusalem, 1972) p.27.
4. M. Koyanagi, T. Endo, S. Koga, G. Yonezaki, and A. Nakamura: Japan. J. Appl. Phys. 14, 1607 (1975).
5. D. W. Palmer, and J. E. Mercereau: IEEE Trans. Magn. MAG-11, 667 (1975).
6. M. T. Levinsen, R. Y. Chiao, M. J. Feldman, and B. A. Tucker: Appl. Phys. Letters, 31, 776 (1977).
7. T. Endo, M. Koyanagi, K. Shimazaki, G. Yonezaki, and A. Nakamura: Atomic Masses and Fundamental Constants 5 (Plenum Press, New York - London, 1976) ed. J. H. Sanders and A. H. Wapstra, p. 464.

Superconducting Resonators: High Stability Oscillators and Applications to Fundamental Physics and Metrology

S. R. Stein
Frequency & Time Standards Section
National Bureau of Standards
Boulder, CO 80303

J. P. Turneaure
High Energy Physics Laboratory
Stanford University
Stanford, CA 94305

ABSTRACT

Superconducting oscillators have achieved better frequency stability than any other device for averaging times of 10 l.c.s. to 1000 l.c.s. This high stability results from the use of solid niobium resonators having Q factors greater than 10^{10}. Such oscillators have direct applications as clocks and spectrally pure sources. They may also be used for accurate measurements of many physical quantities and to perform a variety of experiments on fundamental constants, relativity, and gravity waves.

INTRODUCTION

The best frequency stability ever reported is 3×10^{-16} which was achieved by a superconducting cavity stabilized oscillator. Although superconductivity is not fundamental to such performance, it results in two very important advantages beyond the usual qualities which characterize a high stability resonator: The very high Q which is obtainable (up to 10^{11}) implies both extremely narrow linewidth and low intrinsic noise; the small effective nonlinearity of the superconductor makes it possible to use relatively high power levels. The combination of these properties and others makes it possible to optimize the stability of oscillators from extremely short times to many days.

A variety of different types of superconducting resonators are in use: lumped circuit, stripline, cavity and Fabry-Perot. They span a frequency range from DC to near 100 GHz. Plated superconductors can be employed, but bulk niobium fired in a high vacuum furnace has produced the highest Q. Many different types of oscillators have also been developed or proposed for use with superconducting resonators. These include use of the superconducting resonator in the feedback path of an amplifier to produce an active oscillator and use of the superconducting resonator in some type of phase bridge to produce a control signal for stabilization of an independent oscillator. The factors which currently limit the performance of superconducting oscillators are not fundamental. The long-term stability is probably determined by ambient influences such as temperature and mechanical stability of critical components. The spectral purity is primarily limited by vibrations and stored energy in the resonator.

Because of their exceptional frequency stability, superconducting oscillators outperform other frequency standards in a number of applications. For example, they have the best stability of any device over the range of averaging times needed for long baseline interferometry. In addition, they should provide unsurpassed spectral purity for use in the synthesis of far-infrared frequencies. Superconducting oscillators are also interesting for a variety of fundamental physics experiments. It is possible to use them to test for time variations in the fundamental constants, to search for gravity waves, and to test the metricity of gravity. Since many physical quantities can be transduced to a frequency, high stability oscillators are excellent candidates for state-of-the-art metrology. Superconducting oscillators are particularly well-suited for some temperature, length and electrical parameter measurements.

SUPERCONDUCTING RESONATORS

A variety of different types of superconducting resonators may be used for high stability oscillators and for applications to fundamental physics and metrology. They include lumped circuit, stripline, cavity and Fabry-Perot resonators. The choice of a resonator type for a particular application involves the consideration of many factors. In general, however, a resonator with a high quality factor (Q) and a very stable resonant frequency is desirable. For applications in which a physical quantity is transduced to a frequency shift, the resonator frequency must be sensitive to the physical quantity, yet insensitive to other conditions. The factors which affect the Q and the frequency of a resonator are discussed below.

Resonator Q. Superconducting resonators with high Q have been studied most extensively for application to particle accelerators which use resonators in the range of 100 MHz to 10 GHz. In this frequency range, resonators made of niobium, Nb_3Sn and lead have all achieved unloaded Q's (Q_o) greater than 10^9. Niobium resonators have been most extensively studied. They have achieved high Q_o's in the range of 10^{10} to 10^{11} for TM_{010}-mode cylindrical cavities at 9 GHz[1] and 1.3 GHz[2], for TE_{011}-mode cavities at 3 GHz[3] and 9 GHz[4], for a helical cavity at 90 MHz[5], and for a re-entrant cavity at 400 MHz[6]. Much less work has been performed on Nb_3Sn resonators. Nonetheless a TE_{011}-mode cavity at 9 GHz has achieved a Q_o of 6×10^9[7]. Although there has been extensive work on lead resonators, Q_o's larger than 10^9 have been limited to TE_{011}-mode cavities which have nearly zero electric field on their surface. A Q_o of 4×10^{10} has been reported for a lead TE_{013}-mode cavity at 12 GHz[8].

An understanding of the relationship of the Q_o to the resonator mode and the operating temperature and frequency comes from the study of the various types of losses in the resonator as well as the mode characteristics.

First, a superconducting resonator whose dielectric is vacuum is considered. In this case, the losses are determined by the magnitude of the magnetic field, H, at the surface of the superconductor. The dissipated power density, p, in the superconductor is

$$p = \frac{1}{2} RH^2, \qquad (1)$$

where R is the total surface resistance of the superconductor. For a particular resonator and mode the unloaded Q is related to the surface resistance by the geometrical factor:

$$Q_o = \Gamma/R. \qquad (2)$$

The geometrical factor is a measure of the ratio of the resonator-stored energy to the integral of the magnetic field squared over the surface of the resonator.

The geometrical factor varies over a wide range depending on the resonator type. A right-circular-cylindrical cavity with its diameter equal to its length has $\Gamma = 750\Omega$ for the TE_{011}-mode and

$\Gamma = 300\Omega$ for the TM_{010}-mode.[9] A Fabry-Perot resonator can have a high Γ of $1 \times 10^4\Omega$[5], whereas a helical cavity can have a low Γ of 5Ω.[5] Thus, the choice of a resonator type can strongly influence Q_o.

The surface resistance of superconductors[10] is approximated by the following relation for temperature less than one-half the superconducting transition temperature (T_c) and for frequency (ω) well below the superconducting energy gap:

$$R = R_o \, \omega^{1.7} \exp\left(-\frac{\Delta_o T_c}{T}\right) + R_{res}, \qquad (3)$$

where R_o is a constant and $\Delta_o = \Delta(0)/kT_c$ is the reduced energy gap. The first term of Eq. 3 is the surface resistance (R_s) of the superconducting state, and the second term is the residual surface resistance (R_{res}) which represents other types of losses at the surface. R_{res} as low as $5 \times 10^{-10}\Omega$ at 90 MHz[5] and $3 \times 10^{-9}\Omega$ at 9 GHz[1] has been reported. Although very low R_{res} can be achieved, it is important to note that at higher temperatures and frequencies R_s, the first term in Eq. 3, may dominate the total surface resistance because of its strong exponential temperature dependence. For example, $R_s = 2.6 \times 10^{-5}\Omega$ at 4.2 K and $R_s = 2.5 \times 10^{-9}\Omega$ at 1.3 K for niobium at 10 GHz. Thus at 10 GHz, a superconducting niobium TM_{010}-mode cavity must be operated below 1.3 K to reach a Q_o of 10^{11}. The parameters for the first term in Eq. 3 are given in Table I for niobium, Nb_3Sn and lead. These parameters should be adequate for the frequency range 100 MHz to 100 GHz.

Table I

Superconductor	Δ_o	T_c (K)	R_o (Ω)
Nb[11]	1.88	9.25	7.13×10^{-22}
Nb_3Sn[7, 16]	2.1	18.	5.05×10^{-22}
Pb	2.05	7.2	5.15×10^{-22}

If the volume of a superconducting resonator is filled with a dielectric, the loss tangent, δ of the dielectric must also be considered. The loss tangent of dielectrics at low temperature has not been extensively investigated. Bagadasarov, et al. have recently reported a very low dielectric loss tangent of 2.4×10^{-8} for sapphire at 2 K[13]. This measurement was made on a solid sapphire 3 GHz TM_{010}-mode cavity coated with lead. This work is continuing.

Resonator Frequency. The frequency of a superconducting resonator is a result of the resonator geometry and mode, the dielectric within the space of the resonator, and the RF properties of the superconductor. Although the characteristics which determine the resonant frequency are relatively well-fixed, they are nonetheless sensitive to the operating conditions of the resonator and also to external conditions. These sensitivities are discussed below.

(a) Resonator Stored Energy. The frequency of a superconducting resonator is shifted from its value at zero stored energy by electromagnetic radiation pressure[1] and by the nonlinear superconducting surface reactance[14]. The frequency shifts for both of these effects are to lower frequencies and are proportional to the stored energy. The electromagnetic-radiation-pressure frequency shift is dependent on the resonator geometry, the mode and the elasticity and size of the structure forming the resonator. If a resonator and its structure are scaled by a factor α, while the stored energy is scaled by α^3 (same energy density), then both the resonator frequency and electromagnetic-radiation-pressure frequency shift scale as $1/\alpha$. Thus, the radiation-pressure fractional frequency shift remains constant. On the other hand, the nonlinear-surface-reactance frequency shift depends only on the resonator geometry and mode. If a resonator is scaled by a factor α, while the stored energy is scaled by a factor α^3, then the resonator frequency scales as $1/\alpha$ and the nonlinear-surface-reactance frequency shift scales as $(1/\alpha)^2$. Thus, the fractional frequency shift due to the nonlinear surface resistance scales as $1/\alpha$.

The frequency shift due to the electromagnetic radiation pressure is smallest in a cavity with a massive structure. For example,[15] a circularly cylindrical TM_{010}-mode 8.6 GHz massive niobium cavity (the length of the cavity and the thickness of the walls are about equal to the cavity radius) has a stored-energy coefficient for the fractional frequency shift of -1.5×10^{-6} J^{-1}. This coefficient is probably dominated by the nonlinear surface reactance contribution. On the other hand, a weakly supported resonator has a large frequency shift due to the electromagnetic radiation pressure. For example, an unsupported helical niobium resonator at 130 MHz[16] has a very large stored-energy coefficient for the fractional frequency shift of -5.5×10^{-3} J^{-1}. This coefficient is dominated by the electromagnetic radiation pressure.

The frequency shift coming from the stored energy has two effects on resonator frequency stability. First, power fluctuations in the source result in fluctuations of stored energy and thus the resonant frequency. Second, the nonlinear coupling of the electromagnetic resonator mode to a mechanical mode of the resonator structure by the electromagnetic radiation pressure produces ponderomotive oscillations above some threshold stored energy[16]. These oscillations involve the transfer of energy between the electromagnetic and mechanical resonators, and they result in very large frequency and amplitude modulation of the electromagnetic resonator. For most applications, the resonator is operated well below the threshold for ponderomotive oscillations.

(b) Resonator Temperature. The frequency of a superconducting resonator is shifted from its absolute zero temperature value by both thermal expansion and the temperature dependence of the superconducting surface reactance. The frequency shifts for both of these effects are downward with increasing temperature. The frequency shift due to the thermal expansion is dependent only on the structure forming the resonator. If the structure of a resonator is scaled by a factor α, the fractional frequency shift due to thermal expansion would remain constant. On the other hand, the fractional frequency shift due to the temperature dependence of the superconducting surface reactance would scale as $1/\alpha$. The temperature coefficient of the frequency shift places some requirement on temperature stability in order to achieve a particular level of frequency stability.

An example of the temperature coefficient for the fractional frequency shift is given in Fig. 1 for a circularly cylindrical TM_{010}-mode solid niobium cavity at 8.6 GHz[17]. There are three contributions to the temperature coefficient: the temperature dependence of the surface reactance (skin depth), the lattice thermal expansion and the electronic thermal expansion. The skin depth contribution has a very strong temperature dependence which is exponential ($\exp(-\Delta_0 T_c/T)$). The electronic thermal expansion contribution has about the same functional form as the skin depth contribution, but its magnitude is much less (2%). The lattice thermal expansion contribution has a much weaker temperature dependence which is T^3. At 8.6 GHz, for the TM_{010}-mode niobium cavity, the skin depth and thermal expansion contributions cross at 1.2 K. The use of a dielectric structure for the resonator, rather than a structure formed with a superconductor, can yield a lower thermal expansion contribution to the temperature coefficient for the fractional frequency shift[18].

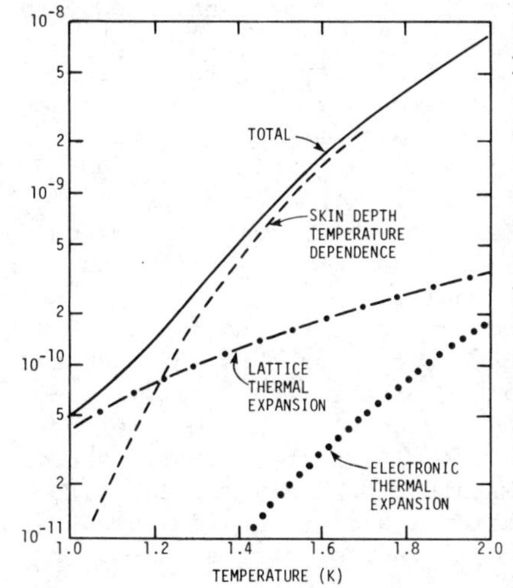

Fig. 1. Temperature coefficient of $\delta\nu/\nu$ for a niobium cavity.

(c) Gravity and Acceleration. On Earth, the largest frequency shift of this type is the result of the resonator structure self-weight, which places the resonator structure itself under strain. Again a massive structure is desirable to reduce the frequency shift on the surface of the Earth. An estimate of the fractional frequency shift of a massive TM_{010}-mode niobium cavity at 8.6 GHz supported at one end with its axis vertical in the gravitational field on the surface of the Earth[19] gives a fractional frequency shift of from 10^{-8} to 10^{-9}, which corresponds to an acceleration coefficient for the fractional coefficient for the fractional frequency shift of from 10^{-9} to $10^{-10}\,s^2m^{-1}$. This frequency shift is relatively large, and it makes the resonant frequency quite sensitive to tilt with respect to the gravitational field. A substantial reduction in the acceleration coefficient could be made by supporting the TM_{010}-mode cavity at its mid-plane so that the upper half is under compression and the lower half under tension. Strain in the resonant structure is also produced by linear acceleration and rotation of the resonator and by the gradient of the gravitational force. Although these effects are small for an Earth-fixed resonator, they can become quite important for resonators in free-falling satellites and space probes.

(d) Radiation. Penetrating radiation, such as cosmic radiation, affects the dielectric characteristics within a resonator, whether it is a vacuum or a solid dielectric. Thus, the radiation causes a frequency shift. The magnitude of the frequency shift due to radiation appears to be small. It has been measured by bringing a 5 mCi Co^{60} source close to a solid niobium cavity with a vacuum dielectric[19]. When the Co^{60} source was brought from a large distance to within 12 cm of the niobium cavity, the fractional frequency shift was less than 1×10^{-12}.

(e) External Pressure. Substantial frequency shifts can result from pressure applied externally to a resonator. This pressure may be the result of the hydrostatic pressure of a liquid or gas or the magnetic pressure due to the ambient magnetic field. The hydrostatic pressure is usually reduced by operating the resonator in a vacuum, and the magnetic pressure is reduced by shielding the resonator from the external magnetic field, and under these conditions the fractional frequency shift is very small.

(f) Vibrations. Mechanical vibrations of a resonator modulate its resonant frequency, which of course directly affects the spectral density of phase fluctuations. The nature of the mechanical vibrations and their affect on the resonant frequency depends on the geometry and mode of the resonator and on the elasticity, geometry, and density of the resonator structure. Although probably small, a frequency shift proportional to the vibrational energy can come about through the anharmonicity of the lattice potential of the resonator structure.

(g) Structural Changes. To this point, it has been assumed that the resonator structure is unchanged except for its single-valued dependence on parameters such as pressure, temperature, etc. It is possible, however, for the atomic arrangement of the resonator structure to change. The energy to change the atomic arrangement can either come from residual stress within the resonator structure created when it is fabricated, or it can come from external sources such as temperature, pressure, gravity or radiation. It is probably not possible to analyze the affect of these energy sources on structural changes and the consequent frequency jumps or drift. The best measure of the affect of these energy sources on structural changes may come from direct observation of the fractional frequency drift in superconducting resonators, which has been observed to be as low as 8×10^{-15}/day[20].

OSCILLATOR DESIGN

Because of the tremendous stability potential of superconducting resonators, a variety of techniques have been used to construct superconducting oscillators. The goals of this research have varied. Some oscillators have been constructed to illustrate feasibility, some to accomplish modest stability goals for further research on superconducting resonators, and others to achieve the ultimate frequency stability over some range of Fourier frequencies or averaging times. As a result, the achieved frequency stability for each technique is probably a poor indication of its capabilities. Instead of making such a comparison, this paper will outline some of the advantages or disadvantages of each method from the point of view of achieving the best possible frequency stability. The techniques discussed here use the superconducting resonator in three different ways--as the sole resonator of an oscillator circuit, as an auxiliary resonator to stabilize a free-running (noisy) oscillator, or as a filter which provides no feedback to the source.

Figure 2b illustrates how an oscillator may be realized using a superconducting resonator and a unilateral amplifier. Oscillation can occur when the amplifier gain exceeds the losses and the total phase shift around the loop is a multiple of 2π rad. Automatic gain control or limiting is necessary in order to produce oscillations at the desired power level. The resonator may be used in either transmission or reflection but the transmission mode is preferable because the insertion loss of the resonator suppresses spurious modes of oscillation which do not lie in its pass bands. This technique has received considerable attention because of its simplicity[21, 22]. The only element which needs to be located in the dewar is the superconducting cavity which can be connected to the room temperature amplifier by long lengths of transmission lines. However, this virtue is its major detraction when state-of-the-art frequency stability is desired. Changes in the phase length of the transmission lines produce proportional frequency shifts. If $\Delta\phi$ is the phase change from any source, the fractional frequency shift is

$$\Delta\nu/\nu = \Delta\phi/2Q_L \, . \qquad (4)$$

The phase changes due to such factors as thermal expansion and vibrations are sufficiently large in a cryogenic system that they totally dominate the short term stability and drift of such an oscillator.

One solution to this problem is to use an amplifier which functions in the same low temperature environment as the resonator and is connected to it by short rigid transmission line. It has been proposed to use a cryogenic travelling wave maser in a unilateral amplifier design[23]. Alternatives are tunnel diode amplifiers and varactor diode parametric amplifiers. Both of these devices function by generating a negative conductance at the resonator frequency. Since they are bilateral, they can simply be connected to the superconducting cavity through an impedance-transforming network as shown in Fig. 2a. When the negative conductance of the amplifier exceeds the positive load conductance of the resonator, oscillation results. Jiminez and Septier have demonstrated the feasibility of the tunnel diode superconducting oscillator[24] and further work is being done by Braginski et al.[25, 26]. One of the authors (Stein) is developing a superconducting non-degenerate parametric oscillator[24]. The major advantage of the tunnel diode oscillator is that it requires only dc bias power for operation. On the other hand, there are several disadvantages. Shot noise in the tunnel junction limits currently available tunnel diode amplifiers to an effective noise temperature of 450K at 9GHz[28]. In addition, the very low operating voltage limits the theoretical output power to 1mW at 10 GHz from commercially available devices (having peak current less than 20 mA). If other problems were solved

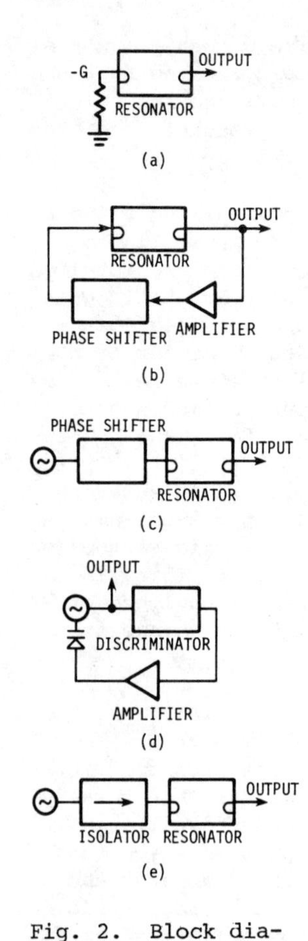

Fig. 2. Block diagram illustrating several superconducting frequency sources: (a) Negative resistance oscillator. (b) Loop oscillator, (c) stabilized voltage-controlled oscillator, and (e) passive filter.

these two difficulties could limit the frequency stability of the tunnel diode superconducting oscillator. In contrast, cooled parametric amplifiers have demonstrated 20 K noise temperatures and room temperature nondegenerate parametric oscillators have produced more than 100 mW at 9 GHz[28, 29].

The most widely studied technique for realizing a superconducting oscillator has been the stabilization of a free-running oscillator with a superconducting resonator[24]. One technique for accomplishing this, called cavity stabilization, is shown in Fig. 2c. The oscillator is injection-locked by the power which is reflected from the superconducting resonator. The stabilization factor, which is the ratio of the free-running oscillator frequency fluctuations to the cavity-stabilized oscillator frequency fluctuations, is given by the ratio of the Q of the superconducting cavity to the Q of the free-running oscillator[30]. Since the frequency fluctuations of an oscillator are inversely proportional to its resonator Q,[31] the best possible performance of a cavity stabilization system reduces to that of an oscillator built with the superconducting cavity as its only resonator. There are two major disadvantages to this technique. First, room temperature oscillators such as klystrons and Gunn-effect devices have extremely high noise temperatures; second, the frequency offset from the center of the resonance is proportional to the line length between the oscillator and the cavity just as in the loop oscillator.

The most successful superconducting oscillator technique to date is the use of active feedback to stabilize a voltage controlled oscillator[32]. The superconducting resonator is the frequency-sensitive element of a discriminator which generates an output voltage proportional to the frequency difference between the oscillator and the center of the superconducting cavity resonance[33]. Although this system also has a long path length between the room temperature oscillator and the superconducting cavity, it is possible to design the discriminator so that the dependence of the oscillator frequency on this path length is greatly reduced. This is accomplished by using phase modulation sidebands on the carrier frequency to provide the reference for locating the plane of the detuned short of the superconducting cavity. Despite the fact that this technique also uses a noisy room temperature oscillator, its performance is not limited by this fact. This is true because in such a system it is possible to greatly multiply the phase vs. frequency slope of the resonator by using external amplifiers. In this way the frequency fluctuations of the free-running oscillator may be reduced until the performance level is determined by the microwave detectors.

Figure 2e shows a superconducting resonator being used to filter the output of an oscillator. This application is particularly important when the oscillator is to be used as a source for frequency multiplication. For example, present state-of-the-art quartz crystal oscillator may be multiplied to 0.5 THz before the carrier is lost in the phase noise pedestal. However, if the same oscillator is filtered by a passive superconducting cavity with loaded Q equal to 2 x 10^9 it could in principle be multiplied to 100 THz[34].

The conclusion of the above discussion is that two types of superconducting oscillators appear most promising for improved stability: stabilization of a VCO and the all-cryogenic oscillator. The fundamental limitations of the two devices are similar, so the most important differences at this time are the practical problems of implementation: The VCO stabilization system has all the critical elements outside the dewar where they are readily available for adjustment and experimentation, but they are necessarily sensitive to problems of temperature fluctuations and vibration. On the other hand, the active oscillator is compact and totally contained in the highly controlled cryogenic environment. Such an oscillator will, however, present some new technical difficulties such as heat dissipation and device parameter fluctuations.

THEORETICAL OSCILLATOR NOISE PERFORMANCE

For the purpose of this discussion, the frequency fluctuations of any oscillator based on a superconducting resonator can be separated into two categories: Statistical fluctuations around the center of the resonance and the perturbations of the resonant frequency itself. The second category determines the ultimate performance level for any oscillator which is relatively independent of its design. It includes temperature, power level, and mechanically-induced frequency shifts which were discussed in a previous section.

If the center of the resonance is sufficiently constant, then other sources of noise will determine the ultimate stability of the superconducting oscillator system. The frequency fluctuations about the center of the resonance are highly dependent on the design of the particular oscillator; however, a lower limit corresponding to the case where all noise sources are filtered by the resonator can be determined. If the perturbing noise is white, then the phase of the oscillator does a random walk[35, 36, 37]. The one-sided spectral density of the phase fluctuations, in a form appropriate for microwave resonators, is given by

$$S_\phi(f) = \left(\frac{\nu_o}{f}\right)^2 \frac{kT}{2P_a\, Q_E\, Q_L} \qquad (5)$$

where ν_o is the operating frequency, k is the Boltzman constant, T is the absolute temperature, P_a is the power dissipated in the load, and Q_E and Q_L are the external and loaded Qs respectively. In the case of the superconducting cavity with $Q_E = 10^{10}$, $Q_L = 5 \times 10^9$, $P_a = 10^{-3}$W, $\nu_o = 10^{10}$ Hz, and T = 1K,

$$S_\phi(f) = 10^{-20} \text{ Hz}/f^2.$$

The active element in a practical oscillator will dominate the thermal noise. In this case T must be interpreted as the effective noise temperature of the device. Such noise temperatures vary from approximately 20 K for varactor parametric amplifiers to more than 10^4 K for a transferred-electron device.

Another important limitation on the stability of a superconducting oscillator is additive noise which results from a white noise voltage generator at the output of the oscillator. In an ideal oscillator the additive noise is due to output buffer amplifiers or a user device. The spectral density of the phase fluctuations is

$$S_\phi(f) = kT'/2P_a, \qquad (6)$$

where T' is the effective noise temperature of the circuitry which sees the output of the oscillator. If the effective noise temperature is 300 K and the available power is 10^{-3} W, then

$$S_\phi(f) = 2 \times 10^{-18}/\text{Hz}.$$

In this case the additive noise dominates the oscillator spectrum for Fourier frequencies greater than .07 Hz. Under the same conditions, the rms fractional frequency fluctuations are given by

$$\sigma_y(\tau) = \left[\left(\frac{8.3 \times 10^{-21}}{\tau^{\frac{1}{2}}}\right)^2 + \left(\frac{4.3 \times 10^{-20} f_h}{\tau}\right)^2\right]^{\frac{1}{2}}, \qquad (7)$$

where f_h is the noise bandwidth of the measurement system.

Equations (5) and (6) show that both the random walk of phase and the additive noise can be reduced by increasing the available power. However, this technique is limited by several factors. The nonlinearity of the resonator couples amplitude and phase modulation and may ultimately limit the stability. If this is not a problem, then at some field level the resonator breaks down. Finally, high power levels may exceed the dynamic range of the user device such as a mixer in a super-heterodyne receiver.

SUPERCONDUCTING OSCILLATOR FREQUENCY STABILITY

The superconducting cavity stabilized oscillator (SCSO), which has been extensively investigated by the authors [33, 38], exemplifies the potential of superconducting cavities to achieve very high frequency stability. An ensemble of three nearly independent SCSOs was used to evaluate their performance. Each SCSO employed a massive niobium TM_{010}-mode 8.6 GHz cavity operating at about 1.2 K with a temperature stability of 10 μK. The unloaded Qs of these superconducting cavities ranged from a few times 10^9 to 7×10^{10}. The frequency stability of these SCSOs is discussed below. The potential frequency stability of SCSOs has not yet been fully realized; for short term stability, circuits utilizing a superconducting cavity as a transmission filter would be more appropriate.

Frequency Domain Measurements. The spectral density of phase fluctuations $S_\phi(f)$ has been measured as a function of Fourier frequency for the SCSO. The results of these measurements[39] for the 8.6 GHz SCSOs are summarized in Fig. 3. There are several aspects of S_ϕ which are important to discuss. First, for Fourier frequencies between 100 Hz and 50 kHz, the baseline for S_ϕ is white and it is 10^{-12} rad^2 Hz^{-1}. This level is the result of the noise characteristic of the 1 MHz amplitude detector used in the SCSO and it does not represent the limit of what may be achieved with a superconducting resonator. For frequencies below 100 Hz, the baseline of S_ϕ makes a transition to flicker frequency noise ($S_\phi \propto f^{-3}$). Superimposed on the baseline for Fourier frequencies below 500 Hz are what appear to be bright lines. These bright lines have been attributed to mechanical vibrations transmitted to the cavity from the floor.[1] The peak in S_ϕ at a Fourier frequency of 250 kHz comes from the specific character of the SCSO feedback frequency response which has an open loop gain of one at about 300 kHz. Thus, there is not enough feedback gain in this region to improve the phase fluctuations of the Gunn oscillator which is stabilized by the superconducting cavity. For the highest Fourier frequencies, S_ϕ is dropping because of the characteristics of the Gunn oscillator itself.

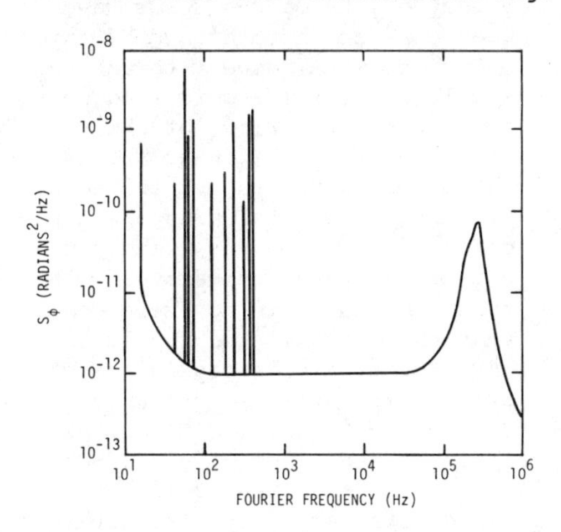

Fig. 3. Spectral density of phase fluctuations S_ϕ as a function of Fourier frequency for an 8.6 GHz superconducting-cavity stabilized oscillator.

Time Domain Measurements. The two-sample Allan variance for the 8.6 GHz SCSOs has been measured[39] as a function of sampling time with a noise bandwidth of 10 Hz, and is shown in Fig. 4 by the open circles. The figure has three regions of interest. First, for sampling times (τ) less than 30 s, the Allan variance (σ^2_y) has the form $\sigma_y = 10^{-14}/\tau$. For the noise bandwidth of 10 Hz, σ_y is dominated by the low frequency bright lines in S_ϕ rather than its white component. Second, σ_y reaches its noise floor of 3×10^{-16} for sampling times

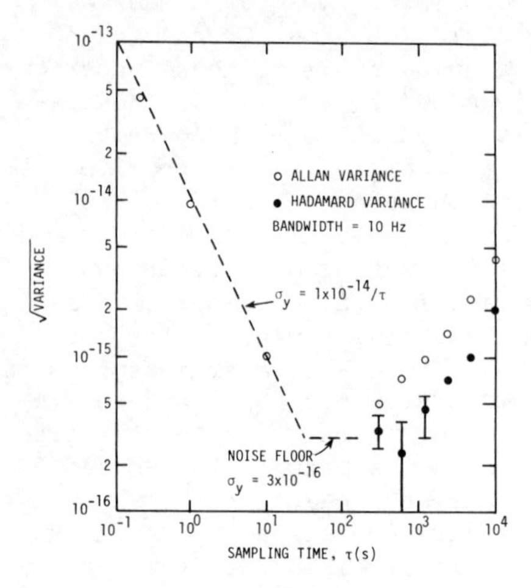

Fig. 4. The Allan and Hadamard variances as a function of sampling time, τ. The data are for a superconducting-cavity stabilized oscillator operating at 8.6 GHz.

between 30 and 100 s. Third, for $\tau > 100$ s, σ_y is dominated by drift in the SCSO, and therefore increases. The presence of drift is demonstrated by computing the second difference variance using three samples. This variance, represented by dots in Fig. 4, removes some of the drift and extends the noise floor to 1000 s.

Frequency Drift. The frequency drifts of the SCSOs have been measured both among the three SCSOs and with respect to an ensemble of cesium atomic frequency standards for periods of about twelve days. The linear drifts of the SCSO's with respect to the cesium frequency standards are typically about 10^{-14} day^{-1} and are both positive and negative.[20] At this time there is no evidence that the drift is the result of any structural changes in the cavities. The drift is probably accounted for by external influences on the cavity and the electronic circuits. In addition to linear drift, the frequency of the SCSOs has a daily variation with an amplitude of a few times 10^{-14}. These periodic variations are correlated with variations of tilt in the floor in which the SCSO dewar is mounted.

APPLICATIONS

Although all of the uses of superconducting cavities result from their low electromagnetic losses, substantially different advantages can be gained from this one property. There are applications where the superconductivity serves primarily to reduce the large amounts of power which would be dissipated in conventional devices; these applications are not discussed here. A second group uses the very high Q to realize exceptional filter characteristics such as narrow bandwidth and high ratio impedance transformation. Low loss also means that practically any object placed in a superconducting cavity will dominate its performance; thus many materials properties may be accurately transduced to a frequency with high resolution and low noise. As a result of their state-of-the-art stability, superconducting oscillators are beginning to find a wide variety of applications either directly as clocks, as components of oscillator systems, or in a variety of physics experiments involving clocks and time. Finally, there are some special applications which are rather unique and are discussed separately from the others.

RF Superconductivity. One of the first applications of superconducting resonators was to provide an RF power source of adequate stability to measure many of the properties of the superconducting resonator itself. In this application, a voltage-tunable microwave oscillator excites the superconducting resonator. In combination with a dc phase bridge, the superconducting resonator produces a frequency discriminator voltage which locks the voltage-tunable oscillator to the frequency of the superconducting resonator. This stabilization technique, which is also applied to superconducting accelerators and superconducting particle separators, avoids the need to provide rather expensive and sometimes cumbersome frequency synthesis of the RF power source from a high-quality quartz crystal oscillator.

Filters. Perhaps the most obvious application of high Q superconducting cavities is as narrow band filters. Below approximately 100 MHz, lumped element resonators are preferred, but at higher frequencies various types of cavities are used: helical structures below 500 MHz; quarter-wave reentrant structures between 500 MHz and a few GHz; low-order mode structures from a few GHz to 10 GHz or more; and Fabry-Perot resonators at higher frequencies. In addition to the obvious advantage of very narrow bandwidth, for example, 0.1 Hz at 10 GHz, cavities have very pure resonance modes. The geometry can often be chosen so that spurious modes are separated by an octave or more. In contrast, quartz crystal filters have spurious modes usually spaced only 0.1 percent from the desired mode.

Superconducting cavities can be mechanically tuned over a wide range, but the instability introduced by the tuning mechanism counteracts the most important property of the resonators. Very fine, stable tuning can be achieved by controlling the temperature of the resonator--the total fractional tunability is on the order of 10^{-10}. A second method which yields a tuning range of 10^{-4} is optical tuning by the photodielectric effect. This has been realized by placing a high resistivity semiconductor wafer in the gap of a quarter-wave reentrant cavity where the RF electric field is very high[40]. When a light beam is directed on the semiconductor its dielectric constant changes, thereby producing a large shift in the cavity resonant frequency. Step changes in the frequency can be made in less than 10 ms but the presence of the semiconductor degrades the band-width of the filter: a Q of 10^6 at 1 GHz has been achieved for such a device.

The narrowband, tunable filters which can be realized with superconducting cavity filters may be applicable to some communications or radar receivers. However, it is more likely that they will be used in highly specialized devices. For example, a superconducting resonator may be used to obtain strong coupling between an evaporated Josephson junction and an electromagnetic field. This is particularly valuable when the junction is used as an oscillator. If such a junction is not incorporated into a resonant structure, the large shunt capacitance shorts out high-frequency voltages. The theoretical maximum output power from a Josephson junction is $0.58\ I_o V$, where I_o is the critical current and V is the bias voltage.[41] Theoretically, a typical junction can output about 10^{-7} W at X band, but only about one percent of this is observed from waveguide-coupled evaporated junctions. In addition to performing the required impedence matching to the junction, the superconducting resonator also provides narrow band filtering of the output signal. Unfortunately, this property detracts from the major advantage of the Josephson junction oscillator--it has a tuning sensitivity 10^5 times greater than conventional oscillators.[42] In order to preserve this property, it will be necessary to use superconducting cavities with extremely agile frequency tuning.

Superconducting resonators may also be useful as high-ratio impedance transformers for use with superconducting antennas. This application also has the disadvantage of narrow bandwidth. Superconducting antennas are not considered advantageous today over conventional devices.[42]

Transducers. Superconducting cavities make attractive transducers for a variety of quantities because they introduce negligible perturbations and their output is usually in the form of a frequency which is easily and accurately measured. Since the frequency of a cavity resonator is approximately inversely proportional to one of its dimensions, it is very natural to use such a resonator to measure changes in that length. The resolution is limited by the frequency instability of the resonator; the best achieved stability of superconducting cavity oscillators corresponds to a random noise level of 3×10^{-16} cm. This performance can be improved in principle by designing the resonator so the controlling dimension is very small. It has been predicted that a quarter-wave reentrant cavity would permit the resolution of 10^{-17} cm for a one-second integration time.[43] However, it has not been demonstrated that the necessary Q can be obtained with such a resonator design. There is an interest today in superconducting cavity length transducers for the detection of gravity waves. In such an experiment, the cavity itself could be used as the antenna, but more likely, two cavities would be coupled to a traditional bar antenna in a way which would cancel a substantial fraction of the frequency noise in the exciting oscillator. By appropriate design, the resonator dimensions and therefore frequency can be made sensitive to extremely small changes in Earth's gravity. A gravimeter with a sensitivity of 10^{-10} g and a drift of 5×10^{-8} g/day has been reported.[44]

Substantial use has been made of superconducting cavities to measure a variety of properties of materials at low temperature. The technique is to place a sample within the resonator and to measure either the frequency or Q as a function of the parameters of interest. The advantage of this technique is that it does not require ohmic contacts and can be used with randomly shaped, powdered or liquid samples. Dielectric constants very near unity and loss tangents as small as 10^{-9} can be measured because of the high stability and very low losses of the superconducting cavity itself. Many semiconductor properties have been measured including relaxation time, lifetime, Fermi level, trap ionization energy, trap density, capture cross section, free carrier density and trap population.[45]

Some properties of liquid helium have also been studied this way. The thermal expansion has been measured with a sensitivity, in terms of fractional density change, of 4×10^{-9}.[46] This sensitivity is sufficient to yield quantitative data concerning the dispersion relation for thermal phonons in liquid helium. The damping of oscillations of the liquid He through a small orifice can be studied by using the frequency of a superconducting cavity to sense the level of liquid helium within it.[47] This data has been used to study the quantization of vorticity in superfluid helium.

Several interesting and useful devices can be implemented using superconducting cavities. A thermometer can be made for the temperature range from .25 to .6 K by filling a cavity with He^3 vapor in equilibrium with the bulk liquid.[46] Changes in density, which are reflected in frequency changes, are interpreted in terms of the temperature of the gas. The accuracy of such a thermometer is estimated to be 0.2 percent. A nuclear radiation detector can be made by placing a properly doped semiconductor crystal on the stub of a quarter-wave reentrant cavity.[48] Below 70 K, the charge carriers created by the absorption of radiation are trapped for very long times at sites in the forbidden band. As a result, the frequency of the cavity shifts proportionally to the total absorbed dose. A detector for low levels of light can be implemented in a similar way only in that case the frequency shift results from the photodielectric effect.

Oscillators and clocks. The excellent spectral purity and medium term stability (up to about one day) of superconducting oscillators have numerous applications. These are divided into three categories in the following discussion: oscillators used as components of instruments; oscillators used as clocks to provide timing functions for complex instrument systems; and oscillators used to perform experiments based directly on clock performance.

Oscillators with very good spectral purity (short-term stability) are important elements of many instruments and measurement techniques. One example is the use as the source oscillator for frequency multiplication from microwave to infrared or higher frequencies. The process of multiplication by an integer n increases the phase noise power by n^2. This creates a severe practical problem because once the integrated white phase noise becomes comparable to 1 rad, it is no longer possible to identify the coherent signal component.[34] Traditional multiplication schemes require the use of several intermediate steps with independent oscillators at each step to reduce the phase noise to an acceptable level. Superconducting oscillators can be used in principle to accomplish high order multiplication in a single step. They have two important advantages in this application: they can operate at frequencies at least as high as 10 GHz and theoretically can produce a signal whose spectral purity is limited by the characteristics of the multiplier. For example, it has been predicted that a state-of-the-art commercial 5 MHz quartz crystal oscillator may be multiplied to .5 THz before the carrier is lost, but the same signal when filtered by a 10 GHz superconducting cavity with $Q = 10^{10}$ can be multiplied directly to 100 THz.

One of the authors (Stein) is testing a superconducting-cavity parametric oscillator at 9.2 GHz for use at the National Bureau of Standards.[27] Also Viet Nguyen at the University of Paris IX (Orsay) is constructing an SCSO for the purpose of multiplication. Both of these devices employ a TM_{010}-mode niobium resonator with two coupling ports. One of the two ports is used by the oscillator or stabilization circuit, and the second port is a transmission port for the useful output power. Both of these devices should have substantially reduced $S_{\phi}(f)$ for higher Fourier frequencies than that shown for the SCSO in Fig. 1. The improvement comes from the filtering character of the cavity at the transmission port.

Certain types of radar also depend critically on the spectral purity of their local oscillator. Return signals from nearby stationary clutter mix with the phase-noise sidebands and limit the signal-to-noise ratio of the true Doppler signal. For example, if a 1 GHz radar has target velocity detection down to 40 m/s and Doppler bandwidth of 10 KHz, then in order to achieve 80 dB sub-clutter visibility it is necessary to use a local oscillator whose phase noise is more than 120 dB below the carrier for Fourier frequencies greater than 200 Hz.[49]

A third application for short-term stable oscillators is as flywheel oscillators in atomic frequency standards. Since the time domain frequency stability, $\sigma_y(\tau)$, cannot improve faster than $1/\tau$, the medium term performance of the standard is limited by the flywheel oscillator if its stability is worse than that of the atomic frequency discriminator at the attack time of the feedback loop. Quartz crystal oscillators do not degrade the performance of current atomic standards, but if expected improvements in these standards are made, then improved flywheel oscillators will be needed and superconducting oscillators are a possible candidate. Flywheel oscillators are also used for autotuning hydrogen masers, a process whereby cavity-pulling and spin exchange frequency shifts are simultaneously reduced.[50] Present autotuning systems utilize a pair of masers, one of which could be replaced by a superconducting oscillator, thereby realizing a significant cost reduction.[51]

Superconducting oscillators can also be used to provide time for complex instrumentation and measurement systems such as radio astronomy and radar ranging. The desirability of superconducting oscillators for these applications range from cost savings to the potential for significant improvements in performance.

Very long baseline interferometry (VLBI) using independent clocks may have the following clock performance requirements for certain types of experiments: initial 1 μs synchronization of the start of recording; total time error of less than 1 ns over a five-hour observation period to insure that all the data have the same initial offset error; and sufficient coherence to guarantee that the recorded signals can be cross-correlated.[52] For a 10 GHz system and an observation time of one hour, the coherence requirement is met by an oscillator with a noise floor $\sigma_y = 5 \times 10^{-15}$. The required noise floor decreases inversely with both the operating frequency and the observation time. The requirements are approximately met by both hydrogen frequency standards (masers and passive devices) and superconducting oscillators. The masers are very expensive (~$250 K) whereas the other two devices appear to cost only about one third as much and may have the possibility of extending the frequency for some types of VLBI to the region of 100 GHz. One of the authors (Turneaure) has analyzed the phase coherence of two standard SCSOs at 8.6 GHz for application to VLBI experiments with 30-and 60-minute cycling times.[39] For 30 min., the rms phase error is 31 m rad; and for 60 min, it is 87 m rad. Turneaure at Stanford is currently constructing an SCSO for VLBI at Owens Valley Radio Observatory.

Various types of navigation systems need state-of-the-art clocks. Since both the navigation requirements and the techniques are somewhat flexible, it is difficult to place fixed requirements on clock performance. Typical performance goals are discussed here for two navigation systems.

The NASA Deep Space Net (DSN) utilizes a network of radar stations to track spacecraft which have left earth orbit. The current capability of the system is approximately 5 m range resolution and 1 μrad angular resolution. Planners foresee the need for approximately an order of magnitude improvement in resolution for some missions which will be flown in the early 1980s such as Jupiter Orbiter (JOP).[53] The clock stability requirements depend on the method of range measurement. One approach which has been suggested is to replace most of the coherent (two way) Doppler ranging with non-coherent (one way) Doppler measurements. The latter technique, also called wideband VLBI, has the advantage that it substantially reduces the tracking time for accuracy comparable to current coherent tracking methods. However, it places the most stringent requirements on clock performance. When daily calibrations are used, the frequency must be constant to 1.5×10^{-14} over one day. Weekly calibrations are preferable to reduce cost and operating time but the frequency stability requirement becomes 3.2×10^{-16} over one day.[53]

An experiment has been proposed to use the DSN to detect gravitational waves. The passage of a gravitational wave pulse past the earth and spacecraft produces an identifiable signature in the range information. This experiment places the strictest stability requirements on the frequency standards. To do a feasibility experiment the frequency stability must be 3×10^{-16} for the duration of the experiment, 40 s to 4000 s, which can be achieved with a state-of-the-art superconducting oscillator system. A desirable experiment would require frequency stability of 3×10^{-18}.[54] It is probable that a superconducting oscillator using a cavity with $Q = 10^{11}$ would eventually be capable of reaching this performance level.

The Global Positioning System (GPS) is a multisatellite system intended to provide earthbound navigation with a precision of about 10 m. Each satellite carries high stability clocks which need to keep time to 10 ns over 10 days. Several hydrogen frequency standards are being developed for this purpose. Depending on future system requirements, superconducting oscillators could become desirable for this application.

There are several fundamental physics experiments which are based directly on superconducting oscillator performance. Two of these, a red shift experiment and a fundamental constants experiment, are performed by comparing the frequency of a superconducting oscillator to the frequency of an atomic standard based on a hyperfine transition such as cesium or hydrogen. In the red shift experiment one looks for a term in the frequency ratio that has a period of one solar day. Because of the Earth's rotation in the Sun's gravitational field, various theories of gravity predict

$$\frac{\nu_{\text{hyperfine}}}{\nu_{\text{superconducting}}} = A[1 + 10^{-12}(\Gamma_o - \tfrac{1}{2}T_1)\cos(2\pi\tau/1 \text{ solar day})] \qquad (8)$$

where Γ_o and T_1 are zero for any metric theory of gravity such as the general theory of relativity.[54] With available standards it would in principle be possible to set an upper limit of 10^{-3} on $\Gamma_o - \frac{1}{2}T_1$. The Eotvos experiment already places a limit of 10^{8} on Γ_o and may place a limit on T_1.

If instead of analyzing the data for diurnal effects, they are fit to a linear drift model, then the same experiment may be analyzed to yield an upper limit on the time rate of change of the fine structure constant. The ratio of the frequencies of a hyperfine standard to a superconducting oscillator is

$$\frac{\nu_{\text{hyperfine}}}{\nu_{\text{superconducting}}} = Bg\left(\frac{m}{M}\right)\alpha^3 \qquad (9)$$

where m is the mass of the electron, M is the mass of the nucleus, and g is the gyromagnetic ratio of the nucleus; B is a constant. By comparing a superconducting oscillator to a cesium standard for 12 days it has been determined that $(1/\alpha)(d\alpha/dt)$ is less than 4×10^{-12}/year with 68 percent probability.[20] The quality of this experiment was determined by the data link connecting the laboratories where the standards were located. Significant improvements may result from a comparison of superconducting oscillators and hydrogen standards in the same laboratory. Although astronomical and geophysical measurements have set a tighter upper limit, they have the disadvantage of averaging possible changes over periods on the order of 10 percent of the age of the universe.

A laboratory experiment has been proposed to use a superconducting resonator excited by a superconducting oscillator to measure the Lense-Thirring effect--the dragging of inertial frames by a rotating mass. A toroidal superconducting waveguide is centered on the rotation axis of an axisymmetric object. The wave travelling in the same direction as the rotating body takes less time to complete one round trip than the counterrotating wave. As a result, the interference pattern rotates around the waveguide. For a 5000 Kg mass with 50 cm radius rotating at an angular velocity of 2×10^{3} rad/s, the angular velocity of drag has been estimated to be 2×10^{-20} rad/s.[56] The required waveguide Q is 5×10^{16} and the stability of the exciting oscillator must be 1×10^{-17}. This stability is probably achievable, but it is impossible to say if such high Qs can be achieved. There are many other problems at least as difficult as these, making this experiment extremely problematic.

<u>Special Devices</u>. Occasionally superconducting cavities can be used to solve some unusual problems. One example of this is the

reduction of the "cavity phase shift" problem in cesium beam frequency standards. At the present time, the most significant limitation in the accuracy of cesium beam frequency standards is the extent to which the phase difference of the RF fields in the two interaction regions can either be nulled or measured. If superconducting cavities were used, the variations of the microwave phase across the apertures of the cavities would become very small and it would be possible to measure the intercavity phase shift with high precision. This approach is not being tried at the present time because of the difficulty of using cryogenic cavities in an atomic beam device and because there is another promising technique with fewer difficulties.

Another novel suggestion is a superconducting cavity gyroscope. One possible configuration is the toroidal microwave cavity which was described earlier in the discussion of the Lense-Thirring effect. The position of the nodes in such a cavity experiences a phase shift proportional to the rotation speed of the cavity and the frequency of the exciting radiation. This is just the Sagnac effect which is used today in laser gyroscopes. Because of the difficulty in optical detection of fringes, laser gyroscopes are constructed today in the form of ring oscillators. These devices must be biased in order to overcome a dead zone at low rotation rates; systematic errors introduced by the bias degrade the performance of the gyroscope. The superconducting version of such a gyroscope is less sensitive by a factor of 10^4 to 10^5 because of the wavelength difference. However, this is offset by the fact that it is possible to detect a much smaller fraction of a fringe at 10 GHz than at visible frequencies. Consequently, it may be possible to construct a Sagnac-type gyroscope (using superconducting cavities) which operates in the interferometric mode with no dead zone and is competitive in sensitivity with the best mechanical gyroscopes.

CONCLUSIONS

There are desirable applications for superconducting cavities with Qs up to 10^{17} and for superconducting oscillators with stabilities as good as 3×10^{-18}. The best Q achieved to date is 5×10^{11} and the best stability is 3×10^{-16}. Since it is generally possible to find the center of a resonance line to one part per million, it is reasonable to expect a stability of 2×10^{-18} will be achieved using present technology within ten years. On the other hand, Q improvement to the 10^{17} level will require major advances in superconducting technology. The likelihood of it happening within the next ten years, if ever, cannot be reasonably assessed.

REFERENCES

1. J. P. Turneaure and N. T. Viet, Appl. Phys. Letters 16, 333 (1970).
2. J. P. Turneaure, IEEE Trans. Nucl. Sci. NS-18, 166 (1971).
3. P. Kneisel, O. Stoltz and J. Halbritter, J. Appl. Phys. 45, 2296 (1974).
4. H. Diepers and H. Martens, Physics Letters 38A, 337 (1972).
5. R. Benaroya, B. E. Clifft, K. W. Johnson, P. Markovich and W. A. Wesolonski, IEEE Trans. Mag. MAG-11, 413 (1975).
6. P. H. Ceperley, I. Ben-Zvi, H. F. Glavish and S. S. Hanna, IEEE Trans. Nucl. Sci. NS-22, 1153 (1975).
7. B. Hillenbrand and H. Martens, J. Appl. Phys. 47, 4151 (1976).
8. J. M. Pierce, J. Appl. Phys. 44, 1342 (1973).
9. J. Benard, private communication.
10. D. C. Mattis and J. Bardeen, Phys. Rev. 111, 412 (1958).
11. J. P. Turneaure and I. Weissman, J. Appl. Phys. 39, 4417 (1968).
12. P. Kneisel, H. Rupter, W. Schwarz, O. Stoltz and J. Halbritter, IEEE Trans. Mag MAG-13, 496 (1977).
13. K. S. Bagdasarov, V. B. Braginskii and P.I. Zubietov, UTP Letters 3, (1977).
14. J. Halbritter, Externer Bericht 3/68-8 (Kernforschungszentrum, Karlsrake, (1968).
15. J. P. Turneaure, unpublished.
16. P. H. Ceperley, Ph.D. Dissertation (Stanford University, 1971).
17. C. Lyneis and J. P. Turneaure, private communication.
18. V. B. Braginskii, private communication.
19. S. R. Stein and J. P. Turneaure, Electron. Lett. 8, 321 (1972).
20. J. P. Turneaure and S. R. Stein, on Atomic Masses and Fundamental Constants, Vol.5, eds. J. H. Sanders and A. H. Wapstra (Plenum, NY, 1976) p.636.
21. M. S. Khaikin, Instrum. and Experimental Tech. No. 3, 518 (1963).
22. Nguyen T. Viet, C. R. Acad. Sci. Paris 269B, 347 (1960).
23. W. H. Higa, NASA Technical Memorandum 33-805, (1976).
24. J. J. Jiminez and Septier, in Proc. 27th Annual Symposium on Frequency Control (Elec. Ind. Assoc., Wash., D.C., 1973) p.406.
25. I. I. Minakova, G. P. Minina, V. N. Nazarov, V. N. Panov and V. D. Popel'nynk, Vestnik Moskovskogo Universiteta. Fizika, 31, 67 (1976).
26. V. B. Braginskii, I. I. Minakova, and V. I. Panov.
27. S. R. Stein, in Proc 29th Annual Symposium on Frequency Control (Electronic Ind. Assoc., Wash., D.C., 1975) pp. 321-327.
28. M. Uenohara, in Handbuch der Physik 23, (Springer-Verlag, Berlin, 1967), pp.1-4.
29. W. G. Matthei and M. J. McCormick, Proc. IEEE 53, 488 (1965).
30. K. Schuneman and B. Schiek, Electronic Letters 7, 618 (1971).
31. L. S. Cutler and C. L. Searle, Proc. IEEE 54, 136 (1966).
32. S. R. Stein and J. P. Turneaure, Proc. IEEE 63, 1249 (1975).
33. S. R. Stein and J. P. Turneaure, in Proc. 27th Annual Symposium on Frequency Control (Elec. Inc. Assoc., Wash., D.C.,1973), p.414
34. F. L. Walls and A. Demarchi, IEEE Trans. Instrum. Meas.IM-24, 210(
35. James A. Barnes et al., IEEE Trans. Instrum. Meas. IM-20, 105 (19

36. E. Hafner, Proc. IEEE 54, 1/9 (1966).
37. K. Kurokawa, An Introduction to the Theory of Microwave Circuits (Academic Press, New York, 1969), pp. 380-397.
38. S. R. Stein, Ph.D. Dissertation, Stanford University (1974).
39. J. P. Turneaure, unpublished.
40. J. L. Stone, W. H. Hartwig and G. L. Baker, J. Appl. Phys. 40, 2015 (1969).
41. Todd I. Smith, J. Appl. Phys. 45, 1875 (1974).
42. 1976 NAVY Study in Superconductive Electronics (ONR Report No. NR319-110) Arnold H. Silver, ed., (Office of Naval Research, Arlington, VA, 1976).
43. G. J. Dick and H. C. Yen in Proc. Applied Superconductivity Conference Annapolis, Maryland, 1972) pp. 684-686.
44. B. I. Verkin, F. F. Mende, A. V. Trubitsin, I. N. Bonderenko, and V. D. Senenko, Cryogenics, 519 (1976).
45. James J. Hinds and William H. Hartwig, J. Appl. Phys. 42, 170 (1971).
46. J. E. Berthold, H. H. Hanson, H. J. Mares, and G. M. Seidel, Physical Review B 14, 1102 (1976).
47. Walter Trela, Ph.D. Dissertation, Stanford University, 1966.
48. C. W. Alworth and C. R. Haden, J. Appl. Phys. 42, 166 (1971).
49. D.L.H. Blomfield, private communication.
50. H. E. Peters, T. E. McGunegal and E. H. Johnson, in Proc. 15th Annual Symposium on Frequency Control (Electronics Inc. Assoc., Wash., D.C., 1968) pp. 464-492.
51. Victor S. Reinhardt, Donald C. Kaufmann, William A. Adams, John J. DeLuca and Joseph L. Saucy, in Proc. 30th Annual Symposium on Frequency Control (Electronics Ind. Assoc., Wash., D.C., 1976) pp. 481-488.
52. Wilfred Konrad Klemperer, Proc. IEEE 60, 602 (1972).
53. D. W. Curkendall, private communication.
54. F. B. Estabrook, private communication.
55. Clifford M. Will, in Second Symposium on Frequency Standards and Metrology (Copper Mountain, 1976) pp. 519-530.
56. Vladimer B. Braginskii, Carlton M. Caves and Kip S. Thorne, Physical Review D 15, 2047 (1977).

BIASING FOR PRECISE FREQUENCY IN A FREQUENCY-AGILE JOSEPHSON JUNCTION

Michael R. Daniel, M. Ashkin and M. A. Janocko
Westinghouse Research and Development Center, Pittsburgh, PA 15235

ABSTRACT

Analog computer studies of a Josephson junction show that the microwave-induced steps in the dc current-voltage curve may be used as ideal (zero impedance) voltage sources with which to bias a second junction. Biased on one of these steps, the second junction is a microwave oscillator of precisely determined frequency. Because the biasing junction impedance → 0 on a step, it contributes no noise to this second junction. A conceptual design using Dayem bridges fabricated as an integral part of a coplanar waveguide structure is proposed.

INTRODUCTION

Although one of the well-established features of a Josephson junction is the very accurate relation between its frequency of oscillation and the dc bias potential, little if any work has been reported on developing a junction as a practical microwave oscillator of precise frequency. The extreme sensitivity of frequency f to voltage V in a junction, $f/V = 2e/h = 0.483591$ GHz/μV, necessitates ways of providing a very accurate bias voltage. One of us (M. A. Janocko) suggested that the dc steps in the I-V curve of a junction irradiated by a microwave source provides an ideal and highly stable voltage source with which to bias a second junction. The stability of this voltage source would of course depend on the frequency stability of the microwave source. This suggests the use of a superconducting microwave cavity oscillator for which a Q factor of 10^{10} is attainable and hence the voltage stability of the biasing junction would be of the order of 1 part in 10^{10}. The second junction then acts as the microwave source of precisely-determined frequency. Additionally, the linewidth of the radiation emitted by a junction, whether due to thermal or shot noise, is proportional to R_D^2, where R_D is the dynamic resistance on the I-V curve at the point of bias.[1] Since biasing is to be on a step then $R_D \to 0$ for the biasing junction, and this would contribute no noise to the second junction.

SIMULATION RESULTS

The results presented in this section were obtained from an analog computer simulation of the well known resistively-shunted model (RSJ model). Details of this simulation will be presented elsewhere;[2] it was done for a junction biased, for example purposes, with an arbitrary but convenient irradiation frequency of 3 GHz.

ISSN: 0094-243X/78/214/$1.50

Self Biasing

It was initially thought that a single junction could act in a self-biasing configuration and simultaneously as a radiation source. However, this resulted in a very complicated spectrum due to the highly nonlinear behavior of a junction causing mixing of the 3 GHz frequency with the natural junction frequency. Substantial frequency components every 3 GHz were present even several octaves beyond the Josephson frequency (determined by the bias point).

External Biasing of a Junction by a Complementary Junction

For two junctions coupled by a series inductance, [this would be the likely configuration in a practical circuit wherein a dc bias potential must be coupled from the first junction (J_1) to the second (J_2)] the output from J_2 was stepped in frequency as a function of the dc bias of J_1 - Figure 1. However, the spectral output shown in Figure 2 was complicated by frequency modulation due to rf coupling between the junctions. Fortunately the inclusion of a single circuit, a low-pass filter between J_1 and J_2, overcame the difficulty. Figure 3 shows the marked improvement in spectral purity obtained with a filter. In this particular case the filter was such that it cut out frequencies above 1/10 GHz at an attenuation rate of 40 dB per decade of frequency. It isolated J_2 from the 3 GHz injected into J_1 and thus J_2 behaved as a single junction but biased by J_1.

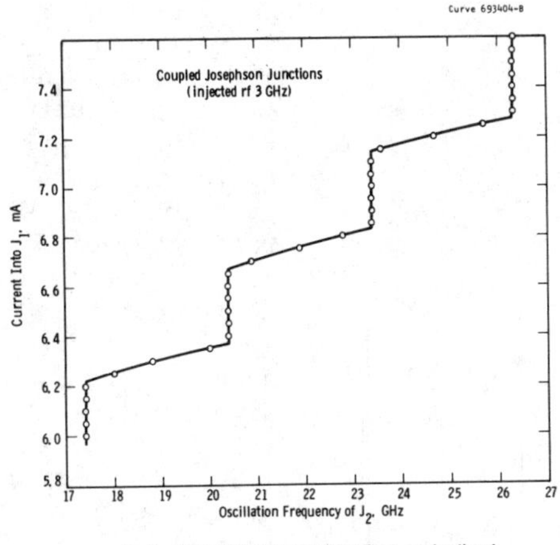

Fig. 1 – The oscillation frequency of junction J_2 as a function of the dc current into J_1

The harmonic content of the rf power in J_2 (see Table 1) was estimated from the spectral response shown in Figure 3. If this junction, with a resistance of 0.025Ω, were coupled to a 50Ω waveguide, a 27 dB mismatch loss would result and only 3.7 x 10^{-11} watts would be detected in the waveguide. This level of power is what is typically obtained experimentally.

The simulation showed the dc bias step height to be determined by the J_1 junction resistance R_1 and to be independent of the second junction resistance R_2. The step height decreased as R_1 increased, but up to X band frequencies R_1 = 0.1Ω was a convenient circuit value. It was noticed that for R_2 > 0.2Ω inadequate damping occurred and the circuit broke into oscillation which was superposed on the Josephson oscillation. Changing the values of the inductance L and the junction capacitance values, C_1 and C_2 did nothing to suppress these oscillations, and thus 0.2Ω is a practical upper limit for R_2 for the present circuit. The two junctions linked by an inductance, L, essentially form a SQUID in which the magnetic flux, ϕ,

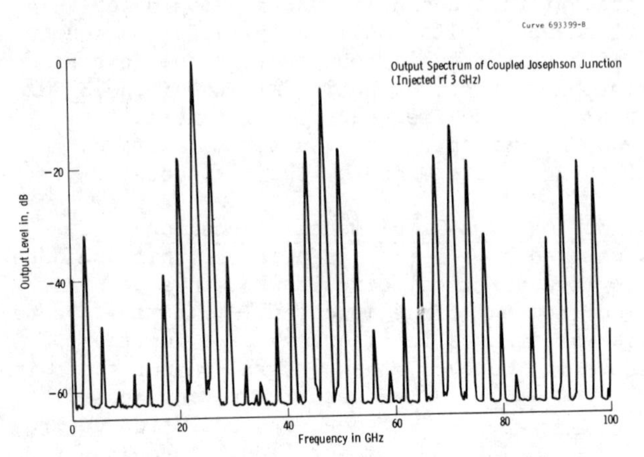

Fig. 2 – The spectral response of junction J_2 when biased on the 8th step of junction J_1

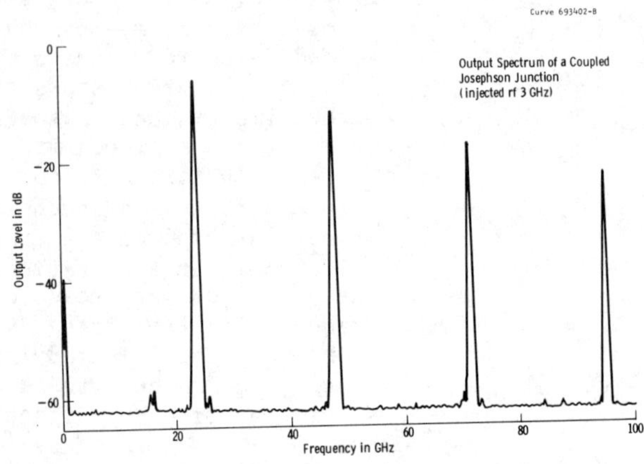

Fig. 3 – The spectral response of junction J_2 when biased on the 8th step of junction J_1 and isolated from it by a low pass filter

enclosed by the circuit of J_1, J_2 and L is quantized in integral multiples of the flux quantum $\Phi_o = 2.067854 \times 10^{-15}$ Webers. At temperature T the mean square fluctuation in flux $(\delta\Phi)^2$ which arises from current fluctuations in L is given by $(\delta\Phi)^2/2L^2 = kT/2$ where k is Boltzmann's constant and the equation results from the equipartition of energy theorem. The voltage separation between steps in a dc I-V curve is just $\Phi_o f$ at frequency f, and if we argue that $\delta\Phi$ should not exceed $1/2\ \phi$ for a step structure to exist, then at 4 K, L $\gtrsim$ 20 nH. We saw no evidence of noise for L $\lesssim$ 10 nH from the simulation results. It is likely that the effects of current noise in L were unintentionally removed by the use of the low-pass filters which were used to record the dc I-V curves and were essential to separate the ac and dc voltages for recording purposes. Hence, there are grounds for accepting 20 nH as an upper limit to the inductive coupling between J_1 and J_2 for operation at 4.2 K above which noise should destroy the step structure generated by J_1. This has not been verified by the present analog simulation, and the opportunity for further work exists. There should be no noise limitation on the values of R_1 for the reasons already stated. Shot and thermal noise due to R_2 will contribute to frequency broadening in J_2 and can be calculated using the equations on page 475 of reference 1, for example. The junction capacitance values, taken to be in the range 0.1 to 1 pF, have little effect upon performance since their reactive impedances are much greater than R_1 or R_2.

Table 1

Energy Spectrum of a Coupled Junction Biased on the 8th Step, -- See Figure 3

Harmonic	Energy, nW	Percentage of Total Energy
First	18.4	72.0
Second	5.3	20.8
Third	1.3	5.0
Fourth	0.4	1.6

CONCEPTUAL DESIGN

The inherent broadband nature of the device can be met by the use of a coplanar waveguide (CPW). Such a waveguide will function satisfactorily probably up to 18 GHz. The use of a CPW then suggests a Dayem bridge type of junction for several reasons. The ground planes and center conductor of a CPW are all one surface and thus a thin film Dayem bridge can readily form an integral part of such a structure and can be fabricated in sequence with the waveguide. Forming such an integral part of the structure, the Dayem bridge can be expected to couple satisfactorily to the CPW. The fabrication technique can comprise standard thin film and photolithographic processes used for microcircuit technology. As an example it is suggested that a sputtered Nb film on sapphire could form the basis for a device.

The necessary rf isolation between the junctions is achieved by a section of high impedance (say 100Ω) CPW between them. Calculation shows (Figure 4) that the two-stage mismatch formed by 0.1Ω junctions terminating a 1 cm length of 100Ω CPW gives more than 30 dB of isolation over most of the range 1 to 18 GHz. Some degree of matching to the junctions can be achieved using tapered sections of waveguide. Preliminary experiments incorporating these suggestions are presently in progress. Figure 5 shows a plan view of a CPW with junctions, which, because of the CPW symmetry, requires two junctions for each function -- bias and oscillator.

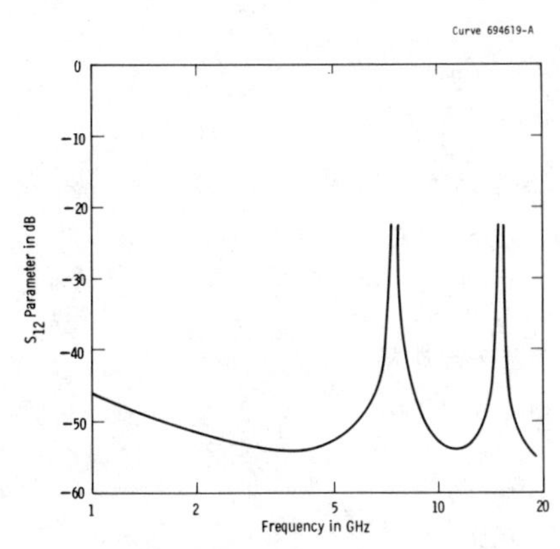

Fig. 4-Attenuation characteristics of a 350 mil section of 100Ω coplanar waveguide terminated on both ends by 0.1Ω junctions and 0.5 pF of capacitance

Amplification of the oscillator junction output is now possible at liquid helium temperatures with a GaAs FET.[3] Employing a low ionization dopant of only a few millivolts, such FET's are capable of giving 15 dB of gain with 1 to 2 dB of noise at X-band. In a grounded

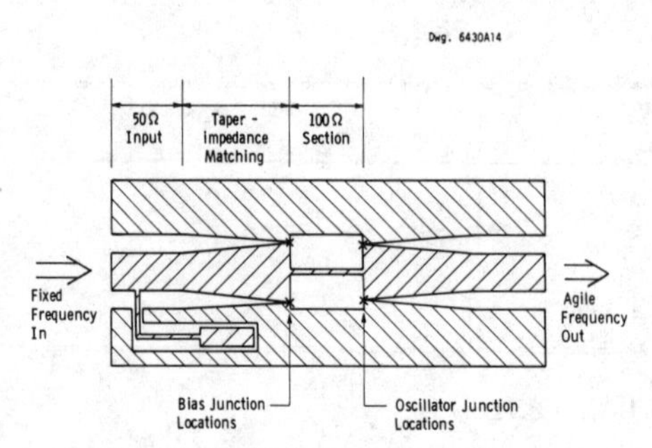

Fig. 5—Conceptual design of the coplanar waveguide with junction location

gate configuration, two stages of amplification would raise a nW level of power from an 0.1Ω impedance junction to a μW level into a 50Ω waveguide.

REFERENCES

* The analog simulation work reported here was supported by the Office of Naval Research.

1. R. Adde, IEEE Trans. on Microwave Theory and Techniques, MIT-25, 473 (1977).

2. Michael R. Daniel, M. Ashkin and M. A. Janocko, submitted to J. Appl. Phys., (1978).

3. R. Zuleeg, private communication.

SUPERCONDUCTING MICROSTRIP FILTERS

R. A. Davidheiser
TRW Systems, Redondo Beach, Ca. 90278

ABSTRACT

Control of the non-zero radiation resistance inherent in metal strip transmission lines results in high-current densities and low-transmission quality factors. Superconducting metals reduce the ohmic losses to approximately the dielectric losses associated with a sapphire substrate. Quality factors of 100,000 are possible. Applications of this effect in communication systems are discussed for the cases of resonant filters and frequency sources.

INTRODUCTION

Metal strip transmission lines have found use in many circuits in the microwave and millimeter frequency range and have been thoroughly characterized in the literature. Planar and electrostatic in character, their simplicity has allowed analysis for a broad range of functions and geometries. Control of circuit impedance values is greatly simplified compared to that of metal and dielectric waveguides. In addition the size and ease of fabrication of metal strip circuits are attractive.

The most common of metal strip lines, microstrip, supports pure TEM modes only in the limit of zero substrate thickness. A microstrip line of finite thickness couples energy to non-TEM modes whose energy content is expressed as a change in the effective dielectric constant of the transmission line [1]

$$\varepsilon_{eff}(h/\lambda) = \varepsilon_{eff}(o) + 2.36(h/\lambda)^{1.33} \qquad (1)$$

where h is the substrate thickness and λ is the free space wavelength. These non-TEM modes couple to free space, and thus at an open termination the radiation resistance of a microstrip line is non-zero and equal to[2]

$$R_{rad} = 240\pi^2 (h/\lambda)^2 F(\varepsilon') \simeq 600\pi^2 (h/\lambda) \frac{1}{\varepsilon} . \qquad (2)$$

For the special case of a 50Ω line one half wave in length open-circuited at both ends the quality factor with respect to radiation to free space is thus

$$Q \text{ (radiation)} = 6.6\text{X}10^{-3} (\lambda/h)^2 \varepsilon . \qquad (3)$$

ISSN: 0094-243X/78/219/$1.50

The quality factor associated with ohmic losses may be approximated by the expression

$$Q\ (\text{ohmic}) = 800(h/\lambda)R_s^{-1}. \qquad (4)$$

For a metal surface resistance, R_s, of $2.2\text{X}10^{-2}\ \Omega$ (copper, 10 GHz, 293K) equations (3) and (4) imply a maximum Q of 230.

The Bardeen-Mattis expression for the BCS microwave surface resistance at finite temperatures is[3]

$$R_s \propto \omega^2/T \exp -[\Delta(T)/k_B T] .$$

Additional residual losses have resulted in surface resistance limits at 10 GHz of $10^{-5}\Omega$ for electroplated lead and $10^{-7}\Omega$ for vacuum-fired niobium.[4] Equations (3) and (4) now lead to maximum Q values of $4\text{X}10^4$ and $8.4\text{X}10^5$ respectively for lead and niobium microstrip.

The dielectric loss tangent of the substrate material induces further loss. Loss tangents of 10^{-4} for quartz and 10^{-5} to 10^{-7} for sapphire are reported. Adding ohmic, radiative, and dielectric losses, quality factors of $3\text{X}10^4$ to $4\text{X}10^4$ are expected for superconducting lead half wavelength resonators at 10 GHz. Niobium resonator Qs could vary from $8.9\text{X}10^4$ to $7.8\text{X}10^5$.

MICROSTRIP FILTERS

The theoretical insertion loss of a microwave filter at centerband is[5]

$$L(\text{dB})^{\text{center}} = 4.34(Q_L/Q_o) \sum_{i=1}^{n} g_i$$

$$\simeq 5n(Q_L/Q_o)$$

where Q_L is approximately the inverse of the filter's percentage bandwidth, Q_o is the unloaded resonator quality factor, n is the number of filter poles, and g_1 through g_n are approximately one. The attenuation characteristic of a Tchebycheff type filter outside the passband is well known[6] and a 10 pole, 0.1 dB ripple, 0.50% passband filter at 10 GHz results in the frequency characteristic shown in figure 1a. Assuming resonator Qs of $1\text{X}10^5$, the insertion loss between 10.00 and 10.05 GHz is 0.2 dB. Figure 1b shows the metal pattern on the upper surface.

The frequency response of the type shown in figure 1 results from proper coupling between each of the resonators. The coupling parameter may be optimized by variation of the resonators' lateral separation. To determine the coupling coefficients (expressed as Q_{ext}) as a function of the ratio of lateral separation to substrate

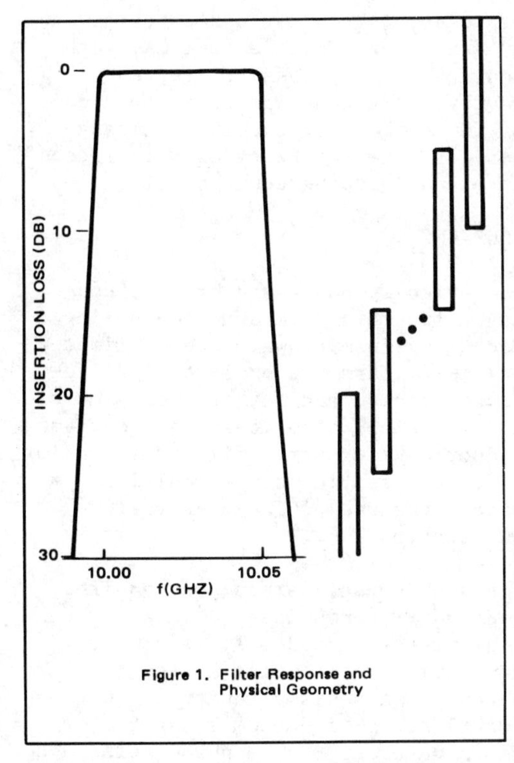

Figure 1. Filter Response and Physical Geometry

thickness (S/h), two port single-pole microstrip resonators were fabricated for use at 1 GHz. The first set was fabricated on .050" thick Epsilam 10 (alumina-filled teflon-coated with 1.4 mil copper sheets, ε=10) the second on .025" Epsilam 10. The insertion loss of the coupled resonators at their resonant frequency is[7]

$$I.L. = 10 \log \left[\frac{Q_{ext}^2}{4Q_L^2}\right]. \quad (5)$$

The insertion loss and loaded Q of each of the filters was measured and Q_{ext} determined from equation (5). The unloaded Q, $Q_o = 1/(1/Q_L - 2/Q_{ext})$, and coupling coefficient, $\beta = Q_o/Q_{ext}$, were also determined. As expected Q_o of the resonators is approximately proportional to the substrate height. Q_{ext} is a function of S/h alone. Input VSWR data was used for the plots of β; the agreement with values of β derived from (5) is good for small β, but discrepancies appear as Q_{ext} approaches $2Q_L$. Theoretical coupling based on Laplacian[8] and Green's function [9] techniques is also plotted in figure 2.

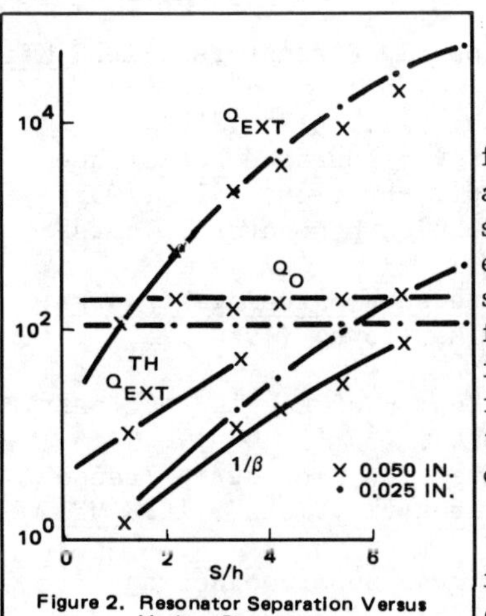

Figure 2. Resonator Separation Versus Various Parameters

FREQUENCY SOURCES

A microwave oscillator, constructed from a superconducting microstrip filter and a microwave amplifier, should have a spectral purity (short-term stability) equivalent to quartz and atomic frequency sources. This property arises primarily from the filter's high Q. The long-term frequency stability of the superconducting microstrip oscillator may be superior to the quartz or superconducting cavity oscillator.

Contributions to the long term frequency drift of the superconducting cavity oscillator include: thermal expansion of the cavity, temperature-dependence of the London penetration depth, expansion of the cavity due to electromagnetic radiation pressure, gravitational strain, and plastic flow of the metal walls.[10] Each of these mechanisms, with the exception of thermal expansion, contribute less to frequency drift than in the microwave cavity. Frequency drift associated with a temperature fluctuation of 10^{-5}K is 1 part in 10^{14}. Changes in the London penetration depth affects the frequency only to second order through the effective substrate height and equation (1).

CONCLUSIONS

The high quality factors theoretically possessed by superconducting metal strip resonators may allow thin film superconducting metals to find at least two applications in microwave and millimeter-wave communication systems. Low-loss filters are employed for frequency isolation, low-noise receiver inputs, and for frequency multiplexing in channelized receivers. The small size, cost, and extreme frequency selectivity of superconducting microstrip filters may allow large-scale multiplexing schemes to replace time domain multiple access (TDMA) systems; and facilitate the use of inexpensive low-data-rate (hence low-power) ground stations.

Low-phase noise and frequency-stable oscillators are required for both digital-phase shift-keyed and FM communication links. The lack of an acceptable source in the microwave and millimeterwave frequency range has forced the development of complex and costly frequency multiplication and frequency synthesis circuitry. Multiplication from 5 MHz for quartz oscillators and from 1 GHz for surface acoustic wave oscillators is necessary. The phase noise and frequency stability of these multiplied signals are at least an order of magnitude inferior to those projected for the superconducting metal strip oscillator.

1. W. A. Getsinger, "Microstrip Dispersion Model", IEEE MTT-21, 34 (1973).
2. E. J. Denlinger, "Radiation from Microstrip Resonators", IEEE MTT, 17, 235 (1969).
3. D. C. Mattis, and J. Bardeen, Phys. Rev. 111, 412 (1958).
4. J. P. Turneaure and I. Weissman, "Microwave Surface Resistance of Superconducting Niobium", J. of Appl. Phys. 39, 4417 (1958).
5. S. B. Cohn, "Diss. Loss in Multiple-Coupled Resonator Filters", Proc. IRE 47, 1342, (1959).
6. G. L. Matthaei, Microwave Filters, Impedance-Matching Networks, and Coupling Structures, McGraw-Hill, New York, (1964).
7. Ibid., pg. 661.
8. F. Gauthier and M. Besse, "Graphical Design of Coupled Microstrip Lines", The Microwave Journal, 36 (1974).
9. T. G. Bryant and J. A. Weiss, "Parameters of Microstrip Transmission Lines and Coupled Pairs of Microstrip Lines", IEEE MTT 16, 1021, (1968).
10. A. H. Silver (Editor), "1976 Navy Study on Superconducting Electronics", ONR #NR319-110, (Aug. 1976).

MILLIMETER WAVE SUPERCONDUCTING DEVICES

P. L. Richards*
Department of Physics,
University of California, Berkeley, CA 94720

ABSTRACT

Significant progress has been made during the past two years in the development of millimeter wave superconducting devices. The present status of bolometric detectors and mixers, Josephson-effect mixers, quasiparticle mixers, and Josephson-effect parametric amplifiers will be reviewed.

INTRODUCTION

The history of high frequency analog superconducting devices has been one of an embarrassment of riches. Many types of devices have been explored which have worked more or less well. In the last few years, however, it has become increasingly clear that the most useful devices will probably be those that perform the conventional functions of detection, mixing and amplification. The author[1,2] and others[3] have published reviews which describe many of the complexities of this subject. The material in these reviews remains generally valid. The purpose of the present article is therefore to report on progress made since 1976.

SUPERCONDUCTING BOLOMETERS

A complete description has been published of the development of superconducting bolometers at Berkeley. Large area (0.16 cm^2) slow (τ = 83 ms) sensitive (NEP + 1.7 $\times$ 10^{-15} W/$\sqrt{Hz}$) bolometers have been made for use as square law (power) detectors at millimeter and submillimeter wavelengths. The thermometer is a superconducting Al film operated at its transition temperature of ~ 1.3K, which is AC biased and read out through a bridge circuit and a step-up transformer. The radiation is absorbed in a Bi film on the back side of a sapphire substrate which also supports the thermometer film. The submillimeter absorbtivity has been shown to be 53 ± 5 percent compared with a theoretical value of 50 percent. This bolometer was the first one reported to approach the theoretical noise limit set by thermodynamic energy fluctuations. Recently similar composite bolometer structures with semiconducting thermometers[5] have given comparable results at He^4 temperatures and significantly better results at He^3 temperatures. The status of development of these

*Work supported by the U.S. Office of Naval Research.

ISSN: 0094-243X/78/223/$1.50

devices is summarized in Table I.

Table I Performance characteristics of 4 × 4mm composite bolometers

Thermometer	Superconducting Al	Ge:Ga	Ge:In:Sb
Bath Temperature (K)	1.27	1.2	.35
Thermal Conductance (W/K)	2×10^{-8}	6×10^{-8}	1.7×10^{-8}
Time Constant (ms)	83	25	6
Electrical NEP (W/$\sqrt{Hz}$)	1.7×10^{-15}	3×10^{-15}	6×10^{-16}

Exploratory experiments have been carried out recently[6] at Berkeley to investigate the use of the superconducting bolometer as an efficient thermal heterodyne mixer. Such a device is able to operate at any RF frequency that can be coupled into it. The IF bandwidth is limited to the thermal relaxation frequency. It should therefore be similar in operation and applications to the InSb hot electron mixer[7] which has found uses in radio astronomy at near-millimeter wavelengths despite its limited IF bandwidth of ~ 3MHz. Another possible application is to near-millimeter radar where the bandwidth required is small. The thermal relaxation frequency can be selected over a wide range. A bandwidth of a few MHz can be obtained with a thin film structure immersed in superfluid He[4], or impedance matched to a substrate.[8] In structures such as point contacts and variable thickness bridges, where the excited quasiparticles escape into the surrounding metal, much higher relaxation frequencies and therefore IF bandwidths are expected.[9] Thermal mixing has often been seen in heating dominated Josephson structures.

For a low noise mixer it is necessary to apply a large enough local oscillator power to obtain a power conversion efficiency $\eta = P_{IF}/P_S > 0.1$. The impedance swing from the superconducting to the normal state is in fact large enough to achieve $\eta \simeq 1$, if thermal runaway can be avoided. Thermal runaway occurs in a current-biased superconductor because the power dissipated by the bias increases with increasing temperature. The best results that have been achieved thus far[6] with Al film mixers in superfluid He are $\eta \simeq 0.04$ and $B_{IF} \simeq 1$MHz. These values are not yet competitive, but no firm limit has been reached.

JOSEPHSON EFFECT MIXER

The use of a Josephson-effect mixer with external local oscillator as a harmonic mixer for frequency comparison[10,11] is the only case in which a Josephson-effect high frequency device has ever been used in a practical application for which it was the optimum device. Recently there has been considerable hope that point contact Josephson-junction mixers operated with an external local oscillator will serve as practical, efficient, low noise receivers for near-millimeter wavelengths.

Table II shows experimental results for mixer noise temperature obtained by several groups at a variety of frequencies.[12-15]

Table II Mixer noise temperature T_M of several point contact Josephson-effect mixers compared with the theoretical prediction of 40 T_{eff}

Frequency (GHz)	Temperature (K)	40 T_{eff} (K)	Experimental T_M (K)
36	1.4	66	54
120	6	266	120
130	4.2	209	180
452	4.2	465	400-1000

The Goddard work[13] at 130 GHz is of particular interest because it employs point contacts which can be recycled repeatedly without a significant change in their characteristics.

In order to understand the optimization of this device quantitatively,[16] an analog junction simulator has been used to compute the detailed experimental parameters of an equivalent circuit which is chosen to represent the actual mixer as closely as possible. The resistively-shunted junction model is imbedded in a series resonant RF input circuit with signal and local oscillator source resistance R_S. The DC bias circuit is decoupled at RF frequencies by an inductor. Noise is represented by an equivalent quasiparticle noise current source which has the form $I_n^2 = 2kTB/R$ in the thermal limit when $2kT>h\nu$ and $I_n^2 = h\nu B/R$ in the photon limit when $h\nu>2kT$. Here B is the bandwidth and R the junction shunt resistance. Quasiparticle shot noise is not important in this device since $h\nu>eV$. There is some uncertainty about the need for an additional pair shot noise term in this simulation. The point contact junction is not well-enough characterized to answer the question theoretically and the experimental evidence appears to be contradictory.[12,17]

In the simulation, the output is represented by an equivalent circuit consisting of an IF current generator in series with the dynamic resistance of the junction. One important result of the simulations is a value for the equivalent output noise current $I_{no}^2 = \beta^2 I_n^2$ in the thermal limit and $\gamma^2 I_n^2$ in the photon limit. If the input coupling is weak ($R_S \gtrsim 4R$) then β^2 and γ^2 are nearly identical functions of the normalized frequency $\Omega = h\nu/2eI_cR$ over the range $0.1<\Omega<1$. Experimental measurements of $\beta^2(\Omega)$ for weakly coupled junctions at 36[12] and 130 GHz[4] lie within a factor 2 above the calculation for $0.1<\Omega<1$. The similarity of $\beta^2(\Omega)$ and $\gamma^2(\Omega)$ shows that the noise in the photon limit can be represented by replacing the ambient temperature T by an effective temperature $h\nu/2k$ when $h\nu>2k$.

For efficient mixer operation it is necessary to couple the junction relatively tightly to the input RF circuit, so that $1<(R_S/R)<4$. In this case the mixer performance becomes a complicated function of nearly all of the mixer parameters. The simulator was used to explore the parameter spare to predict conversion efficiency and mixer noise temperature for the assumed equivalent circuit. Conversion efficiencies $\gtrsim 0.3$ are available for $0.1<\Omega<1$. The best value of single-sideband mixer noise temperature computed was $T_M \simeq 40T_{eff}$ $(0.1<\Omega<1)$ where T_{eff} is the larger of the ambient temperature or $h\nu/2k$. The theoretical predictions are compared with experiments in Table II. Although these experimental results represent the lowest mixer noise temperatures reported in this frequency range, the margin of improvement over cooled conventional Schottky diode mixers is not very wide. It is of importance for the future Josephson mixers to obtain better results. The simulations do not preclude the possibility of $T_M < 40T$, if a more favorable mixer circuit is discovered. It may be that this has occurred in the work at 120 GHz.[13]

QUASIPARTICLE MIXERS

The I-V curve for quasiparticle tunneling in superconductor-insulator-superconductor (SIS), superconductor-insulator-normal metal (SIN), and superconductor-insulator-semiconductor (super-Schottky) junctions has sharp enough curvature that their use in classical mixers and detectors is very promising. The super-Schottky is the most extensively explored device of this type. The mixer noise temperature of $T_M = 6K$ and Video NEP $= 5 \times 10^{-16}$ W/$\sqrt{Hz}$ observed at X-band by the Aerospace group[18] are an order of magnitude better than has been observed with any other diode. As with the conventional Schottky diode, the limit to the operating frequency is set by the discharge of the junction capacitance through the series (spreading) resistance in the semiconductor. The cut-off frequency has been quite low in the super-Schottky diodes used thus far. Super-Schottky diodes with very thin single crystal Si as the semiconductor layer are now being fabricated[19,20] which are expected to extend the operating range of these very promising devices to higher frequencies.

Preliminary experiments have begun at Berkeley and Boulder[21] to explore the properties of classical mixing in SIS and SIN tunnel junctions. The SIS case has larger curvature in the I-V curve, but is complicated by the Josephson effect. It is not promising to use an SIS tunnel junction as a Josephson mixer with an external local oscillator because junctions with sufficient current density to eliminate the hysteresis on the I-V curve are not yet available, and the hysteretic Josephson mixer is rather noisy.[22] Only a moderate critical current density is required, however, to make a classical mixer at millimeter wavelengths. It appears to be sufficient to make the capacitative reactance of the shunt capacitance at the operating frequency large compared with the normal resistance of the junction. This occurs at 36 GHz in Pb-Pb junctions with a current density of $\sim 300 A/cm^2$. Another requirement is that the junction resistance be large enough to match easily to the RF and microwave circuits. Junction resistances in the neighborhood of 100Ω are available with junction dimensions of a few μm.

When an increasing local oscillator power is applied to an SIS junction, Josephson steps move out to larger voltages. When the Josephson steps appear near the knee in the quasiparticle tunneling curve, which is the DC bias point chosen for this mixer, then strong Josephson mixing effects are seen. This occurs when the swing in the local oscillator voltage is wide enough to include zero voltage. For smaller amounts of local oscillator power, the I-V curve in the neighborhood of the bias point is dominated by photon-assisted tunneling effects, and essentially classical mixing phenomena are seen. Although the mixer noise in this mode appears to be quite low,[21] detailed measurements have not yet been made.

EXTERNALLY PUMPED PARAMETRIC AMPLIFIER

Josephson-effect parametric amplifiers with both internal[23,24] and external pump have been explored. There has been more progress with the externally pumped devices recently so this discussion will focus on them.

We first consider the four-photon doubly degenerate parametric amplifier or SUPARAMP invented by Chiao which has $\omega_S + \omega_I = 2\omega_P$. Detailed theoretical investigations of this device have been carried out by Chiao and his coworkers at Berkeley[25,26] based on the assumption that the Josephson element is RF voltage biased. All of the early experimental work, however, was carried out with series arrays of Dayem bridges[25] or point contacts[27] which were RF current biased. Efforts have been made to produce an adequate theory of the current-biased device.[27] Significant gain has been observed at frequencies as high as 86 GHz,[28] but the gain is accompanied by amplified junction noise which appears to be related to phase instabilities that occur when the RF current exceeds I_c.[29]

A new experimental approach has been made by the Gothenberg group[30] who have operated a SUPARAMP constructed from a series array of tunnel junctions. This device is used in a mode in which each junction is RF voltage biased by the parallel junction capacitance. A magnetic field in the plane of the junctions is used to adjust the amplifier to an operating point with large (non-reentrant) gain. Very good agreement is found between the performance of this amplifier and the theory of the voltage-biased SUPARAMP.[25]

An entirely separate line of development of externally pumped Josephson parametric amplifiers has been pursued by the group at the Technical University of Denmark. This work started with the discovery (made using a junction simulator) that when a tunnel junction is pumped at twice the Josephson plasma resonance frequency ω_J, then a parametric oscillation is excited at ω_J. The properties of a three-photon singly degenerate parametric amplifier with $\omega_P = \omega_S + \omega_I$ were then calculated.[31] Significant gain has recently been observed at X-band using a single Josephson tunnel junction.[32]

The use of tunnel junctions in Josephson high frequency devices can be considered to be a favorable development because they are well characterized and have minimal heating effects. It can be anticipated that the heavy investment in Josephson-effect computers will make stable reproducible tunnel junctions available with a wide variety of junction parameters.

REFERENCES

1. P. L. Richards, Semiconductors and Semimetals, R. K. Willardson and A. C. Beer, eds. (Academic Press, New York, 1977) V12, p. 395.
2. P. L. Richards, SQUID, H. D. Hahlbohm and H. Lübbig, Eds. (deGruyter, Berlin, 1977) p. 323.
3. R. Adde and G. Vernet, Superconductor Applications SQUIDS and Machines, B. B. Schwartz and S. Foner, Eds. (Plenum Press, New York, 1976), p. 248.
4. J. Clarke, G. I. Hoffer, P. L. Richards and N-H Yeh, J. Appl. Phys. 48, 4865 (1977).
5. N. S. Nishioka, D. P. Woody and P. L. Richards, Appl. Opt. (to be published).
6. P. L. Richards and T. M. Shen (to be published).
7. T. G. Phillips and K. B. Jefferts, Rev. Sci. Inst. 44, 1009 (1973).
8. S. B. Kaplan, C. C. Chi, D. N. Langenberg, J. J. Chang, S. Jafary and D. J. Scalapino, Phys. Rev. B14, 4854 (1976).
9. M. Tinkham, M. Octavio, and W. J. Skocpol, J. Appl. Phys. 48, 1311 (1977).

10. D. G. McDonald, A. S. Risley, J. D. Cupp, K. M. Evenson and J. R. Ashley, Appl. Phys. Lett. 20, 296 (1972).
11. T. Blaney, and D. J. E. Knight, J. Phys. D6, 936 (1973).
12. J. H. Claassen, Y. Taur, and P. L. Richards, Appl. Phys. Lett. 25, 759 (1974).
13. Y. Taur and A. R. Kerr (this conference)
14. J. H. Claassen and P. L. Richards, J. Appl. Phys. (to be published).
15. T. Blaney (this conference)
16. J. H. Claassen and P. L. Richards, J. Appl. Phys. (to be published).
17. G. Vernet and R. Adde, Appl. Phys. Lett. 19, 195 (1971).
18. M. McColl, R. J. Pedersen, M. F. Bottjer, M. F. Millea, A. H. Silver and F. L. Jr., Vernon, Appl. Phys. Lett. 28, 159 (1976); M. McColl, M. F. Millea, A. H. Silver, M. F. Bottjer, R. J. Pedersen and F. L. Jr., Vernon, IEEE Trans. MAG-13, 221 (1977).
19. T. Van Duzer, and C. L. Huang, IEEE Trans. ED-23 579 (1976); J. Maah-Sango and T. Van Duzer (this conference).
20. L. B. Roth, J. A. Roth, and P. M. Schwartz (this conference).
21. P. L. Richards, T. M. Shen, R. E. Harris, and F. L. Lloyd (to be published).
22. Y. Taur, J. H. Claassen, and P. L. Richards, Appl. Phys. Lett. 24, 101 (1974).
23. H. Kanter, IEEE Trans. MAG-11, 789 (1975); J. Appl. Phys. 46, 4018 (1975).
24. A. N. Vystavkin, V. N. Gubankov, L. S. Kuzmin, K. K. Likharev, V. V. Migulin, and V. K. Semenov, IEEE Trans. MAG-13, 223 (1977).
25. P. T. Parrish, and R. Y. Chiao, Appl. Phys. Lett. 25, 627 (1974); R. Y. Chiao, and P. T. Parrish, J. Appl. Phys. 46, 4031 (1975); M. J. Feldman, P. T. Parrish, and R. Y. Chiao, J. Appl. Phys. 47, 2639 (1976).
26. M. J. Fledman, J. Appl. Phys. 48, 1301 (1977).
27. Y. Taur and P. L. Richards, J. Appl. Phys. 48, 1321 (1977); IEEE Trans. MAG-13, 252 (1977).
28. M. Bottjer, H. Kanter, R. Pedersen, and F. L. Vernon, Bull. Am. Phys. Soc. (Ser II) 22, 375 (1977).
29. R. Y. Chiao, M. J. Feldman, D. W. Petersen, B. A. Tucker, and M. T. Levinson (this conference).
30. S. Wahlsten, S. Rudner, and T. Claassen, Appl. Phys. Lett. 30, 298 (1977); J. Appl. Phys. (to be published).
31. N. F. Pedersen, M. R. Samuelsen, and K. Saermark, J. Appl. Phys. 44, 5120 (1973).
32. O. H. Sorenson, J. Mygind and N. F. Pedersen (this conference).

JOSEPHSON MIXERS AT SUBMILLIMETRE WAVELENGTHS: PRESENT EXPERIMENTAL STATUS AND FUTURE DEVELOPMENTS

T. G. Blaney
Division of Electrical Science, National Physical Laboratory, Teddington, TW11 OLW, U.K.

ABSTRACT

The use of Josephson mixers with an external local oscillator for sensitive heterodyne reception at submillimetre wavelengths is discussed. Recent experiments, primarily at 450 GHz (0.66 mm wavelength), together with our present theoretical knowledge, allow realistic estimates of future receiver performance to be made. Although some practical problems remain, the probability of building field-usable receivers with very competitive sensitivity now appears good. A discussion is also given of the use of the high-harmonic mixing properties of Josephson devices for frequency measurement in the far infrared.

INTRODUCTION

The submillimetre-wavelength region (i.e. wavelengths from 1mm to a few tens of microns) is of increasing scientific interest, although it has yet to attract any large-scale commercial applications. Submillimetre measurement techniques are making an increasing impact in areas such as stratospheric monitoring, the diagnosis of plasmas in nuclear fusion research and materials research. However, both existing and potential future applications, such as short-range radars and satellite - satellite communications, are still greatly hindered by the performance limitations of existing devices and systems.

The Josephson effects, as presently known, reach their upper frequency limit in the submillimetre region. However, they can still make a useful contribution, particularly at the longer submillimetre wavelengths. This paper is concerned with two areas of current interests at NPL:

i) Heterodyne receivers using a Josephson device as a frequency mixer with an external local oscillator. The main requirements for such receivers at present are in astronomy (particularly from platforms above the earth's atmosphere), the detection of radiation in submillimetre laser-scattering diagnostics of plasmas, and in general high-resolution spectroscopy. The particular attractions of Josephson mixers are the low receiver noise temperatures, the high intermediate-frequency (i.f.) bandwidths and the low local-oscillator (l.o.) power requirements which appear to be possible.

ISSN: 0094-243X/78/230/$1.50

ii) Direct frequency measurement of submillimetre laser sources by high-harmonic mixing with a microwave frequency standard. As techniques for frequency measurement move towards the visible region [1], there is a need for simpler schemes and hence improved accuracy at the longer wavelengths of the far infrared. The remarkable harmonic-generating capability of Josephson devices offers both flexibility and the potential for this improved accuracy which is important for the development of frequency and length standards.

HETERODYNE RECEIVERS WITH AN EXTERNAL LOCAL OSCILLATOR

A study has been made at NPL to assess the potential of Josephson receivers for application in submillimetre astronomy. In particular, tests were carried out to measure the performance of an experimental receiver at 450 GHz (0.66 mm wavelength). A full account of the experiments will be published elsewhere [2], but the brief details are given below.

Adjustable niobium point contacts were mounted across a conventional hollow waveguide, which was small enough to allow propagation in only one mode at 450 GHz. The waveguide was fed from the radiation signal and l.o. sources via a horn (Figure 1), the horn and waveguide being formed inside a solid copper block.

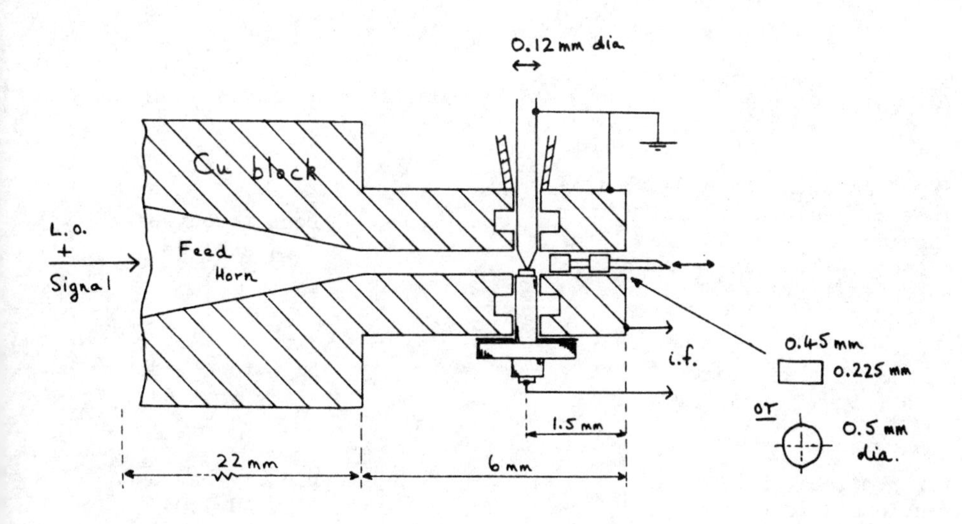

Fig. 1 Schematic mount of 450 GHz mixers

Two types of waveguide were used: (a) circular cross-section (0.5 mm dia. in the straight section) and (b) rectangular cross-section (0.45 x 0.225 mm). The lead-through holes for the point-contact wires (0.125 mm dia.) were provided with quarter-wavelength choke sections to reduce loss of radiation power from the waveguide. A movable reflecting plunger, to optimize the radiation-coupling efficiency was situated behind the mixer contact. The apparatus was operated immersed in liquid helium in a cryostat provided with windows so that radiation could be directed straight into the mouth of the horn. The l.o. and signal radiations were combined with a dielectric beam-splitter outside the cryostat. The l.o. was a 0.663 mm - wavelength (452 GHz) laser (optically-pumped $CH_2:CF_2$), and receiver noise temperatures were measured with 77K and 300K thermal sources. The i.f. output from the mixer was amplified by a room-temperature amplifier (20-100 MHz, noise temperature ~ 110K).

Some of the receiver characteristics are noted in Table I.

Table I Measured receiver characteristics at 450 GHz

Best overall receiver noise temperature, double sideband (including i.f. amplifier noise and all input losses except that of the beam-splitter)	2100K ($\equiv 6 \times 10^{-20}$ WHz^{-1})
Overall conversion loss (signal to i.f.)	Down to 11dB (± 1dB)
Input matching loss, excluding loss at beam-splitter. (Estimated using RSJ model for effect of l.o. on device characteristics)	Down to between 4 and 6dB
Intrinsic mixer conversion loss (signal to i.f.)	Down to ~ 5dB (Estimated)
Mixer noise temperature, single sideband	Typically between 350 and 1000K
Optimum l.o. power required (at external window of cryostat)	Few tens of nW to ~ 1uW, depending on mixer resistance

The best receiver noise temperature was taken with a rather lossy sapphire cryostat window, and this figure would reduce to about 1400K with our more usual polyethylene windows. There appeared to be no significant improvement in receiver noise performance when the operating temperature was reduced from 4.2K to $<$2.2K.

The resistively-shunted-junction (RSJ) picture of mixing behaviour [3] agrees with the experimental results in a broad qualitative way, and gives reasonable quantitative agreement as regards optimum l.o. power requirement and mixer conversion efficiency (although the model [3] tends to overestimate conversion loss by a few dB in comparison with the experiments). There is, however, little understanding of the junction noise levels at the i.f.

A peculiarity of the receiver is that at low values of l.o. power, the system exhibits a non-heterodyne (broad-band) response to thermal signal sources, although with reasonable precautions, this spurious response is usually negligible when l.o. power is increased to optimum level. The saturation properties of the mixer have not been studied in detail, but rough estimates indicate that the receiver will start to saturate at signal levels in the 10^{-10} to 10^{-9} W range.

Some much less exhaustive tests at 0.337 mm wavelength [2] with niobium junctions verify that the RSJ predictions of mixer conversion loss at this shorter wavelength are reasonably good.

RECEIVER PERFORMANCE IN THE FUTURE

On the basis of a pragmatic mixture of practical experience and those parts of the theory which appear to be reliable, we have made predictions of the overall receiver noise temperatures which can now be expected in the 1 mm to 0.3 mm wavelength range. Our main assumptions are: (a) that Josephson devices with similar (or superior) properties to those of point contacts be used; (b) that mixer conversion efficiency be as predicted by the RSJ model with $R_{dyn} = R$ (see ref. 3); (c) that junction i.f. noise be equivalent to that of a resistor of temperature 100K; (d) that an i.f. amplifier of noise temperature 30K be used; and (e) that input coupling efficiency of 10% or better be achieved at 0.3 mm, rising at longer wavelengths. Fig. 2 gives the likely range of performances for Nb and Nb_3Sn junctions, which are assumed to have $I_J R$ products of 80% of those for ideal tunnel junctions[4]. (This appears to be readily achievable with niobium point contacts). No predictions are given for wavelengths shorter that 0.3 mm, as the mixer theory and/or the predictions of input-coupling efficiences are presently too uncertain there.

Receivers based on adjustable point contacts, which can be used in a reasonably well-controlled environment, are possible now. As regards the construction of rugged field-usable receivers, the main practical problem is providing a suitable

Josephson device. The main device requirements are: (a) a resistance of $\lesssim$ 10 ohms; (b) a resistance-capacitance time constant of $\gtrsim 10^{-13}$ to 10^{-12} s; (c) to be made from materials of high energy gap and of near-ideal $I_J R$ product; (d) to be capable of fabrication in a geometry suitable for radiation signal coupling; and (e) to be mechanically and electrically stable and temperature-cyclable. Although yet to be demonstrated in practice, several viable routes to such devices appear to be open.

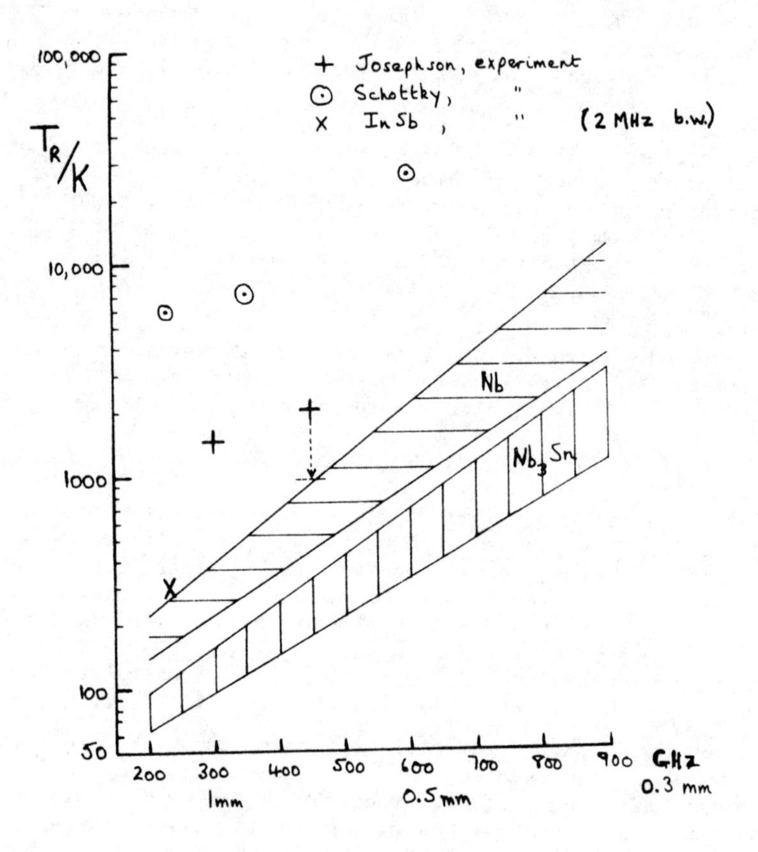

Fig. 2 Double sideband receiver noise temperatures. The shaded areas are realistic projections for Josephson-based receivers. Present values for Josephson receivers [2, 18] and Schottky diode receivers are also given.

Firstly, there is a good probability that point contacts of sufficient stability and reproducibility and of suitable geometry can be made (some preliminary work on this problem is in progress at NPL). Secondly, simple theoretical considerations show that if evaporated-film tunnel junctions are made sufficiently small in area ($\gtrsim 1\ (\mu m)^2$) and have a high enough Josephson current density ($> 10^9\ Am^{-2}$), they may well meet the requirements (see e.g. refs. 5 and 6). Thirdly, semiconductor-barrier junctions of small area ($\gtrsim 1\ (\mu m)^2$) may also meet the requirements[7]. The situation with constriction microbridges is much less clear, and they presently appear less promising than the devices already mentioned. In the longer term, arrays of junctions may have a contribution to make[8], but both theoretical and practical uncertainties make them much more speculative.

Assuming Josephson device resistance is not too low, reasonably efficient radiation coupling down to 0.3 mm wavelength appears possible. Our experience with horn-fed fundamental-mode waveguides at 0.66 mm indicates that mounts of this type can almost certainly be fabricated for half this wavelength, although only experiment will show what coupling efficiency can ultimately be achieved by this method. Viable alternatives are other forms of resonant cavity (e.g. the biconical cavity [9]) or small antenna systems, which are particularly suited to evaporated-film Josephson devices (e.g. double-slot dipole antennas [10]).

The l.o. power requirements are such that they can probably be met by harmonic-generation techniques using lower-frequency oscillators such as klystrons, backward-wave oscillators or solid state sources.

At present, the only useful alternative heterodyne mixer of high i.f. bandwidth for these wavelengths appears to be the Schottky barrier diode [11]. A good direct comparison between the latter and the Josephson device has yet to be made. However, it seems likely that the Josephson devices will maintain an advantage as regards overall receiver noise temperature, and will always require less l.o. power by several orders of magnitude. The Schottky diodes retain the advantage of ambient-temperature operation (although cooling may improve performance) and may prove to have a better dynamic range than the Josephson device which is essentially limited to relatively low signal levels.

HIGH-HARMONIC MIXING

Using a harmonic mixer, an unknown (or imprecisely known) frequency f_L is related to a lower standard frequency f_S by

$$f_L = N f_S \pm \Delta f$$

where N is the harmonic generated by the mixer, and Δf is a relatively small and readily-measured "beat" frequency. For comparison of far-infrared lasers with microwave standards, three main demands may be made of the mixer:

(a) It should operate with high values of N.

(b) The beat, at Δf, should have a high signal-to-noise ratio, preferably high enough to allow phase-locking between the sources of f_L and f_S. (It is also desirable that the power required at f_L and f_S be as small as possible).

(c) It should operate to high values of f_L.

For values of f_L up to at least 4 THz (~ 70 um wavelength), niobium point-contact Josephson mixers excel in the first two criteria. Harmonics (N) up to 825 (at 891 GHz [12]) and 401 (at 3.8 THz [13]) have been observed with useful signal-to-noise ratio. As an example of the flexibility offered by this high-N capability, the simple experimental scheme in Figure 3 can be used to measure virtually any laser frequency up to several terahertz without change of components [14]. It is also possible under some conditions to phase-lock the microwave oscillator to the laser to give virtually a continuous monitor of laser frequency on the counter. The system has operated satisfactorily with only a few tens of microwatts of laser power [14].

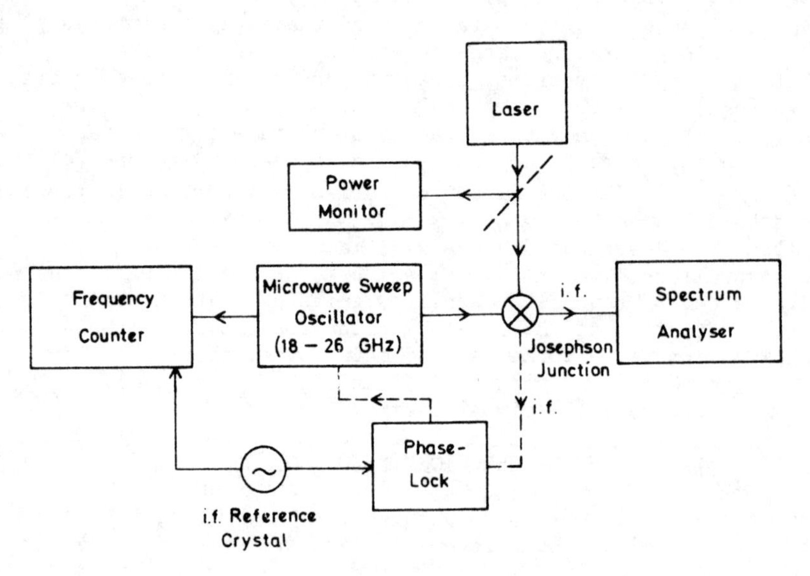

Fig. 3 Simple scheme for measuring and/or monitoring laser frequencies [14]

The main alternative device for harmonic mixing in this part of the spectrum is the Schottky barrier diode (point contact or planar devices), although even with source powers considerably higher than those required by the Josephson device, such diodes are unlikely to operate with values of N much beyond 30 to 40 [15, 16].

Room-temperature metal-insulator-metal point contacts can also be used, and although only useful to relatively small N ($\lesssim$ 13), will operate to beyond 190 THz [1].

Josephson harmonic mixers of this sort are mainly of interest in metrological laboratories. In the near future, Josephson point contacts will be used to measure laser frequencies around 4THz as part of new determinations of the frequency of the methane-stabilized He-Ne laser at 88 THz (3.39 μm wavelength). The simplification offered by the high-N capability of the Josephson device should contribute substantially to reducing the uncertainty in the latter frequency (presently about 5 parts in 10^{10}) by at least an order of magnitude.

In the longer term, it is desirable to push Josephson harmonic-mixing techniques to higher frequencies, with a particular aim of being able to make direct frequency "jumps" from microwave standards to the important region around 30 THz (10 μm wavelength). This will probably require the use of the highest energy gap materials in fabricating Josephson devices, although device heating effects may well prove to be as important a limitation as any intrinsic deterioration of the Josephson effects with increasing frequency. It seems likely that the point contact will remain the most useful device for this type of work. It is also worth noting that an extension of this sort to higher frequencies also requires parallel improvements in the spectral purity of the "base" frequency sources at f_S.

Josephson techniques may also play a part in the measurement of difference frequencies in the visible or near-infrared, as time-domain measurements become established in those regions. A non-Josephson device will probably generate the difference frequency, but the sensitivity and flexibility of a Josephson harmonic mixer could well be used to measure it.

ACKNOWLEDGEMENTS

The author is grateful to his experimental collaborators at NPL, particularly N R Cross, R G Jones and D J E Knight, and to the European Space Research and Technology Centre (Noordwijk, Holland) who supported the work on heterodyne receivers.

REFERENCES

1. K M Evenson, D A Jennings, F R Petersen and J S Wells, Laser Spectroscopy III Ed. J L Hall and J L Carsten (Springer, Berlin 1977) pp 56-68.
2. T G Blaney, N R Cross and R G Jones, to be published
3. Y Taur, J H Claassen and P L Richards, Rev. Phys. Appl. 9 263-8 (1974)
4. V Ambegaokar and A Baratoff, Phys. Rev. Lett. 10 486-6 and 11, 104 (1963)
5. D G McDonald, E G Johnson and R E Harris, Phys. Rev. B 1028-31 (1976)

6. J Niemeyer and V Kose, Appl. Phys. Lett. 29 380-2 (1976)
7. M Schyfter, J Maah-Sango, N Raley, R Ruby, B T Ulrich and T Van Duzer IEEE Trans. on Magnetics MAG-13 862-5 (1977)
8. P L Richards Semiconductors and Semimetals Vol 12, Eds. R K Willardson and A C Beer (Academic Press: New York 1977) p. 395
9. J J Gustincic Proc. S.P.I.E. 105 40-3 (1977)
10. A R Kerr, P H Siegel and R J Mattauch, IEEE MTTS-International Microwave Symposium Digest 1977 IEEE Cat. No. 77 CH 1219 - 5 MTT pp 96-8
11. T G Blaney "Radiation Detection at Submillimetre Wavelengths", to be published in J.Phys. E: Sci. Instrum.
12. T G Blaney and D J E Knight J.Phys.D: Appl. Phys. 7 1882-6 (1974)
13. D G McDonald, A S Risley, J D Cupp, K M Evenson and J R Ashley Appl. Phys. Lett. 20 296-9 (1972)
14. T G Blaney, D J E Knight and E K Murray Lloyd, "Frequency measurements of some optically-pumped laser lines in CH_3OD", to be published in Opt. Commun.
15. H R Fetterman, B J Clifton, P E Tannenwald, C D Parker and H Penfield, IEEE Trans. on Micr. Th. and Tech. MTT-22 1013-5 (1974)
16. B F J Zuidberg and A Dymanus, Appl. Phys. Lett. 29 643-5 (1976)
17. M Tinkham, M Octavio, and W J Skocpol, J. Appl. Phys. 48 1311-20 (1977)
18. J Edrich, D B Sullivan and D G McDonald, IEEE Trans. on Micr. Th. and Tech. MTT-25 476-9 (1977)

MIXING WITH JOSEPHSON JUNCTIONS IN THE FAR INFRARED

R. Adde and G. Vernet
Institut d'Electronique* Université Paris-Sud, 91405, ORSAY, France

ABSTRACT

The main problems encountered in order to realize a heterodyne mixer in the far-infrared (≃ 1 THz) are surveyed: junction characteristics, optimum performance, coupling efficiency at the signal and i.f. frequency.

INTRODUCTION

We are concerned here with Josephson heterodyne mixers operating in the short submillimeter wavelength region, i.e. at frequencies in the terahertz range and above. Then the waveguide technology currently in use at lower frequencies becomes increasingly difficult and quasi-optical systems appear more feasible. We survey the main problems encountered to realize a mixer (fundamental or harmonic, with external or internal local oscillator (LO)) in the terahertz range and report the existing results in this area.

Few results have been obtained up to now in this frequency range. Blaney[1] was the first to report on the performances of a Josephson mixer with external LO at 891 GHz. Most of the remaining work is concerned with high-order harmonic mixing for Far Infra-Red (FIR) synthesis developed at NBS,[2] NPL[3] and Orsay,[4] a kind of experiment where enough signal power is available from the FIR sources (a few mW). More recently we reported on mixing at 891 GHz using the internal Josephson oscillation[5] and in the last few months the group of Tinkham published results on the performance of point contacts in the FIR.[6,7]

JJ FOR FAR INFRARED MIXERS

Superconducting point contacts are still the preferred type of Josephson Junction (JJ) for high frequency performances. This is due to their very small area which minimizes the shunt capacitance and also to their easier coupling to the FIR radiation. Heating effects are still low at 1 THz due to their 3-dimensional structure, at least in high resistance junctions. The realization of these point contacts is an important aspect of a FIR mixer. The most complete results come from the recent systematic study made at Harvard[6,7] leading to reproducible and relatively stable point contacts with characteristics not far from ideal ($R_N I_c \simeq 2$ mV and $R_N \simeq 50 - 100\ \Omega$).

Most point contacts for high frequency (hf) applications are made of niobium. The methods of preparation use chemical etching, electropolishing, and anodization. In the air atmosphere Nb oxidizes and

*Laboratoire associé au CNRS.

ISSN: 0094-243X/78/239/$1.50

the exact nature of the oxide layer is not well defined, and may include several types of Nb oxides (NbO, Nb_2O_5...), interstitial oxygen atoms, etc. By comparison, a detailed study of the growth and characterization of the Nb oxide layer in Nb-NbO_x-Pb tunnel junctions by the Grenoble LETI group has shown the importance of a) realizing an amorphous Nb_2O_5 oxide layer and b) preventing contamination in order to obtain good Josephson tunneling properties and large current densities. We have observed that such junctions in which the oxide barrier has been punctured present an IV characteristic very similar to our point contacts for hf applications. We believe that our method, leading to high performance point contacts in the FIR, gives in one way or another a shunted oxide barrier between the two NB electrodes, fulfilling the requirement observed for tunnel junctions. The difficulty for contacts consists in a precise characterization of the interface existing at the time of the experiment. It is generally observed that the junctions must be exposed to atmosphere as short a time as possible and placed in He atmosphere or vacuum to prevent a too-thick oxide layer from forming.

Actually all the best contacts for the FIR are still made "in situ". A burn-in process[9] is widely employed to "form" the contact. The junctions are first adjusted mechanically to give a high resistance contact (500-1000 Ω), and the operation is controlled using a low voltage bias. In this way there is no severe voltage breakdown when the contact is formed, which would give junctions with low resistance, strong I_c and weak gap structure. These are characteristic of severe shorts which make the contact behave like a microbridge. The burn-in process which follows consists of applying controlled voltages or pulses to the contact. Small voltage breakdowns are created across the interface between the Nb electrodes, which decrease the resistance and lead to the appearance of a supercurrent and strong gap structure in the VI characteristic. The art consists of obtaining a large $R_N I_c$ with a moderately high resistance (50-100 Ω) to help matching to the high frequency signals and to the IF chain. Therefore the critical currents are in the 20-50 μA range. One may speculate on the nature of these point contacts. Are they rather microbridges or shunted tunnel junctions? It would be interesting to apply this method to the preset contacts of Taur and Kerr[10] which have excellent performances at 100 GHz but have too low of an $R_N I_c$ product (0.6 - 0.8 mV) for operation near 1 THz.

Whatever the progress to date in the performances of point contacts, their relative instability remains a drawback. A real step forward in hf detection would be made with the availability of suitable tunnel junctions. Moreover this would separate the problems relative to the junction and to the antenna. However it is necessary to realize non-hysteretic junctions ($\beta_c = 2e\, R^2\, C\, I_c/h < 1$). We are working in this direction with "shunted" tunnel junctions in collaboration with the LETI Grenoble group. It will be necessary to use junctions (e.g. Nb-NbO_x-Pb) of micron dimensions and large current densities ($\simeq 10^4$ A/cm^2) to fulfill the above conditions and have good performances at 1 THz.

EXPECTED PERFORMANCES OF A FIR JOSEPHSON HETERODYNE RECEIVER[11,17]

If all conditions required for optimum performances are met, we estimate that extremely low receiver noise temperatures of a few tens of K may be realized at 1 THz. It is not realistic to consider such an ideal situation and we give here a more realistic point of view of what can be reached in a reasonable future. Such estimates are based on the RSJ model (even though the IV characteristic of a FIR contact differs appreciably from that of this model), and neglect the effect of the superconducting energy gap. In the THz range, analytical calculations may be made with the Resistively Shunted Junction (RSJ) model based on the voltage source model of the JJ.

a) Conversion efficiency: At frequencies $\Omega = \omega/\omega_c \gtrsim 1$ ($f_c = 2eR_NI_c/h \simeq 1$ THz for $R_N \simeq 50\Omega$ and $I_c \simeq 40$ µA), the optimum conversion efficiency of a Josephson mixer may be simply approximated by:

- external L.O.: $\eta_{opt} = (C_{hf}/16\ \Omega^2)(R_d/R_N)$ (1)

where C_{hf} is the input coupling efficiency for the signal and R_d is the dynamic resistance. Looking at real junctions VIC (e.g. ref. 6,7) with the above parameters shows that the VIC under L.O. level to give optimum efficiency becomes such that $R_d \simeq R_N$. Therefore $\eta \simeq C_{hf}/16\ \Omega^2$.

- internal L.O.: $\eta = C_{hf}/16\ \Omega^2$ (2)

Therefore the optimum conversion efficiency at 1 THz with a quite ideal Nb contact ($\Omega \simeq 1$) is $\eta \simeq 0.06\ C_{hf}$. This result shows clearly that the high conversion efficiencies observed at lower frequencies will not be reached in the THz range even with excellent Nb point contacts. Moreover, the coupling coefficient, which is far from being optimized, will actually further decrease the performances.

b) Noise temperature: The physical origin of internal noise in JJ which are not sandwich tunnel junctions is still under question. Measurements of the voltage noise or of the Josephson current linewidth in point contacts[12,5] and at least some types of microbridges[13,14] indicate that the noise level is better explained on the basis of a "current shot type noise" model than by assuming only the Johnson noise related to the shunt resistance. We have recently proposed[15] a tentative explanation of this result based on strong coupling inside the junction between the high frequency Josephson current and a dissipative mechanism.

Noise measurements in Josephson mixers are usually interpreted assuming Johnson noise and introducing a noise parameter $\beta (\gtrsim 1)$ to take into account down-converted noise at harmonics of the L.O. and Josephson frequencies.[16] The importance of these processes become small in the far infrared ($\beta \doteq 1$ if $\Omega \gtrsim 1$).

Taking the favorable case of Johnson noise, the minimum mixer input noise temperature is then both for external or internal L.O.

operation:

$$T_M = T\eta = 16\ \Omega^2\ T/C_{hf} \tag{3}$$

where T is the temperature of the mixer.

The contribution of the IF amplifier to the receiver noise temperature referred to the input of the mixer is $T_{IF}/\eta\ C_{IF}$ where C_{IF} is the coupling efficiency at the IF ($\simeq$ 1 for junctions with $R_N \simeq 50\ \Omega$). The total receive input noise temperature (not including antenna and background noise) is

$$T_R \simeq 16\ \Omega^2\ (T + T_{IF})/C_{hf} \tag{4}$$

with the above quite ideal Nb junctions ($\Omega \simeq 1$, $R_N \simeq 50\ \Omega$) which gives at 4 K:

IF amplifier (4.5 GHz)		T_R (K)
cooled paramp:	$T_{IF} \simeq 20$ K	$\simeq 400/C_{hf}$
cooled FET:	$T_{IF} \simeq 60$ K	$\simeq 1000/C_{hf}$

Equation 4 and the above numbers indicate clearly which parameters are most important. A good FIR receiver should have:

- ideal junctions with Ω as small as possible
- low T_{IF}
- efficient coupling at the signal frequency

Except if a maser IF amplifier is used ($T_{IF} \simeq 2$ K) the noise contribution of the IF amplifier may be larger than that of the mixer, and cooling a junction under 4 K will help mostly to improve the junction HF properties and lower Ω.

MIXER WITH EXTERNAL OR INTERNAL L.O.

The estimated performances given above show that comparable receiver noise temperatures may be expected in the FIR with both modes of operation. The very simplicity of the internal local oscillator mode (the frequency of the Josephson oscillation is directly proportional to the dc bias voltage across the junction) makes it attractive for a receiver operating in a wide range of frequencies. However the attractiveness of this mode of operation is reduced with conventional point contacts because of the wide linewidth[5] of the Josephson current (a few GHz at 1 THz with $R_N \simeq 10\ \Omega$). The reason is that the voltage fluctuations across the junction frequency modulate the Josephson current. For many applications a 10^{-3} frequency resolution is a good criterion, and we intend to develop structures allowing the low frequency ($\lesssim$ 1 GHz) voltage noise components across the junction to be shunted without deteriorating the performances with an IF chain in the 4.5 GHz range.

The mixer with external L.O. has therefore received much more attention since its selectivity is only determined by the linewidth of the external L.O.; the amplitude and frequency of the L.O. may be adjusted externally, leaving the junction dc bias as another independent parameter to optimize the mixer performances. Microwave oscillators may be used if the junction is operated as a harmonic mixer but they quickly give complicated spectra to analyze due to the high number of sidebands which are created. Optically-pumped cw FIR lasers seem to be an attractive source as a L.O. for JJ mixers since a very low level of L.O. power is required (10^{-6} watt or less for a matched junction) and hundreds of lines are available in the FIR region. However a performant mixer requires a stable L.O. amplitude, which is not a simple task with these FIR lasers. The power level of FIR lasers is strongly dependent on the frequency matching between the CO_2 pump laser and the absorption line of the molecule. Therefore a frequency stabilization of the CO_2 laser is required. The simplest scheme involves frequency-locking of the CO_2 laser frequency on the absorption line frequency of the molecule using an absorption cell containing a) the same molecule if there is overlapping of the two lines, b) another suitable molecule if the above condition is not fulfilled. The whole unit consisting of a CO_2 laser ($\simeq$ 20 watts cw), a cw FIR molecular laser and the frequency stabilization system constitutes actually a rather "bulky" oscillator which needs to be made in a more compact configuration for many applications.

INTERMEDIATE FREQUENCY AMPLIFIER

Since Eq. 4 requires a low T_{IF} to obtain a low-noise receiver and make optimum use of the sensitivity of the JJ, a cooled parametric amplifier seems appropriate in such a receiver. However our actual experience with a 4.5 - 5 GHz He cooled paramp shows that much caution must be taken. We have used a He cooled circulator between the paramp and JJ to insure a high level of isolation. However, we have found that a rather strong level of the second harmonic of the pump frequency ($\simeq$ 20 GHz) is created in the paramp and reaches the junction. The JJ is highly non-linear in the microwave region and many harmonics of this spurious signal are created giving back many sidebands at the IF. This makes the signal more difficult to identify and degrades the performance of the mixer. We have studied cooled filters to eliminate this effect.

FIR COUPLING

In the millimeter and submillimeter wavelength region down to 0.5 mm, the signal is coupled to the mixer with waveguide techniques. At shorter wavelengths the realization of a reduced height waveguide for impedance matching to the junction ($\simeq$ 50 Ω in favorable cases) becomes increasingly difficult. The technological problems seem too difficult to be well resolved and it seems natural to use quasi-optical techniques. The coupling is a general problem for all types of mixer diodes in this frequency range (JJ, Schottky, MIM); and studies are now under development for finding antenna structures optimizing the FIR coupling to the mixer. In many radar or radiometric applications the incoming beam may be considered as a laser beam, parallel or focussed which illuminates an antenna. If the mixer is of the point contact type, this antenna has more or less a conical shape with the two extreme situations being the thin dipole antenna and the conical antenna. There is actually no precise data concerning the optimization of radiation coupling in the FIR and only general properties may be given.

In a first approximation one may neglect the presence of an active device in the antenna. The optimization of the system requires 1) optimizing the impedance of the antenna with that of the junction. The reactive part of a point contact hf impedance is usually small[17] which means antenna structures with small reactance components should be selected. If the antenna resistance R_a is close to R_N, the effective impedance seen by the junction is $(R_N R_a / R_N + R_a)$. This strongly increases Ω and deteriorates the mixer performances (Eqs. 1 and 4). The optimum antenna impedance must therefore be somewhat larger than R_N. 2) The incident signal beam must fit as well as possible the radiation pattern of the antenna. This results from the reciprocity theorem which states that the antenna pattern for reception is identical with that for emission. This means that with cylindrical antenna structures such as whiskers or conical contacts, the excitation should fill a toroidal radiation pattern. A toroidal excitation means with polarized beams (e.g. laser beams) a polarization transformer must be realized. Another possibility is to break the azimuthal symmetry of the radiation pattern with reflectors.[18] However focussing the reflected beams in phase on the junction is increasingly difficult when λ decreases.[18] The zenithal angular width of the incident beam must also be fitted to the antenna.

A JJ is an active non-linear device which makes the problem much more complex. It is useful to operate these junctions in a wide-band structure to prevent the appearance of hysteresis in the IV curve. Therefore the junction-creating harmonics of the signal ω_S makes the antenna operate as a receiver at ω_S and an emitter, not only at ω_S, but also $2\omega_S$. Only detailed antenna calculations may show the importance of these effects and the consequences on the optimization of the radiation pattern.

Finally an efficient coupling structure with Josephson junctions using thin film structures would be the two-slot radiator antenna which has been already employed successfully at 100 GHz.[19]

CONCLUSION

The remarks about FIR coupling indicate that it is in this area that the effort must be made now to improve the performances of a FIR receiver at $\simeq$ 1 THz. Several groups are working in this direction. Due to the actual limitations of the hf coupling efficiency the realization of a heterodyne receiver using Nb contacts with a noise temperature of 10^4 K at 1 THz would be an important step. In parallel, the realization of preset Nb contacts with near-ideal performances would greatly simplify the design of the mixer and improve its mechanical stability. Heating effects[20] have been shown to be severe in point contact alloys (Nb_3Sn) and this disadvantage outweighs the gain in $R_N I_c$ so that Nb may remain the better material. Finally the future of FIR Josephson receivers will be certainly strongly related to the realization of thin film junction structures: (shunted) tunnel junctions, junctions with Si membranes. These planar structures are less sensitive and one could make full use of the properties of high T_c materials.

REFERENCES

1. T. C. Blaney, Rev. Phys. Appl. 9, 279 (1974).
2. D. G. McDonald, A. S. Risley, J. D. Cupp, K. Evenson and J. R. Ashley, Appl. Phys. Lett. 20, 296 (1972).
3. T. G. Blaney and D. J. E. Knight, J. Phys. D 7, 1882 (1974).
4. J-M. Lourtioz, R. Adde, G. Vernet and J-C. Hénaux, Rev. Phys. Appl. 12, 487 (1977).
5. G. Vernet, J-C. Hénaux and R. Adde, IEEE MTT 25, 473 (1977).
6. D. A. Weitz, W. J. Skocpol and M. Tinkham, Appl. Phys. Lett. 31, 227, (1977).
7. D. A. Weitz, W. J. Skocpol and M. Tinkham, Phys. Rev. Lett. 40, 253 (1978).
8. J. C. Villegier, Thèse Docteur-Ingénieur, Université de Grenoble, (1977) (unpublished).
9. H. C. Tolner, C. D. Andriesse and H. A. Schaeffer, Infrared Physics 16, 213 (1976).
10. Y. Taur and A. R. Kerr (to be published).
11. R. Adde and G. Vernet, in Superconductor Applications (Edited by B. B. Schwartz and S. Foner), Plenum Press, New York (1977).
12. H. Kanter, Phys. Rev. Lett. 25, 588 (1970).
13. S. Decker and J. E. Mercereau, Appl. Phys. Lett. 27, 466 (1975).
14. P. Crozat, G. Vernet and R. Adde, Appl. Phys. Lett. 32, April (1978).
15. G. Vernet (private communication).
16. J. H. Claassen, Y. Taur and P. L. Richards, Appl. Phys. Lett. 25, 759 (1973).
17. P. L. Richards, in Semiconductors and Semi-metals, Willardson and Beer, Vol. 12, Academic Press, New York.
18. H. Krautle, E. Sauter and G. V. Schultz, Infrared Physics 17, 477 (1977).
19. A. R. Kerr, P. H. Siegel and R. J. Mattauch, IEEE-MTT (1977) Int. Microwave Symposium Digest, IEEE Cat. No. 77-CH-1219-5 MTT.

MICROWAVE PARAMETRIC AMPLIFIERS USING EXTERNALLY PUMPED JOSEPHSON JUNCTIONS

O. H. Soerensen, J. Mygind, and N. F. Pedersen
Physics Laboratory I, The Technical University of Denmark
DK-2800 Lyngby, Denmark

ABSTRACT

Externally pumped parametric amplifiers are discussed. Theory and experiments on the singly degenerate parametric amplifier based on a Josephson junction are presented. Advantages and limitations of the singly degenerate and doubly degenerate parametric amplifiers are discussed. Some plans and proposals for future research are presented.

INTRODUCTION

One of the areas in which Josephson junctions offer a potential for future use is as active elements in microwave parametric amplifiers. Extrapolations from X-band measurements indicate that amplifiers with working frequencies as high as ~300 GHz may be feasible.

The results obtained at X-band include an internally pumped[1, 2] (self-pumped by the Josephson ac-voltage) parametric amplifier using a point contact junction, an externally pumped, doubly degenerate amplifier (DDA) using either an array of microbridges[3] or tunnel junctions[4], and finally an externally pumped, singly degenerate amplifier (SDA) using a single tunnel junction as the active element[5]. At 35 GHz preliminary results have been reported for doubly degenerate amplifiers using an array of microbridges[6] and a single point contact[7].

In this paper we shall discuss the externally pumped parametric amplifier with main emphasis on the singly degenerate version, and comparisons are made with the properties of the doubly degenerate mode of operation. Some recent experimental results on the SDA are presented and discussed. Of particular interest is the coupling between the active device and the external circuits. Also the problem of thermal saturation of the amplifier will be briefly mentioned. Finally, we outline some of the problems to which we shall devote our efforts in the future and as a closing remark we suggest a scheme for a wholly integrated, externally pumped parametric amplifier.

THEORETICAL CONSIDERATIONS

The basic configuration of a Josephson junction parametric reflection amplifier is sketched in Fig. 1. The signal at frequency ω_s and (possibly) also the pump at frequency ω_p is supplied to the junction via a transmission line of impedance Z_o. A dc bias current, I_o, is also provided for. The junction has normal state

ISSN: 0094-243X/78/246/$1.50

resistance, R, capacitance, C, and a critical current, I_c.

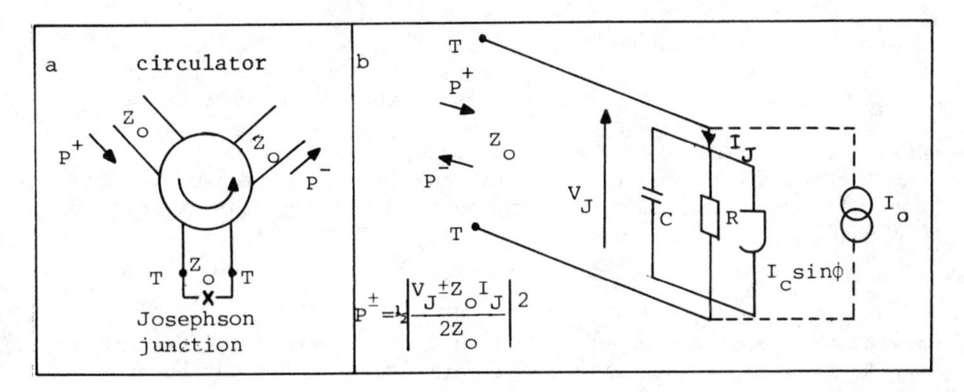

Fig. 1. a. Josephson junction reflection amplifier.
b. Equivalent diagram at terminals T.

Under certain conditions the input admittance of the junction at frequency ω_s can become negative such that the signal gain defined as the power reflection coefficient, Γ, at ω_s is greater than one.

The theory for the DDA mode has already been published[3] whereas the results for the SDA mode has not yet appeared[8].

The DDA may be characterized by the conditions (i) $I_o = 0$ and (ii) $2\omega_p = \omega_s + \omega_i$ with $\omega_p \sim \omega_s \sim \omega_i$, where ω_i is the idler frequency. The SDA is similarly characterized by (i) $I_o \neq 0$ and (ii) $\omega_p = \omega_s + \omega_i$ where $\omega_s \sim \omega_i \sim \frac{1}{2}\omega_p$, and further that the junction plasma frequency ω_R must be approximately equal to ω_s. With $\omega_o = (2eI_c/\hbar C)^{\frac{1}{2}}$ we have that $\omega_R = \omega_o(\cos\phi_o)^{\frac{1}{2}}$ where $\phi_o \simeq \sin^{-1}(I_o/I_c)$.

Within the framework of the shunted junction model it may be shown that the power reflection coefficient at the signal frequency for the SDA is given by

$$\Gamma = \frac{\{\eta^2(1-\sigma^2) + \eta^2 A^2 - B^2\}^2 + 4\eta^4 A^2 \sigma^2}{\{\eta^2(1+\sigma)^2 + \eta^2 A^2 - B^2\}^2} \quad . \tag{1}$$

Here, $\eta = \hbar\omega_p/4eRI_c$ is the ratio of half the pump frequency to the characteristic frequency, $\omega_c = 2eRI_c/\hbar$, of the junction; $\sigma = R/Z_o$ measures the impedance mismatch between the junction and the waveguide. The term A^2 is defined by

$$\eta^2 A^2 = \{(\omega_p/2\omega_o)^2 - (\omega_R/\omega_o)^2\}^2$$

and reflects the detuning from the plasma frequency, ω_R. Finally, $B = J_1(\phi_p)\sin\phi_o$ is the term responsible for the parametric amplification. Here, J_1 is the first order Bessel function and ϕ_p is the junction phase modulation at the pump frequency. For the sake of clarity the effects of the phase-dependent conductance has been neglected in eq. (1) but may readily be included. By inspection of eq. (1) it is apparent that in order to achieve gain the denominator must be made small, which may be realized if the dc bias and pump power are adjusted such that the $J_1(\phi_p)\sin\phi_o$ term outbalances the internal and external losses represented by the $(1+\sigma)$-term and the detuning given by the A-term. We also point at the importance of having a small value of η, i.e., a low reduced frequency if gain shall be realized at reasonably low power levels.

The threshold condition for the onset of subharmonic oscillations[9] is obtained from eq. (1) by setting the denominator equal to zero

$$J_1(\phi_p)\sin\phi_o = \eta((1+\sigma)^2 + A^2)^{\frac{1}{2}} \tag{2}$$

showing that observation of subharmonic oscillations (at frequency $\frac{1}{2}\omega_p$) indicates that conditions for parametric amplification are satisfied.

Although the SDA and the DDA appear as two rather different modes of operation they are very similar. Whereas the conditions for gain in the SDA-mode is derived by an expansion of the $\sin\phi$-term maintaining only the first harmonics at ω_p, ω_s, and $\omega_p-\omega_s$ the latter with amplitude $J_1(\phi_p)\sin\phi_o$, the DDA-mode results from an expansion including 2nd harmonic of ω_p and the sidebands $\omega_s = 2\omega_p-\omega_i$, and $\omega_i = 2\omega_p-\omega_s$ with amplitudes $J_2(\phi_p)\cos\phi_o$. In fact eq. (1) may be transformed into the corresponding expression for the DDA by the following substitutions

1) $I_o = 0$, i.e., $\sin\phi_o = 0$, $\cos\phi_o = 1$

2) $\omega_p \rightarrow 2\omega_p$

3) $J_1\sin\phi_o \rightarrow J_2\cdot\cos\phi_o = J_2\cdot 1$

With respect to experimental convenience the SDA has the advantages of a lower pump-power requirement (provided the pump line is matched), of a pump frequency far removed from the signal frequency preventing pump return power from saturating the succeeding receiver stage, and of being tunable by means of a dc-current. The DDA on the other hand has the advantages that no dc-bias leads are necessary, the signal, idler and pump frequencies are all matched by the same circuit, and by symmetry no even harmonics are present. In both modes a magneti field may be used for tuning purposes.

An upper frequency for the SDA is readily estimated. In order to be able to reach the parametic regime at reasonable pump power levels, the Q of the plasma resonance ($Q = \omega_o RC$) must be of order 2 to 5 and not greater if hysteresis in the response should be avoided. Since, theoretically

$$\omega_o Q = 2eRI_c/\hbar = 2e/\hbar\cdot\pi/4\cdot 2\Delta \tag{3}$$

where 2Δ is the energy gap of the superconductor, we find that with $Q \simeq 2$, the maximum attainable plasma frequency is $\sim$200 GHz for Sn junctions and $\sim$500 GHz for Pb junctions. Such high plasma frequencies will require high supercurrent densities ($\omega_o^2 \propto I_c/C$), which in turn requires small area junctions in order to maintain reasonably high junction impedances.

SOME EXPERIMENTAL AND NUMERICAL RESULTS

In this section a brief account will be given on some recently published X-band experiments with the SDA. The main result was that with a single tunnel junction a net gain of 16 dB in a 4 MHz bandwidth was obtained. The noise temperature of the device was fairly high in these preliminary experiments, i.e., of the order 3000 K.

Typical recordings of the detected power at the signal, idler, and half pump frequencies are shown in Fig. 2a, b, and c. The dashed curve in Fig. 2a is the theoretical response calculated with parameters relevant for the experiment. The only free parameter in the calculation was the absolute pump power level which was adjusted to bring the peaks to just about coincide with the experiment.

The input signal level was fairly high in this particular experiment ($\sim 3 \cdot 10^{-14}$W) and an appreciable saturation of the amplifier had taken place (the 16 dB net gain quoted above was achieved with a lower input signal level). The horizontal dotted line in Fig.2a represents the input signal level applied to the receiver. In this experiment we used a Sn-O-Sn tunnel junction with dimensions 100 by 200 μm, a maximum supercurrent density of order 20A/cm^2, and normal state resistance 0.18Ω.

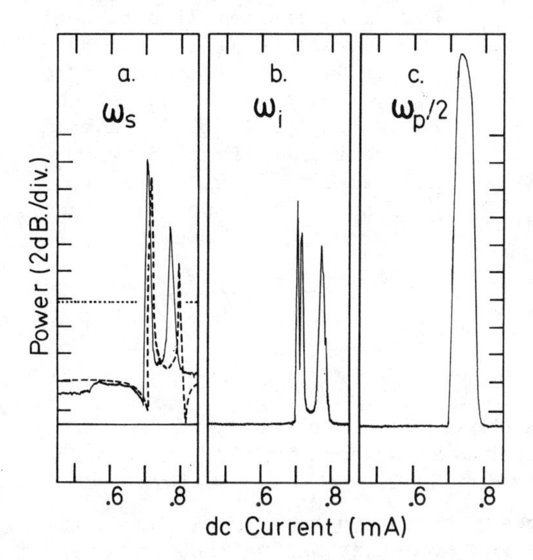

Fig. 2. The received power in a 100 kHz bandwidth centered at frequencies ω_s, ω_i, and $\omega_p/2$. The dotted line indicates input power level, and the dashed curve is a theoretical calculation with parameters relevant to the experiment.

The junction was mounted across the low impedance end of a two-step binomial transformer as shown in Fig. 3c. The correct termination of the transformer should be of order a few Ohms.

The dotted curve in Fig. 3a shows the transmission through two transformers mounted end to end. The bandwidth is about 2 GHz and the insertion loss at center band is $\sim$2dB.

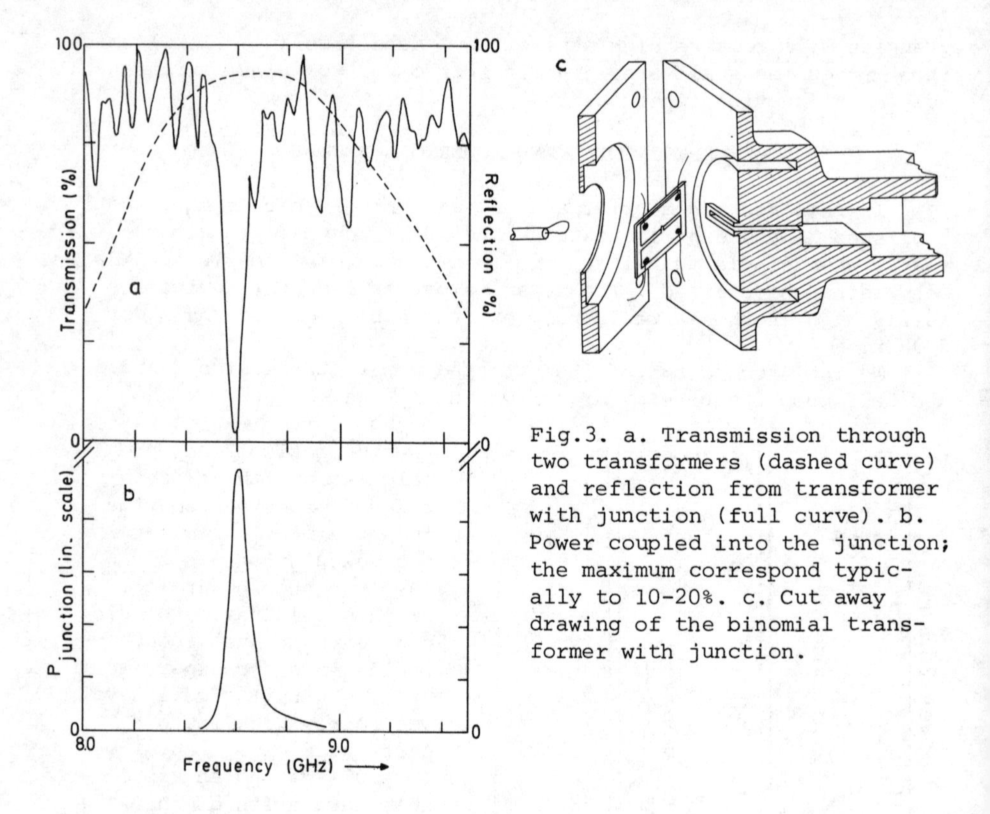

Fig.3. a. Transmission through two transformers (dashed curve) and reflection from transformer with junction (full curve). b. Power coupled into the junction; the maximum correspond typically to 10-20%. c. Cut away drawing of the binomial transformer with junction.

The solid curve in Fig. 3a shows the power reflected from one transformer terminated with a junction as shown in Fig. 3c. Fig. 3b shows the power available to the junction, $P_{junction}$, determined by measuring the attenuation necessary to keep the magnitude of the 1st rf-induced step constant as the frequency was varied. The peak in $P_{junction}$ and the dip in the reflection are in precise correspondence. We do not understand why the coupling is so narrow a band but suspect that spurious wave guide resonances was involved[10].

The narrow band coupling served as an input filter limiting the incoming thermal radiation from room temperature and thus reducing the thermal saturation of the amplifier. Calculations have shown that for the experiment in Fig. 2 the signal saturation was by far the most important.

The effect of thermal saturation in the SDA-mode is illustrated in Fig. 4. Here the power gain is calculated with the same parameters as in Fig. 2 but without signal input. Trace a shows the gain without incoming thermal noise (T = 0 K). Trace b shows the gain with a noise input corresponding to T = 100 K. Finally, trace c shows the gain with a noise input at T = 300 K.

The calculations presented here are very similar to those of Feldman,[11] and a detailed account will be given elsewhere[12].

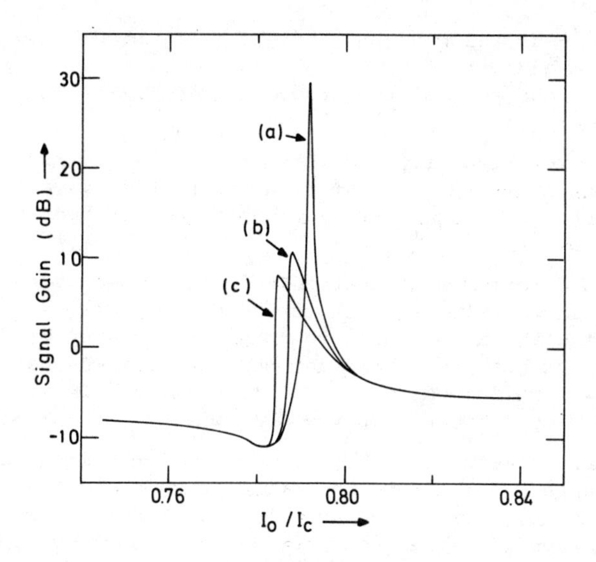

Fig. 4. Calculated signal gain vs. dc bias current with incoming thermal radiation at temperature (a) 0 K, (b) 100 K, and (c) 300 K. Only the gain peaks at low bias currents are shown.

PARAMETRIC EFFECTS ON rf INDUCED STEPS

Some very recent and not yet fully understood results should be mentioned. We have observed subharmonic oscillations not only at zero voltage but also at the rf induced steps. It turned out that as the pump power was increased above the threshold for subharmonic generation at zero voltage thresholds were reached for subharmonic generation at the n = 1, 2, 3, ... rf induced steps. With applied pump power at 17 GHz we have followed this so far as to the n = 7 step without any noticeable decrease in emitted power at 8.5 GHz. With an applied signal near 8.5 GHz parametric effects were seen bearing much resemblance to the behaviour observed at zero voltage. Idler conversion was found on each step, and a net signal gain ($\sim$2 dB) was achieved on the n = 1 step.

PLANS FOR THE FUTURE

In the near future we shall perform experiments at frequencies above X-band.[13] An experimental rig prepared for 35/70 GHz work is near completion. Our intention is to study the SDA at 35 GHz (pump at 70 GHz) and the DDA at 35 and 70 GHz using small area tunnel junctions prepared using evaporation masks fabricated by photolithographic methods. Dimensions of the order 10 by 10 μm and maximum plasma frequencies from 50 - 100 GHz have been achieved. Microwave impedance transformers for 35 and 70 GHz have been made and tested (at 70 GHz the waveguide height is reduced to 70 μm at the low impedance end).

We shall outline some of the problems which we feel need further consideration:

i) to compare experimentally and theoretically standard properties such as gain, bandwidth, dynamic range, stability, and internal noise generation of parametric Josephson devices operated in the SDA and DDA modes.

ii) to study in detail the saturation effects caused by signal power and thermal radiation applied to these low-power handling devices. A limited thermal saturation possibly has a stabilizing effect (cf. Fig. 4), but it remains an open question whether incoming noise power influences the properties in the SDA mode as strongly as found by Feldman[11] in the DDA mode.

iii) to elaborate on methods to bridge the impedance mismatch between the Josephson device and the signal/pump circuitry. This can be done either by increasing the effective junction impedance using compact arrays of series-connected junctions or by reducing the impedance of the external waveguide/stripline with the aid of tapers, transformers, and resonant structures.

iv) to look into the question of using thin film microwave circuitry with tunnel junctions as integrated parts. In this context the high-frequency properties of the high current density Pb-In-(Au) alloy tunnel junctions should be studied.

In conclusion we suggest one such scheme employing integrated microwave circuit technology: the design of a fully integrated parametric amplifier operating somewhere in the frequency range 100 - 300 GHz and using a Pb-alloy tunnel junction (-array) as the parametric device. Apart from impedance transformers, filters, and couplers, the thin film circuit should contain a second high-current density tunnel junction which biased on an internal geometrical resonance step or injection-locked to harmonics of an external oscillator should provide the necessary pump power.

The work was supported in part by the "Statens Naturvidenskabelige Forskningsråd", grant No.: 511-8048.

REFERENCES

1. A.N. Vystavkin, V.N. Gubankov, L.S. Kuzmin, K.K. Likharev, V.V. Migulin, and V.K. Semenov, IEEE Trans.Magn. MAG-13, 233 (1977).
2. H. Kanter, J.Appl.Phys. 46, 4018 (1975).
3. M.J. Feldman, P.T. Parrish, and R.Y. Chiao, J.Appl.Phys. 46, 4031 (1975).
4. S. Wahlsten, S. Rudner, and T. Claeson, Appl.Phys.Lett. 30, 298 (1977).
5. J. Mygind, N.F. Pedersen, and O.H. Soerensen, Appl.Phys.Lett. 32, 70 (1978).

6. R.Y. Chiao and P.T. Parrish, J.Appl.Phys. 47, 2639 (1976).
7. Y. Taur and P.L. Richards, J.Appl.Phys. 48, 1321 (1977).
8. The full derivation will be given in a future communication.
9. J. Mygind, N.F. Pedersen, and O.H. Soerensen, Appl.Phys.Lett. 29, 317 (1976).
10. In recent experiments we have inserted a 8 dB (cooled) attenuator vane in the waveguide. All sharp resonances were quenched and a fairly broad band coupling resulted.
11. M.J. Feldman, J.Appl.Phys. 48, 1301 (1977).
12. O.H. Soerensen, N.F. Pedersen, and J. Mygind, to be published (Appl.Supercond.Conf. 1978).
13. In collaboration with M.T. Levinsen of the H.C.Ørsted Institute, Univ. of Copenhagen and a graduate student B. Dueholm.

PROGRESS ON MILLIMETER WAVE JOSEPHSON JUNCTION MIXERS

Y. Taur* and A. R. Kerr
NASA Goddard Institute for Space Studies
Goddard Space Flight Center, New York, N. Y. 10025

ABSTRACT

Preset, recyclable Nb point contacts are tested as low-noise Josephson mixers at a signal frequency of 115 GHz. The best result achieved is a mixer noise temperature (single sideband) of 120 K with unity conversion efficiency (SSB) for a junction at 6 K. Variation of mixer properties with temperature and other parameters is presented.

INTRODUCTION

Josephson junctions are suitable for applications in mm-wave heterodyne receivers due to their non-linear characteristics and low power requirements.[1] Previous investigations have shown very promising results at 36 and 130 GHz, using low temperature adjusted point contacts.[2,3] A reasonably satisfactory theoretical model has also been developed for Josephson mixers.[4] However, in order to use them in practical receivers such as on a mm-wave telescope, the mechanical instability problem with point contacts must be solved. We have successfully fabricated recyclable Nb point contacts which show desirable mechanical properties and excellent mixer performance at 115 GHz.

FABRICATION

Our point contact is made with a Nb wire and a Nb foil between two substrates as shown in Fig. 1(a). The tip of the Nb wire is etched electrochemically to a sharp point of radius $<0.25\ \mu m$. The substrate material is Corning 8260 glass which matches the thermal expansion of Nb.[5] Strip-line choke patterns on the substrates are fabricated by photo-lithographic techniques to avoid leakage of signal power into the output cable. To assemble a junction, the substrate with Nb foil is first glued to the base substrate; then the substrate with Nb wire is moved in with a differential micrometer. When making the contact, we adjust for a junction resistance $\approx 30\ \Omega$ at room temperature before gluing the whisker substrate to the base.

The assembled junction is mounted in a reduced-height waveguide as shown in Fig. 1(b). A backshort behind the junction can be adjusted for best coupling to the RF source. The junction is located by a spring wire in a slot in the surface of the mixer block. A choke section on the foil substrate is in contact with an indium-coated surface of the mixer block to ensure a well-defined RF and dc ground. The intermediate frequency (IF) output and dc bias are connected to the end of one substrate by a spring contact. Such a configuration avoids any rigid constraint on the glass structure, and

*Also with Columbia Radiation Laboratory, Columbia University, New York, N. Y. 10027

ISSN: 0094-243X/78/254/$1.50

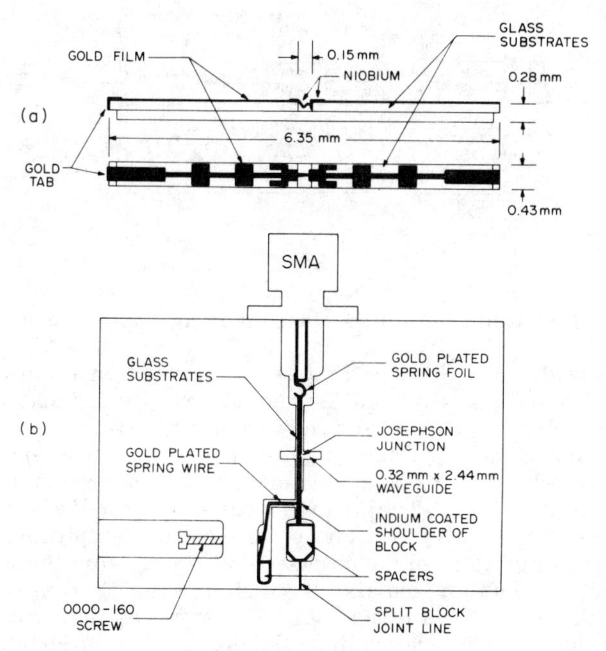

Fig. 1. (a) Side- and top-view of an assembled Josephson point contact. (b) Junction mixer mount, the substrate lies sideways in the machined slot.

allows it to expand and withstand mechanical shocks without deforming the junction.

The junction is sealed in He exchange gas so that its temperature can be controlled between 1.7 K and 9.0 K. The junction resistance usually increases by ~15% from 300 K to 4.2 K, probably because of a slight pressure change in the contact area. Such a resistance change, however, is reversible and reproducible, and the superconducting I-V characteristics stay the same between cycles. One of the junctions has been cycled between room and liquid He temperatures more than ten times without significant change over a period of one month.[6]

TEST SYSTEM

The block diagram in Fig. 2 shows the system for our mixer tests. The signal source is a calibrated gas-discharge noise tube with a waveguide iris filter (center frequency 115.6 GHz, 3 dB bandwidth 2.5 GHz) in front of it. The local oscillator (LO) power is supplied by a klystron tube and coupled to the signal line via a 10-dB directional coupler. The RF power is fed to the junction via a stainless-steel waveguide and a four-to-one stepped transformer to match to the reduced-height mixer waveguide. The IF output passes through a lumped resonant filter (center frequency 1.4

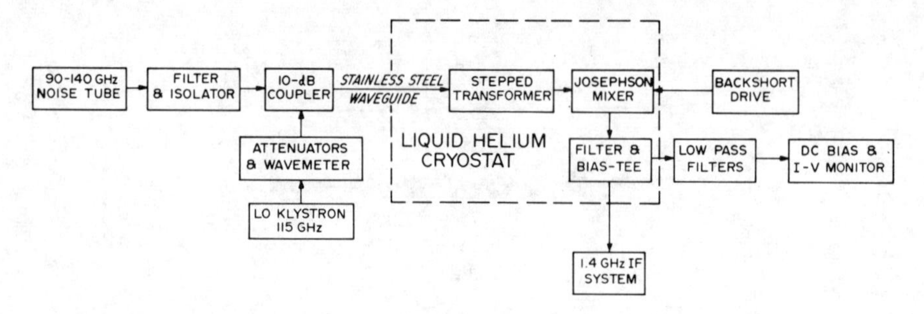

Fig. 2. Block diagram of the Josephson mixer test system.

GHz, 3 dB bandwidth 0.2 GHz) to a measuring system at room temperature. Leads for dc bias and I-V curve measurement are filtered and connected to the IF line with blocking capacitors and inductors.

The losses of the input and output transmission lines are carefully measured separately so that one can determine the conversion efficiency of the Josephson mixer alone. The 1.4 GHz radiometer/reflectometer IF system (bandwidth 0.05 GHz) consists of an isolator, amplifier, and detector.[7] It is calibrated to read the output noise power from the junction in absolute temperature units, and also the match at the IF port of the mixer. When the signal noise tube is turned on, we observe an increase in the junction IF noise output, which determines the conversion efficiency. The mixer noise temperature, referred to the input flange of the waveguide transformer, can then be determined from the junction IF noise temperature, conversion efficiency, and RF input noise temperature seen by the junction.

RESULTS

To minimize the mixer noise temperature of a junction, we vary the temperature, backshort position, LO power, and dc bias. For most junctions, the best result is usually obtained when the LO power incident on the junction is about 2 to 3 nW, and when the dc bias is slightly below half of the voltage of the first LO induced step. The results from our best junction are shown in Fig. 3, where the mixer noise temperature is plotted against backshort distance in units of waveguide wavelength λ_g for several operating temperatures. At 4.2 K, this junction has a resistance (R) of 55 Ω and a critical current (I_c) of 20 μA. There is a negative slope region at low voltages on the I-V curve due to our dc voltage bias of the junction.[8] The negative resistance disappears, however, when the junction is heated to 6 K, where its critical current becomes 12 μA.

Since the signal frequency is centered at 115.6 GHz and the IF is at 1.4 GHz, the LO frequency can be either at 114.2 GHz for upper sideband (USB) mixing or at 117.0 GHz for lower sideband (LSB) mixing. In both cases, the mixer noise temperature has a minimum as the backshort is varied at a fixed temperature. Figure 3 shows that the lowest mixer noise temperature is obtained at T = 6 K. In addition, the optimum backshort position shifts with temperature consistently. This arises because the junction behavior (RI_c product) has a rather strong dependence on temperature. Furthermore, USB mixing seems to yield slightly better results at all temperatures than LSB mixing. We believe that this is caused by the frequency-dependent impedance of the choke and waveguide structures, rather than being an intrinsic property of Josephson mixers.

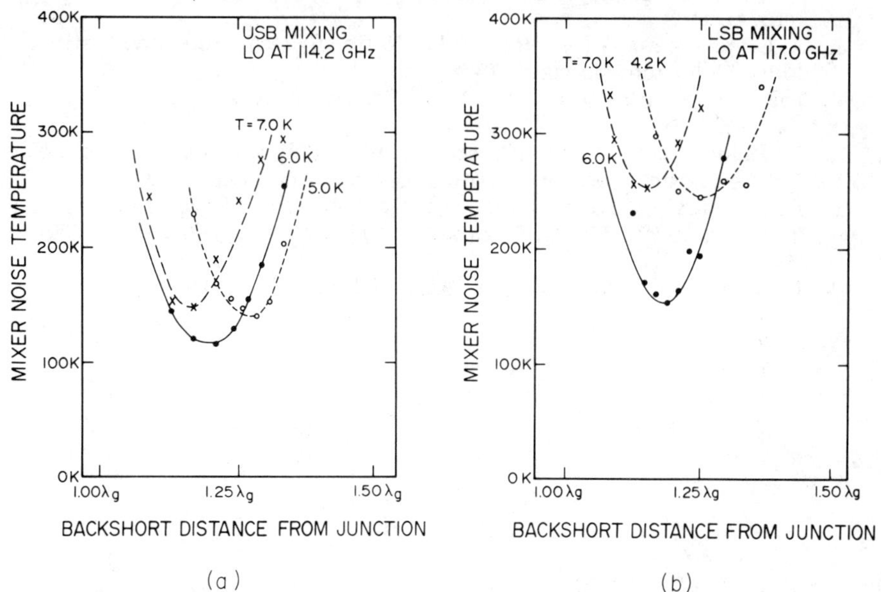

Fig. 3. Mixer noise temperature vs. backshort position for (a) upper sideband and (b) lower sideband mixing at several temperatures. Each data point is obtained by adjusting the LO power and dc bias for the lowest noise temperature at a fixed backshort setting.

The best mixer noise temperature we have obtained is 120 K (±10 K) with unity conversion efficiency (±10%) for USB mixing at 6 K. The power mismatch between the junction and the 50 Ω IF system is less than 5 percent. These results are within a factor of two from the theoretically predicted performance of Josephson mixers at this frequency.[4] Further improvement on the mixer noise temperature may be possible if the relationship between the junction parameters and the mixer embedding circuit can be better understood.

CONCLUSION

Our results represent significant progress toward realization of Josephson mixers in a practical mm-wave heterodyne receiver. Such a system, consisting of a low-loss input waveguide and a cooled IF amplifier following the Josephson mixer, is now under construction. Possible saturation problems can be solved by using a narrow-band cooled filter in front of the junction. We expect to obtain a single sideband system noise temperature of less than 220 K, which should make it the best coherent receiver at this frequency.

REFERENCES

1. P. L. Richards, F. Auracher, and T. Van Duzer, Proc. IEEE 61, 36 (1973).
2. Y. Taur, J. H. Claassen, and P. L. Richards, Appl. Phys. Lett. 24, 101 (1974).
3. J. H. Claassen and P. L. Richards, National Radio Science Meeting, Boulder, Colorado, January 1978.
4. Y. Taur, J. H. Claassen, and P. L. Richards, Rev. Phys. Appl. 9, 261 (1974).
5. R. A. Buhrman, S. F. Strait, and W. W. Webb, Technical Report No. 1555, The Material Science Center, Cornell University (1971).
6. Y. Taur and A. R. Kerr, paper submitted to Appl. Phys. Lett.
7. A. R. Kerr, IEEE Trans. Microwave Theory Tech. MTT-23, 781 (1975).
8. Y. Taur and P. L. Richards, J. Appl. Phys. 46, 1793 (1975).

PHASE INSTABILITY NOISE IN JOSEPHSON JUNCTIONS*

R.Y. Chiao, M.J. Feldman, D.W. Peterson, B.A. Tucker
Department of Physics, University of California, Berkeley, CA 94720

M.T. Levinsen
Physics Lab. I, H.C. Ørsted Institute, University of Copenhagen

ABSTRACT

We explain a noise phenomenon seen in microbridge SUPARAMPs in terms of phase instability in the junction voltage waveform associated with rf-induced nulls in the supercurrent. We show that the same instability occurs between all rf-induced steps. Experimental verification is presented.

INTRODUCTION

In our work with the SUPARAMP (Superconducting Unbiased PARametric AMPlifier)[1,2] we have often observed a sharp rise in the noise output closely associated with the parametric process. An example is shown in Fig. 1. The signal (ω_s), pump (ω_p), and idler (ω_i) frequencies are related by $2\omega_p = \omega_s + \omega_i$. This triplet, with a monochromatic signal gain of 14 dB, sits upon a symmetric broadband noise background. This "noise rise" has a peak output temperature of 10^6 K, implying a noise temperature of 5×10^4 K for the amplifier. The gain and noise rise were observed to occur almost simultaneously over a narrow range of pump power (0.5 dB). This was the pump power for which the supercurrent is suppressed to zero.

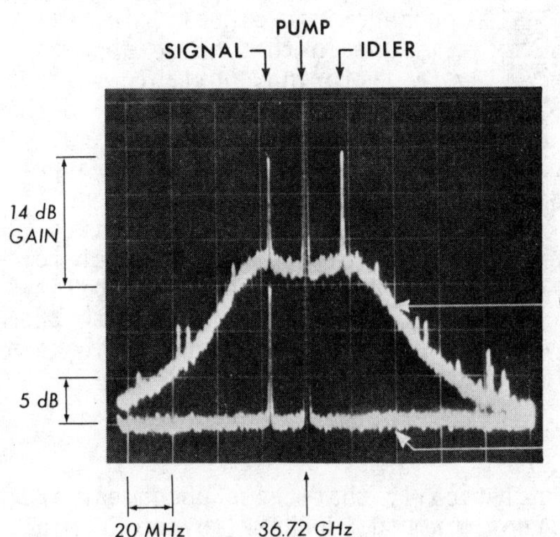

Fig. 1. Input and output spectra of a microbridge SUPARAMP.

*This work was supported in part by NASA under grant NGL 05-003-272 and in part by NSF under grants AST 75-13511 and AST 72-05136.

ISSN: 0094-243X/78/259/$1.50

This noise cannot be due to amplified thermal input.[3] The temperature of the input thermal radiation is only 8 K (the circulator and a 20 dB input attenuator were both at 2 K). In other cases, we have seen noise rise without any monochromatic gain.[4] Rather, the noise rise indicates the presence of intrinsic internal noise.[5]

The indium VTB used to obtain Fig. 1 was 300 nm square and less than 150 nm thick, and had banks of thickness 500 nm.

NOISE IN THE BAK ANALOG

We observed much the same behavior by simulating the SUPARAMP using a Bak[6] RSJ analog. The impedance of our analog is much lower than the source impedance at all frequencies. Thus the pump current through the device is perfectly sinusoidal, whereas the voltage waveform can develop many harmonics. We shall call this the <u>current-clamped</u> situation. In the <u>voltage-clamped</u> situation the voltage waveform is a pure sinusoid, but the current waveform may contain many harmonics. Which of these two cases applies in practice depends on the impedance seen by the Josephson element at the harmonics of the pump frequency. If there is a large inductance in series with the Josephson element, as for microbridges, current clamping occurs. If there is a large parallel capacitance, as for oxide barrier junctions, voltage clamping occurs. The voltage-clamped case has been treated analytically.[7] The current-clamped case, however, must be solved numerically.[4] We did this with the Bak analog and also by digital computations.

A spectrum analysis of the output of our analog showed large amplitude broadband noise coincident with the pump-induced nulls in the supercurrent. This occurred even in the absence of any applied broadband noise or monochromatic signal. In the time domain this noise appeared as a jittering, unstable voltage waveform (see Fig. 2) due to the onset of a voltage spike and, more importantly, sudden reversals in the sign of the reactance (i.e. 180° phase shifts of the voltage with respect to the current). This instability was closely associated with the appearance of monochromatic gain as in our experiments.

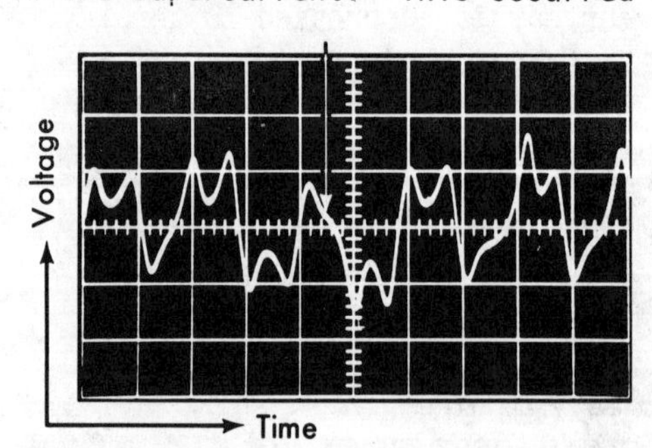

Fig. 2. Unstable voltage waveform from analog.

PERTURBED LIMIT-CYCLE SOLUTION

We have been able to show analytically that these phenomena are intrinsically connected in a current-clamped SUPARAMP, by performing a linear instability analysis of the RSJ model equation

$$\frac{d\phi}{dt} + \sin\phi = \alpha\cos\Omega t \qquad (1)$$

(normalized units), where α and Ω are the pump current amplitude and frequency. The phase difference ϕ across the Josephson element is expressed as $\phi = \phi_o + \phi_1$ where ϕ_o is the large-amplitude limit-cycle solution, phase-locked to the pump current, obeying the periodicity condition $\phi_o(t + nT) = \phi_o(t)$ where $T = 2\pi/\omega$ is the pump period. ϕ_1 is a small perturbation. The linearized equation for ϕ_1 is

$$\frac{d\phi_1}{dt} + \phi_1 \cos \phi_o(t) = 0. \quad (2)$$

The solution at discrete times $t = nT$ is $\phi_1(nT) = \phi_1(0)\, e^{-nAT}$,

where $A \equiv \frac{1}{T} \int_0^T \cos \phi_o(t)dt = \langle\cos \phi_o\rangle$.

By inspection $-1 \leq A \leq 1$. For a stable limit cycle A must be positive. The special case $A \rightarrow 0$, however, deserves special attention: a perturbation takes an infinite time to die away. The addition of the slightest noise will prevent the voltage waveform from finding its limit cycle. Also the phase of the limit-cycle waveform changes abruptly as A passes through zero (see below). Large phase jitter or FM noise results. This is the origin of the noise rise.

SECOND LIMIT-CYCLE SOLUTION

As noted in Ref. 4, if $\phi_o(t)$ is a limit-cycle solution of Eq. 1 then $\phi_o'(t) = \pi - \phi_o(-t)$ is also a solution. The primed voltage waveform $d\phi_o'/dt$ is the time-reverse of the unprimed. Clearly $A' = -A$ so that only one solution can be stable. Therefore, when A crosses zero the system must switch to the other solution. Were the voltage in phase with the current (i.e. resistive), it would be an even function of time, so no phase change would result. But since the voltage waveform is of predominantly reactive phase for $\Omega \ll 1$ due to the Josephson inductance, there is an abrupt phase change of nearly π with respect to the current upon passage through $A = 0$.

Equation 1 can be modelled by a rigid pendulum immersed in a viscous fluid (so that it is always at terminal velocity), under the influence of an applied torque $\alpha \cos \omega t$. ϕ is the angle of the pendulum with respect to its equilibrium position. For small α the pendulum oscillates about $\phi = 0$ and $A \equiv \langle\cos \phi\rangle \lesssim 1$. A decreases if we increase α. After A crosses through zero, gravity cannot overcome the applied torque and the pendulum rapidly goes through its inverted position, making a complete turn (analogous to a voltage spike). But now that A is negative, the pendulum is top-heavy on the average. This is an unstable situation. The pendulum must invert: its phase changes by π.

RF-INDUCED NULLS OF THE SUPERCURRENT

Digital computer solutions of Eq. 1 indicate that $A = 0$ at the rf-induced nulls of the supercurrent. To show this analytically, we generalize Eq. 1 by adding a dc bias current i_{dc}. A limit-cycle solution for the mth step has the form $\phi_o(t) = m\Omega t + \Phi_o(t)$ where $\Phi_o(t+nT) = \Phi_o(t)$. If we now inject a small extra amount of dc current Δi_{dc} at t = 0, the small response, ϕ_1, obeys the linear equation $d\phi_1/dt + \phi_1 \cos \phi_o(t) = \Delta i_{dc}$. The solution at discrete times t = nT is

$$\phi_1(nT) = e^{-nTA} \left[\frac{e^{nTA} - 1}{e^{TA} - 1} E \Delta i_{dc} + \phi_1(0) \right] \tag{3}$$

where $E \equiv \int_0^T \exp \left\{ \int_0^t \cos \phi_o(t')dt' \right\} dt.$

If A > 0, we see by inspection that there is no time-average change in voltage. But if A = 0, then

$$\phi_1(nT) = n E \Delta i_{dc} + \phi_1(0). \tag{4}$$

The phase now grows linearly in time: a dc voltage change results.[8] This means that A = 0 must correspond to the top or bottom of an rf-induced step.

The rf-induced nulls of the supercurrent are the special case m = 0 and i_{dc} = 0. Therefore A = 0 at these nulls; we expect phase-instability noise in an unbiased, current-clamped Josephson junction whenever microwave radiation causes the supercurrent to go through a zero. Experimentally, whenever we see a large noise rise from a VTB SUPARAMP, the supercurrent is in fact suppressed to zero. Note that Taur and Richards[4] found by digital computer that monochromatic gain should occur near (but not exactly at) the supercurrent nulls, reminiscent of Fig. 1.

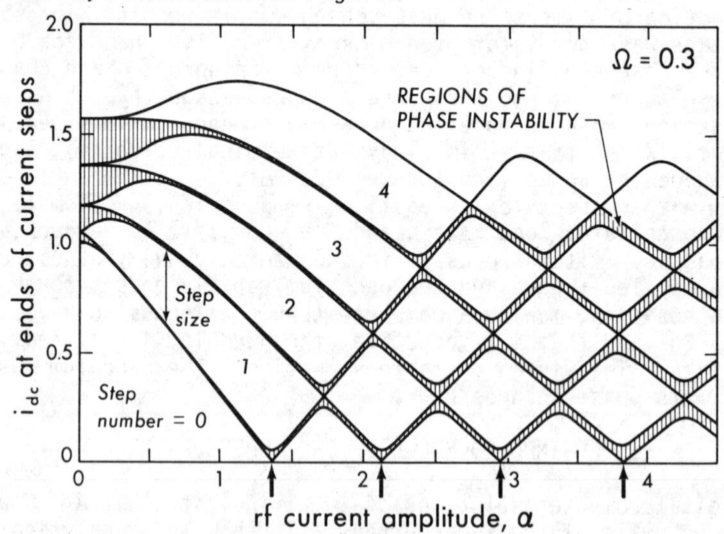

Fig. 3. Schematic of rf-induced steps. In unshaded areas A > 0. In shaded areas and on boundaries A = 0. Arrows indicate points of noise rise in the SUPARAMP.

Belykh et al.[9] have shown that the regions of the i-v curve between steps correspond to unsynchronized Josephson oscillations. No limit cycles exist here; rather, solutions are of a quasi-periodic nature. If we generalize our definition of A to an infinite time average, then A = 0 between the steps as well as at their extremities. In confirmation of this prediction, we have seen a noise rise from an indium VTB (Ω = 0.2) in these regions. The maximum noise occurred halfway between steps. The analog simulator also shows phase instability noise in these regions.

These results are summarized in Fig. 3, adapted from Ref. 10. A noise rise is seen on the analog throughout the shaded areas and along the boundaries.

VOLTAGE-CLAMPED SUPARAMP: NO NOISE RISE

For a voltage-clamped SUPARAMP both voltage and current are determined unambiguously. The phase cannot be unstable so there can be no noise rise. The analysis of Ref. 7 is then applicable. This limit can be approached by using small-area oxide barrier junctions. The capacitance shorts out high harmonics, and a close approximation to voltage-clamping should be achievable. Wahlsten et al.[11] have reported a 10 GHz SUPARAMP which uses a series array of such junctions. Their results are in excellent agreement with the SUPARAMP saturation theory,[3] with a noise temperature as low as 30 ± 20 K. They reported no extraneous noise rise associated with gain.

REFERENCES

1. P.T. Parrish and R.Y. Chiao, Appl. Phys. Lett. 25, 627 (1974).
2. R.Y. Chiao and P.T. Parrish, J. Appl. Phys. 47, 2639 (1976).
3. M.J. Feldman, J. Appl. Phys. 48, 1301 (1977).
4. A symmetric noise rise without monochromatic gain was also reported by Y. Taur and P.L. Richards, J. Appl. Phys. 48, 1321 (1977), for a point contact paramp.
5. We have also seen large noise outputs associated with parametric processes from thin films with no microbridges. This is presumably due to a flux-flow nonlinearity (see V.N. Gubankov, K.K. Likharev, and N.B. Pavlov, Sov. Phys. Solid State 14, 2721 (1973)) along the edges of the film. This effect is distinguishable from phase-instability noise in that it is very weak for thick films (as in the VTB), nonexistent for point contacts (see Ref. 4), and is not associated with the supercurrent nulls.)
6. C.K. Bak, Revue de Physique Appliquée 9, 15 (1974).
7. M.J. Feldman, P.T. Parrish, and R.Y. Chiao, J. Appl. Phys. 46, 4031 (1975).
8. After ϕ_1 becomes large our analysis is no longer valid.
9. V.N. Belykh, N.F. Pedersen, and O.H. Soerensen, Phys. Rev. B16, 4853 (1978).
10. K.K. Likharev and V.K. Semenov, Radio Engineering and Electronic Physics 16, 1917 (1971).
11. S. Wahlsten, S. Rudner, and T. Claeson, to be published.

DIFFERENTIAL HETERODYNE FREQUENCY CONVERSION AT 0.4 mm WAVELENGTH

B.T. Ulrich*, R.W. van der Heijden, and J.M.V. Verschueren
Physics Laboratory and Research Institute for Materials
University of Nijmegen, Toernooiveld, Nijmegen,
The Netherlands

ABSTRACT

A Josephson-effect differential heterodyne receiver converts a variable signal frequency f_S, near a fixed local oscillator f_L, to a fixed intermediate frequency $f_I = f_S - f_L - f_Q$, by mixing between $f_D = f_S - f_L$ and a voltage-tunable variable quantum oscillation $f_Q = 2eV/h$ provided by a double-junction resistive interferometer. At $f_L = 0.7$ THz provided by a far infrared formic acid laser, with $f_S = f_L + f_D$ provided by a Doppler shift, power conversion efficiencies η_Q from $f_D \rightarrow f_I$ up to 0.2 were observed. Linewidths of the quantum oscillation $\delta f_Q \sim (2k_BT/\pi L)^{1/2}$ (2eR/h) down to $\sim$ 0.5 Hz were observed for $R = 3.3 \times 10^{-9}\ \Omega$ interferometer resistance, and $L \sim 10^{-9}$ H interferometer inductance, demonstrating an intrinsic frequency resolution of 1 in 10^{12} at high conversion efficiency for the technique.

INTRODUCTION

A high-resolution voltage-tunable heterodyne receiver combines the advantages of a Josephson-effect mixer pumped by an external local oscillator with the advantage of voltage tunability via the Josephson voltage-frequency relation. We consider in turn the mechanism of differential frequency conversion and its experimental realization.

PRINCIPLE OF OPERATION

In a single Josephson junction, the internal quantum oscillation can serve as local oscillator. However, the line is broadened by Johnson-Nyquist voltage fluctuations in the intrinsic junction shunt resistance to widths of order 0.5 GHz. An external local oscillator, such as provided by a far infrared laser, has a much narrower line width, but a different intermediate frequency is required for each signal frequency. In differential heterodyne fre-

*Physics Department, Univ. of Geneva, 1211 Geneva 4, Switzerland.

ISSN: 0094-243X/78/264/$1.50

quency conversion, a two-step frequency conversion is used. First, an external local oscillator is used to mix with the signal, producing a difference frequency f_D. Second, a narrowband quantum oscillation at f_Q provided by a Josephson-junction structure mixes with f_D to produce a frequency within the bandwidth of an intermediate-frequency amplifier at fixed frequency.

The narrowband quantum oscillation is provided by a resistive double-junction quantum interferometer, or "resistive d.c. SQUID", in which two Josephson junctions are connected in parallel at one end by a superconductor, and at the other end by a low resistance R. When the junctions are biased in parallel, then the voltages V_1 and V_2 across the junctions can differ by the voltage $V = V_2 - V_1$ across R. The internal quantum oscillations then beat at frequency $f_Q = 2eV/h$. Provided R is small, the quantum oscillation line at f_Q can be extremely narrow, of order $\delta f_Q \sim (2k_BT/\pi L)^{1/2}$ (2eR/h), where L is the inductance of ring containing the two junctions. For $R = 10^{-9}\Omega$, T = 4.2 K, and with a typical $L \sim 10^{-9}$ H, the line width is $\delta f_Q \sim 0.1$ Hz.

In the two-step conversion process, the resistive double junction interferometer simultaneously serves as mixer with an external local oscillator at f_L and signal at f_S, with conversion to a difference frequency $f_D = f_S - f_L$ with conversion efficiency η_e analogous to that for a single Josephson-junction mixer. This frequency f_D mixes with the internal quantum oscillation frequency f_Q to produce a fixed intermediate frequency $f_I = f_- = f_S - f_L - f_Q$ with conversion efficiency η_Q. The two-step process is

$$f_S \overset{\eta_e}{\rightarrow} f_D = f_S - f_L \overset{\eta_Q}{\rightarrow} f_I = f_- = f_D - f_Q \, .$$

The special aspect of the present technique is the conversion from f_D to $f_I = f_- = f_S - f_L - f_Q$. This voltage-tunable conversion process makes it possible to choose f_I at the lowest possible frequency consistent with the required bandwidth, and where low noise amplifiers are available.

EXPERIMENT

A Josephson-effect differential heterodyne receiver was constructed to measure the conversion efficiency η_Q at signal frequency 693 GHz, and to study the voltage tunable properties of such a receiver.

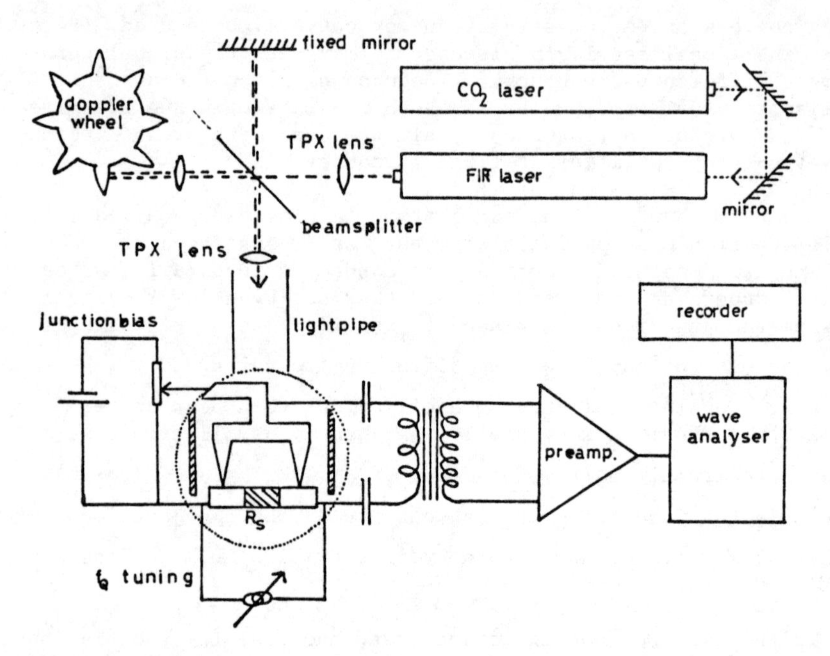

Fig. 1. Experimental arrangement for differential heterodyne receiver.

The double Josephson-junction resistive quantum interferometer used two point contacts formed on opposite ends of a "U" shaped niobium wire, as shown schematically in Fig. 1. The points were pressed by a differential screw mechanism against two solder (PbSn) electrodes melted onto opposite sides of a 30 μm thick copper sheet of 0.5 cm^2 area which served as the resistive part R of the quantum interferometer. At 4.2 K, the resistance $R = 3.3 \times 10^{-9}\ \Omega$ was measured via the voltage-frequency relation for the quantum oscillations at $f_Q = 2eV/h = 2eIR/h$, with the imposed current I measured with a microammeter. The interferometer was current-biased at finite voltage to observe oscillations in the interferometer critical current. The oscillating voltage across the interferometer was amplified by a transformer-coupled low-noise amplifier, and the frequency measured with a counter to determine f_Q, or the frequency spectrum measured with an audio frequency spectrum analyzer during the frequency conversion experiments. Estimated interferometer inductance of order 1×10^{-9} H lead to an expected linewidth $\delta f_Q \sim$ 0.3 Hz. Measured linewidth was 0.5 Hz.

The output of a formic acid far infrared laser at $f_L = 693$ GHz served as local oscillator. Part of the laser beam was split off and Doppler-shifted to form the signal at $f_S = f_L + f_D$ as illustrated in Fig. 1. The Doppler shift f_D provided by a translating mirror, or

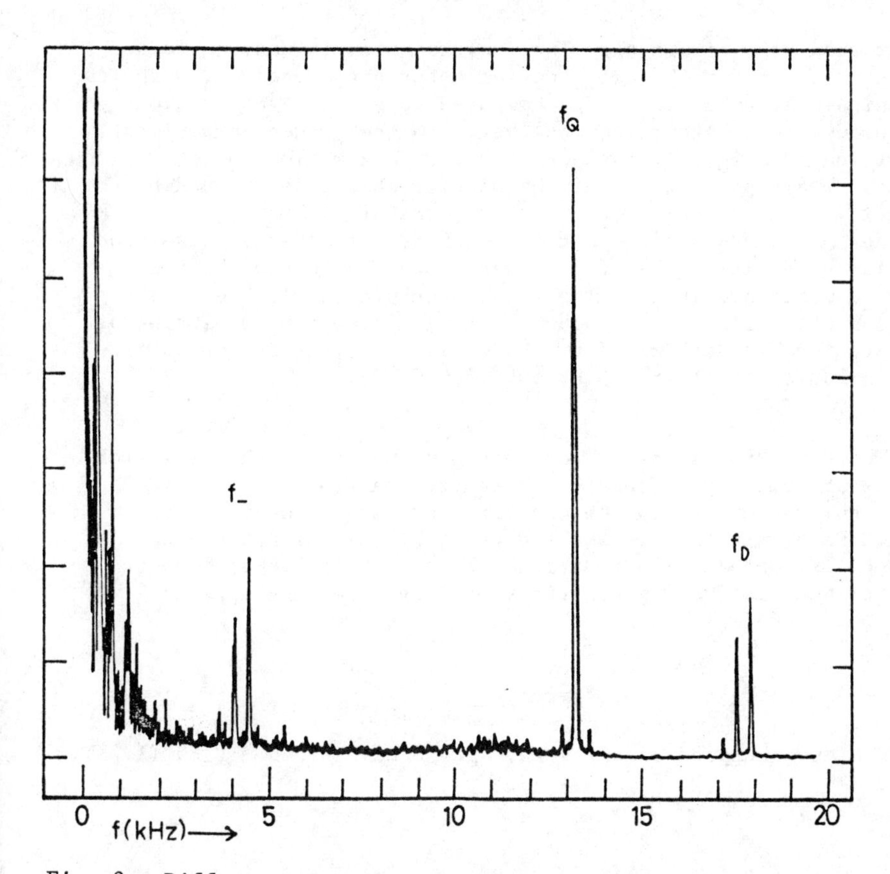

Fig. 2. Differential heterodyne frequency conversion at 693 GHz. The signal was provided by Doppler-shifting part of the local oscillator line. Mixing between the local oscillator and signal in the junction gave the difference signal f_D at 18 kHz. Structure in f_D is due to uneven reflectivity of the teeth on the Doppler wheel. The quantum oscillation f_Q in the double junction resistive interferometer at 13 kHz mixes with f_D to down-convert the signal further to f_-, preserving the structure due to the Doppler wheel. An audio frequency spectrum analyzer served as intermediate frequency amplifier, and was used to measure the voltage spectrum across the junction. The audio frequency spectrum illustrated has not been corrected for the frequency response of the coupling transformer which falls off for frequencies above about 10 kHz.

by a Doppler wheel[1] was adjustable from a few Hz to 50 kHz. The purpose of the present experiment was to observe the magnitude of η_Q, rather than to optimize η_e. The conversion efficiency η_Q is a figure of merit for the differential heterodyne receiver, with its convenience of voltage tuning, compared to a single-conversion Josephson-effect heterodyne receiver. In the present experiment, conversion from f_D to f_I was observed, with a maximum η_Q of 0.2 observed. The typical width of the Doppler shift, determined by the stability of the Doppler-shift method, was $\delta f_D \sim 70$ Hz, when f_D was of order a few kHz. Differential heterodyne frequency conversion is illustrated in Fig. 2, for a case in which the Doppler wheel produced structure in f_D. The same structure is visible in the down-converted line at f_-. The intrinsic frequency resolution of the technique is determined by the width of f_Q, giving an intrinsic resolution of $\delta f_Q = 0.5$ Hz out of 0.7 THz.

We thank Professor P. Wyder for his support of this research and for stimulating discussions. We are grateful to Dr. N. Nieuwenhuijzen of Utrecht Observatory for the use of the Doppler wheel. Part of this work was supported by the Stichting voor Fundamenteel Onderzoik der Materie (FOM) with financial support from the Nederlandse Organisatie voor Zuiver Wetenschappelijk Onderzoek (ZWO).

REFERENCE

1. N. Nieuwenhuijzen and C. Veth, Opt. & Laser Tech. 3, 89 (1971).

JUNCTIONS - TYPES, PROPERTIES, AND LIMITATIONS*

M. Tinkham
Harvard University, Cambridge, Mass. 02138

ABSTRACT

The types of Josephson junctions are surveyed, with emphasis on tunnel and constriction configurations as limiting examples. The limits on the intrinsic high-frequency performance set by capacitive and heating effects are analyzed theoretically, and the latter compared with experimental data on VTB's and point contacts. Nonequilibrium effects more subtle than those accounted for by simple heating are briefly considered. Finally, the properties of point contacts which apparently show nearly ideal intrinsic frequency-dependent behavior out to far infrared frequencies of twice the gap frequency are summarized.

INTRODUCTION

Although Josephson[1] originally envisioned his effect in terms of two massive superconductors weakly connected by electron tunneling through a thin insulating barrier, it is now clear that the qualitative features of the Josephson effect arise for many other forms of weak-link coupling. For example, the weak link can take the form of a point contact, a short narrow constriction in a thin film (a microbridge or "Dayem bridge"), a VTB (Variable Thickness Bridge, in which the bridge is thinner than the banks), an SNS junction (in which the superconducting pairs "tunnel" through a thin normal metal layer), a proximity effect bridge (similar to SNS junction, except that the "normal" metal may be a superconductor just above or below its T_c), an S-Se-S junction (a tunnel junction using the lower barrier of a semiconductor to allow the barrier to be thicker), or even a weak segment induced in a homogeneous film strip by a localized magnetic field or by a creation of a localized excess quasiparticle population by tunnel injection or optical irradiation. Given all these configurations, as well as choices of material and physical dimensions, it is clear that there is much scope for optimization of the junction chosen for a given application, especially when ease of fabrication and durability are also taken into consideration. Since the following papers will address the latter issues, I shall concentrate my attention on the properties and limitations of several of the major junction types as set by the basic physics of the situation.

For the purposes of our discussion, the conventional RSJ (Resistively Shunted Josephson Junction) model provides an adequate framework. In this model, one considers an ideal Josephson element with critical current I_c shunted by a normal resistance R and a capacitance C. Thus the total current is given by

*Research supported in part by NSF, ONR, and JSEP.

ISSN: 0094-243X/78/269/$1.50

$$I = I_c \sin\phi + \frac{V}{R} + C\,\frac{dV}{dt} \quad , \tag{1}$$

where ϕ is the phase difference of the superconducting wave function across the link, which increases with time with the instantaneous frequency

$$\omega = \frac{d\phi}{dt} = \frac{2eV}{\hbar} \quad . \tag{2}$$

The capacitance C in (1) depends critically on junction geometry, and bears no universal relation to the other parameters I_c and R. On the other hand, for ideal junctions, either tunnel[2] or constriction,[3] I_c and R are not independent, but are related by

$$I_c R = \frac{\pi\Delta(T)}{2e} \tanh \frac{\Delta(T)}{2kT} \quad , \tag{3}$$

which is proportional near T_c to $(T_c - T)$, and approaches the limit $\pi\Delta(0)/2e$ or $0.24T_c$ (mV) at $T = 0$, assuming the BCS result $\Delta(0) = 1.76kT_c$. Since the Josephson coupling energy is only $-(\hbar I_c/2e)\cos\phi$, in practice I_c must be greater than about 1 μA to avoid partial loss of coherence due to noise. For junctions showing the ideal I_cR product, this corresponds to requiring $R < 1000\ \Omega$. On the other hand, R should be greater than ~1 Ω for efficient matching to circuits and radiation.

In discussing the limits and optimization of junctions, one must distinguish two broad classes of electronic applications: computers and sensitive detectors. Present computer development emphasizes "latching" logic, in which the Josephson element operates in a hysteretic mode. This is normally obtained with a tunnel junction with sufficient capacitance to satisfy the Stewart-McCumber criterion,[4] although the hysteresis induced by self-heating effects in metallic weak links provides another possible route to easily switched, if somewhat slower, bi-stable devices. By contrast to these latching logic elements which rely on hysteresis for operation, hysteresis in detectors is usually an undesirable feature. Thus capacitance and heating effects become twin obstacles to be overcome, particularly if performance is to be pushed to the highest possible frequencies. Since my own experience centers on understanding and trying to circumvent these high-frequency limits, I shall use that goal as the organizing principle in the discussion which follows.

HIGH-FREQUENCY LIMITS

It is now generally recognized that the usefulness of Josephson devices at high frequencies is limited by three distinct considerations: (1) The I_cR product defines a characteristic frequency $\omega_c = 2eI_cR/\hbar$ of order the energy gap frequency ω_g, above which resistive shunting of the ideal nonlinear inductive Josephson element increasingly degrades performance. Since the I_cR product scales with T_c, this consideration favors high T_c materials, independent of junction type. (2) The RC time constant defines a characteristic frequency $\omega_{RC} = (RC)^{-1}$ above which capacitive shunting becomes more serious than resistive shunting. This constraint depends critically on junction type, being most serious in tunnel junctions. (3) Nonequilibrium effects due to the power dissipated tend to increase as $(\hbar\omega/2e)^2/R$, since the ac voltage level V_1 must increase roughly as $\hbar\omega/2e$ to give a useful nonlinear response. This constraint also depends critically on junction type, and is most serious in metallic weak links. While considerations (1) and (2) give power-law rolloffs, consideration (3) leads to an exponential cutoff at high voltage levels.

CAPACITIVE LIMITS

At sufficiently high frequencies, capacitive shunting always dominates. However, so long as $\omega_{RC} = (RC)^{-1}$ exceeds the intrinsic characteristic frequency $\omega_c = 2eI_cR/\hbar$, capacitive shunting will be less than resistive shunting up to $\omega_c \sim \omega_g \sim 10^{12} - 10^{13}\ \text{sec}^{-1}$, above which the performance rapidly falls off in any case. Thus, capacitive shunting is not a major limitation on performance so long as

$$C \lesssim \frac{\hbar}{2e(I_cR)R} \approx \frac{1\text{pF}}{T_c(\text{K})R(\Omega)} \quad . \tag{4}$$

This is recognized also as the Stewart-McCumber criterion[4] for non-hysteretic operation of a Josephson device. In the second form of (4), the result is reexpressed in practical units, assuming I_cR is near its low temperature limiting value. As noted above, R is constrained to the general range of 1-1000 Ω by consideration of noise and external coupling, so (4) sets a limit on C for non-hysteretic or high-frequency operation. For example, for the typical values T_c = 10 K and R = 10 Ω, we find $C \lesssim 10^{-2}$pF, which implies very small physical size, and correspondingly high critical current densities.

In carrying this analysis further, we shall focus on two limiting cases: a) tunnel junctions and b) constriction weak links. These are limits in that the former has an insulating barrier between the two superconductors, while the latter is of homogeneous superconducting metal, so that the barrier is purely geometrical. Oxide barriers typify (a); microbridges and VTB's exemplify (b), but clean point contacts should also fall into this category. Most other forms of links, such as dirty point contacts and SNS junctions, fall into an intermediate category, in which the barrier material is intermediate between an insulator and a strong superconductor.

TUNNEL JUNCTIONS

In a tunnel junction, both C and 1/R scale in proportion to the area, so that the RC time constant is independent of the area, and depends only on the barrier properties. The capacitance per unit area is 0.09 ε/d (in pF), where d is the thickness (in cm) of the barrier, whose dielectric constant is ε. In contrast to this d^{-1} dependence of C, the tunnel conductance per unit area falls exponentially with d. As a result, the RC product is reduced by going to a thin barrier, and a correspondingly high tunnel current density. Combining (3) and (4), we obtain the condition

$$J_c = \frac{I_c}{A} \geq 1.5 \times 10^{-5} \frac{\varepsilon T_c^2}{d} \text{ (amp/cm}^2\text{)} \quad (5a)$$

or the equivalent condition in terms of resistance

$$R \leq \frac{15.6\ d}{A \varepsilon T_c} \ (\Omega) \quad (5b)$$

where A is the area of the junction in cm^2, and I_c is the low-temperature limiting value. Taking the typical values $\varepsilon = 3$, $T_c = 8K$, and $d = 2 \times 10^{-7}$cm, these conditions yield

$$J_c \geq 1.4 \times 10^4 \text{amp/cm}^2 \quad \text{and} \quad RA \leq 1.3 \times 10^{-7}\ \Omega\text{-cm}^2 \ . \quad (5c)$$

Since J_c depends exponentially on d, useful values of d lie in a narrow range. Also, useful values of T_c lie within about a factor of 2 of 8K. Thus these numerical estimates should give the right order of magnitude in most cases of interest. Tunnel current densities satisfying (5c) have been reported by several groups.[5] Note that (5c) implies a junction area of only $10^{-8}cm^2$ (i.e., 1 μm × 1 μm) to allow a resistance as large as 10 Ω in a junction without capacitive limitation.

Since the control of capacitive effects forces one to high current density, thin-barrier, small-area junctions, it is clear that to some degree one is approaching a constriction geometry as the barrier thickness decreases. An important parameter for monitoring this convergence is λ_J, the Josephson penetration depth, which characterizes the strength of the superconductivity in the barrier in essentially the same way as the penetration depth λ does in the bulk metal. Neglecting d/2λ compared to unity, it is given by

$$\lambda_J = \left[\frac{c\Phi_o}{16\pi^2 J_c \lambda}\right]^{\frac{1}{2}} \quad (6)$$

where $\Phi_o = hc/2e = 2.07 \times 10^{-7}$ gauss-cm^2 is the fluxoid quantum. As J_c is increased by decreasing the barrier, λ_J decreases. For the minimum J_c found in (5c) and the typical value $\lambda = 5 \times 10^{-6}$cm, one finds $\lambda_J = 3.8$ μm, almost two orders of magnitude greater than λ. It is also large enough to justify our assumption that the tunnel current is uniform over the area of a 1 μm square junction. If the barrier is further reduced, so that J_c approaches values typical of bulk superconductors, λ_J decreases and approaches λ.

It follows from the dependence of λ_J on J_c in (6) that there is an upper limit on the critical current through a tunnel junction before deterioration of its performance due to non-uniform supercurrent density. As shown in the early work of Ferrell and Prange,[6] and in more detail by Owen and Scalapino,[7] in a junction whose width greatly exceeds λ_J, the total supercurrent is limited to $2\lambda_J J_c$ times the perimeter of the junction. Thus for a circular junction of radius r_1, r_1 must be limited to $\sim 2\lambda_J$ if one expects a nearly uniform supercurrent density of $\sim J_c$. For larger radii, I_c increases only as the perimeter, while C and 1/R increase as the area, so that $I_c R$ should fall as r_1^{-1}, with corresponding degradation of performance. Using (6), we find the critical current of this "maximal junction" to be

$$I_c \approx \pi(2\lambda_J)^2 J_c = \frac{c\Phi_o}{4\pi\lambda} \quad . \tag{7}$$

Note that the barrier-dependent properties J_c and λ_J have dropped out, leaving a result depending only on the value of λ in the metal electrodes. Again taking $\lambda = 5 \times 10^{-6}$ cm as typical of most superconductors at low temperatures, we find $I_c \sim 30$ mA. The corresponding value of R is $\sim 0.008 T_c (\Omega)$, or 0.06 Ω for $T_c = 8$K. Since we have given only a qualitative argument and have no experimental data to test this conclusion, these numerical results should be treated with caution, but it appears that such a limit should exist.

CONSTRICTION GEOMETRIES

In constriction geometries, whether point contact or microbridge, the critical current densities are typically 4 orders of magnitude greater than even in the high J_c tunnel junctions described above, and the transverse dimensions correspondingly reduced. Accordingly, junction capacitance is usually not a serious limitation, at least for frequencies up to $\omega_c \sim \omega_g$.

Since (7) depends only on the metal of the electrodes and not on the tunnel barrier, one might expect it to apply also to a constriction weak link. One can in fact carry out an argument similar to that above, restricting the radius a of the constriction to be $\lesssim \lambda$ and noting that the effective length of the constriction including the distance required for the current to fan out into the banks can be no less than the radius a, if the material is homogeneous. Within the accuracy of this approximate analysis, one finds the same result (7). Thus, for both tunnel and constriction weak links, it appears that there should be an upper limit on I_c and lower limit on R before deterioration of performance sets in. Since practical considerations normally dictate a higher resistance level in any case, this limit appears to be primarily of academic interest.

HEATING EFFECTS

The above discussion indicates that it should be possible to fabricate tunnel junctions as well as constriction weak links with sufficiently high current density that capacitive shunting would not seriously degrade the high-frequency performance implicit in the I_cR product. But when one tries to use the non-linear properties of Josephson devices at high frequencies, this implies correspondingly high voltage levels V_1 of order $\hbar\omega/2e$, with the attendant dissipation of power of the order of

$$P \approx V_1^2/R \approx (\hbar\omega/2e)^2/R \quad . \tag{8}$$

This power input into the electronic system creates a non-equilibrium distribution of quasi-particles and phonons, limited by the speed with which the excess energy can be removed. Since diffusion of these "hot" electrons is the fastest energy removal process, the finite-voltage performance of the device is governed by the geometric constraints on this diffusion. The best performance results from metallic geometries in which the "hot" electrons diffuse out in three dimensions. Examples are point contacts and VTBs. In this case, the eventual extraction of the energy to the He bath occurs over a large area, and is no bottleneck. Conventional planar microbridges are characterized by 2-dimensional electronic diffusive cooling. In this case the distance traveled during the transfer of energy from electrons to phonons and then across the interface to He or substrate enters logarithmically in the temperature rise in the link itself. In 1-dimensional strip geometries, the bottleneck of the transfer of energy by phonons to the environment plays a more critical role in determining the temperature rise. As shown by Skocpol, Beasley, and Tinkham,[8] the excess energy is electronically carried along the strip for the thermal healing length η, typically ~5 μm, before escaping through the surface as phonons. From these considerations, SBT concluded that heating effects would limit the usefulness of planar bridges to roughly ω_g, or $V \sim I_cR$, but that 3-dimensional geometries should be useful to higher voltages, if made with very narrow constrictions (radius $a < \xi$).

These semi-quantitative conclusions were refined by Tinkham, Octavio, and Skocpol[9] (TOS). They showed that, in 3-dimensional constriction links, the critical current should fall as

$$I_c(P) = I_c(0)e^{-P/P_o} \tag{9}$$

where

$$P_o = \frac{K(T_c)T_c\xi(0)\Omega}{\sqrt{2}}\,(1 - t_b^2)^{\frac{1}{2}} \tag{10}$$

where K is the metallic thermal conductivity, $\xi(0)(1-t)^{-\frac{1}{2}}$ is the coherence length, and Ω is the solid angle through which the diffusive electronic cooling occurs. P_o is typically ~10 μW, falling to zero as $t_b = T_b/T_c \rightarrow 1$. ($T_b$ is the bath temperature.) This prediction was confirmed experimentally by comparing the dependence of the amplitude of harmonic steps induced in a tin VTB by X-band microwaves with the expected Bessel function dependence as shown in Fig. 1. Note that over 180 steps are observed, corresponding to Josephson currents at 1.8 THz and operation at over 3mV. Other measurements of Octavio also gave a good confirmation of the predicted temperature dependence of P_o.

In addition to limiting the voltage or frequency to which the devices operate, nonequilibrium heating effects are manifested in two other ways: a decrease in the energy gap and an increase in Johnson noise. Insofar as one can characterize the extremely nonequilibrium situation at the center of the link by a temperature, the model of TOS (consistent with much older work[10] in a different context) shows that the maximum temperature should be

$$T_m = [T_b^2 + 3(eV/2\pi k)^2]^{\frac{1}{2}} \tag{11}$$

which reaches ~15K in Octavio's tin VTB and ~70K in Nb point contacts studied by McDonald and others. Thus, the center of the contact is actually far above T_c, and the device becomes a form of SNS link. Since the structure on the I-V curve at the energy gap and its subharmonics is dominated by the value of the gap in the material about ξ from the center, where T is slowly varying and much closer to T_b than to T_m, the downward shifts in measured gap are small; they are well-fitted by this model, however. Such gap shifts in a tin VTB are shown in Fig. 2, as measured by Octavio, et al.[11] The increase in Johnson noise is more dramatic, because so much of the resistance of the link comes from the narrow neck which is near T_m. In fact, under a plausible simplifying approximation, one expects the noise temperature to be simply the average $\frac{1}{2}(T_b + T_m)$. This prediction has been supported by measurements on noise rounding of Josephson steps of Nb point contacts by Weitz, Skocpol, and Tinkham,[12] as illustrated in Fig. 3. The fact that the measured noise is significantly below that expected from a shot noise model supports the presumption that these contacts act more like metallic constrictions than like tunnel junctions.

These many consistent bits of experimental evidence show that the simplification of treating the disequilibrium induced by Joule heating by simple macroscopic heat flow analysis is remarkably successful in accounting for the major effects which limit device performance. In particular, the crucial importance of using metallic geometries which facilitate the diffusive escape of nonequilibrium quasiparticles from the center of the weak link is highlighted.

It is also appropriate to remark that tunnel junctions normally are not troubled by heating problems because the tunnel resistance can be kept large compared to the normal resistance of the film, a degree of freedom which is absent in constriction links made of homogeneous metal. Thus, for any given tunnel conductance per unit area, the density of nonequilibrium electrons injected by tunneling into the electrodes can be kept low by simply using sufficiently thick films. Even in the high-current density tunnel junctions described above, the current densities and dissipation levels are orders of

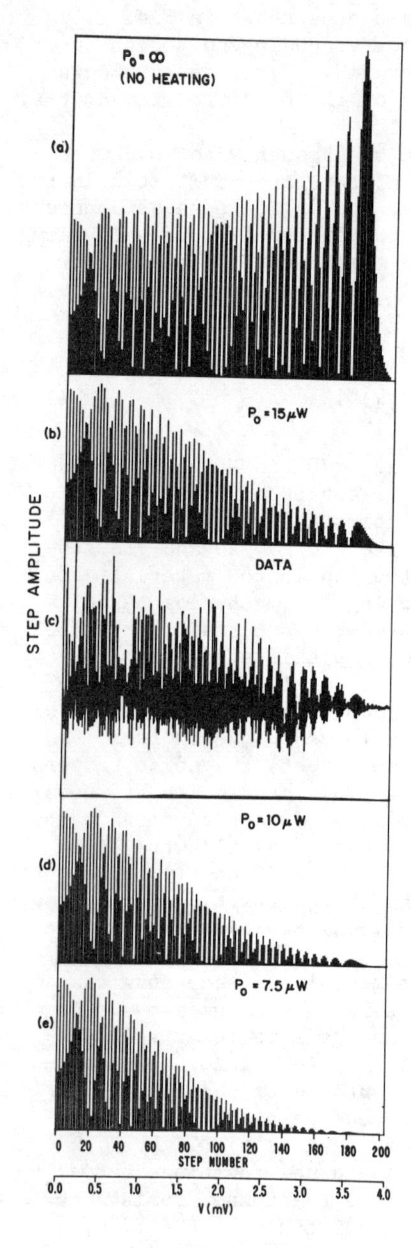

Fig. 1. Step amplitudes at 10 GHz for different values of P_o compared with experimental step amplitudes for a variable thickness tin bridge with R = 0.33 Ω.

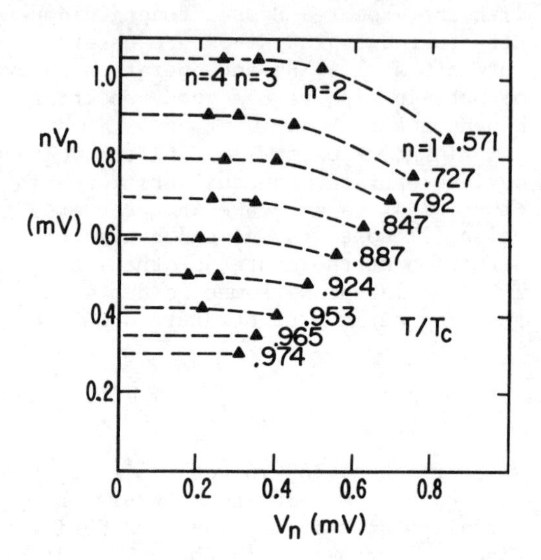

Fig. 2. Gap values inferred from subharmonic gap structure observed at $V_n = n^{-1}(2\Delta/e)$ in 0.33 ohm tin VTB. (See ref. 11.) Values of $nV_n = 2\Delta/e$ are depressed by heating, which increases as V_n^2.

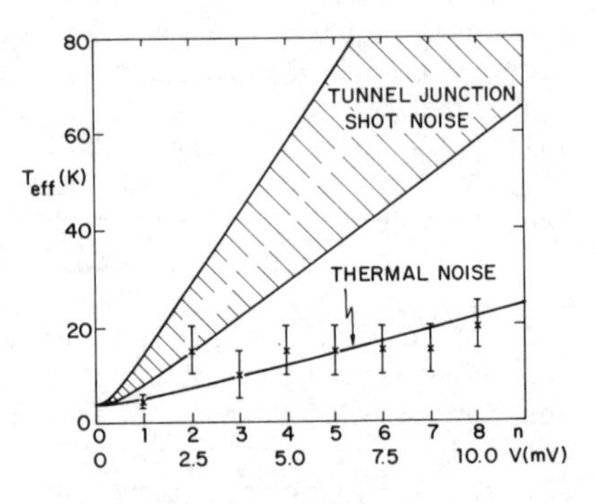

Fig. 3. Noise temperatures inferred from rounding of 496 μm-laser-induced steps on I-V curve of Nb point contact. (See ref. 12.) Thermal noise refers to $\frac{1}{2}(T_b+T_m)$, with T_m given by (11).

magnitude lower than in constriction links.

OTHER NONEQUILIBRIUM EFFECTS

In addition to these rather gross "heating" effects, which can be modeled by a suitable time-average temperature profile through the junction, there are other nonequilibrium effects which depend on the relaxation occurring within a Josephson period. Such effects have been analyzed recently by Likharev and Yakobsen[13] and by Baratoff and Kramer[14] using the time-dependent Ginzburg-Landau (TDGL) theory. Their analysis provides an explanation of the so-called "excess current" of $(0.5 - 0.8)I_c$ observed in the I-V curves of metallic weak links. (This "excess current" refers to the nearly constant excess of current in the superconducting state over that in the normal state at the same voltage, for voltages above the gap voltages. No such excess current is predicted by the simple RSJ model.) At relatively low voltages (or Josephson frequencies) such a time-average supercurrent arises because the quasi-particle occupation numbers lag the time-varying energy gap because of the slow rate of inelastic scattering τ_E, giving rise to the "foot" observed in I-V curves.[15] However, in the excess current regime at higher voltages, the order parameter ψ (or gap) itself is varying so rapidly that $\tau_{GL}(\ll\tau_E)$ comes into play. This causes ψ in the center of the link to be exponentially small, but near the ends, ψ is sustained by proximity to the banks. This allows a lower voltage to drive a given current than if the whole bridge were normal, or equivalently, there is an excess current for a given voltage.

This excess current is not observed with tunnel junctions, and remains one of the few features of I-V curves which appears to distinguish the low-capacitance tunnel junction from the metallic weak link. The explanation in terms of TDGL theory given above would also account for the absence of the excess current in tunnel junctions, since there can be no "proximity effect" superconductivity in the non-metallic barrier region adjacent to the film electrodes.

Conceptually, one can consider a continuous succession of weak links between two massive superconductors, starting with a very narrow short constriction in homogeneous metal, and going to progressively wider constrictions with a tunnel barrier spanning the neck, with the barrier thickness increasing from zero at such a rate that all the links have the same resistance and critical current as the original constriction. Given this concept, it seems intuitively plausible that as the tunnel resistance increases, the magnitude of the excess current should decrease roughly in proportion as the fraction of the resistance of the link which still comes from the metallic part. For the more realistic geometry of tunneling between overlapping segments of two one-dimensional film strips, one might expect the excess current to remain absent, so long as the tunnel currents never exceeded the critical current level in the film strips.

Although the VTB configuration has been shown[9,11] to work up to twice the gap frequency in tin, to reach the highest far infrared frequencies, one must resort to cat-whisker-type point contacts. These have been extensively studied by Weitz[12,16] in our laboratory, using an optically-pumped gas laser FIR source focussed on the contacts by a lens to observe radiation-induced steps up to 12 mV, corresponding to ~6 THz. I shall summarize his results very briefly. He found that good performance required that the contacts be "clean." This is accomplished by etching both point and anvil immediately before installation, and by making the contact and burning it in while it is immersed in liquid helium. Successful FIR contacts could be identified by study of the dc I-V curves. The hallmarks are 1) near theoretical I_cR product, and 2) a sharply defined slope change (at least a factor of 2) at the energy gap voltage. If the ratio of these slopes is denoted S, it is found that the maximum useable frequency scales up roughly linearly with S. Moreover, the resistances were typically large (50-150 Ω), facilitating coupling to radiation, and minimizing the heating due to power dissipation, V^2/R. With these contacts, extrinsic limits on high-frequency performance were so well controlled that one could discern the theoretically expected intrinsic frequency dependence of the Josephson effect in the vicinity of the gap frequency and above.[16,17] Thus, although such contacts suffer from practical drawbacks, they have enabled us to establish with some confidence what factors govern the ultimate capabilities of Josephson devices for operation at very high frequencies.

REFERENCES

1. B. D. Josephson, Phys. Letters 1, 25 (1962); Advan. Phys. 14, 419 (1965).
2. V. Ambegaokar and A. Baratoff, Phys. Rev. Lett. 10, 486 (1963); 11, 104 (1963).
3. L. G. Aslamazov and A. I. Larkin, JETP Letters 9, 87 (1969).
4. W. C. Stewart, Appl. Phys. Lett. 12, 277 (1968); D. E. McCumber, J. Appl. Phys. 39, 3113 (1968).
5. See, for example, J. Niemeyer and V. Kose, Appl. Phys. Lett. 29, 380 (1976).
6. R. A. Ferrell and R. E. Prange, Phys. Rev. Lett. 10, 479 (1963).
7. C. S. Owen and D. J. Scalapino, Phys. Rev. 164, 538 (1967).
8. W. J. Skocpol, M. R. Beasley, and M. Tinkham, J. Appl. Phys. 45, 4054 (1974).
9. M. Tinkham, M. Octavio, and W. J. Skocpol, J. Appl. Phys. 48, 1311 (1977).
10. R. Holm, Electric Contacts (Springer-Verlag, Berlin, 1967), p. 31; F. Kohlrausch, Ann. Phys. (Leipzig) 1, 132 (1900).
11. M. Octavio, W. J. Skocpol, and M. Tinkham, IEEE Trans. Magn. MAG-13, 739 (1977).
12. D. A. Weitz, W. J. Skocpol, and M. Tinkham, Appl. Phys. Lett. 31, 227 (1977); and to be published.

13. K. K. Likharev and L. A. Yakobsen, Zh. Eksp. Teor. Fiz. 68, 1150 (1975) [Sov. Phys.-JETP 41, 570 (1976)].
14. A. Baratoff and L. Kramer, preprint.
15. M. Octavio, W. J. Skocpol, and M. Tinkham, Phys. Rev. B17, 159 (1978), and references cited therein.
16. D. A. Weitz, W. J. Skocpol, and M. Tinkham, Phys. Rev. Lett. 40, 253 (1978), and to be published.
17. W. J. Skocpol, this Proceedings.

ADVANCED IMAGING TECHNIQUES FOR SUBMICROMETER DEVICES

T. G. Blocker, R. K. Watts, and W. C. Holton
Texas Instruments Incorporated, Dallas, Texas 75265

ABSTRACT

As the dimensions of electronic circuits are reduced in the quest for higher performance and greater density, the limits of conventional imaging techniques are being reached. New imaging methods using electron beams and X-rays rather than optical radiation are being developed in many laboratories. This paper gives an overview of work in our laboratory on high resolution electron beam and X-ray lithography. The GaAs field effect microwave transistor is highlighted as an example of a device for which the increased cost of advanced imaging techniques is offset by the enhanced performance obtained.

OPTICAL LITHOGRAPHY

Looking first at the left-hand side of Figure 1, we see that conventional optical lithography consists in generating an oversize, typically 10X, circuit pattern on a glass plate called a reticle. The reticle is then reduced to final size and stepped to fill a master mask with the desired number of identical circuit patterns. Modern step-repeat cameras have interferometrically controlled stages for increased stepping precision, reduction lenses capable of producing images in photoresist of 1.2 μm features, and environmentally controlled housings.

If the mask is to be used in contact printing, copies are made because the contact produces defects in the mask and reduces its useful life. If the mask is to be used in projection printing, where the separation between mask and wafer is large, or in proximity printing, where there is a small gap between mask and wafer, the master mask may be used rather than a copy. Typical electronic devices require six to twelve masks for fabrication of the various levels. The patterns of the levels must be properly oriented and registered among themselves on the wafer. In general, the registration tolerances of all levels will not be equal, e.g., some will require much more precise registration than others. Since the patterns of the levels cannot be registered more accurately than the overlay precision of the set of masks used to image them, this overlay precision or mask "stacking" is an important quantity. Existing step-repeat cameras have image placement precision of ±0.25 μm (2σ) over 15 cm, leading to stacking errors for an N-mask set of $\pm 0.25N^{\frac{1}{2}}$ μm over this large distance. New cameras under development should soon offer even better placement precision. Either the reticle or the final mask is often now generated by an electron beam pattern generator, as also indicated in the figure. In this case, the stacking is determined by variations such as residual magnetic fields across the mask, stability of deflection electronics, and stage-positioning accuracy. Commercially available machines provide stacking errors about half

ISSN: 0094-243X/78/280/$1.50

as large as for the step-repeat camera.

Optical projection printers are of two types. In the full field type, the mask pattern is imaged on the wafer with unit magnification (1:1). In the reduction type, the mask is oversize, and a reduced image (typically 10:1, 4:1, or 2:1) is stepped over the wafer with a registration operation preceding exposure of each field. This second type of printer is similar to the step-repeat camera. Some characteristics of these systems are collected in Table I. For imaging in resist a value of the modulation transfer function of at least 0.4 to 0.5 is required. To achieve the high resolutions shown, a large numerical aperture is required, leading to a small depth of focus. Thus the requirement of wafer flatness becomes stringent. Field distortion as a fraction of image field size is very small, generally near 10^{-5}. Throughput of optical exposure systems is high. For automatic contact printers, values of 150 wafers/hour are possible. For the full field projection printer, typical throughputs are 60 to 100 wafers/hour, and for step-repeat projection, about 10-60 wafers/hour.

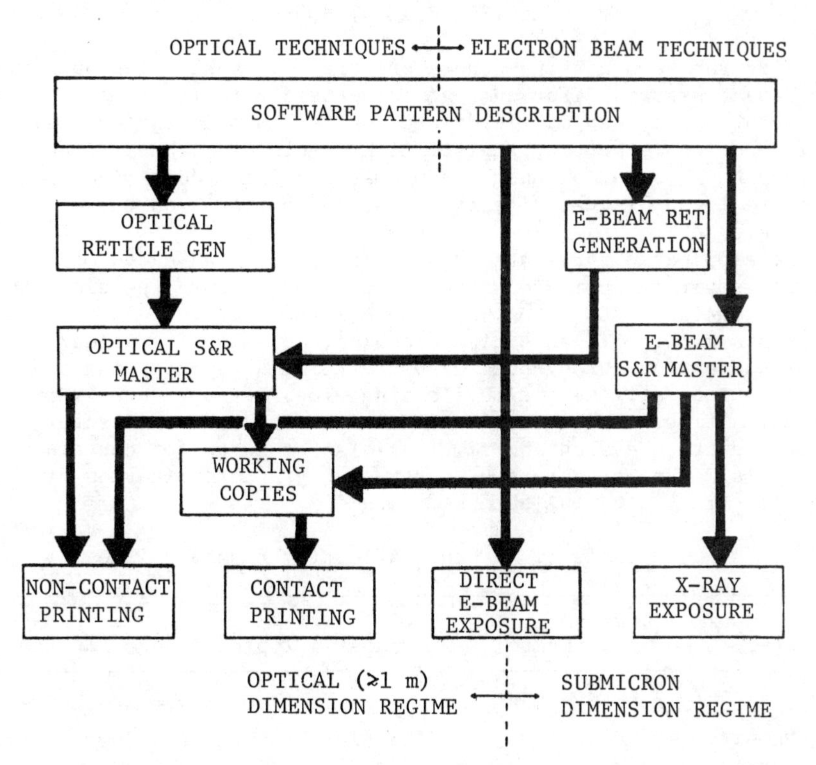

Fig. 1. Pattern definition techniques for microelectronic devices.

Table I Optical Projection Printers

Printer	Reduction	Resolution (μm)	Field (cm)	Field Distortion (μm)	Depth of Focus (μm)
Perkin Elmer 140	1:1	2.0	8.3x10	1	±5.5
Canon FPA 141	4:1	0.9	1.0 x 1.0	< 1	±1.5
Mann DSW	10:1	1.25	1.0 x 1.0	<0.1	Auto Focus
Thomson CSF Slice Stepper	10:1	1.25	1.0 x 1.0	<0.1	-

ELECTRON BEAM LITHOGRAPHY

Electron beam pattern generators are comparable with optical step-repeat systems. For current integrated-circuit geometries in manufacture, the stepped fields are smaller than in the optical case, the resolution is better, registration precision is better, and throughput is worse. We now concentrate on the middle-right portion of Figure 1 and discuss direct exposure of patterns on the wafer by the beam.

Development efforts at Texas Instruments on advanced microwave devices requiring submicron structures have employed the electron pattern generator EBM I to image the critical levels. In order to use such a machine in an optimal fashion, one should consider the factors affecting throughput: exposure time, stage-stepping and settling time, alignment time, loading time, set-up and pattern evaluation time. Exposure time is usually the dominant factor, although pattern evaluation time may also be large for certain critical levels. The parameters of interest for a determination of exposure time are listed in Table II.

Table II Factors Affecting Exposure Time

Parameter	Symbol	Units	Typical Range for EBM I
Inverse Speed	E	μsec/μm	.2 - 20
Line Density	D	lines/μm	2 - 20
Beam Current	i	nA	.5 - 30
Resist Sensitivity	S	μC/cm^2	.5 - 50

The relationship between these parameters can be derived by considering the raster exposure of a geometry. As sketched in Figure 2, a beam of current, i, is scanned at an exposure, E, with a spacing $\Delta = 1/D$ in a resist of sensitivity, S. Using the proper units, the relationship is

$$E \cdot D = S/i \quad . \tag{1}$$

Since $E \cdot D$ is the exposure time/unit area, the total exposure time can be calculated from $(E \cdot D)fA$, where A is the total area to be patterned and f is the percent area coverage of the pattern. For highest throughput, one should use the minimum $E \cdot D$ consistent with the resist sensitivity, resolution requirements, and electronics limitations. From Equation 1, the $E \cdot D$ product can be decreased by higher resist sensitivity (lower S) or higher beam current (brighter gun).

The required dose to expose a resist characterized by a sensitivity S is a complicated function of thickness, beam voltage, substrate material and development parameters. Its value can depend on the scan density (proximity effect), a particular pattern edge acuity requirement, or by the nonuniform energy distribution in the beam (aberration effects).

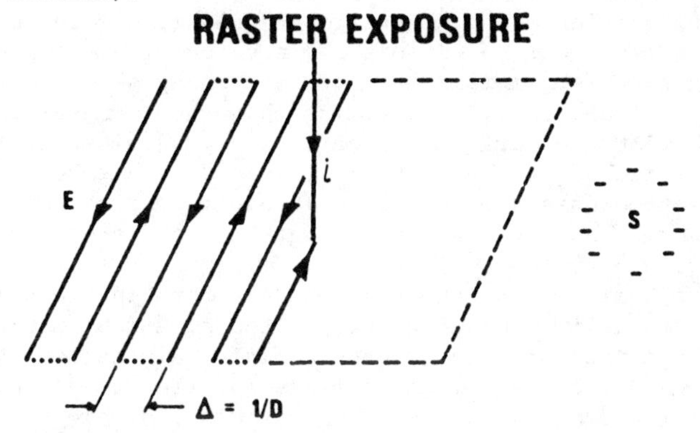

Fig. 2. Writing beam scanning resist of sensitivity S to expose parallelogram pattern element.

The maximum probe current is a complicated function of cathode material, voltage, and temperature, and is specified by beam brightness, β, the aberrations of the electron optics system employed, and the cone angle, α, of the final aperture in the system. There is an optimum α for which the probe current is a maximum. In practice, an $\alpha < \alpha_{opt}$ is chosen because of the need to minimize deflection coil aberrations to achieve maximum field size.

Typical values of probe current and effective beam diameter for EBM I are 25 nA at 5000 Å and 1 nA at 2500 Å. The effective beam diameter is determined by electron back scatter from the substrate in the resist. The minimum exposure E that can be employed is limited by the response of the deflection coils, the drive electronics, and computer data rate (clock frequency). E's of .2-20 μsec/μm are available on EBM I.

The minimum D that one can use is limited on the low side in principle by the beam diameter d - i.e., D cannot be less than 1/d for a geometry to be filled in. In practice, D must range from 2/d to 4/d to ensure that the minimum geometries are cosmetic with smooth edges - the value chosen is strongly dependent on the resist and the pattern topology. The minimum E is determined by the beam deflection electronics. The beam current is strongly dependent on beam diameter,

$$i_{max} \propto d^{8/3} \qquad (2)$$

so that for submicron patterning, writing time becomes long.

Such considerations have played an important role in the choice of the fabrication sequences and resist selections for devices with submicron features under development at Texas Instruments. As an example, we consider the GaAs FET. For this device, a hybrid process was implemented using conventional optical lithography for all levels except the e-beam defined gate level which involved the most stringent resolution and registration requirement. The motivations for this approach were to exploit the previously developed process for >2 μm gate FETs, minimize the process changes necessary to exploit e-beam lithography, to optimize throughput and allow easy comparisons with the performance of optically defined devices (made with optical techniques but with lower yield than e-beam) and evaluate the electron dose damage (if any) resulting from the e-beam pattern generation. Nominal gate lengths of the e-beam defined devices were 0.5, 0.75 and 1.0 μm. To minimize the effects of charging from the semi-insulating Cr doped substrate on the alignment, the mesa photomask was modified to put the alignment marks on interconnecting epi regions. The source-drain metallization photomask was modified to include alignment marks for the gate level. The Au-Ge-Ni source-drain metallization previously developed allowed acceptable secondary electron contrast against the GaAs background. Somewhat thicker Au metallization was employed and more attention was given to achieving minimum roughness and eutectic composition "structure." Under machine conditions where EBM I would accept high-quality etched marks in silicon to a criterion of ±500 Å per alignment mark leg, these procedures allowed ±2000 Å registration per leg to be routinely achieved over the 80 x 80 mil^2 (2 x 2 mm^2) field of the bar. The e-beam gate level consists of the gates themselves and the gate pads which were typically greater than 4 x 4 mil^2 (100 x 100 mm^2). These pads presented a potential throughput limitation since the maximum beam diameter (and hence beam current) is determined by the smallest geometry, L_{min}, which is the gate length of 0.5 μm in this case. As discussed above the beam diameter, d, used in writing rectilinear geometries, is chosen to be $d \simeq L_{min}/4$. In order to get reasonable exposure times for the bond pads, a line scan was used to generate the smallest gate, and the exposure setting was used to control the geometry. Although this approach requires a test-pattern calibration and some software modifications (as described below), it results in a 10X improvement in throughput. A nominal 4000 Å diameter beam with beam currents in the 20 nA range was used to expose the gate level. The gate patterns were tapered at the mesa edge to reduce gate resistance and to prevent constriction of the resist at the step. The resist used was polymethyl methacrylate (PMMA) with a nominal thickness of 7500 Å and the developed pattern served as the lift-off mask for 4000-5000 Å of aluminum gate metallization. The GaAs was lightly etched prior to deposition to improve metal adhesion and to recess the gate below the epitaxial surface.[1] Figure 3 is a SEM of the gate region.

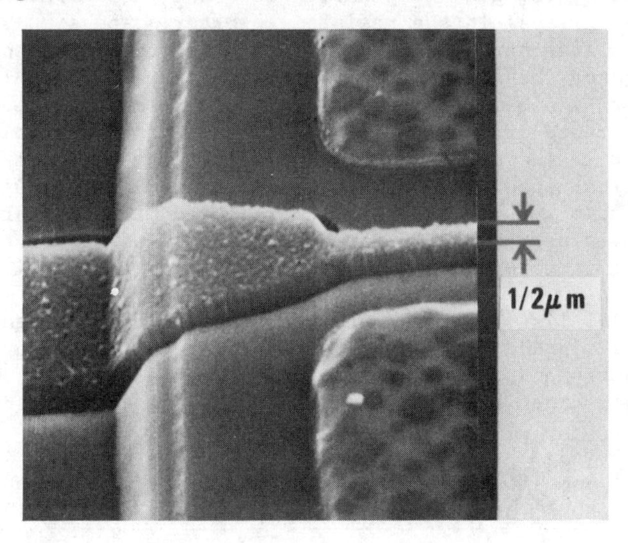

Fig. 3. SEM photo of gate region of FET showing 0.5 μm gate length and mesa edge.

The nominal slice used for the e-beam runs was slightly greater than 1 cm^2 which was dictated by the holder geometry and other tooling. This allowed 588 small-signal devices (twenty-five 80 x 80 mil^2 bars) to be patterned in a single pump down. The exposure time required per slice, including stage step and alignment, was six minutes. A typical "run" consisted of first patterning a GaAs pilot with the same resist thickness with an exposure series of a test pattern to determine the correct exposures for the pads and the various gates - a typical exposure ratio for the pad, 1 μm, 0.75 μm (all raster rectangle-generated) and 0.5 μm (line-generated) gates were 1:1.2:1.2:2. After setting these exposure changes in the program the live slices were initiated. Each slice was immediately developed and inspected after exposure so that "mid course" corrections could be made if necessary. In practice about $\pm$10% control of the gate lengths was achieved. Much of this is due to the difficulty of spinning thick (∿7500 Å) resist on small (1 cm^2) slices which have topology variations due to previously processed levels.

X-RAY LITHOGRAPHY

X-ray lithography is analogous to optical proximity printing. The shorter wavelength, 4 Å to 50 Å, necessitates a thin membrane mask and makes diffraction effects negligible, which are troublesome in the optical case. The exposing radiation diverges from a small source; it is not collimated as in the optical printer. The energy of radiation, 3 keV to 0.25 keV, is much less than the 10 to 20 keV used in electron lithography. Therefore, the ultimate resolution, which is set by the path length of electrons (photoelectrons produced by the X-rays) in the resist is greater. The simultaneous exposure of a large patterned area offers a throughput advantage which becomes greater the smaller the pattern features. Most electronic circuits with minimum feature size L_{min} have a level-to-level registration tolerance of $L_{min}/3$ to $L_{min}/5$. Achieving such tolerances consistent with the high resolution possible involves solving difficult mechanical problems. However, magnetic bubble circuits[2] have a smaller ratio of registration tolerance to minimum feature size, and as smaller semiconductor devices are designed, increasing use of self-aligned structures promises relief from these restrictions here also.

Table III lists some radiation sources which have been used in lithography. In addition, synchrotron radiation has been used.[3] At Texas Instruments, aluminum radiation has been largely employed, although some experiments with Mo and Cu sources have also been carried out.

Table III. Some Line Radiation Used for X-ray Lithography

Source	Wavelength (Å)	Energy (keV)
Pd L	4.37	2.84
Rh L	4.60	2.70
Mo L	5.41	2.29
Si K	7.13	1.74
Al K	8.34	1.49
Cu L	13.4	0.925
C K	44.6	0.278

Since diffraction effects can be neglected, the projection parameters of interest are given by simple geometrical considerations. Associated with the finite diameter of the spot on the anode from which the radiation originates is a blur or shadow σ in the image. For a mask-wafer gap g and source-mask spacing D, the relation is

$$\sigma = (g/D)b \tag{3}$$

where b is the focal spot size. Typical values are $g = 20\ \mu m$, $D = 40$ cm, $b = 3$ mm, $\sigma = 0.15\ \mu m$. The resolution obtained in resist depends on the value of σ, but also on resist contrast, and the edge profile of the pattern elements on the mask. Typical absorber thickness is 0.5 μm for Al radiation. Ideally, the edges should be vertical. Such ideal profiles are achievable by electroplating, and very nearly ideal profiles can be obtained by reactive ion milling using a refractory metal milling mask.[4] Also associated with the point source projection is a runout ρ or slight magnification of the image,

$$\rho = (R/D)g. \tag{4}$$

R is the radius of the wafer on which the image is formed. For $R = 2.5$ cm, $\rho = 1.25\ \mu m$. Control of the gap g is important so that changes in ρ from one process level to another do not limit level-to-level registration.

In our work, two aluminum radiation sources are used. One is a low-power source consisting of a modified electron-beam evaporator and supplying a flux in vacuum of $\phi = 8/D^2$ mW/cm^2. Focal spot size is $b = 1$ mm. The other is a high-power generator with focal spot size of $b = 1$ mm or $b = 3$ mm. It supplies a flux in helium atmosphere of up to $270/D^2$ mW/cm^2. The most sensitive resist employed to date is a negative resist of sensitivity 1 mJ/cm^2. Exposure times of less than 1 minute are obtained with the high-power source at long working distances $D \sim 50$ cm. It is more difficult to obtain high sensitivity in positive resists, and the most sensitive of these is less sensitive by a factor of 40.

In earlier work with polymer membrane mask substrates precision of level-to-level registration was limited by the dimensional instability of the membrane. This problem arises in part from the viscoelastic nature of polymers and in part from the small Young's modulus which allows the material to be easily strained by the metal absorber pattern. To avoid these difficulties we are developing a SiC membrane mask. SiC is deposited on a silicon wafer at high temperature by a chemical vapor deposition process. The center of the wafer is etched away to leave a membrane. The parameters of the deposition process can be controlled to optimize the tension in the membrane and the optical transparency. Figure 4 is a photograph of such a mask.

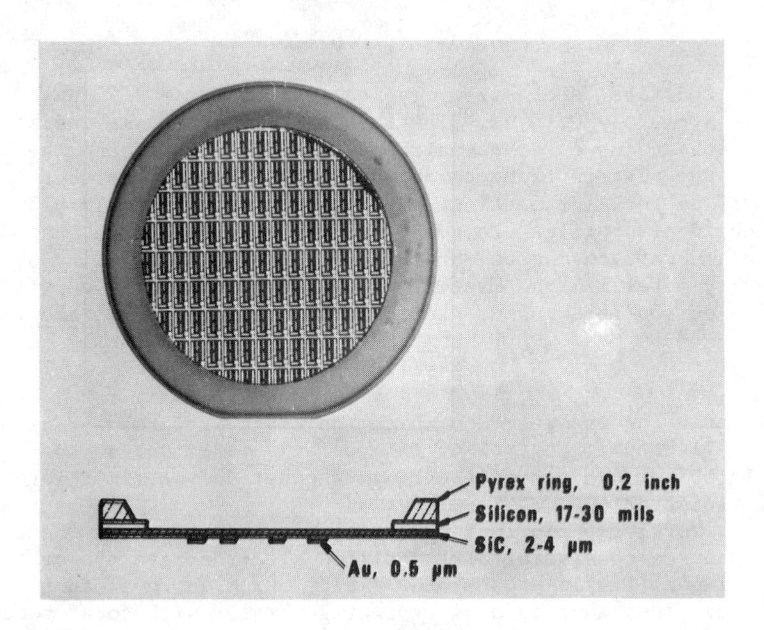

Fig. 4. SiC Mask. Outer diameter is two inches.

In summary, advanced imaging techniques are being developed for devices with submicrometer geometries. These techniques will permit many devices which are currently laboratory curiosities to impact advanced systems requiring high density memory and high-speed logic. It is also conceivable that the existence of these advanced imaging techniques will stimulate totally new device concepts which can uniquely exploit their advantages and avoid their disadvantages.

REFERENCES

1. H. M. Macksey, R. L. Adams, D. N. McQuiddy, Jr., D. W. Shaw, and W. R. Wisseman, IEEE Trans. El. Dev. ED-24, 113 (1977).
2. R. K. Watts, H. M. Darley, J. B. Kruger, T. G. Blocker, D. C. Guterman, J. T. Carlo, D. C. Bullock, and M. S. Shaikh, Appl. Phys. Letters 28, 355 (1976).
3. E. Spiller, D. E. Eastman, R. Feder, W. D. Grobman, W. Gudat, and J. Topalian, J. Appl. Phys. 47, 5450 (1976).
4. R. K. Watts, D. C. Guterman, and H. M. Darley, Proc. SPIE 80, 100 (1976).

HIGH RESOLUTION ELECTRON BEAM LITHOGRAPHY AND APPLICATIONS TO SUPERCONDUCTING DEVICES

A. N. Broers and R. B. Laibowitz
IBM Thomas J. Watson Research Center
Yorktown Heights, New York 10598

ABSTRACT

High resolution electron beam lithographic techniques are essential for the fabrication of a wide variety of present and future electronic devices and integrated circuits. Advanced structures have been made for silicon, bubble and superconducting technologies. The e-beam lithography techniques used to obtain the highest possible resolution will be reviewed and our recent results using electron-beam produced contamination layers will also be presented. These high resolution techniques have been used to make superconducting Nb nanostripes with widths of 25-50 nm. The I-V characteristics for these bridges will be presented along with transmission electron micrographs of the structures.

INTRODUCTION

The advantages of scanning electron beam fabrication for thin film device manufacture are that it offers higher ultimate resolution than UV light methods, and that high resolution patterns can be written directly without the need for expensive and vulnerable masks. Recent efforts on this technology have concentrated on reducing cost per exposure, on improving the precision with which it is possible to overlay patterns, and on perfecting electron resists and processes for the fabrication of semiconductor devices. These efforts have produced experimental integrated circuits with higher density than had previously been possible, and led to cost reductions in manufacturing environments where quick turn-around is of great importance. The full resolution potential of electron beam has not been of interest and dimensions below 1μ have only been explored in a few instances.

At dimensions below 1 μ, electron–scattering effects must be taken into account in obtaining satisfactory profiles in the developed resist patterns, and in compensating for the differences in effective exposure that occur for dense and non-dense areas of a pattern. This latter effect is known as the proximity effect.[2] When dimensions reach 0.25 μ, scattering effects become so serious that complex patterns can only be formed in very thin resist layers and significant corrections have to be made for proximity effect.[3] The smallest metallic lines formed using conventional resists and normal solid substrates have been about 1000 Å in

: 0094-243X/78/289/$1.50

width. (See for example ref. 4.) When smaller dimensions are required it is necessary to use substrates that are thin compared to the mean-free-path of the electrons and very thin resist layers.[5] The smallest structures have been made using molecular layers of hydrocarbon contaminants, formed on the sample surfaces in oil-pumped vacuum systems, as the resist. In the last two years these specialized techniques have been perfected to the extent that it is now possible to make simple structures with dimensions down to about 100 Å. These samples can be handled readily and it is possible to make electrical contact to them. They are of interest in exploring the physical nature of electrical conductivity, particularly in superconducting structures such as Josephson devices, and in this paper we describe some preliminary results obtained with microbridges fabricated in this manner. We also describe preliminary results obtained with PMMA electron resist which indicate that this resist is also capable of very high resolution.

ELECTRON BEAM LITHOGRAPHY

The instrument most suited to the study of the ultimate resolution of electron beam fabrication is the scanning transmission electron microscope (STEM). The thin substrates required in order to minimize electron back-scattering have the added advantage in this instance that the samples can be examined 'in-situ' in the transmission electron microscope mode. The STEM we have used was built at the IBM T. J. Watson Research Laboratory and uses three magnetic electron lenses and a lanthanum hexaboride cathode electron gun. The resolution of the instrument is below 5Å in scanning transmission and the beam diameter was kept below 20 Å for all experiments described. In no instance was there any indication that the beam diameter limited resolution. The beam current can be continuously varied up to 10^{-6} ampere by reducing the strength of the demagnifying electron lenses and the beam diameter can be maintained below 100 Å for currents up to 20 nanoamperes. Accelerating potentials of about 50 KV were used. The electron column is "slaved" to a flying-spot scanner for the generation of patterns other than arrays of lines or dots.

PMMA Exposure

PMMA was the first resist developed specifically for electron beam exposure, and, while many recently discovered electron resists have higher sensitivity, none has better resolution.[7] Fig. 1 shows 300 Å wide lines in 225 Å thick gold-palladium produced by ion etching through a PMMA mask. The resist layer was spin-coated onto a Au-Pd coated Si_3N_4 window substrate of the type described in U. S. Patent 3,971,860. Exposure was made under flying spot scanner control. The sample was etched with argon ions. Details of exposure and etching conditions

Fig. 1: S.T.E.M. bright field image of portion of zone-plate pattern showing 300 Å wide Au-Pd lines fabricated using PMMA and ion etching. (Au-Pd lines are dark).

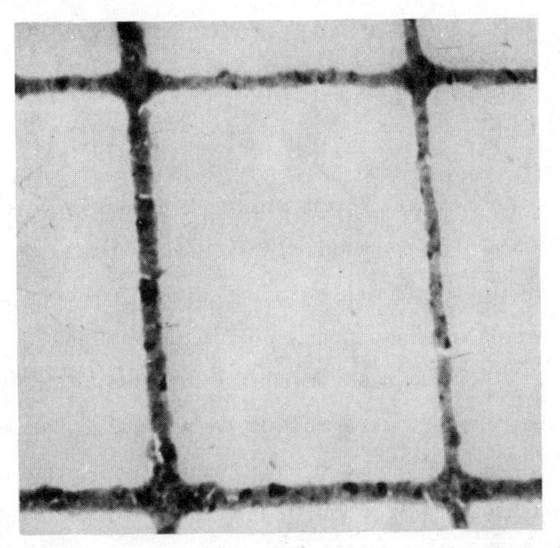

Fig. 2: Au-Pd lines fabricated on carbon substrate using contamination resist and ion etching. Narrowest lines are 80 Å wide.

are to be published.[8] Thinner lines were not resolved in this pattern presumably because the resist was too thick. It will be difficult at this resolution with PMMA and ion etching to fabricate structures from materials with lower sputtering rates than Au-Pd because PMMA itself has a relatively high sputtering rate. Nevertheless this result indicates that resolution beyond that previously considered possible with conventional resist can be obtained under suitably well-controlled conditions.

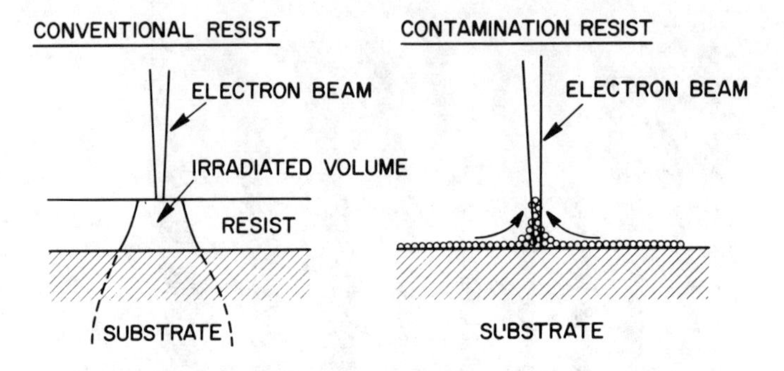

Fig. 3: Comparison of conventional electron resist and contamination resist processes. Aspect ratio can be better in contamination resist process although sensitivity is three orders of magnitude lower.

Contamination Resist

Contamination has been used since 1964 as a resist for etching processes.[9] It offers higher resolution for a given thickness than conventional resist layers because the protecting layer is built up mono-layer at a time, thus reducing spreading in the final resist pattern (see fig. 3). Initial work with the STEM and argon ion etching produced Au-Pd lines 80 Å wide and 100 Å on carbon substrates.[6] (Fig. 2). Subsequently membrane substrates of the type shown in fig. 4 have been used because they allow electrical contrast to be made to the samples. Resolution in the example shown in fig. 2 was limited by imperfections in the blanket metal film. A subsequent unpublished modeling study by Parikh and Broers indicates that resolution is also set by the range of the low-energy secondary electrons in the metal film. This shows why it would be difficult with this method to produce metal structures smaller than about 50 Å even if uniform metal coatings were available. It is possible to make lines of contamination less than 20 Å in width but they are not thick enough to survive ion etching of electrically continuous metal films. Such films generally have to be greater than 100 Å thick.

Significant disadvantages of contamination resist are its very low sensitivity ($\sim 10^{-1}$ C/cm^2) and the depletion effect which occurs when complex patterns have to be formed.

APPLICATIONS TO SUPERCONDUCTIVITY

Sample Preparation

While there are several possible superconducting devices and structures that can be fabricated using the above-described high resolution lithography, only the work concerned with constructing very narrow bridges (nanostripes or nanobridges) between two relatively large banks will be described. A schematic of our preferred sample structure for linewidths below 500 Å is shown in Fig. 4. As discussed above the window substrate is important both for reducing backscattered electrons which degrade resolution and for observing the samples in the STEM. Examples of membranes are thermally grown SiO_2, Si_3N_4 and carbon. A disadvantage of such a substrate is its fragility.

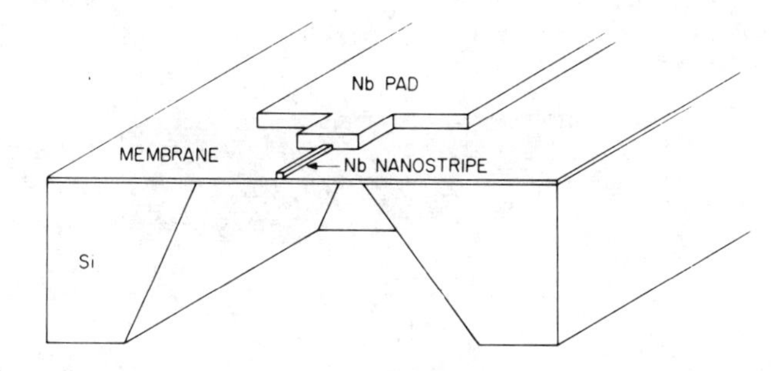

Fig. 4: Schematic diagram of window substrate showing Nb nanostripe.

As is also shown in Fig. 4 Nb was chosen as the material for the stripe, and the Nb thickness was generally about 200 Å. The length of the bridge is determined by the pad separation. These lengths can be seen in the micrographs that follow. Our initial results are on bridges about 0.8 μ long; standard e-beam lithography is used to make the mask for these pads. The width of the stripe is determined by the width of the contamination resist and by the conditions used in the subsequent ion milling.[8]

Fig. 5: Micrograph of a 120 Å gold stripe between gold pads. The irregular pad edges are caused by the large grain size in the film and the lift-off used to form the pads.

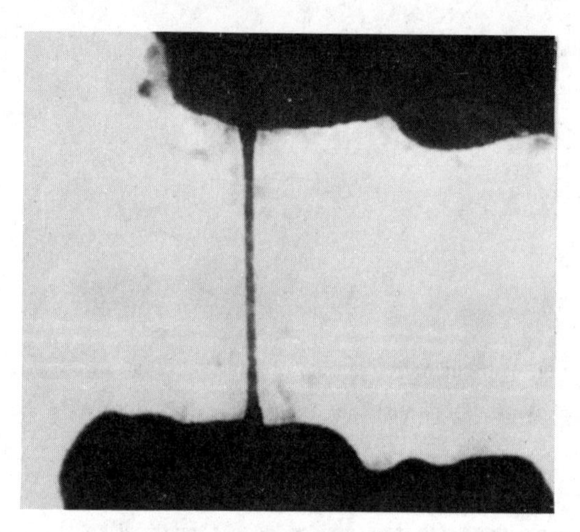

Fig. 6: Micrograph a Nb line which is about 250 Å wide and 200 Å thick.

At several stages in the fabrication process, the samples are examined in the STEM; some final micrographs are shown in Fig. 5 and 6 in which the dark areas are metallic. Fig. 5 shows a Au stripe and pads. The width of the Au stripe is about 120 Å, and the irregular shape of the pad edges is due to the large grain size in the Au film combined with the characteristics of the electron beam lift-off technique used to form the banks. A Nb bridge about 250 Å wide is shown in Fig. 6. This is the narrowest Nb line we have successfully measured.

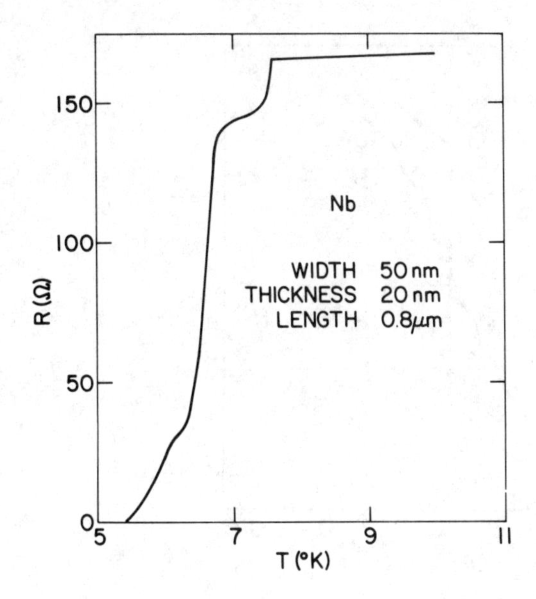

Fig. 7: Resistance vs temperature for a Nb line measured at a current of 0.1 μ A.

Results and Discussion

Four terminal i-v measurements with and without microwaves were obtained on these Nb samples. In addition resistance measurements as a function of temperature in the region at the transition temperature were also taken. As it is difficult to make contact directly to the nanostripe, we often see a complex R vs temperature curve due to the series resistances of the pads. An example of such a curve is shown in Fig. 7 which also shows the sample dimensions. The highest transition temperature is usually that of the large connecting pads; a second transition possibly caused by intermediate pads is often seen and finally as has been observed in past work; the weak link itself generally has the lowest transition temperature. The bank Tc in Fig. 7 is somewhat lower than bulk Nb, presumably due to impurity contamination. Higher bank Tc's have been obtained.[10-12] It is interesting to observe that above the bank Tc the resistance is still slightly decreasing with decreasing temperatures which may be the first observation of fluctuation phenomena in a one-dimensional Nb stripe. Sample geometries where the stripe resistance is a much larger part of the total resistance are needed to effectively study such phenomena.

The I-V curves for these structures generally show intermediate states when their zero voltage critical current is exceeded. This might be expected as these are long bridges in which heating, impurities, or defects can cause such effects.[12] A full description of the switching

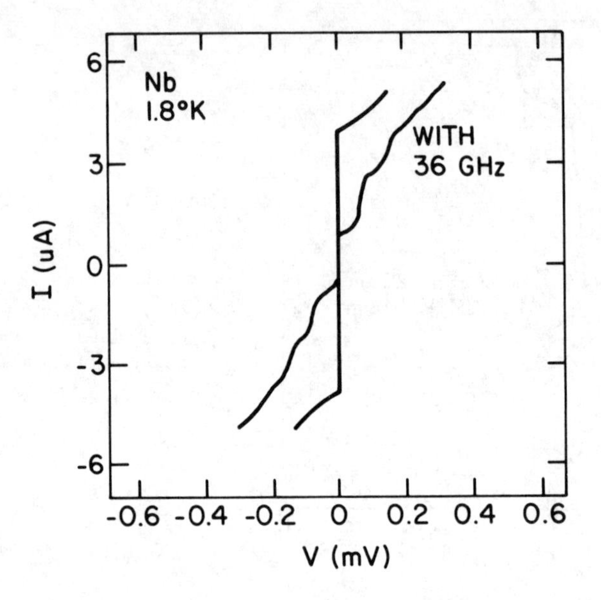

Fig. 8: I-V curves for a Nb nanobridge with and without 36 GHz microwaves applied.

behavior of these nanobridges must await further studies in which the sample geometry is controllably varied. Such experiments can only be accomplished using high resolution techniques. An example of an I-V characteristic with and without microwave radiation is shown in Fig. 8. The device is biased up to its first intermediate state. Three microwave-induced steps can be observed, and the spacing corresponds to the Josephson relation 2 eV=hv. Steps have not been generally observed above the first transition, but in the nanobridges studied in this paper they are observed over the entire temperature range below the bridge Tc. This result and the high normal resistances of these nanobridges ($\gtrsim$ 100 Ω) may enable applications using bridges to become practicable.

ACKNOWLEDGEMENTS

The authors gratefully acknowledge helpful discussions and expert collaboration with the following: J. Harper, ion milling; W. Molzen and J. Cuomo, substrate development and preparation, metallization, sputter etching; J. Viggiano and J. Yeh, measurements; S. Raider and R. Drake, Nb metallization; W. Grobman, H. Luhn and T. Donohue, electron-beam masking; J. Kuran, bonding; and M. Hatzakis for advice about PMMA resist processing.

REFERENCES

1. H. N. Yu, R. H. Dennard, T. H. P. Chang, C. M. Osburn, V. Dilonardo, H. E. Luhn, J. Vac. Sci. and Tech. 12, 1297 (1975).

2. T. H. P. Chang, J. Vac. Sci. and Technol., 12, 1271 (1975).

3. Parikh, M., Proc. 14th Symposium on Electron Ion and Laser Beam Technology. To be published in J. Vac. Sci. and Tech. (1978).

4. A. N. Broers, and M. Hatzakis, "The Scanning Electron Microscope in the Fabrication of Microminiature Electronic Components", Septieme Congres International de Microscopie Electronique, Grenoble, Societe Francaise de Microscope Electronique, Paris, France, 1970.

5. T. Sedgwick, A. N. Broers and B. J. Agule, J. of Electrochem. Soc. 119, 1769 (1972).

6. A. N. Broers, W. Molzen, J. Cuomo and N. Wittels, Appl. Phys. Lett., 15, p. 98 (1976).

7. M. Hatzakis, J. Electrochem. Soc. 116, p. 1033 (1969).

8. A. N. Broers and J. Harper, To be published.

9. A. N. Broers, Proc. 1st Internat. Conf. on Electron and Ion Beam Science and Technology, R. Bakish, Ed., John Wiley, New York, p. 181 (1964).

10. R. B. Laibowitz, Appl. Phys. Let., 23, 407 (1973).

11. E. P. Harris, and R. B. Laibowitz, IEEE Trans. Mag.,MAG-13, 724 (1977).

12. R. B. Laibowitz, A. N. Broers, J. M. Viggiano, J. J. Cuomo and W. W. Molzen, Bull. Am. Phys. Soc. 23, 357 (1978).

13. W. J. Skocpol, M. R. Beasley and M. Tinkham, J. Appl. Phys. 45, 4054 (1974).

FABRICATION OF MICROBRIDGE JOSEPHSON JUNCTIONS USING ELECTRON BEAM LITHOGRAPHY[†]

J.E. Lukens, R.D. Sandell and C. Varmazis
State University of New York, Stony Brook, N.Y. 11794

ABSTRACT

We describe the fabrication of microbridge Josephson junctions using a small research-oriented electron beam lithography system. Data is presented on the dimensional dependence of the resistance and critical current of such junctions. Progress toward the development of large coherent arrays of these junctions and the potential properties of such arrays is discussed.

INTRODUCTION

It appears that for some technical applications of Josephson junctions, junctions formed by fabricating a narrow constriction (microbridge) between two superconducting films may have advantages over the more common tunnel junction. These microbridge junctions, however, must have dimensions of the superconducting coherence length ξ or less for optimum performance. Since ξ is usually only a few thousand Å or less, submicron fabrication techniques are required. In contrast tunnel junctions can be made using standard industrial photolithographic technology.

One of the goals of our own research program at Stony Brook is to develop coherent arrays of Josephson junctions. Since one possible use of such arrays is as a broadly tunable microwave source, it was decided at the start to use microbridges instead of tunnel junctions since the microbridges are nonresonant and do not present the problem of making oxide barriers with very high critical current densities and long-term stability. The technique used to fabricate the microbridge arrays must, in addition to having a resolution of a few thousand angstroms, be able to produce large numbers of nearly identical junctions. Since so little was known about coherence among Josephson junctions, it also seemed essential to use a fabrication method with great flexibility so that it would be relatively easy to change junction parameters and circuit configurations to incorporate the results of experiments. Finally, since part of the reason for the study of arrays is their potential for technical applications, it is desirable that fabrication methods be used which will eventually be compatible with industrial thin film circuit techniques. While many ingenious methods have been used to make single microbridges almost since the discovery of the Josephson effect, it seemed that the only fabrication method which met the criteria listed above was electron beam lithography (EBL).

[†]Work supported by the Office of Naval Research.

ISSN: 0094-243X/78/298/$1.50

The general techniques of EBL are discussed elsewhere in these proceedings[1] and summarized in Fig. 1.

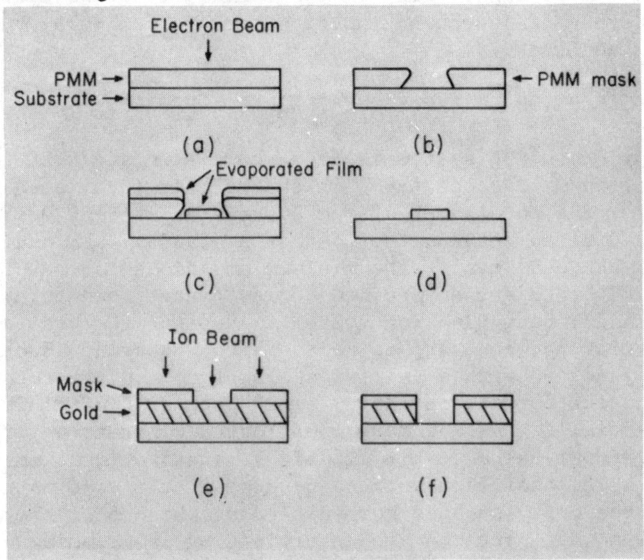

Fig. 1. EBL Processing Techniques. For liftoff processing, PMM resist is spun onto the substrate and exposed by an electron beam with the pattern of the desired metal film (a). Mask is developed (b). Metal film is evaporated (c). Note that undercutting in resist due to electron scattering makes the metal film discontinuous at mask edge. Finally, unexposed resist is removed leaving patterned metal film. The effective polarity of the resist can be reversed by depositing the metal film under the PMM then exposing the PMM with the negative of the final pattern (e). The developed PMM is then used as a mask for ion milling which removes the exposed metal (f).

The reader is referred to these papers for a detailed description of this fabrication method. The EBL facility in our research group at Stony Brook is, however, less sophisticated and operates with considerably less technical support than is common of these industrial systems. Much of the Josephson effect research is now at a rather awkward stage where fundamental work, of the type best done at universities, is still needed, but progress is being hindered by the lack of sophisticated fabrication facilities more common in industry. One purpose of this paper is to discuss our experience using EBL at Stony Brook for the past three years and hopefully to give some feeling for the practicality of operating such a system in connection with a small research program. Further, we will try to indicate the limitations of small EBL systems and discuss the areas of Josephson junction fabrication where the large effort required to operate such a system seems likely to be worthwhile. We will also survey the properties of microbridges produced using EBL and discuss the prospect for its application in the fabrication of coherent arrays. For a more general review of fabrication of different types of Josephson junction see, for example, the review by T. Van Duzer[2].

DESCRIPTION OF THE STONY BROOK EBL FACILITY

The Stony Brook EBL system is based on a standard ETEC scanning electron microscope (SEM) which has been interfaced to a mini-computer. The total cost of the system (about $150,000) is about equally divided between the cost of a standard SEM and the extra cost of EBL equipment. The minimum additions to the SEM required for EBL work are a specimen current meter to determine the exposure and a beam blanking system to deflect the beam out of field when moving between various parts of the pattern. Such a system in conjunction with a pair of analog function generators can make EBL masks for microbridges. The computer control is, however, essential for making more complicated circuits or arrays. The computer system has a 16-bit × 16-bit resolution in a 2mm field. This permits high resolution work to be done at the same magnification as course work, such as current leads, which fill the entire field. The exposure patterns are determined by programming a sequence of rectangles using the interactive FORTRAN language. The system then exposes the interior of the rectangle point by point with programmable resolution and dwell time (fixed within a given rectangle). The rather slow speed of the system (30μs minimum between points) makes it necessary to change the beam current and resolution between the small, high resolution and larger, course parts of the pattern. (Exposing the full field at maximum resolution would take about 34 hours.) Changes in beam current shift the center of the field and require that the subsequent pattern be reregistered to align it with the high resolution parts. The SEM has a manual registration system which permits this alignment to within about 1000Å. The registration system is also essential for patterns requiring more than one layer.

MICROBRIDGE FABRICATION USING THE EBL SYSTEM

Most of our studies of microbridges and microbridge arrays have been done using planar indium bridges made using EBL fabricated masks and the lift-off technique. The bridges are presently made using the computer-controlled system described above. However, masks for earlier bridges were fabricated using a standard SEM, scanning the beam with two triangle wave function generators for the X and Y axes. Simple homemade electronics were used to deflect the beam out of the field in areas, such as the gap between the banks of the bridge, which were not to be exposed. Single bridges made using this technique are just as good as those made with the computer-controlled system. The major advantage of the computer system is the ability to make large arrays of bridges and the connecting circuitry in a precisely reproducible way. The details of bridge fabrication using both systems has been reported elsewhere[3]. A few of the characteristics and limitations of the technique are worth emphasizing.

Electrons scatter in the resist material and from the substrate; this produces an undercut pattern in the resist mask. While this undercutting aids in the lift-off of the metal film evaporated through the mask, it also limits the resolution of the pattern to about 50% of the resist thickness. Since our microbridges are 1000Å thick, the dimensions for reproducible bridges are limited to between 1000Å and 2000Å. Another effect of the electron scattering is to make the correct exposure of a given area, such as the bridge, dependent on the exposure of other areas within several microns[4]. While this effect has been corrected for in large systems with powerful computers, at the moment it seems impractical to do so in the type of small system discussed here. Thus the bridge dimensions, while reproducible, are different than the programmed dimensions, so properly exposing small bridges is more an art than a science. This problem becomes especially severe for dimensions less than about 1/3 micron. Also, for dimensions less than this, rather tight control of the resist (PMMA) has been required. As a result of these problems it has been our experience that casual users of the EBL system can fairly easily make masks with 1/3 micron dimensions. The bridge in Fig. 2(a) is typical of what can be rather easily made. The fabrication of smaller dimensions has been practical only when one person has had essentially full-time responsibility for operation and maintenance of the system.

EBL is also well-suited to the fabrication of variable thickness bridges (VTBs). We have made such bridges using several techniques[5]. Perhaps the most versatile is a two-layer technique in which a narrow line, which will be the bridge, is first masked and metalized. A second mask for the banks is then made on top of the line; the oxide is ion milled off of the first layer and the metal for the banks deposited. A VTB made in this way is shown in Fig. 2(b).

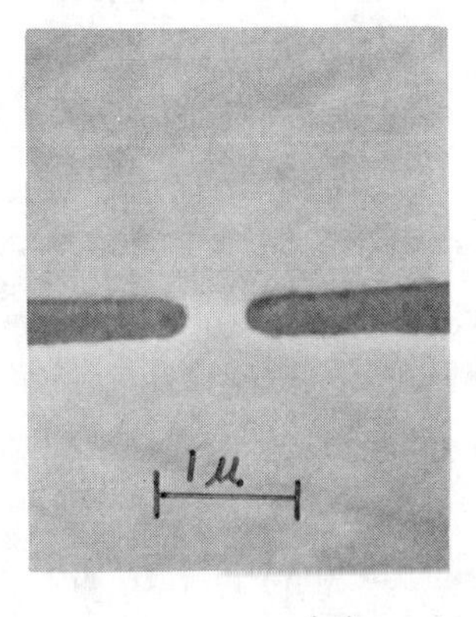

(a)

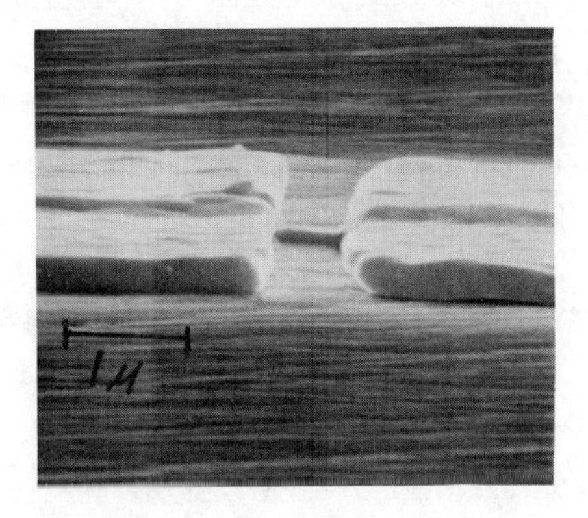

(b)

Fig. 2. (a) Planar In bridge made using EBL and liftoff masking. Film thickness is about 1000Å. (b) Variable thickness In bridge (micrograph taken at 80° tilt) fabricated as described in text. The bridge thickness is about 500Å and that of the banks about 2000Å.

This technique has the advantage that different materials can be used for the bridge and the banks, thus making possible a wider range of bridge characteristics. Also, bridge dimensions are more easily controlled than for single layer junctions, since the electron scattering during the exposure of the banks does not effect the initial bridge line. A disadvantage of the technique, however, is that the need for a submicron gap limits the resist thickness and thus the thickness of the banks to several thousand Å where a micron or more might be desirable[6]. One possible way around this problem might be to use x-ray lithography with an EBL fabricated mask[7]. Since x-rays do not scatter in the resist, much thicker resist layers can be used.

There are a number of photolithographic techniques which are quite competitive with EBL for the fabrication of single submicron bridges or even more complex structures. Contact photolithography has been demonstrated to work reliably down to 0.4μ dimensions[8]. This of course requires a mask with the same dimensions, which can be made using EBL and then used many times. Such an approach may well be easier for studies of single bridges than to fabricate all samples using EBL directly. The use of x-ray lithography with an EBL mask gives even higher resolution. Josephson junction bridges and arrays have also been fabricated by projecting the image of a large mask through an optical microscope[9]. Single bridges as small as 0.2μ have been made using this method[10]. The fabrication of submicron tunnel junctions and VTBs has been reported using a photolithographic mask of one micron resolution through which two sequential evaporations are carried out at different angles to give an overlap area whose dimensions are smaller than those of the original mask[11]. There is, however, an EBL technique[12] (contamination EBL) which has been used to fabricate structure of less than 100Å. In this case no resist material is used in order to eliminate the problem of electron scattering. A layer of material (contamination) builds up on the metal film where the electron beam strikes it; this layer then serves as a milling mask. It is not clear how well this can work in standard EBL systems in which drift and noise are often comparable to the 1000Å resolution of the PMMA resist. The smallest structures which have been produced on our own system using contamination EBL have a size of about 500Å. We have, however, made no attempt to reduce substrate scattering.

FABRICATION OF COMPLEX STRUCTURES

The real justification for the use of EBL mask fabrication is in the ease with which complex structures can be made and in the rapid-turn around time possible. For instance, minor changes in a sample might require changing only a couple of program statements. Figure 3 shows two double-bridge structures along with the gold coupling resistors and current and voltage leads.

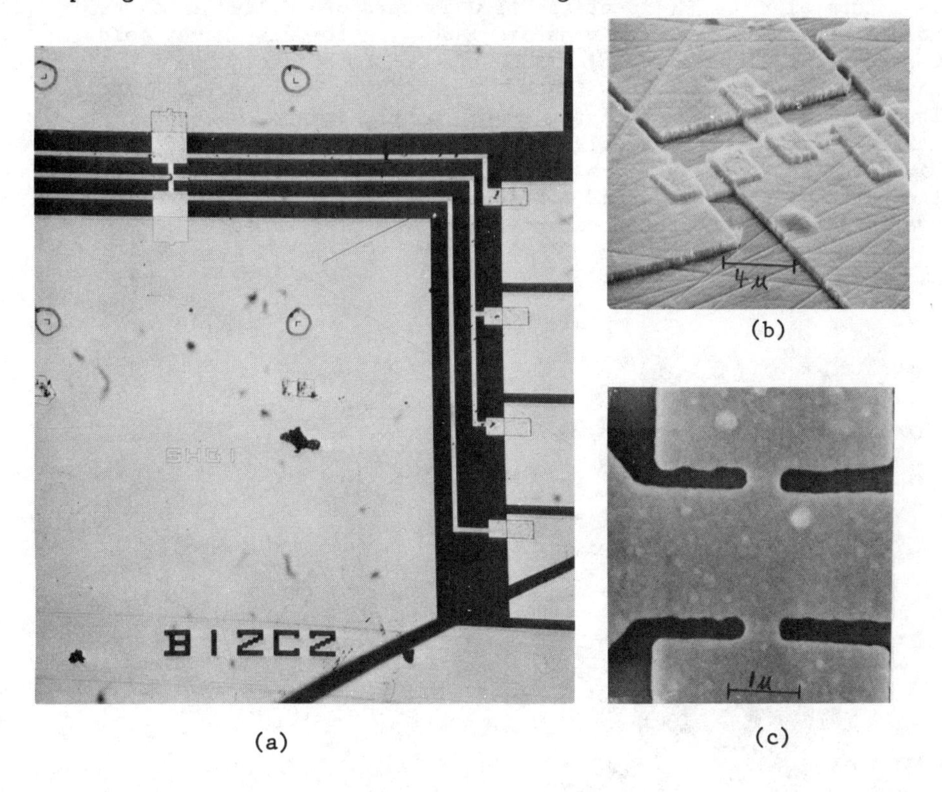

(a) (b) (c)

Fig. 3. (a) Thin film circuit for coupled microbridges, showing individual current and voltage leads and gold contact pads for microwave coupling. Large gold pads were patterned first by ion milling through an EBL mask and include registration marks (circled) to align subsequent layers to within 1000Å. In the second layer the gold resistors (if used) are made using the lift-off technique. The In bridges and leads are patterned in a third layer also using the liftoff technique. (b) Two In bridges used in circuit (a) coupled by gold shunt resistors. (c) Alternate coupling configuration in which bridges, spaced less than 2μ apart, are intrinsically coupled without the use of shunts.

In this case the large gold contact pads were also made using EBL. For the pads the gold film was first deposited, then an EBL mask made for ion milling, which removed the gold in the exposed areas. This procedure was used mainly to change the effective polarity of the resist in order to reduce the area of electron exposure. An added advantage of the technique is that if the PMM mask is post baked after development, the edges are rounded, giving a tapered edge to the final gold pad and reducing the problem of step coverage for subsequent layers. Registration marks for the microbridges and resistors along with identification numbers are milled into the gold at the same time the pads are made. In the next layer gold resistors are deposited using the lift-off technique. The final layer is the indium film with the microbridges and current leads. The total fabrication time for such a sample is about one day. A larger array with twenty-five coupled bridges is shown in Fig. 4. Fig. 4(b) shows the various rectangles (about 100) used to produce the pattern. This array of rectangles is computer-generated in order to check the pattern once it has been programmed.

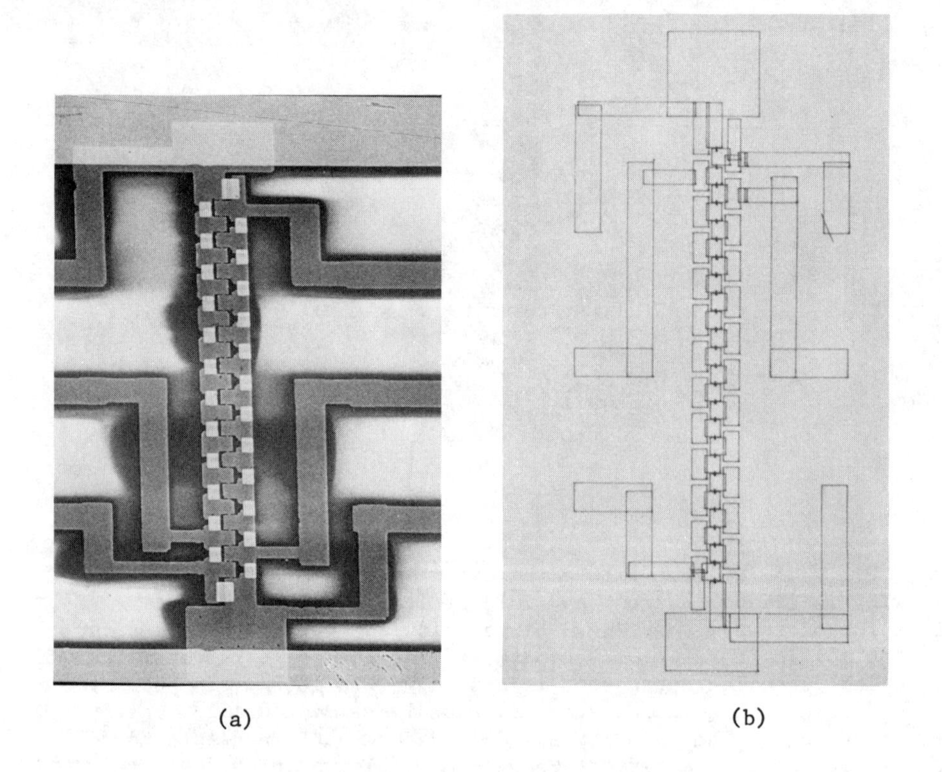

(a) (b)

Fig. 4. (a) An array of twenty-five microbridges coupled by gold shunt resistors with voltage and current leads. (b) Computer plot of programmed rectangles used to produce array pattern.

Once the initial pattern has been tested, the changes (e.g. resistor size) which are needed for a new sample can be made in the program in a matter of minutes.

To summarize our experience, we have found computer-controlled EBL to be an indispensable tool for the fabrication of microbridge arrays. One of the major advantages of the system has been the ability it provides to rapidly make design changes and fabricate new samples as indicated by experimental results without being confined to any fixed sample geometry. In this, we have found the rapid turn-around time provided by immediate access to the system to be of crucial importance.

PROPERTIES OF PLANAR MICROBRIDGE JUNCTIONS MADE USING ELECTRON BEAM LITHOGRAPHY

One of the principle requirements that must be met by microbridges to be used in large coherent arrays is that of highly uniform resistance and critical current. This is necessary to permit all bridges to lock to the same voltage when they have a common bias current. Fortunately the resistance and critical current of planar bridges depend rather weakly on the dimension of the bridge since these properties are determined to a fair extent by the banks. Figure 5 shows an example of two not exceptionally uniform bridges along with a plot of the differential resistance versus current for both.

(a)

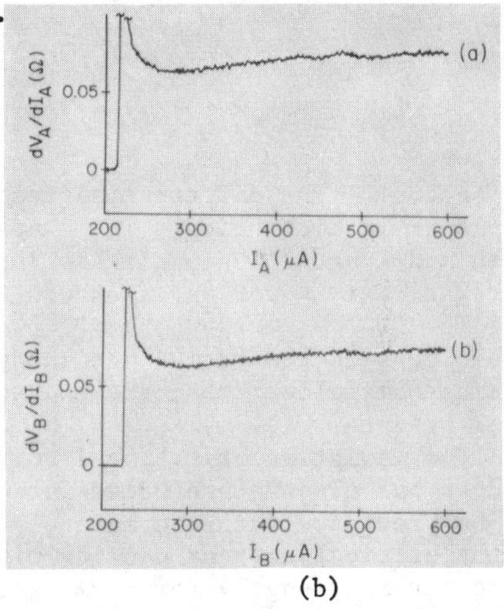

(b)

Fig. 5. (a) Two In microbridges, A and B, coupled by 4μ × 4μ gold shunt resistors (light squares). (b) Differential resistance vs. current for the two bridges. Note that critical currents differ by about 5% while the differential resistances of the bridges are nearly identical.

The differential resistances of the bridges are virtually identical, while the critical currents differ by only about 10μA out of more than 200μA. These differences are small compared to variations caused by changes in film properties. Thus, this sort of reproducibility could probably not be obtained for two bridges evaporated under different conditions. This, however, does not present a problem for array fabrication. A systematic study of the variation in bridge resistance with dimension has been made[13]. The results are summarized in Fig. 6.

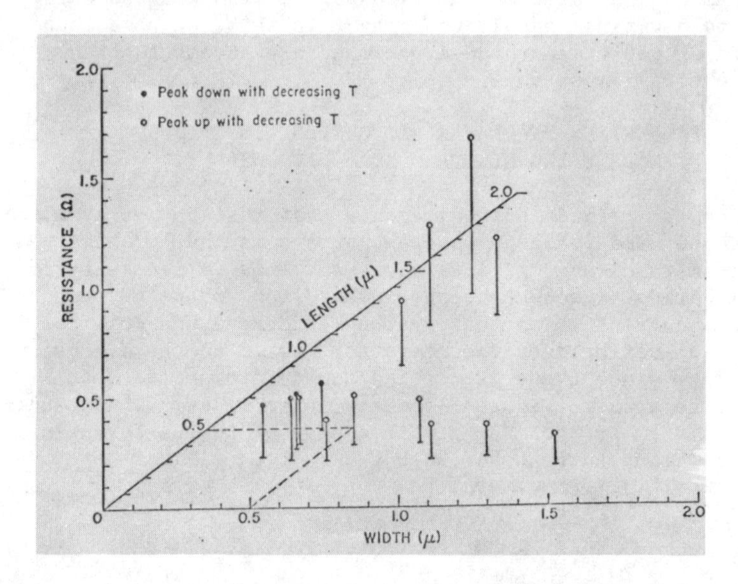

Fig. 6. The differential resistance at low critical current of planar In microbridges (thickness of 1000Å) as a function of width and length. Open circles indicate bridges which become hysteretic at I_c for large enough I_c.

As can be seen the resistance change with bridge dimensions is considerably slower than the changes in the length and width for small bridges.

The variation of critical current with the dimensions of the bridges has also been studied for a range of bridge dimensions and can be predicted rather accurately[14]. Figure 7 shows the change in the critical current density with bridge width (in units of the coherence length) for three bridges of different width. The critical current density of all bridges is enhanced over that of a uniform strip of the same superconductor. Since this enhancement is greater for narrower bridges, the total critical current of the bridge varies more slowly than the bridge width.

The current phase (I-ϕ) relations have been measured[14,15] for planar bridges up to critical currents of 10μA. Small bridges with dimensions of the order of ξ are found to exhibit a sinusoidal (I-ϕ) relationship. It was also found that bridges with width and length much less than ξ had I-ϕ relations which were in excellent agreement with those calculated for VTBs, assuming rigid boundary conditions[16].

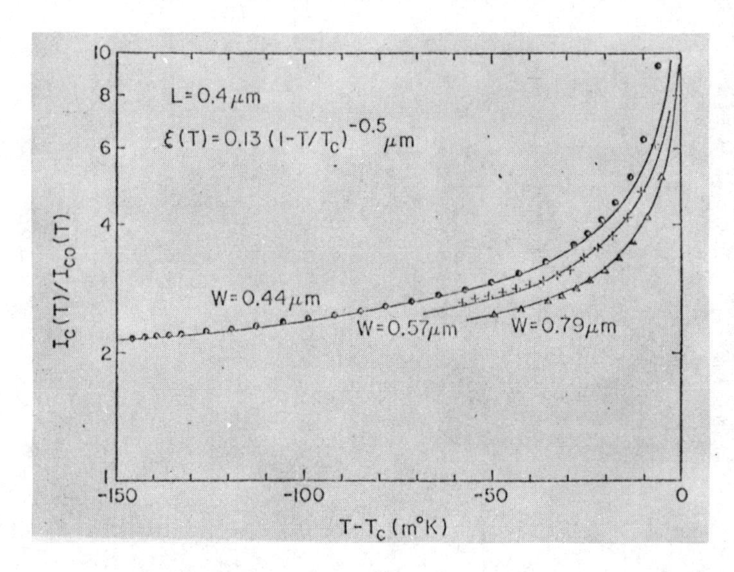

Fig. 7. Enhancement of critical, $I_c(T)$, for microbridges of various widths over the critical current of a long uniform strip of the same width $I_{co}(T)$. Bridges are about 0.4μm long and 0.1μm thick. Solid lines are theory with $\xi(0)$ used as an adjustable parameter (the same for all bridges).

In particular, for a 2.7μ long bridge, the I-ϕ relation became significantly nonsinusoidal when ξ was less than the bridge length but did not become reentrant until the length was between three and four times ξ.

Single microbridges of this type have a rather wide radiation linewidth, typically several hundreds of MHz[17]. Part of the reason for the wide line is inherent in the nonresonant nature of the systems. The intrinsic linewidth is proportional to $\langle V_N^2 \rangle$, the voltage noise at low frequency[18], and $\langle V_N^2 \rangle = R_D^2 \langle I_N^2 \rangle$. R_D is the differential resistance of the junction at the operating point and $\langle I_N^2 \rangle$ the intrinsic current noise. For resonant systems such as tunnel junctions and junctions in cavities, the interaction of the Josephson oscillations with the cavity resonance greatly reduces R_D. In addition to this, extensive measurements of the low-frequency voltage noise of microbridges[13,19] show the current noise to be different than that calculated for tunnel junctions and reportedly observed for several different types of Josephson junctions. The current noise of our microbridges is nearly independent of voltage for I greater than I_c, but the noise increases with I_c and is always significantly greater than the Johnson noise corresponding to the bridge resistance. There are two methods of reducing this noise. One is by simply shunting the bridge with a low resistance such as shown in Fig. 8.

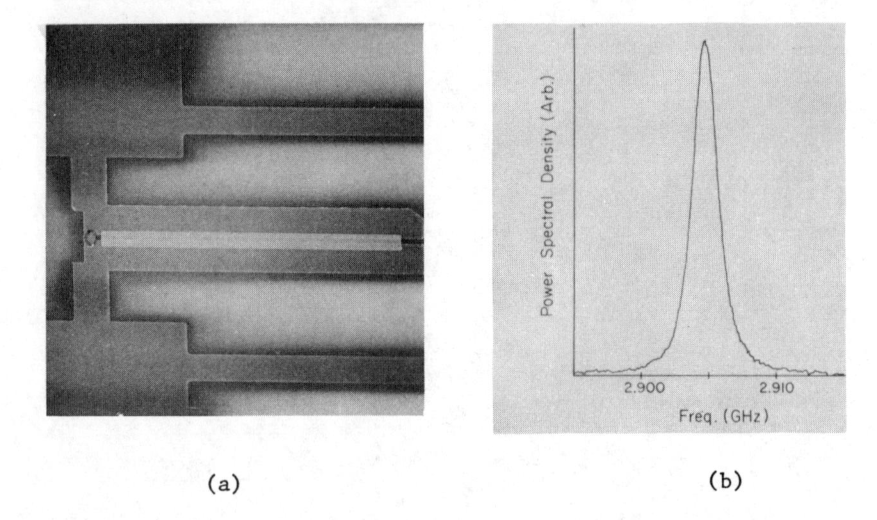

(a) (b)

Fig. 8. (a) Indium microbridge (circled) shunted by .015Ω gold resistor (light narrow strip). (b) Spectral density of radiation from shunted bridge.

Since the bridge noise is much larger than the Johnson noise of the shunt resistor, the linewidth will decrease approximately as R_D^2 and rather dramatic reductions can be achieved. For example, the radiation from an unshunted bridge with $R_D = 0.45\Omega$, reported in Ref. (17), had a linewidth of 1 GHz. A similar bridge, shown in Fig. 8, at the same critical current but shunted with a resistance of 0.015Ω, had a linewidth of only 2 MHz. The disadvantage of this approach of course is that it increases the impedance mismatch between the bridge and the external world. This problem can be overcome by stringing a large number of such junctions together in series in order to match the load. If changes in R_D due to the interaction of the bridges in the array are neglected, this approach will only lead to a reduced linewidth when the shunt resistors are not the dominant current noise source. However, it is desirable in this case to reduce the junction resistance in order to obtain the maximum power from the array. These points are discussed in more detail below.

In order to be useful as a radiation source the junctions of an array must voltage-lock. That is, when the voltages of the junctions are close, they must tend to spontaneously lock together to a common voltage[20]. This is important since there will inevitably be small variations in characteristics of the junctions, and it is not practical to overcome them by independently biasing each of the junctions. We have found two different configurations, shown in Fig. 2a and Fig. 2b, which give voltage locking between microbridges[21]. The technical challenge in the fabrication of coupled arrays such as in Fig. 3 is to obtain junctions with great-enough uniformity and large-enough locking range that the locking range (i.e., the amount by which the current through a single junction can be varied while the bridges still remain locked) exceeds the variation in critical currents among the junctions. It is possible to calculate the locking range approximately using the RSJ model for the bridge. On the basis of these calculations and the attainable junction uniformity, it appears possible to construct large arrays which will operate coherently. In fact this has been demonstrated in the VHF range for proximity effect bridge arrays[22]. For the coupled bridges shown in Fig. 4 a large-enough locking range has already been achieved at low microwave frequencies (to about 5 GHz). We have, however, observed coherent radiation with a power about twice the sum of the powers of the individual bridges even without voltage locking[23]. For example, coherent radiation up to 18 GHz has been obtained for the sample in Fig. 4 even though voltage locking does not extend that far. In fact, a high degree of coherence can be present even when the DC voltages of the two bridges differ by several hundred nanovolts. The limitation on the locking range for this sample appears to be due to the inductance of the bridge shunt resistor loop. There are, however, several techniques (e.g., superconducting ground planes) which should reduce the loop inductance substantially. Further, the locking range of larger arrays should double since each junction is then coupled to two others. While we are presently restricting our studies to linear arrays, it seems likely that two- and three-dimensional arrays would work even better and certainly appear to be technically feasible to fabricate.

There is yet another advantage to voltage-locked arrays for obtaining narrow linewidths. In the region where the junctions are locked, the differential resistance relating the voltage change in the array to a current change in a single junction (e.g., the noise current) is greatly reduced[21]. It appears theoretically that this reduction should be proportional to N, the number of junctions, at least in an ideal array. Coupling this effect with that of the reduced bridge resistance R_J, which can be used for large N (assuming R_J is changed with N such that $NR_J = R_L$, the load resistance), the linewidth $\Delta\nu \propto N^{-2}$. Also the total power available in the fundamental of an array matched to its load should increase as N^2. This occurs because of the constraint that $R_J I_c < V_{DC}$, the operating voltage in order to get the maximum power. Thus the reduction in R_J possible when increasing N allows operation at high I_c and thus high power. Including the effects of line-narrowing, the peak power should increase as N^4. Thus arrays should be vastly superior to single junctions as generators of microwave radiation which can be swept over many octaves. Power increases and line-narrowing consistent with these arguments have been observed for the two coupled junctions, but there is obviously a long way to go. Using values for R_J and I_c which seem attainable with present technology it appears that such an array could have a power of more than a microwatt with a linewidth of a few hundred Hertz.

While the complete control over all junction parameters and coupling geometry which such arrays will require will make full use of the power of computerized EBL systems, it appears that the fabrication of such large arrays is within the capabilities of present laboratory systems. As technology advances over the next few years and our understanding of the mechanisms for coherence increases, it would seem that coherent Josephson systems should be an interesting area for study and possible technical application.

We would like to thank A.K. Jain for performing some of the microwave measurements reported and also P. Mankiewich for the fabrication of the samples discussed.

REFERENCES

1. A.N. Broers, R.B. Laibowitz, paper presented at this Conference; T.Blocker and P.K. Watts, Paper presented at this Conference.
2. T. Van Duzer, SQUID (Walter de Gruyter, Berlin, 1977) p. 63.
3. D.W. Jillie, J. Lukens and Y.H. Kao, IEEE Trans. on Magn. MAG-11, 671 (1975); D.W. Jillie, Ph.D. Thesis, State Univ. of New York at Stony Brook, unpublished.
4. J.H.P. Chang, M. Hatzakis, A.D. Wilson, A.J. Speth, A. Kern, and H. Luhn, IBM J. Res. Develop., 376 (1976).
5. R.D. Sandell, G.J. Dolan and J.E. Lukens, SQUID (Walter de Gruyter, Berlin, 1977) p. 93.
6. M. Octavio, W.J. Skocpol and M. Tinkam, IEEE Trans. Mag-13, 739 (1977).
7. E. Spiller and R. Feder, X-ray Optics (Springer, 1978) .
8. Henry I. Smith, N. Efremow, and P.L. Kelley, J. Electrochem, Soc.: Solid-State Science and Technology, 121, 1503 (1974).
9. P.W. Palmer and J.E. Mercereau, IEEE Trans. Mag-11, 667 (1975).
10. M.D. Feuer, D.E. Prober, and J.W. Cogdell, paper presented at this Conference.
11. G.J. Dolan, Appl. Phys. Lett. 31, 5 (1977).
12. A.N. Broers, W.W. Molzen, J.J. Cuomo, and N.D. Wittles, Appl. Phys. Lett. 29, 9, 596 (1976).
13. T.W. Lee, Ph.D. Thesis, State Univ. of New York at Stony Brook, unpublished.
14. S.S. Pei, Ph.D. Thesis, State Univ. of New York at Stony Brook, unpublished.
15. S.S. Pei, R.D. Sandell, and J.E. Lukens, to be published.
16. K.K. Likharev and L.A. Yakobson, Sov. Phys. Tech. Phys. 20, 950 (1976).
17. C. Varmazis, J.E. Lukens and T.F. Finnegan, Appl. Phys. Lett. 30, 660 (1977).

18. A.J. Dahm, A. Denenstein, D.N. Langenberg, W.H. Parker, D. Rogovin, and D.J. Scalapino, Phys. Rev. Lett. 22, 1416 (1969).
19. T.W. Lee and J.E. Lukens, Bull. of the Amer. Phys. Soc. 23, 357 (1978).
20. D.W. Jillie, J.E. Lukens, Y.H. Kao, and G.J. Dolan, Phys. Lett. 55A, 381 (1976).
21. R.D. Sandell, C. Varmazis, A.K. Jain and J.E. Lukens, paper presented at this Conference.
22. D.W. Palmer and J.E. Mercereau, Phys. Lett. 61A, 135 (1977).
23. C. Varmazis, R.D. Sandell, A.K. Jain, and J.E. Lukens, to be published.

ALTERNATIVES IN FABRICATING SUBMICRON NIOBIUM JOSEPHSON JUNCTIONS

G.M. Daalmans and J. Zwier
Department of Applied Physics, Delft University of Technology
Delft, The Netherlands

ABSTRACT

This paper deals with the fabrication of submicron niobium microbridges and submicron niobium tunneljunctions. Several techniques are described to fabricate niobium microbridges. The most important of them is a technique in which niobium is evaporated through a bakable submicron mask. The technique to fabricate submicron tunneljunctions is basically a shadow evaporation technique. In contrast to earlier attempts it is a reproducible method suitable for large scale application.

INTRODUCTION

Submicron niobium bridges have been prepared for the first time by Laibowitz[1] in 1973. These bridges were of the uniform-thickness type. They were fabricated by evaporating niobium on a silicon substrate coated with a P.M.M. layer in which a microbridge pattern was made with an electron beam and subsequently lifting off the resist pattern. This technique requires a low substrate temperature in order to avoid plastic deformation of the resist pattern by heating during the evaporation run. Due to the strong gettering action of niobium it is necessary to evaporate on a heated substrate in order to obtain the best niobium layers. Therefore one obtains the best niobium bridges if it is allowed to warm up the substrate during the niobium evaporation run. We developed different techniques to achieve this goal.

As far as we know Niemeyer[2,3] was the first who made submicron lead or indium-lead tunneljunctions. Jutzi[4] made tunneljunctions with one niobium and one lead electrode with a contact area of a few square microns. The technique used by Niemeyer was a shadow evaporation technique; Jutzi made use of conventional photolithography. The shadow evaporation technique has the advantage over the photolithographic technique that it is not necessary to open the evaporation chamber between the two evaporation runs. If there are no means in the evaporation chamber to clean the base layer of the junction before the barrier formation, it is necessary to keep the vacuum chamber closed and to make use of the shadow evaporation technique. That is our actual situation. We improved the shadow evaporation technique of Niemeyer in that sense that we made it more reproducible, easier to perform, and applicable for large quantities of niobium junctions.

SUBMICRON MICROBRIDGES

If one wants to make thin film patterns there are two alternatives; A. one starts with a good quality niobium thin film and makes a pattern in the film or B. one evaporates the niobium through a mask which is in close contact with the substrate.

A. Patterning niobium films

One starts with a niobium-coated substrate and puts electron

ISSN: 0094-243X/78/312/$1.50

resist (P.M.M.) on it, prebakes it for half an hour at 180°C and brings the substrate into an electron beam imaging system[5]. In this system a copper mask, containing the enlarged pattern desired, is imaged by an electron beam of 15 keV onto the substrate. Expose the resist with a dose of about 10^{-4} Coul./cm^2. Under the influence of such an amount of charge the molecular chains are chopped in shorter pieces[6] and are soluble in a mixture of isopropyl alcohol and methyl isobutyl ketone. So we obtain a positive resist image of the mask pattern (figure 1). Subsequently the substrate is installed in the evaporation chamber and a thin layer of chromium is evaporated on top of it. The thickness of the chromium layer must be at least one half of the thickness of the niobium layer. Afterwards the resist pattern is lifted off in trichloroethylene and a chromium mask equal to the desired niobium pattern remains on the niobium layer. The lift-off of a P.M.M. mask exposed with electrons should be facilitated[7] by a slight undercut (≈ 5°) due to reflecting electrons. This undercut is frequently overshadowed by effects due to misfocusing of the electron beam, vibrations, stray fields and over-exposure. These errors produce a resist pattern with a certain linewidth that means there is a gradual transition from not-exposed resist to fully-exposed resist (figure 2.)

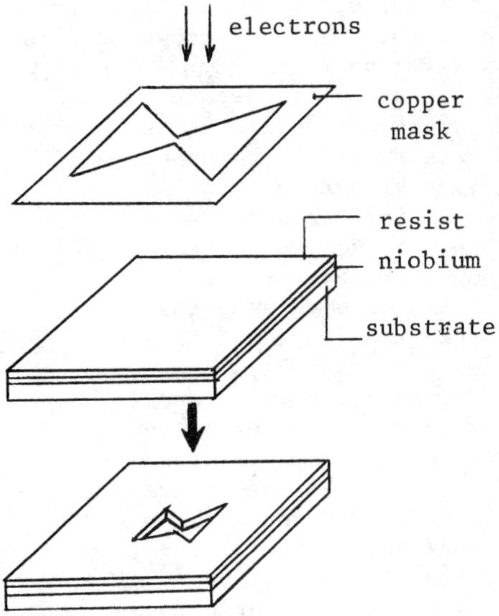

electrons
copper mask
resist
niobium
substrate

Figure 1.

In our case the line widthis about 0.3 μ. In order to copy the chromium pattern in the niobium layer, an ion etch technique is used[8]. Our sample is exposed to a nearly parallel beam of argon atoms. The argon atoms etch the unprotected niobium away as well as the chromium mask. The etch rate of niobium is about twice the etch rate of chromium, so the thickness of the chromium mask must be at least one half of the thickness of the niobium layer in order to keep protection during the entire etch time. The residue of the chromium mask is removed afterwards in a KOH solution. With thin niobium layers (200 and 300 Å) we found that the niobium pattern is indeed a one-to-one copy of the chromium mask. Moreover the yield is very high, almost 100%.

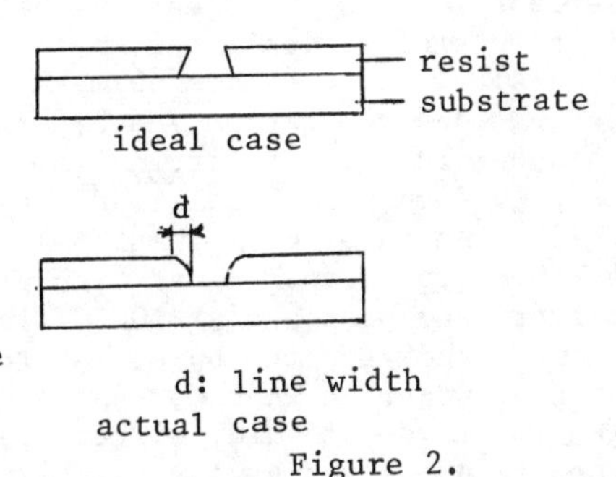

Figure 2.

B. Evaporation through sub-micron masks

1. Evaporate a thick chromium (or aluminum) layer on a substrate and evaporate on top of it a thin layer of aluminum (or chromium). The thickness of the bottom layer is determined by the thickness of the niobium pattern desired and is at least twice the thickness of the niobium layer. The top layer has a thickness of less than 1000 Å. Spin on top of the double layer a P.M.M. layer of about 0.3 micron thick and prebake the sample at 180°C for half an hour. Expose with electrons as in the foregoing section. Develop the resist pattern and postbake at 100°C for half an hour. The result is shown in figure 3a. The next step is to etch away the unprotected upper layer with chemical (wet) etch (figure 3b) In order to keep the un ercut negligible it is necessary to use an upper layer thinner than 1000 Å. Remove the resist pattern with tri and clean the sample thoroughly (fig. 3c) Finally we etch away the thick groundlayer of the bimetallic sandwich again with chemical etch. Undercut is now unavoidable but at the same time very useful (fig. 3d). The undercut can be one micron without losing pattern definition, so lift-off of rather thick niobium layers can be achieved easily. The result is a bakable metallic mask with a resolution power equal to the resolution power of the P.M.M. mask. The yield of this more complicated procedure is mainly determined by the quality of the resist pattern and therefore can be very high. The method developed here can be very useful too for those who want to make structures of niobium alloys, which are very difficult to handle.

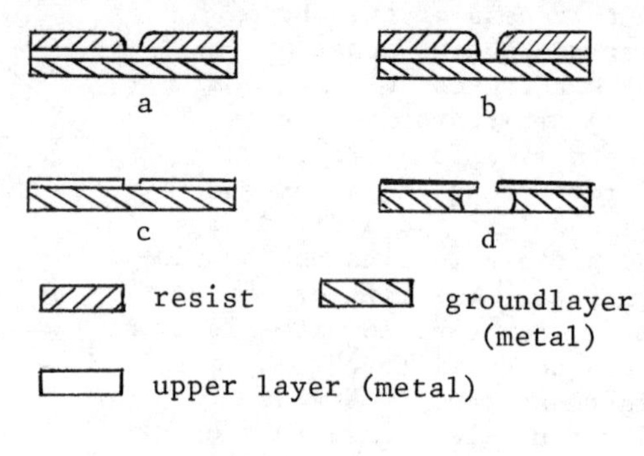

Figure 3.

2. The metallic mask prepared in the preceeding section generates flat bridges. It is proved[9] that heating effects are very serious in those bridges. The heating problem can be overcome partly by making variable thickness bridges (thin bridge, thick electrodes). It is possible to make these variable thickness bridges during one evaporation run by using a shadow evaporation technique as shown by Dolan[10] for indium. We adapted these methods for niobium.

Evaporate a thick ground layer of chromium or aluminum and a thick upper layer of silver, gold or copper on a substrate. Spin on a P.M.M. layer and expose with electrons as before. Develop the pattern and postbake at 100°C. Install the sample in the ion etch chamber and etch the upper layer away. It is essential to etch without undercut in order to save the pattern definition. By using an upper layer of gold, silver or copper with a much higher etch rate than the resist layer, it is even possible to obtain a thick metallic mask with a smaller linewidth (figure 4a). Remove the resist pattern and etch the ground layer away chemically. Again we appreciate the associated undercut (figure 4b). The evaporation mask obtained in this way has only one difference compared with the mask of the foregoing section and that is the thickness of the top layer. The thick top layer enables us to make variable thickness micro-bridges during one evaporation cycle if we rotate the sample during the evaporation in such a way that the overhanging gold, silver or copper parts screen the substrate in the bridge region totally for incoming niobium atoms. The first results are encouraging and seem to indicate that we will succeed in preparing variable thickness niobium bridges in this way.

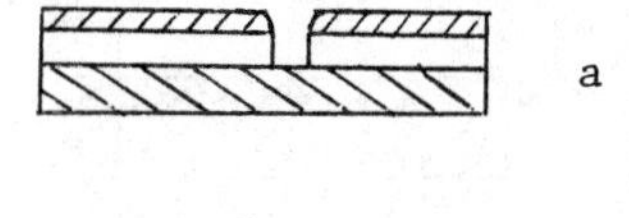

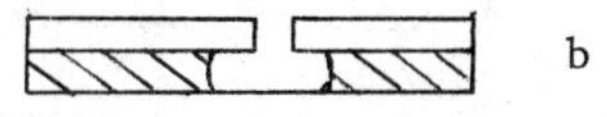

silver, copper or gold

resist

aluminum or chromium

Figure 4.

SUBMICRON TUNNEL JUNCTIONS

As mentioned in the introduction we do not have sputter-etch apparatus in our evaporation chamber. That makes it very difficult to work with resist masks, because in that case we have to open the evaporation chamber between the two evaporation runs and so we lose the control over the barrier formation. The only alternative left was the shadowing technique of Niemeyer. He used the shadow of a glass fiber crossed over a scratch in the substrate. We used the same method but found the results not reproducible enough. Moreover making one junction took too much time because fibers less than one micron in diameter are not easy to handle.

In our method proposed by one of us (J. Zwier) a thick layer of Al or Cr is evaporated on a silicon substrate. Afterwards a P.M.M. layer is spun on top of it and the sample is baked at 180°C for half an hour. The resist is exposed with electrons and the same mask as before is used. After development the same resist pattern is obtained as in figure 1. The sample is mounted in the vacuum chamber and a thin layer of niobium is evaporated on it. The resist pattern is lifted off in tri. The result is a niobium bridge pattern deposited on the thick chromium or aluminum under-layer. The underlayer is etched away in an appropriate etchbath and the etch time is taken as long as is necessary to make so much undercut that the niobium bridge is free-hanging (figure 5.) These free-hanging bridges can be used as a shadow evaporation mask, which is principally identical with the fiber-scratch mask of Niemeyer. The free-hanging niobium bridge plays the role of the fiber and the scratch is exchanged by the tunnel etched away under the niobium bridge. In order to fabricate a tunnel-junction with this mask we evaporate the two electrodes under different angles of incidence of the niobium vapour (figure 6).

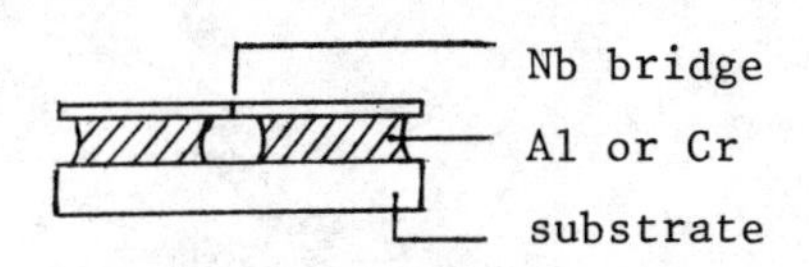

Figure 5.

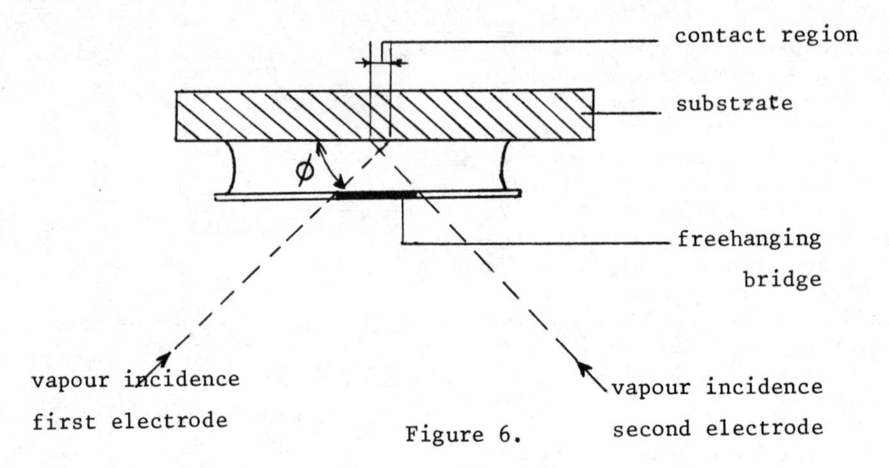

Figure 6.

The dimensions of the contact are determined by the width of the bridge, the thickness of the base layer and the angle of incidence ϕ. With so many degrees of freedom it is easy to obtain junction areas of less than one square micron. The yield is nearly 100% and the smallest contact area obtainable is about 0.1 square micron.

We wish to thank T.M. Klapwijk for careful reading of the manuscript.

REFERENCES

1. R. Laibowitz, Appl. Phys. Lett. 23, 407 (1973).
2. J. Niemeyer, P.T.B.-Mitt. 84, 251 (1974).
3. J. Niemeyer et al, Appl. Phys. Lett. 29, 380 (1976).
4. W. Jutzi, Electr. Lett. 8, 589 (1972).
5. J.E. Mooij et al., Revue Phys. Appl. 9, 173 (1974).
6. I. Haller, I.B.M. Journal 12, 251 (1968).
7. M. Hatzakis, Journ. Electrochem. Soc. 116, 1033 (1969).
8. H. Smith, Proc. of the IEEE 21, 1380 (1974).
9. T.M. Klapwijk et al., Phys. Lett. 47A, 351 (1974).
10. G.J. Dolan, Appl. Phys. Lett. 31, 337 (1977).

submicron tunneljunction

FABRICATION OF SUBMICRON JOSEPHSON MICROBRIDGES USING OPTICAL PROJECTION LITHOGRAPHY AND LIFTOFF TECHNIQUES*

M.D. Feuer, D.E. Prober and J.W. Cogdell
Becton Center, Department of Engineering and Applied Science,
Yale University, New Haven, CT. 06520

ABSTRACT

We describe the fabrication of sub-micron, type II microbridges, using a reflected-light microscope for high-resolution, projection photolithography and metal liftoff techniques for precise pattern transfer. Microbridges as narrow as 2000 Å have been produced. The novel technique of through-the-substrate exposure, which gives extremely sharp edge definition, is discussed, as well as other factors determining resolution and liftoff yield. We present experimental results for a 0.2 μm wide $Pb_{0.86}In_{0.14}$ microbridge with a device resistance of 5.5 Ω. Finally, we consider the general applicability of these fabrication techniques.

INTRODUCTION

Microbridges based on high-resistivity Pb alloys can achieve the high device resistances which are of particular importance for Josephson mm-wave detectors and for lower-noise SQUID magnetometers. In the future, type II materials may also offer higher operating temperatures and frequencies. However, their use in constriction weak links places stringent limits on device size due to the short coherence lengths involved. To effectively utilize type II films in Josephson microbridges we have developed techniques which allow the production of two-dimensional microbridge constrictions as narrow as 0.2 μm. This has been made possible by a novel through-the-substrate projection exposure technique which ensures good lift-off.

The basic procedure developed is illustrated in Fig. 1. It comprises three steps:

(1) Photolithography--A layer of photoresist (PR) is applied to a substrate, then exposed and removed in the regions which are to be metallized in the final device (Fig. 1a, 1b).

(2) Film Deposition--A uniform superconducting film is deposited by vacuum evaporation, overlying both bare and PR-covered areas of the substrate (Fig. 1c).

(3) Liftoff--The PR is dissolved away, removing the metal which overlies it, while the metal in direct contact with the substrate remains (Fig. 1d).

A typical good microbridge obtained with this procedure is shown in Fig. 2 in a SEM micrograph. A 1 μm scale is shown in the figure; the bridge width is 0.2 μm. Data for this microbridge are reported below. Microbridges of this size can be obtained fairly routinely.

*Research supported in part by NSF Grants ENG 76-09253 and 77-10164.

ISSN: 0094-243X/78/317/$1.50

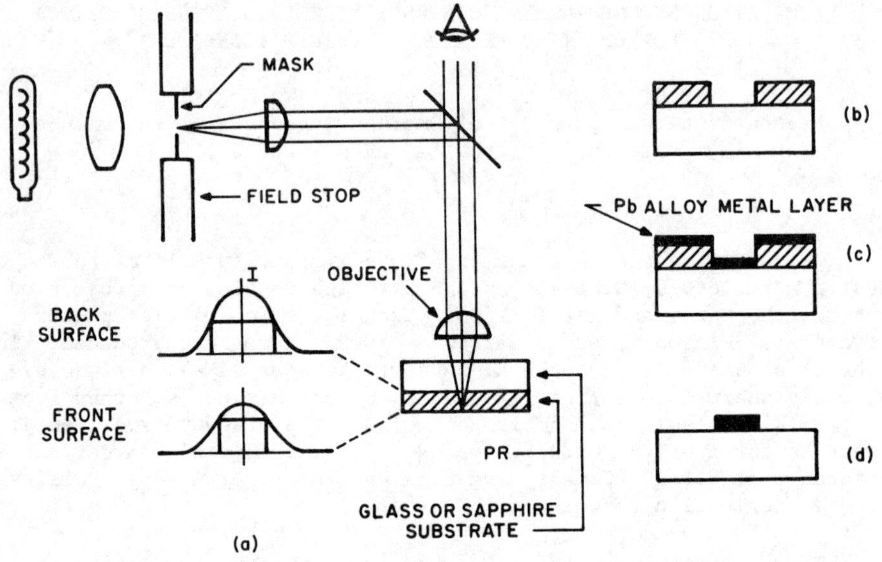

Fig. 1. Schematic of the microscope projection-lithography and lift-off procedure. Example shown is a 0.2 μm width metal line, seen end-on (or the central portion of a microbridge). (a) Projection exposure of the PR in a reflected-light microscope. Note that PR is on the bottom of the substrate. Schematic intensity (I) of exposing radiation vs. position at two heights in PR is also shown, assuming a 0.2 μm pattern and diffraction broadening. Rectangles show width of fully exposed region; (b) After development; (c) Following metallization; and (d) Results of liftoff.

DETAILED LITHOGRAPHIC PROCEDURES

Shipley 1350B positive photoresist is used. It is spin-coated onto glass or sapphire substrates at 10,000 rpm to obtain a thickness of ∿ 0.25 μm. Standard cover glasses are quite satisfactory if handled with care. Baking procedures recommended by Shipley are followed. A first exposure employing contact lithography defines the electrical contact pad areas. A second projection exposure defines the microbridge pattern. This is done with a Zeiss Photomicroscope with vertical illuminator, in a procedure first described by Palmer and Decker[1] (see Fig. 1a). The mask is placed in the field stop of the microscope, so that its image, demagnified by the objective, is focused on the PR layer. A red filter is used for focusing, while a monochromatic blue glass filter is used during PR exposure. Typically a series of 5 patterns are exposed on one slide, with exposure times ranging from 18 to 24 seconds to bracket the optimum exposure. Extra microbridge patterns are later scratched out. Correct focusing is critical; the proper stage height for PR exposure is ∿ 1 μm below the apparent correct focus in the red. This was determined empirically by subsequent SEM inspection of a test series of exposed patterns.

A demagnification factor of 43x is obtained with the 100x oil immersion objective. For sub-micron work, the resolution of masks made from 35mm film (∿ 3-5 μm) is only marginal, and masks made from high-resolution plates are used. It appears that with the better masks, the lower limits on device size are set only by diffraction effects in the microscope.

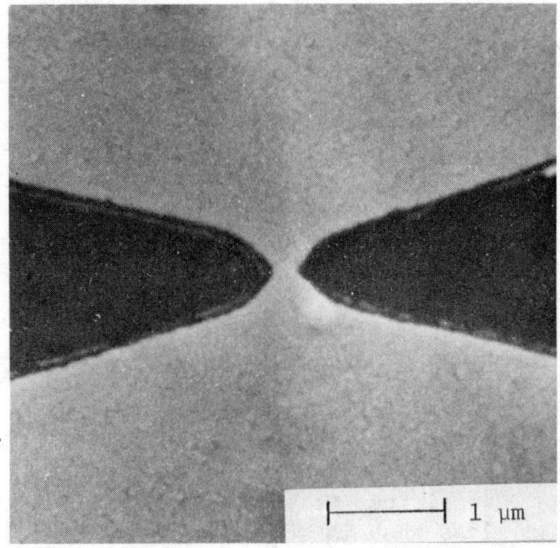

Fig. 2. Microbridge I-7. Width is 0.20μm.

Essential to the success of liftoff is the need for square or undercut PR profiles, so that the metal film is discontinuous at the PR edge and will part cleanly. The major innovation of our procedure is through-the-substrate exposures (back exposures) in which the light passes through the substrate from the back, and impinges on the bottom layer of PR first. This leads to very sharp, possibly undercut, PR profiles, as can be seen in the oblique scanning electron micrograph, Fig. 3. Such profiles greatly increase liftoff yield.

Our explanation of the sharp PR profiles yielded by back-exposure is shown schematically in Fig. 1a. Absorption within the PR reduces the intensity ∿ 30% prior to full exposure as the light passes through the PR,[2] so that the width of the fully exposed region decreases from the initial entry point to the exit point--the PR-air interface. After development, this yields very sharp walls. Essentially, the non-linear development of the PR allows production of PR patterns even smaller than the size suggested by the diffraction limit. Even narrower lines appear feasible, because for lines one has diffraction smearing in only one dimension, instead of two as for microbridges.

A curious result is that focusing and PR patterns are much sharper on the cover glass substrates than on sapphire. This is because the 5 mil sapphire plate, with its higher refractive index, introduces spherical aberration for which our objective has no correction.

FILM DEPOSITION AND LIFTOFF

We have used 250-900 Å films of $Pb_{0.7}Bi_{0.3}$ and $Pb_{0.86}In_{0.14}$ deposited by vacuum evaporation. Pb and Bi have the same vapor pressure and can be evaporated directly. Pb and In have different vapor pressures, however, so that the initial charge must be evaporated to completion for reproducible results. Alloying appears to take place readily and with no special aid.

To achieve good film adhesion necessary for the liftoff process, we employ a glow discharge cleaning and Pb-alloy oxide undercoat.[3] The substrate is cooled to 77 K prior to film deposition to prevent film agglomeration. As seen in Fig. 2, the films are indeed quite smooth. PbIn films are quite resistant to atmospheric exposure and may be stored for months with little degradation. PbBi films, unfortunately, oxidize quickly when exposed to moisture.

The deposited thin films have properties almost identical with bulk material.[4] Below 400 Å, the T_c of PbIn films rises slightly (50 mK at 250 Å), possibly due to scavenging of indium by the oxide layer. Transitions are typically less than 15 mK broad, and different samples agree to within 25 mK.

Fig. 3. Sharp PR profile produced by back exposure, with a 0.3 μm opening.

When the necessary precautions are taken in the first two steps, the liftoff step is satisfyingly simple and straightforward. The substrate is subjected to ultrasonic agitation[5] in acetone, which removes the photoresist and the undesired regions of metal film, leaving a sub-micron microbridge with extremely sharp, clearly defined edges.

We find nearly perfect liftoff, with a yield > 90% even for our smallest microbridges, for through-the-substrate exposures as described above. Top exposures, on the other hand, give reasonable lift-off yields (∿ 50%) only for the larger (0.5 μm) devices.

Current yield for the complete procedure on 0.3 μm width microbridges is about 20-30% (roughly 1 per slide) limited primarily by variations of the optimum exposure time. Limits on film thickness have not been determined; a maximum of 0.25 μm is set by the presently-used photoresist thickness.

RESULTS AND CONCLUSIONS

Microbridges produced range from 0.2 μm to 2 μm in width and from 250 Å to 900 Å in thickness. Critical currents are roughly proportional to $(T_c-T)^{1.3}$, in agreement with the theory of Mooij and Dekker,[6] which calculates the critical current due to supercurrent depairing in a planar, hyperbolic neck. Steps at 9 GHz appear in the I-V characteristics of all devices which are 0.5 μm or less in width. As a function of microwave power, the steps initially show the quasi-periodic dependence predicted by the resistively-shunted-junction (RSJ) model, but at higher power levels, the steps fade away due to self-heating effects. Some data for the $Pb_{0.86}In_{0.14}$ microbridge shown in Fig. 2 are given in Table I. This bridge is 0.2 μm wide and 250 Å thick.

Table I. Material parameters and data for microbridge shown in Fig. 2.

T_c	7.12 K
$\xi(0)$	250 Å
$\frac{dI_c}{dT}$ (ΔT = 0.1 K)	0.25 mA/K
R_{diff}	5.5 Ω
V_{max} for highest step	375 μV
Λ_{qp} - quasiparticle diffusion length	∿ 0.2 μm

The techniques of projection back-exposure and liftoff outlined above allow production of extremely small high-quality Josephson microbridges, which have quite good performance for planar (uniform-thickness) structures and have the desired high resistances. These techniques should be applicable to any soft superconductor which can be patterned with a single lithography step. It may also be possible to apply them to other thin-film materials, such as those used in magnetic-bubble devices or surface-acoustic-wave devices. One limitation on the back-exposure technique is that it cannot be used to obtain resist undercutting with repeated depositions, if the first deposited film is not transparent. In contrast, electron-beam lithography allows one to obtain undercut resist profiles on top of metal films, due to electron scattering in the PMM resist. Still, one can fabricate sub-micron variable-thickness microbridges with optical projection lithography if the first lift-off step is followed by ion milling to thin the bridge region. This is currently being investigated in our laboratory.

ACKNOWLEDGEMENTS

We wish to thank W.H. Henkels, C.J. Kircher, and especially M. Octavio for helpful discussions about fabrication techniques. At Yale, we also thank A. Pooley for the scanning electron micrographs, and P. Dressendorfer and A. Spielberg for mask patterns.

REFERENCES

1. D.W. Palmer and S.K. Decker, Rev. Sci. Instrum. 44, 1621 (1973); S.K. Decker, Ph.D. Thesis, CalTech, 1975.
2. Frederick H. Dill, William P. Hornberger, Peter S. Hauge and Jane M. Shaw, IEEE Trans. ED22, 445 (1975).
3. R.Y. Chiao, M.J. Feldman, H. Ohta and P.T. Parrish, Rev. Phys. Appl. 9, 183 (1974); M. Octavio, private communication.
4. J.E. Evetts and J.M.A. Wade, J. Phys. Chem. Solids 31, 973 (1970).
5. L.D. Jackel, Ph.D. Thesis, Cornell University, 1976.
6. J.E. Mooij and P. Dekker, to be published; G.M. Daalmans, T.M. Klapwijk and J.E. Mooij, IEEE Trans. MAG-13, 719 (1977).

SUPERCONDUCTING MICROBRIDGES AND ARRAYS

P.E. Lindelof, J. Bindslev Hansen and P. Jespersen
Physics Laboratory I, H.C. Ørsted Institute
University of Copenhagen, 2100 Copenhagen
Denmark

ABSTRACT

The current-voltage characteristics and Josephson radiation have been studied for a number of indium bridges as a function of temperatures. Our results show strong deviations from the predictions of the RSJ model, even in the temperature regime where the bridge size is smaller than the coherence length. We ascribe this to non-equilibrium effects in and around the microbridge.

In order to improve the applicability of the Josephson effect, we have investigated coherent arrays of 2, 4 and 10 microbridges. The 10 microbridge array is a novel design which allows the bridges to be parallel-biased at dc (equal voltages) and series-biased at 9 GHz (high output impedance).

MICROBRIDGES

A superconductor with a constriction smaller than the coherence length exhibits the Josephson effect. Small structures are obtained with clean point contacts or using thin-film technology. In the last case it has been customary to distinguish between two types of thin-film microbridges: constant thickness bridges (CTB) and variable thickness bridges (VTB). The distinction is not always experimentally clear-cut. In a VTB the variation in the order parameter is well localized and heat is effectively removed.

The resistively shunted Josephson junction (RSJ) model, incorporating a sinusoidal current-phase relation $I_s = I_0 \sin \phi$, derived on the basis of the Ginzburg-Landau theory with rigid boundary conditions, is central to the understanding of superconducting microbridges. Recently, some attempts have been made to incorporate non-equilibrium phenomena in such a two-fluid description in the I-V characteristics of superconducting microbridges. Three distinct features show up, which, with all probability, are of such origin, namely, the large excess current (shoulder) at small voltages,[1] a saturation of the microwave power of the Josephson radiation,[2] and the subharmonic energy gap structure.[3]

Figure 1 shows a typical I-V curve and the Josephson radiation detected by a superheterodyne receiver (9 GHz). The definition of the quantities plotted in the next figures are shown. R_N is the slope of the I-V characteristic just below T_c.

In Fig. 2 we have plotted R_{min}^{-1}, I_{c1}/I_0 and $P^{1/2}$ vs. $(1-t)\propto I_0$ for 4 representative indium bridges. R_{min}^{-1} is linear in (1-t) and has an intersection with (1-t)=0 which is R_N^{-1}. I_{c1}/I_0 grows as the temperature decreases, but then saturates and is constant over a large temperature interval, i.e. $I_{c1}\propto I_0$. The emitted power of the Josephson radiation grows as $(1-t)^2$ close to T_c but also saturates in the same temperature regime where I_{c1}/I_0 saturates. The saturation value

ISSN: 0094-243X/78/322/$1.50

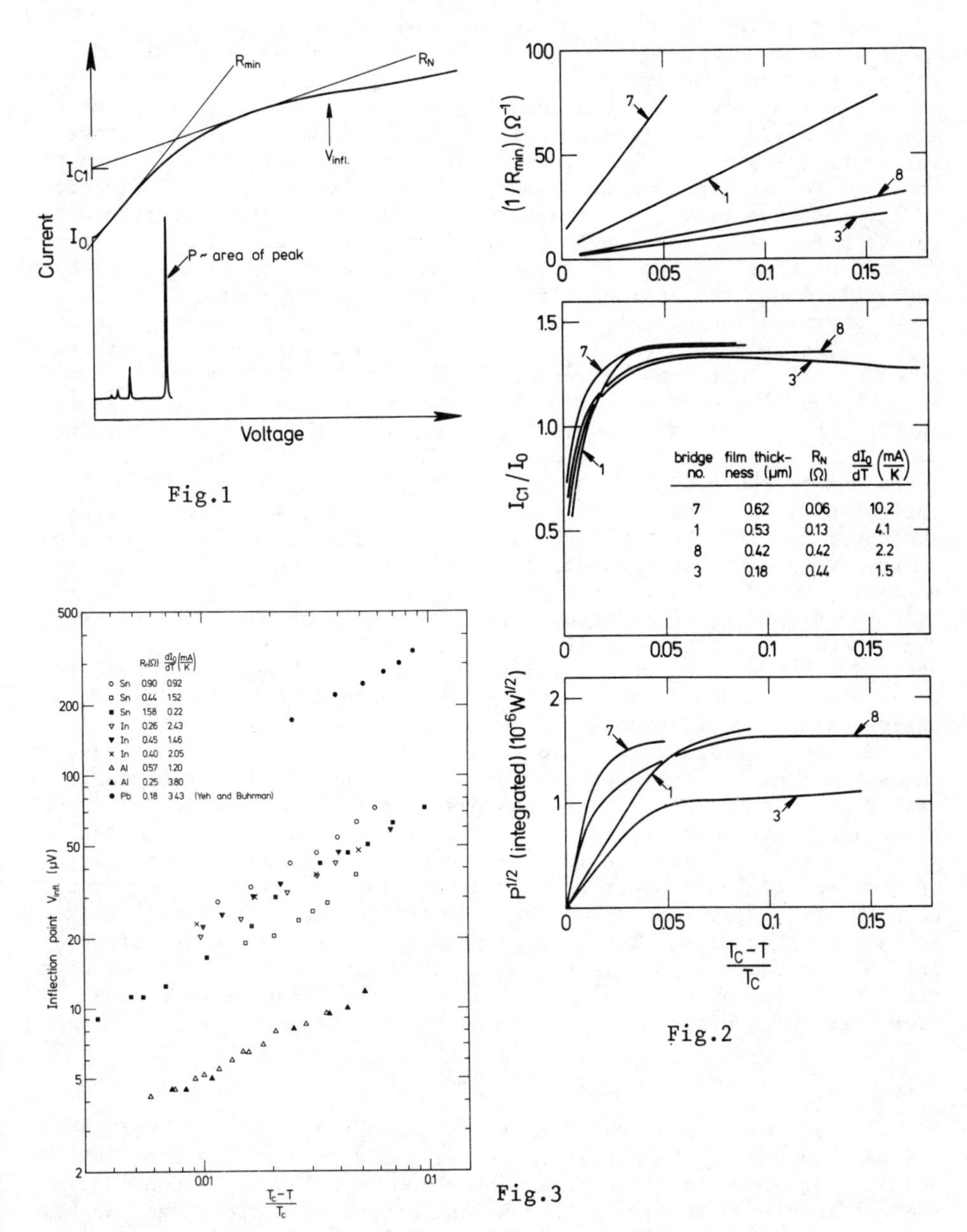

Fig.1 Typical I-V curve showing shoulder and the definition of some quantities.

Fig.2 Plot of R_{min}^{-1}, I_{c1}/I_0 and $P^{\frac{1}{2}}$ vs. (1-t) for 4 indium microbridges.

Fig.3 Inflection point voltage, $V_{infl.}$ vs. (1-t) for a number of Pb, In, Sn and Al microbridges.

of the integrated Josephson microwave power at 9 GHz is typically found to be larger in indium than in tin; in aluminum the emitted power is again much smaller.

Figure 3 shows a double logarithmic plot of $V_{infl.}$, i.e. the voltage where $d^2V/dI^2 = 0$ (see Fig. 1) vs. (1-t). As seen this voltage varies as $(1-t)^{1/2}$ for bridges made of lead, indium, tin and aluminum. The data for lead are taken from ref. 4. There is a clear difference in the position of $V_{infl.}$ at a given reduced temperature for the different materials. This reflects almost with certainty the inelastic relaxation time for the electrons in the different materials, which in turn determines the characteristic times in the superconductor.

Determination of the shoulder in aluminium microbridges presents a special problem because it occurs at such low voltages. It can easily be mixed up with the effect of noise.

We have speculated that perhaps the low voltage shoulder and the saturation of the Josephson radiation could be the effect of the injection of a large number of quasi-particles into the immediate surroundings of the bridge. As the bridge cross-section increases, this current will increase, and by the time the bridge is of the order of the coherence length, it will have a maximum effect on the energy gap. As the temperature is lowered, I_0 increases and the oscillating quasi-particle current will increase as well until the injected current of quasi-particles times their average lifetime is of the order of, say, the number of thermally-excited quasi-particles in a volume of the constriction out to the coherence length. At low voltages this will tend to keep the Josephson oscillation below a certain saturated level as suggested by our experiments.

The third feature which we relate to non-equilibrium phenomena is the subharmonic energy gap structure. This structure is presumably caused by pair-breaking due to phonons created by relaxation of quasi-particles to the singularity of the density of states. The excitation of the quasi-particles can in turn be caused by Josephson radiation or quasi-particles crossing the bridge giving an image picture of the density of quasi-particles of one side of the bridge an energy eV higher on the other side. The density of states on the two sides of the bridge is presumably modulated due to the oscillating electrical potential on the two sides. Another cause of such a modulation can be phonon-trapping effects.

ARRAYS

It has been demonstrated that two microbridges closely spaced in the same superconducting film perturb the I-V characteristics (IVC) of one another and phase-lock their Josephson oscillations. Considering the potential applicability of coherent arrays of microbridges we have focused most of our interest on the phase-locking mechanism. In experiments using tin and indium microbridges we found[5] a maximum critical spacing for the observation of phase-locking phenomena which was about 3.5 μm for indium and 12 μm for tin. Furthermore, we found that the locking could only be observed close to T_c. We concluded that the phase-locking presumably was established through the oscillation of the quasi-particle potential, i.e. over a distance which at low frequencies is Λ, the quasi-particle diffusion length, and at higher frequencies is δ, the skin depth of the quasi-particle potential. When investigating the phase-locking by direct detection of the Josephson radiation from different bridges we have observed both fully in-phase and partially anti-phase coherent behavior. We suggested[5] that this could be related to a low-velocity propagating mode.[6] Such a mode is, however, almost always heavily overdamped. A phase-shift skin depth of the ac quasi-particle potential is a more likely cause. Obviously, a phase-shift also develops if the coupling between the bridges goes via some external impedance elements, but this can hardly be the dominant coupling mechanism in view of the temperature and distance dependencies we find.

We have also looked for coupling between two aluminum microbridges. With inter-bridge spacings ranging from 6 μm to 100 μm, we have so far not been able to detect any phase-locking either in the radiation or in the IVCs. We believe that this is a consequence of the very long

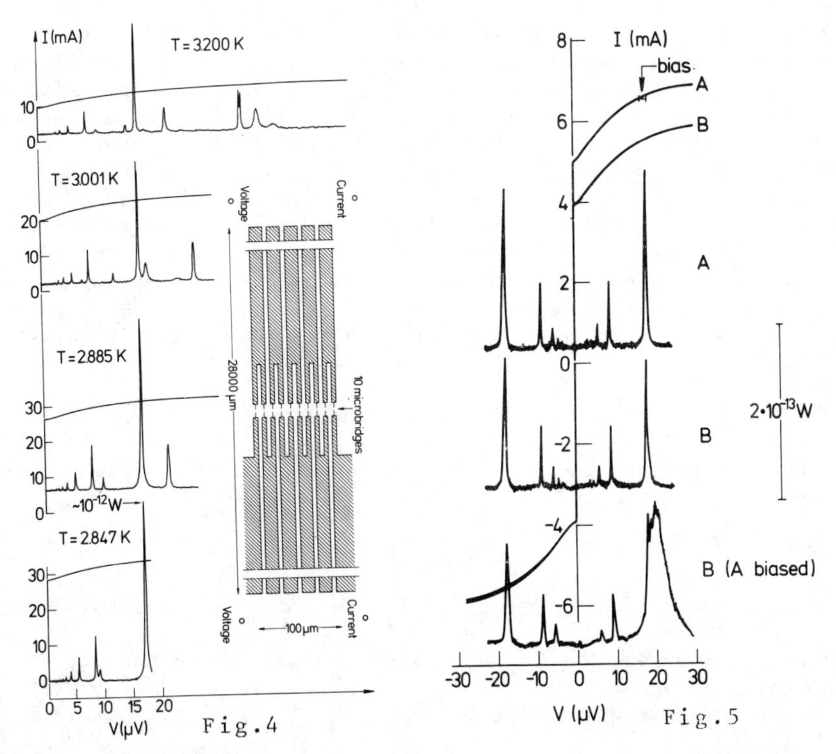

Fig.4 Fig.5

Fig.4 Detected Josephson radiation from an array of 10 In microbridges vs. voltage as well as the IVCs taken at different temperatures. The series/parallel configuration of our array is shown in the insert.

Fig.5 IVCs and Josephson radiation power for two independently biased arrays of 5 In microbridges, A and B. The Josephson radiation when both arrays are biased to radiate into the receiver is shown in the bottom trace (B(A biased)).

inelastic relaxation time in aluminum which presumably limits the frequency range for coupling to such low values that the thermal noise inherent in the bridge extinguishes the coherence phenomenon.

Experimentally we have studied arrays with 4 and with 10 closely-spaced microbridges. The insert in Fig. 4 shows the geometry used. The basic idea is to force all the bridges to have essentially the same voltage (parallel connection at dc) while attempting to force the same microwave current through all the microbridges (series connection at 10 GHz). To improve impedance matching to the waveguide, the sample was mounted at the end of a two-step quarter-wave transformer. Figure 4 shows IVCs of the system at several different temperatures, as well as the radiation at 10 GHz in a bandwidth of 10 MHz. Close to T_c, the five 5 μm wide strips develop phase-slip centers with the result that the voltage we measure is larger than the one across the bridges. However, as the temperature is lowered, the critical current of the strips rises faster than that of the bridges and the radiation peaks from all 10 bridges gradually come to coincide in voltage. In order to probe the coherence of the system another configuration was used that enabled us to bias two sections, A & B, each with 5 bridges independently. In Fig. 5 A and B give the results of the two sections measured independently. B (A biased) shows the total radiation which could be obtained when both sections were biased to radiate simultaneously into our receiver. The broad radiation curve to the left in the figure is caused by an overall change in the voltage across section A due to a change in the current through B. Clearly, the power output is not a simple super-position of the two individually-measured values, but rather corresponds to the output from two coherent oscillators with a phase-difference of about 120°. In this case we did not observe any direct voltage-locking in IVCs.

ACKNOWLEDGEMENT

We are indebted to Ole Eg for preparation of our microbridges and arrays.

REFERENCES

1. H. Højgaard Jensen and P.E. Lindelof, J. Low Temp. Phys. 23, 469 (1976).
2. O.H. Sørensen, J. Mygind, N.F. Pedersen, V.N. Gubankov, M.T. Levinsen and P.E. Lindelof, J. Appl. Phys. 48, 5372 (1977).
3. P.E. Gregers-Hansen and G.R. Pickett, Rev. Phys. Appl. 9, 135 (1974).
4. J.T.C. Yeh and R.A. Buhrman, J. Appl. Phys. 48, 5360 (1977).
5. P.E. Lindelof and J. Bindslev Hansen, J. Low Temp. Phys. 29, 369 (1977).
6. R.V. Carlson and A.M. Goldman, Phys. Rev. Lett. 34, 11 (1975).

STUDY OF THE PROPERTIES OF COHERENT MICROBRIDGES COUPLED BY EXTERNAL SHUNTS

R.D. Sandell, C. Varmazis, A.K. Jain, J.E. Lukens*
Department of Physics, State University of New York,
Stony Brook, New York 11794

ABSTRACT

We have studied the coherent radiation emitted by two microbridges coupled by external shunt resistors. Power spectra data and differential resistance measurements are presented. The relation between the observed linewidths and the differential resistances and considerations for the design of larger arrays are discussed.

PROPERTIES OF COHERENT MICROBRIDGES

Considerable interest exists for the possible application of coherent arrays of Josephson junctions as rapidly tunable microwave generators. As a first step toward that goal, we have studied the properties of a coherent two-junction microbridge array. Using nonresonant coupling, the radiation power spectra from the junctions are measured. We have observed coherent radiation at frequencies from 2-18 GHz and observed, over a factor of two, narrowing of the linewidth of the coherent radiation as compared to the radiation from an individual junction.

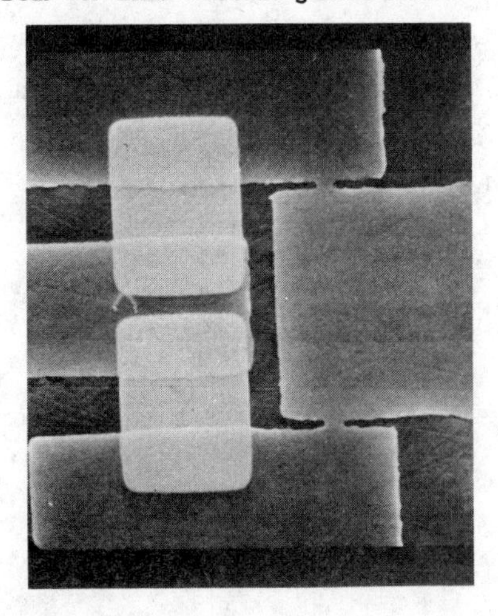

(a)

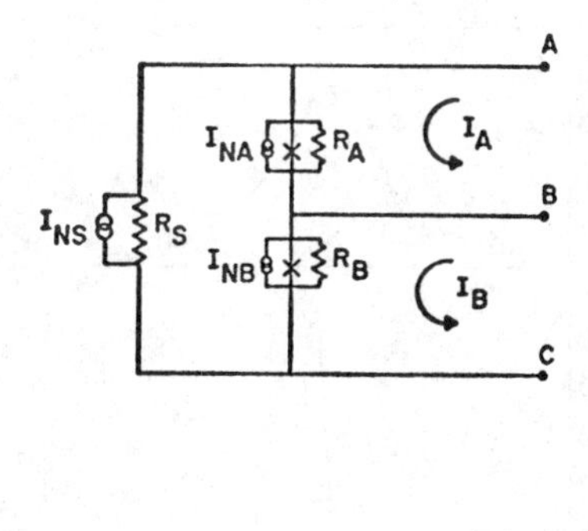

(b)

*Work supported by the Office of Naval Research

ISSN: 0094-243X/78/327/$1.50

Fig. 1. (a) Micrograph of the two-microbridge sample. Lightly shaded rectangles are gold shunts. (b) Equivalent circuit of the sample.

The samples were fabricated using electron-beam lithography[1]. A micrograph of the sample is shown in Fig. 1(a). The Indium microbridges are 0.3μ long, 0.5μ wide and spaced 8μ apart. The lighter rectangles in the micrograph are 1000Å thick gold film resistors which provide the coupling between the bridges. A simple equivalent circuit of the sample is shown in Fig. 1(b). The samples have eight leads to allow independent current-biasing and voltage measurements. We will refer to the current-biasing shown in Fig. 2(b) in which the bias currents are flowing in the same direction as series-biasing. If the direction of one of the bias currents is reversed, the sample is opposed-biased. The two microbridges have nearly identical properties; the critical currents differ by ∿5% and differential resistances differ by less than 2%. The value of the bridge and shunt resistances are all approximately 0.12Ω. The advantage of this type of coupling compared to quasiparticle coupling of closely-spaced bridges[2,3,4] is that the nonequilibrium effects, such as heating, between bridges are eliminated, and direct comparison of the radiation from the individual bridges and the coherent radiation from two bridges can be made. Also, the circuit equations for Fig. 1(b) are easily written down and can be analyzed and solved on a computer. Such an analysis shows that voltage and phase locking of the two bridges does occur for certain values of the parameters.

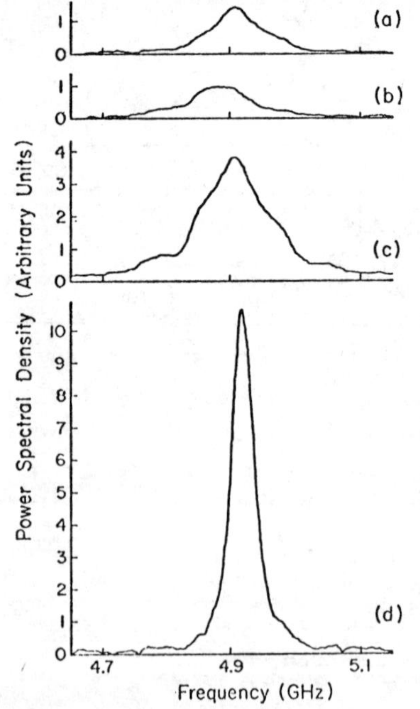

Fig. 2. Radiation power spectra. (a) Radiation from bridge A, bridge B unbiased. (b) Radiation from bridge B, bridge A unbiased. (c) Coherent radiation for opposed-biased bridges. (d) Coherent radiation for series-biased bridges.

The sample is connected across the end of a coaxial line which is brought out of the dewar. This nonresonant coupling allows us to tune and observe the radiation over a wide frequency range.

The observed radiation power spectra are shown in Fig. 2. The data is taken by fixing the bias currents I_A, I_B and sweeping the receiver. Figures 2(a) and 2(b) show the radiation of the individual bridges when the bias current in the other bridge is zero. The microbridges have linewidths of approximately 80 and 95 MHz and critical currents of 220 and 210μA.

Figure 2(c) shows the power spectrum when the bridges are opposed-biased and the currents I_A and I_B are adjusted to give a maximum output. The total integrated power output is ∿2 times the sum of the individual powers, indicating nearly total in-phase coherence. The linewidth (120 MHz) is broader than that of an individual bridge.

In Figure 2(d) the corresponding radiation for series-current biasing is shown. The total integrated power output is the same as for the opposed bias, again indicating in-phase coherence. However, the linewidth is now less than half the linewidth of an individual bridge, and thus the peak power is 10 times greater than the peak power of a single bridge.

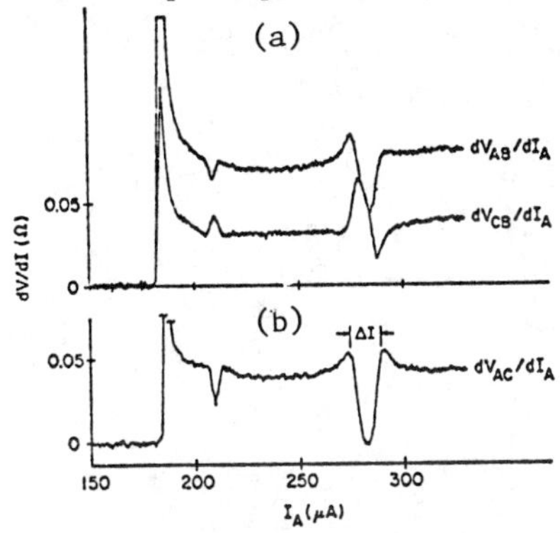

Fig. 3. Derivative measurements dV/dI_A for opposed-current biasing as a function of the bias current in bridge A. (a) Change in voltage in bridge A (top curve) and in bridge B (bottom curve). (b) Change in voltage across both bridges. Voltage locking occurs when $dV_{AC}/dI_A = 0$ and corresponds to the region in (a) where the curves overlap.

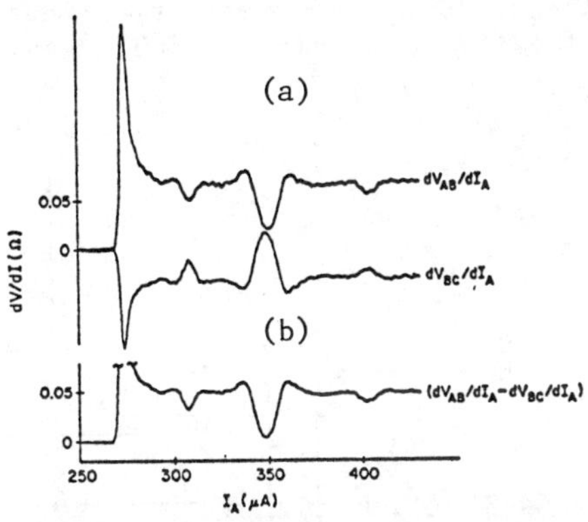

Fig. 4. (a) Similar curves of Fig. 3(a) for series-current biasing. (b) Difference of the change in voltage in bridge A and the change in bridge B as a function of the bias current in A.

Typical $\frac{dV}{dI}$ measurements are shown in Fig. 3 for the opposed bias and in Fig. 4 for the series bias. The data are obtained by fixing the current I_B, applying a small AC current to I_A and sweeping I_A. When the bridges are biased in parallel, there is a range in currents I_A for which the voltages in each bridge are identical. This voltage-locking region is indicated when dV_{AC}/dI_A in Fig. 3(b) is zero, i.e., the range in bias current I_A, for which the change in voltage in bridge A is equal to the change in bridge B. This may also be seen in Fig. 3(a) which shows the voltage changes in bridges A and B respectively as the bias current I_A is swept. In the locking range these two curves overlap. The differential resistance dV_{AB}/dI_A of bridge A, Fig. 3(a), is greatly altered in the neighborhood of the locking region where the frequencies of the two bridges are equal. The figure also indicates evidence of harmonic interactions when the frequency of bridge A is one half that of B.

Figure 4 shows similar dV/dI curves for the series bias. There is incomplete voltage locking in this data, since no range in current I_A exists where the change in the voltages in A and B are equal. This is somewhat surprising since we do observe nearly totally coherent radiation without voltage locking. Again, the differential resistance of each bridge is reduced in the region where coherent radiation is observed. However, in the series-biased case, the differential resistance is much less than the value for parallel-biased bridges.

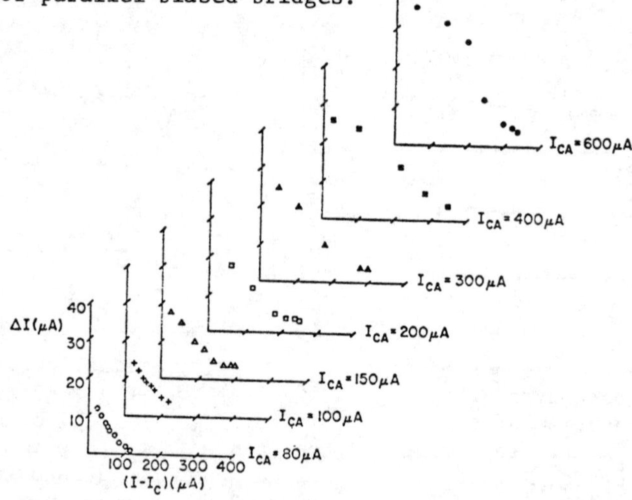

Fig. 5. Voltage-locking range ΔI for opposed biasing vs. bias current at several critical currents. ΔI is defined in Fig. 2(b).

The variation of the range of voltage locking for parallel biasing with bias current and critical current is shown in Fig. 5. This range, ΔI, is taken to be the current interval between the maxima in dV_{AC}/dI_A in Fig. 2(b). The locking falls off at higher bias currents and thus, higher frequencies. We suspect this is partly due to the decrease in the coupling current through R_S as the result of the inductance associated with loop formed by R_S and the microbridges. The locking range increases monotonically with critical current. At larger critical currents, the point at which the locking decreases rapidly is extended to higher bias currents, and locking also occurs at larger bias levels. The upper limit for the locking at high critical currents appears to be due to heating in our planar bridges.

The various linewidths in Fig. 2 may be quantitatively compared to the measured differential resistances. The linewidth is proportional to the square of the noise voltage per Hertz across the bridge $\langle Vn^2 \rangle$. If we assume current noise sources due to the shunt resistor I_{NS} and intrinsic noise currents in the microbridges I_{NA}, I_{NB}, then the linewidth of bridge A with B turned off is given by

$$\Delta\nu_A = C[\langle I_{NA}{}^2 \rangle (\frac{dV_{AB}}{dI_A})^2 + \langle I_{NS}{}^2 \rangle (\frac{dV_{AB}}{dI_{AC}})^2] , \tag{1}$$

when C is a proportionality constant and

$$\frac{dV_{AB}}{dI_A} = \frac{dV_{AB}}{dI_{AC}} = \frac{R_S R_A}{R_A + R_S} .$$

The linewidth can thus be decreased by lowering R_S. If the noise current I_{NA} due to the bridge dominates, the linewidth will decrease as $R_S{}^2$. If, however, R_S is reduced enough, its contribution to the linewidth will dominate since this Johnson noise current is inversely proportional to R_S, and the linewidth will then decrease as R_S. In a single junction sample shunted by an .015Ω external gold film resistor, a linewidth of 2 MHz was obtained. This is the expected linewidth for a total noise current equal to the Johnson noise of the shunt. A similar unshunted bridge would have a linewidth of up to one GHz.

When the bridges are voltage-locked, the large difference in linewidths observed in series-and opposed-biasing may be explained by consideration of the differential resistances in the locking region. The differential resistance of bridge A may be calculated by considering the change in voltage in it, ΔV_{AB}, due to a current change ΔI_A.

$$\Delta V_{AB} = [(dV_{AB}/dI_{AJ})\Delta I_{AJ} + (dV_{AB}/dI_{BJ})\Delta I_{BJ}] , \tag{2}$$

where ΔI_{AJ} and ΔI_{BJ} are the changes in currents through bridges A and B, and $\Delta I_A = \Delta I_{AJ} + \Delta I_{BJ}$. Since only I_A is varied ΔI_{BJ} also equals the change in current ΔI_S through R_S. When the bridges are locked with opposed bias, the voltages are equal and opposite and no DC current flows through R_S, i.e., $\Delta I_S = \Delta I_{BJ} = 0$. The opposed bias differential resistance is then,

$$R_D \equiv \Delta V_{AB}/\Delta I_A = dV_{AB}/dI_{AJ} \tag{3}$$

An analysis of an array of two equal junctions reveals that R_D is ½ the junction resistance and in general for arrays of N equal junctions will decrease as 1/N.

When the bridges are locked in series $\Delta I_S = 2\ V_{AB}/R_S$ and $dV_{AB}/dI_{AJ} = -dV_{AB}/dI_{BJ}$. Solving for $\Delta V_{AB}/\Delta I_A$ gives the series differential resistance,

$$R_{DS} \equiv \Delta V_{AB}/\Delta I_A = R_D \left[\frac{R_S}{R_S + 4R_D}\right]. \qquad (4)$$

The series differential resistance should then be 1/6 R_A since in sample measured $R_A \approx R_S$ and R_D is given by $R_A/2$. We do not observe complete voltage locking in series; however, the dV/dI curves in Fig. 4(a) are approaching a value of .022Ω which is to be compared to $R_A/6$ = .02Ω. The radiation linewidth when the bridges are locked is given by:

$$\Delta\nu = C[(I_{NA}{}^2)(dV_{AB}/dI_A)^2 + (I_{NB}{}^2)(dV_{AB}/dI_B)^2 + (I_{NS}{}^2)(\frac{dV_{AB}}{dI_{AC}})^2]. \qquad (5)$$

In our samples the intrinsic bridge noise dominates; therefore, we neglect the noise current in R_S and assume that the linewidths $\Delta\nu_A$ and $\Delta\nu_B$ of the individual bridges are determined by I_{NA} and I_{NB} respectively. Then using the measured values of dV_{AB}/dI_A and dV_{AB}/dI_B the expected linewidths are ∿125 MHz (opposed-biasing) and ∿31 MHz (series-biasing), which are in reasonable agreement with the measured values of 120 and 40 MHz. If complete locking had been obtained in series, a linewidth of 20 MHz could have been expected since the differential resistance in series should then be 1/3 the value when biased-opposed.

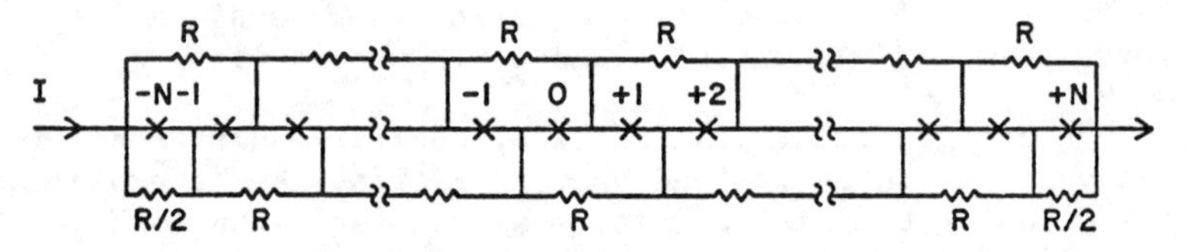

Fig. 6. Schematic of a large series array using nearest neighbor coupling resistors.

The two bridge array has an impedance mismatch of about 10^3 to 1, and thus only 10^{-3} of the power available is coupled into the 50Ω load. For practical microwave generators, one would like to match the impedance of the array to the load. A feasible way of matching is to use a series array of N coherent microbridges to increase the impedance, such as illustrated in Fig. 6. If the number of junctions is chosen such that $NR_J = R_L$ where R_J is the resistance of a single junction and R_L is the load impedance, the maximum power available to the load is approximately 1/4 $I_c{}^2R_L$, independent of the number of junctions. For a 50Ω load and 2mA critical currents, this results in about 50μW of power. While a single bridge matched to the load delivers as much power as a large array of bridges, such a bridge if it could be made would have a small critical current because of the limits of the I_cR_J product. In the above discussion we have also assumed that the operating voltage V_J is $\geq I_cR_J$ so that the power goes into the fundamental frequency. If V_J is less than I_cR_J, then much of the power goes into harmonics of the fundamental oscillations and the power available at the fundamental is independent of I_c. This is a disadvantage for a matched array with a small number of junctions, since the desire for a large I_cR_J means the operating voltage would be less than I_cR_J and much of the power would be lost in higher harmonics.

An advantage of an array is then the ability to match to the load, maintain large critical currents, and still operate at voltages such that $V_J > I_cR_J$. It is desired to adjust I_c such that $I_c \simeq V_J/R_J$ since a further increase in I_c does not give further power output. The power available from a matched array ($\frac{1}{4}I_c{}^2R_L$) then becomes $P = \frac{1}{4}V_J{}^2N^2/R_L$, where $NR_J = R_L$. The power can thus be increased as N^2 by decreasing R_J and increasing N and operating at higher critical currents. This, of course, will be limited by the maximum I_c at which the bridges will operate without detrimental effects such as heating.

Considerations of the linewidth of the radiation reveal additional benefits obtained from large arrays. Since the linewidth of the radiation from a single bridge is proportional to $R_J{}^2$, it is desirable to make R_J small. If R_J is reduced by external shunts and the number of elements of an array is increased to obtain matching $NR_J=R_L$, then the linewidth of the coherent array will decrease as 1/N. This assumes that the intrinsic noise currents of the microbridge and not the noise from the shunt resistor dominate. If the shunt resistor is made small enough, then its noise currents will dominate, and the linewidth will remain constant. The linewidth data for two microbridges shows that the coupling provides an additional line narrowing due to the decrease in the differential resistance in the locking region. Since the simple analysis of a resistively coupled array of identical bridges indicates that this differential resistance R_D decreases as the number of bridges in the array, the linewidth of an array should decrease as $1/N^2$. The peak power of an impedance-matched array is then expected to increase as N^4, a factor of N^2 due to the decrease in linewidth and another N^2 from the increase in total power gained by operating at higher critical currents. If the intrinsic noise of the bridges is greater than that of the shunts even further enhancement may occur.

CONCLUSION

We are encouraged by the results from the two coupled junctions. Coherent radiation with narrowed linewidth and the expected decrease of the differential resistance are observed. We are extending our work to larger arrays using the configuration of Fig. 6. This will test the scaling predictions and no doubt point out new problems. We would like to thank P. Mankiewich for the preparation of samples used in this work.

REFERENCES

1. J.E. Lukens, R.D. Sandell and C. Varmazis, paper presented at this conference.
2. R.D. Sandell, C. Varmazis, A.K. Jain and J.E. Lukens, Bull. APS 23, 94 (1978).
3. P.E. Lindelof and J. Bindslev Hansen, J. Low Temp. Phys. 29, 369 (1977).
4. D.W. Palmer and J.E. Mercereau, Phys. Lett. 61A, 135 (1977).

HIGH-FREQUENCY PROPERTIES OF MICROBRIDGE AND POINT CONTACT JOSEPHSON JUNCTIONS*

W.J. Skocpol
Harvard University, Cambridge MA 02138

ABSTRACT

Recent experiments by the Harvard Superconductivity Group concerning the response of variable-thickness microbridges to high harmonics of microwave radiation and the response of selected Nb point contacts to far-infrared laser radiation are reviewed. Gap structure on the dc I-V characteristics of the various devices can be used to recognize those junctions with the best high-frequency performance. Microbridges which show large numbers of microwave-induced steps to voltages as high as 4.2 mV have been studied, although their poor coupling and low impedance has not yet allowed the observation of steps induced by submillimeter radiation. Selected Nb point contacts which have better coupling and higher resistance have been used to study the steps induced by harmonics of 604 GHz (496 μm) laser radiation to as high as 12.5 mV, and by higher fundamental frequencies up to 2.52 THz (119 μm) at 5.2 mV. Such contacts have been used to measure the intrinsic rolloff of the ac Josephson effect beyond the superconducting energy gap frequency.

INTRODUCTION

The Harvard Superconductivity Group has been studying the high-frequency properties of various types of Josephson junctions for several years with the aim of identifying and, if possible, overcoming those factors which limit their high-frequency performance. Most of our work has relied upon the direct observation and systematic study of ac Josephson effect steps at large voltages induced by microwave and far-infrared radiation. The observation of a constant-voltage step at a voltage V is direct evidence for the existence of components of the Josephson supercurrent at frequencies $f = 2eV/h$; the maximum amplitude of such steps compared to the dc critical current yields information about the strength of the Josephson effect at that frequency under the operating conditions present in the junction.

In the best of our thin-film microbridge Josephson junctions, we have observed more than 200 steps caused by 10 GHz radiation, reaching to approximately 4.2 mV. The performance of Nb cat-whisker point contacts is even better, and the highest voltage step that we have observed is the 12.5 mV tenth step due to 604 GHz (496 μm) laser radiation, while the highest fundamental step that we have been able to observe has been the 5.22 mV step due to 2.52 THz (119 μm) radiation. In this paper, we review some of the results of our investigations.

*Research supported in part by NSF, ONR, and JSEP.

ISSN: 0094-243X/78/335/$1.50

MICROBRIDGES

The most important limitation on the high-frequency performance of microbridges is Joule heating, which raises the temperature in the vicinity of the bridge because of the large power dissipation encountered at the high voltages desired.[1,2] Better cooling may be obtained by using the variable-thickness geometry in which the tiny bridge connects thicker, more massive banks.[3,4] Yet even in this most favorable three-dimensional geometry,[5] heating can still provide important limits to the high-frequency performance.[6]

One of the interesting results of our systematic study of the microwave response of variable-thickness microbridges is that the gap structure on the dc I-V curve of the microbridges can be used to select those bridges which will have good high-frequency performance. In a bridge in which power dissipation leads to a large temperature rise, the gap structure appears at a lower voltage corresponding to the elevated temperature.[4] Figure 1 shows the gap-structure portion of the I-V curves of several tin bridges of varying high-frequency performance as measured by V_{max}, the maximum voltage to which 10 GHz or 32 GHz steps could be observed at low temperatures. In those bridges with the best performance, the gap structure for a given bath temperature occurs at the highest voltages, approaching the value expected for negligible temperature rise above the bath temperature. This structure thus provides a useful rule of thumb for identifying well-cooled bridges. Such bridges tend to have higher impedances and banks at least 1.5 μm thick. The performance that we observe with our best variable-thickness bridges represents approximately an order of magnitude improvement over that of ordinary uniform-thickness microbridges.

Thus far, thin film microbridges have not been successfully used with high fundamental frequencies for the incident radiation. The one exception has been the observation of a step due to 300 GHz carcinotron radiation in an unusually shaped microstrip which had an 80Ω normal resistance.[7] Like other researchers,[8] we have not yet been able to observe a step due to 604 GHz (496 μm) laser radiation. The primary difficulty lies with coupling the radiation into the

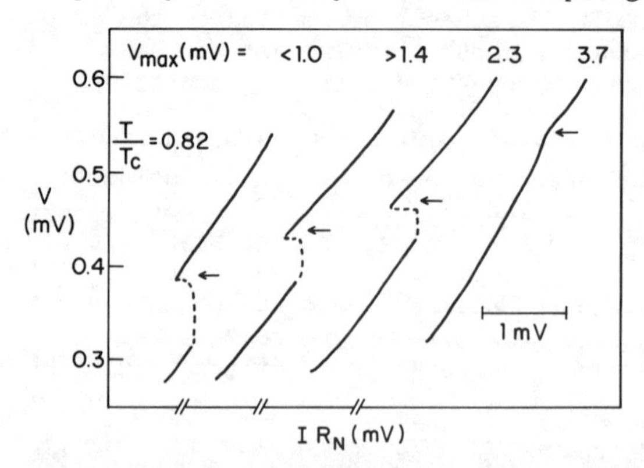

Fig. 1. Gap structure in microbridges differing in high-frequency performance (V_{max}).

bridge. From the 0.1% reductions of the critical current observed at full incident power, it is clear that a first step would be much smaller than the noise on the I-V characteristic. The coupling is poor because we have not yet succeeded in optimizing the thin-film antenna structures or in producing microbridges with resistances substantially exceeding 1Ω. It is possible that progress in this direction can be made in the near future.

POINT CONTACTS

We have been much more successful in working directly at high frequencies using Nb point contacts, which are well-cooled devices with much higher resistance levels. We have analyzed[9] the Josephson response at 604 GHz in some detail and have extended our observations throughout the far-infrared spectrum up to 2.52 THz (119 μm).[10-12]

Our point contacts are made in liquid helium between 75 μm-dia. Nb wire electroetched to a fine point, and a polished and etched Nb flat, using an electrical burn-in technique to reduce the resistance once a high-resistance ohmic contact is established. The detailed structure of the contact region is of course unknown, but all of the experimental data available seem consistent with the contact being a very small metallic constriction.[12] We find that the point contacts vary widely in their degree of response to far-infrared radiation, and that the contacts with the best high-frequency performance can be identified by the pronounced sharpness of the gap structure on the dc I-V characteristic,[10-12] appearing at ~2.8-3.0 mV. Figure 2 shows the gap-structure portion of the I-V curves for several point contacts of varying high-frequency performance, as measured by N_{max}, the highest harmonic of 604 GHz at which a step could be observed. The poorer contacts have less pronounced structure, at lower voltages. As in the microbridges, therefore, the gap structure can be used to recognize good contacts from their dc characteristics. The shape of the gap structure in these high-impedance point contacts is different from the shape in low-impedance microbridges, but Yanson[13] has shown a similar changeover of shape with increasing impedance for metallic shorts through dielectric layers ranging in resistance from 0.1Ω to 22Ω.

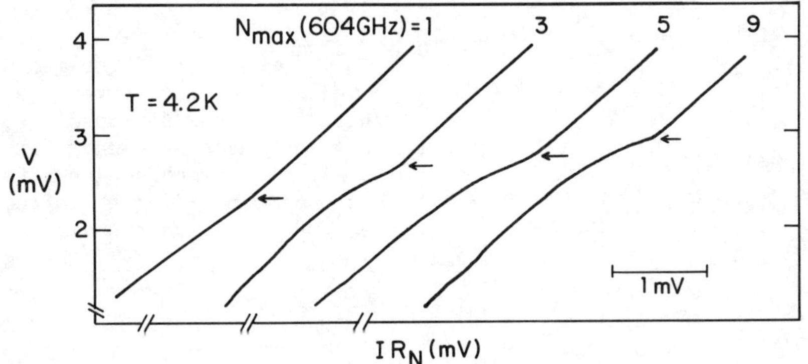

Fig. 2. Gap structure in point contacts of different N_{max}.

Although we are not certain of the factors responsible for washing out the gap structure and degrading the high-frequency performance in some point contacts, we do not believe that heating effects analogous to those in microbridges are primarily responsible. The apparent gap depression in poor contacts is proportionally much larger than the small additional depressions due to heating observed when large amounts of radiation are dissipated in the junction. Moreover, our theoretical model[6] indicates that the heating effects should not be important, unless some aspect of the material in the vicinity of the contact leads to an anomalously low value of the characteristic power level P_o at which reductions of the critical current due to heating become significant. An alternate explanation takes note of the possible formation of rather thick layers of Nb-suboxide on the Nb surfaces. There is evidence[14] that these layers have reduced transition temperatures and hence reduced gap values; contacts formed of this oxidized material might well have less ideal performance.[12]

We have also characterized the performance of our best point contacts as detectors of low levels of incident radiation, using the "resonant" response at voltages near the incipient ac step. Because of low-frequency voltage fluctuations (which could have been in part avoided by using chopping frequencies greater than the 450 Hz used), the measured NEP for 604 GHz radiation was only about 10^{-13} W/Hz$^{\frac{1}{2}}$, with a voltage responsivity of about 10^5 V/W.[11]

For large incident power levels, the high-frequency performance of our best point contacts is sufficiently good to allow study of the expected intrinsic rolloff[15] of the strength of the Josephson effect at frequencies exceeding the energy gap frequency in the material. In Fig. 3 we show the maximum current half-width of the first Josephson step for a number of frequencies in the far infrared, normalized to the dc critical current. The solid dots are the largest values actually observed, and the crosses are our estimates of the corrected step widths in the absence of heating and noise-rounding. The solid curve is the prediction of the voltage-bias version of Werthamer's frequency-dependent theory (most applicable at the higher voltages); the dashed curve is the prediction of the resistively-shunted-junction (RSJ) model, which does not include the

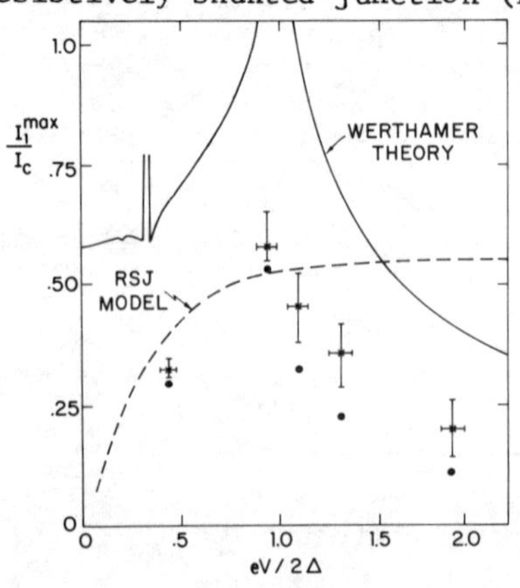

Fig. 3. Dependence of the maximum observed size of the current half-width of the first Josephson step on incident laser frequency (step voltage). The symbols with error bars are corrected for the effects of heating and noise-rounding.

intrinsic gap-related frequency dependence but does include the effects of current bias which become more important at the lower voltages. The data are in reasonable qualitative agreement with the frequency-dependent theory in the appropriate range, although absolute discrepancies of the order of a factor of two or more are observed.

CONCLUSION

Selected Nb cat-whisker point contacts can be used to observe the ac Josephson effect throughout the far infrared at levels limited ultimately by the expected gap-related rolloff. Because these junctions have impedance levels comparable to that of free space and have a built-in long-wire antenna, external coupling to high-frequency radiation is quite effective. Comparable performance has not yet been achieved with thin-film microbridges, but it is possible that careful attention to improving the impedance and coupling characteristics may provide further significant improvements in microbridge performance at high frequencies.

REFERENCES

1. W.J. Skocpol, M.R. Beasley, and M. Tinkham, Revue Phys. Appl. 9, 19 (1974).
2. W.J. Skocpol, M.R. Beasley, and M. Tinkham, J. Appl. Phys. 45, 4054 (1974).
3. T.M. Klapwijk and J.E. Mooij, IEEE Trans. MAG-11, 85 (1975).
4. M. Octavio, W.J. Skocpol, and M. Tinkham, IEEE Trans. MAG-13, 739 (1977).
5. M. Tinkham, this Proceedings.
6. M. Tinkham, M. Octavio, and W.J. Skocpol, J. Appl. Phys. 48, 1311 (1977).
7. B. Kofoed and K. Saermark, Phys. Rev. Lett. 31, 1124 (1973).
8. T.D. Clark and P.E. Lindelof, Phys. Rev. Lett. 37, 368 (1976).
9. D.A. Weitz, W.J. Skocpol, and M. Tinkham, Appl. Phys. Lett. 31, 227 (1977).
10. D.A. Weitz, W.J. Skocpol, and M. Tinkham, Phys. Rev. Lett. 40, 253 (1978).
11. D.A. Weitz, W.J. Skocpol, and M. Tinkham, Proceedings of the 3rd International Conf. on Submillimeter Waves and their Applications, Guilford, England, March 1978 to be published in Infrared Phys.
12. D.A. Weitz, W.J. Skocpol, and M. Tinkham, submitted to J. Appl. Phys.
13. I.K. Yanson, Fiz. Nizkikh Temp. 1, 141 (1975) [Sov. J. Low Temp. Phys. 1, 67 (1975)].
14. W. Schwarz and J. Halbritter, J. Appl. Phys. 48, 4618 (1977).
15. N.R. Werthamer, Phys. Rev. 147, 255 (1966).

LIGHT-SENSITIVE JOSEPHSON JUNCTIONS: TEN YEARS AGO, TEN YEARS FROM NOW.

A. Barone, M. Russo
Laboratorio di Cibernetica del CNR, Arco Felice, Italy

R. Vaglio
Istituto di Fisica, Università di Salerno, Salerno, Italy

ABSTRACT

Recent developments of experimental investigations on light-sensitive junctions are discussed, also in connection to future applications.

In 1968 Giaever[1,2] reported successful experiments on Josephson tunneling through semiconductor barriers and, in particular, the first observations on the light-sensitive properties of junctions employing CdS barrier layers. Thus the present discussion falls at the half-way point in the ten-year perspective of the Conference. Here we report some of our more significant experimental results on the subject to direct the attention of other investigators to the state of the art in this matter. We consider some possibilities for future applications hoping that this can stimulate further ideas and suggestions.

Josephson junctions with light-adjustable barriers offer very attractive characteristics. In particular it is possible to make a sample whose critical current can be continuously increased by light exposure. The modifications of the V-I characteristics induced by light can be considered as stable, though resettable, for all practical purposes [3].

Stimulated in fact by a visit of Ivar Giaever to our Laboratory in 1970 we have developed a rather systematic work on light sensitive Josephson structures towards the following two objectives: 1) Increase of the reliability of the samples; 2) Accurate characterization of the junction behavior. As far as concerns the stability of such structures these are still far from the levels achievable by some oxide-barrier junctions (namely Nb-based junctions or lead alloy electrode junctions). However, more recent results indicate that a significant improvement of the junction properties can be obtained by a suitable choice of the electrode materials [4,5]. We believe that, to some extent, this aspect has been overlooked throughout the literature. The use of In electrodes, for instance, allows the possibility of making rather stable junctions with a very thick (≈1000 Å) barrier, avoiding also the oxidation step which is typically performed in order to fill the pin-holes occurring in the CdS

ISSN: 0094-243X/78/340/$1.50

layer. In this case the stability of the structure should be related to a stable interface which is created by a limited diffusion of the indium into the cadmium sulfide. CdS, with the related problems of the difference in the vapor pressure of its constituents, has to be considered a "difficult" material. However, the reason for choosing this material among other photo-sensitive semiconductors lies mainly in the widespread possibilities it offers for making films of different characteristics for laboratory studies, rather than in the perspective of its actual use in future commercial applications.

Let us now discuss various experimental aspects which characterize light-sensitive Josephson junction behavior.

a) V-I characteristics. Clean voltage-current characteristics are found which can be rather easily interpreted in terms of the fabrication parameters and light exposure conditions [3]. In Fig.1a is shown the sharp gap structure of a Pb-CdS-Pb junction with a thin (≈100Å) semiconductor barrier. The effect of the light on the normal resistance of a Pb-CdS-In structure is reported in Fig.1b. In Fig. 1c and 1d, the V-I curve of a In-CdS-In junction is reported in dark conditions and after light exposure, respectively. We emphasize that the absence of detectable zero-voltage current in dark conditions guarantees the absence of shorts, so that the light-induced zero-voltage current is indeed the dc Josephson supercurrent.

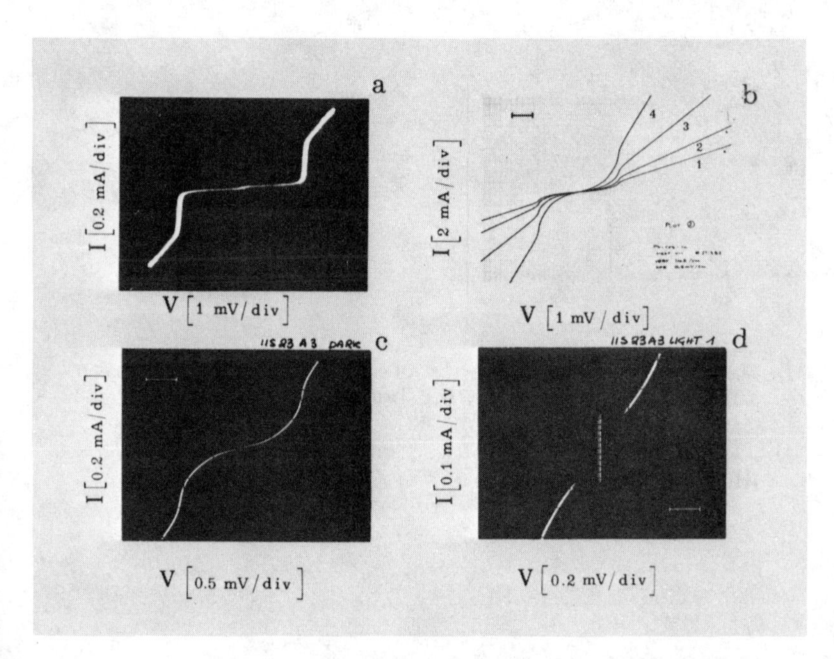

Fig. 1 - Voltage-Current characteristics of light-sensitive tunnel junctions.

b) Critical current vs. external magnetic field. The behavior of light-sensitive junctions in applied fields has been widely investigated. For small junctions a nearly perfect Fraunhofer-like pattern is found [5]. Moreover, increasing the illumination level the maximum light-induced supercurrent I_J^L increases and, correspondingly, the Josephson penetration λ_J decreases. As a result the ratio of the junction dimensions to λ_J becomes larger and larger so that a change from "small" to "large" behavior can be observed on the same sample [6].

c) Double junction behavior. Diffraction and interference phenomena have been investigated on these structures [5]. It is possible in fact to induce an enhancement of the tunneling current at the edges of a junction varying both the light exposure and the transparency (thickness) of the top metal film layer. It is thus rather simple to control the current peaking at the junction edges creating interference conditions as clearly shown in Fig. 2a.

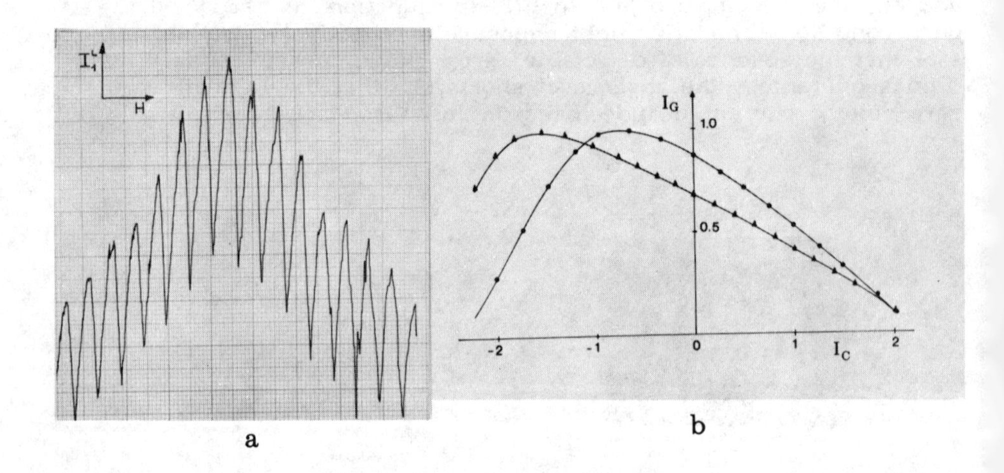

Fig. 2 a) Magnetic field dependence of light-induced Josephson current (interference -like behavior)

b) Josephson gate current vs. control current for two illumination levels.

d) Structural fluctuations [7]. The presence of fluctuations in the effective thickness of the barrier due to randomly distributed inhomogeneities has been observed by accurate magnetic field measurements. For these junctions in fact a background current is present in the

I_J^L vs. H pattern. This can be related to a correlation length which gives a measure of the nonuniformity [8].

e) Resonant modes. Recently, investigations on light-sensitive structures have been extended to a.c. phenomena [9,10]. Fiske steps have been observed and their magnetic field dependence discussed within the framework of Kulik's theory [10]. The junctions behave as low Q ($\sim$10) resonant cavities.

f) Temperature dependence. Measurements of the temperature dependence of both asymmetric (Pb-CdS-In) and symmetric (In-CdS-In) structures have been performed [5,11]. The peculiar behavior in each case can be accounted for in the context of proximity effect theory [11,12].

Whether light-sensitive junctions can or cannot play a rôle in applications is a debatable point; however, we believe that the above-mentioned features and the experimental results collected so far make such structures rather promising. As an example, in the context of applications to digital devices we observe that for a given junction and fixed circuit parameters, "self-resetting", "latching" and "non-latching" modes [13,14] of switching operation can be realized depending upon the particular state of illumination. Furthermore, it is recognized [14] that, due to the low critical current achievable, it is very difficult with semiconductor (non-light sensitive) barriers to obtain a "large" junction behavior; however, light-sensitive junctions can realize a "modifiable" gate-control dependence as in Fig. 2b. In conclusion, we do not claim that light-sensitive structures necessarily have a secure future, rather we feel that the "one more degree of freedom" deriving from the possibility of an external optical input could increase the potential of Josephson junctions and extend perhaps also their use to perform other tasks such as parallel processing in pattern recognition [15].

The authors are grateful to Professor R.D. Parmentier for comments and suggestions.

References:

1. I. Giaever, Phys.Rev. Lett. 20, 1286 (1968).
2. I. Giaever and H.R.Zeller, Phys.Rev. B1, 4278 (1970).
3. A.Barone, P.Rissman and M.Russo, Rev.Phys. Appl. 9, 73 (1974).
4. A.Barone, M.Russo, Phys. Lett. 49A, 45 (1974).
5. A.Barone, G.Paternò, M.Russo, R.Vaglio, in "Proc. LT14" Eds. M. Krusius and M.Vuorio (North Holland, Helsinki, Vol.4, p.88, 1975).

6. A.Barone, G.Paternò, M.Russo, R.Vaglio, Phys. Lett. 53A, 393 (1975).
7. I.K. Yanson, JETP 31, 800 (1970).
8. A.Barone, G.Paternò, M. Russo, R.Vaglio, JETP (in press).
9. M. Russo, Phys. Lett. 61A, 191 (1977).
10. M. Russo, R.Vaglio, Phys. Rev. B (in press).
11. F. Andreozzi, A.Barone, G.Paternò, M. Russo, R. Vaglio (to be published).
12. N.L.Rowell, H.J.T.Smith, Can. J. Phys. 54, 223 (1976).
13. H.H. Zappe, J. Appl. Phys. 44, 1371 (1973).
14. W.Y. Lum, H.W. Chan, T. Van Duzer, IEEE Trans. Mag. 13, 48 (1977).
15. A. C. Scott, IEEE Trans. Systems, Man and Cybernetics SMC-1, 267 (1971)

MILLIMETER WAVE RESPONSE OF Nb VARIABLE THICKNESS BRIDGES*

D. W. Barr and R. J. Mattauch
Department of Electrical Engineering, University of Virginia

Li-Kong Wang and B. S. Deaver, Jr.
Department of Physics, University of Virginia
Charlottesville, Virginia 22901

ABSTRACT

Niobium variable thickness bridges are being fabricated for testing as millimeter wave mixers. Bridges have been fabricated on fused quartz, silicon and gallium arsenide substrates and characterized by measurements of the I-V curve, and dV/dI versus V and by mixing experiments at 80-100 GHz. The bridges on Si and GaAs exhibit small Josephson steps but the response of all the bridges is limited by self-heating, and mixing appears to be primarily bolometric. However the cooling of the bridges on Si is much better than the others.

INTRODUCTION

The high frequency response of metallic superconducting bridges has been shown to be limited by self-heating.[1] However bridges of variable thickness consisting of thin, narrow links joining two much thicker bulk films are less limited by heating than bridges in films of uniform thickness.[2] In order to test their response as millimeter wave mixers we have been fabricating Nb variable thickness bridges on fused quartz, silicon and gallium arsenide substrates. We report here some initial experiments at 80-100 GHz.

FABRICATION

The bridges are of the form shown in Fig. 1 and are located at the center of a microwave choke structure designed to isolate the microwaves from the if and biasing terminals. Junction fabrication begins with the rf sputtering of a Nb film 500 Å thick on the substrate. The film thickness was determined by measuring its optical density using a laser. The films were deposited at ∿400 Å/min and yielded films with a room

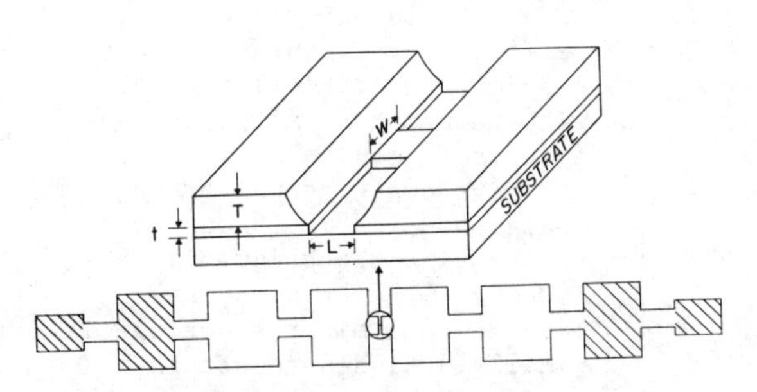

Fig. 1. Bridge and microwave choke. W = 1 μm, L = 0.57 μm, t = 500 Å, T = 5000 Å. The choke structure is Nb with Au pads (crosshatched) overlaid.

ISSN: 0094-243X/78/345/$1.50

temperature to helium temperature resistance ratio of 9. The films exhibited an initial T_c of about 8K although after subsequent processing steps it was decreased to $\sim$5.5 K.

A quartz fiber 1 μm dia. was placed across the film and after a sputter etch to clean the surface, a layer of Nb 5000 Å thick was deposited. The quartz fiber was then removed and photolithography and chemical etching were used to define the bridges and then the choke structure. A lift-off technique was used to fabricate the Au pads on the ends of the choke to provide stable contacts to the film. Our masking technique allowed for simultaneous fabrication of 50 bridges. They were separated using a diamond saw giving individual units 0.8 cm long x 0.5 mm wide. The junctions are extremely stable showing no change in critical current upon cycling between room temperature and helium temperatures, and no precautions appear to be necessary to protect them from burn-out.

MEASUREMENTS OF BRIDGE CHARACTERISTICS

The bridges were characterized by measurements of I-V and dV/dI vs V and by microwave mixing experiments with measurements of P_{if} vs V. Their resistances ranged from 0.5 to 6 ohms. The critical current varied linearly with temperature near T_c. There was the usual subharmonic gap peak in dV/dI at $V = \Delta(t)/e$ which gave an extrapolated T_c the same as that obtained from critical current data. At low temperature the I-V curves were hysteretic.

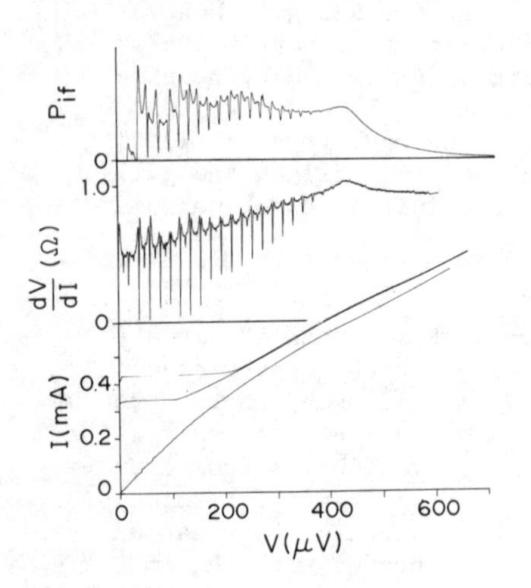

Fig. 2. Data for a bridge on fused quartz substrates showing hysteretic I-V curve without microwaves, and I-V, dV/dI, and P_{if} at 60 MHz with microwave power at 9.90 GHz applied.

Measurements at 9.9 GHz on bridges on fused quartz substrates (Fig. 2) showed Josephson response at harmonics as high as 100-200 GHz and exhibited both Josephson response and bolometric mixing.[3] The peak in dV/dI at V = 430 μV in Fig. 2 indicates the onset of a self-sustaining hot spot[4] and at higher bias P_{if} is due entirely to bolometric mixing and reflects the exponential variation of the critical current with temperature after the hot spot is formed.

The experimental arrangement shown in Fig. 3 was used for testing the bridges at 80-100 GHz. The bridges were mounted in a mixer block specifically designed to obtain a better match to the low impedance bridge. The bridge with its choke was supported in a stripline cavity which positioned the bridge across the

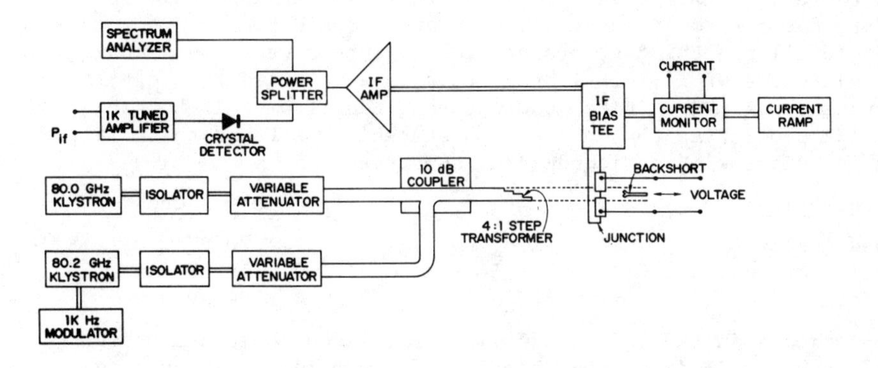

Fig. 3. System diagram for 80 GHz mixing experiments.

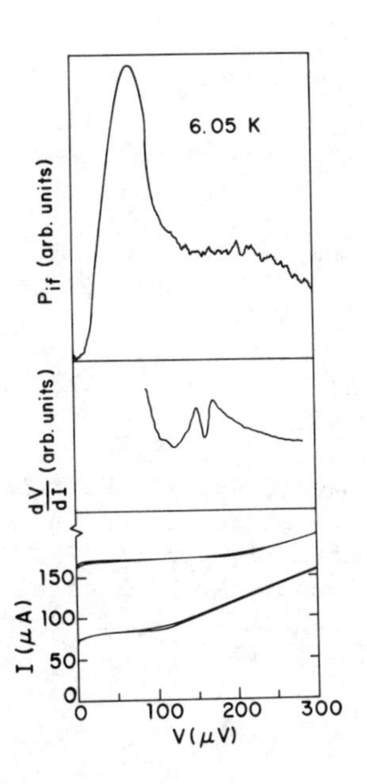

Fig. 4. Data for bridges on Si substrate showing I-V curve with and without microwaves, dV/dI with small Josephson step, and P_{if} at 200 MHz.

height of a quarter-height RG 99 U waveguide. Behind the bridge in the quarter-height guide was a moveable backshort which was adjusted to maximize the power coupled to the bridge.

No Josephson steps or mixing were found at 80 GHz for bridges on fused quartz although the critical current was depressed at reasonable power levels.

Figure 4 shows data at 80 GHz for a bridge on a silicon substrate. The I-V curve was not hysteretic for the temperature shown. There was a small Josephson step in dV/dI but it was not distinguishable in I-V. The P_{if} curve shows a peak at 83 μV but it is likely to be due to bolometric mixing which also peaks at a maximum in the dynamic resistance.

The response of a bridge on a GaAs substrate to 100 GHz is shown in Fig. 5. Again no steps are distinguishable on the I-V curve but the first step and two subharmonics are visible on dV/dI. This bridge was tested as a mixer at 80 GHz and there was mixing response as before for intermediate frequencies up to 800 MHz.

In all the bridges, hysteretic I-V curves were found above a certain value of I_c. This hysteresis is attributed to the voltage dependence of the critical current resulting from self-heating. The I_cR product at the on-set of hysteresis was found to be larger for junctions with semiconducting substrates.

Substrate	$R_n(\Omega)$	I_cR (at hysteresis)
Fused quartz	1.0	220 μV
Gallium Arsenide	.25	750 μV
Silicon	2.75	1200 μV

The increase in the I_cR product in the silicon substrate bridge would seem to indicate that heating in the bridge is decreased since heat conduction by electrons is 2.75 times less than for the bridge on fused quartz but the I_cR product is increased by a factor of 5. Even though a step transformer was used, there was a large mismatch of the microwave power to these very low impedance bridges and apparently there is extraneous heating throughout the mounting structure. This may account for the dominance of thermal effects in these data.

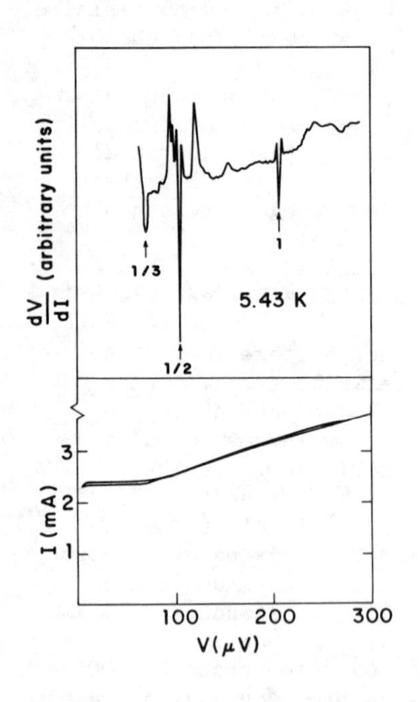

Fig. 5. Data for bridge on GaAs substrate. I-V and dV/dI with 100 GHz microwave applied.

REFERENCES

*Research supported by ONR contract N00014-77-c-0440

1. M. Tinkham, M. Octavio and W. J. Skocpol, J. Appl. Phys. 48, 1311 (1977).
2. T. M. Klapwijk and T. B. Veenstra, Phys. Lett. A47, 351 (1974).
3. Li-Kong Wang, A. Callegari, B. S. Deaver, Jr., D. W. Barr and R. J. Mattauch, Appl. Phys. Lett. 31, 306 (1977).
4. W. J. Skocpol, M. R. Beasley and M. Tinkham, J. Appl. Phys. 45, 4054 (1974).

PROPERTIES OF VARIABLE THICKNESS BRIDGES ON GaAs SUBSTRATES*

C. H. Galfo, Li-Kong Wang and B. S. Deaver, Jr.
Department of Physics, University of Virginia

D. W. Barr and R. J. Mattauch
Department of Electrical Engineering, University of Virginia
Charlottesville, Virginia 22901

ABSTRACT

Structure on the I-V curves of Sn and Nb variable thickness bridges on GaAs substrates are being studied systematically both to characterize the junctions and to search for interactions of the Josephson oscillations with the substrate.

INTRODUCTION

Careful analysis of the profusion of structure on the I-V curves of superconducting microbridges is a useful technique both for characterizing the properties of the bridge itself and for studying its interaction with its surroundings. In the latter category, in addition to structure induced by coupling to electromagnetic modes of the bridge geometry, is structure induced by resonant absorptions in material coated on the bridge and in the substrate.[1-3] Measuring this structure is then a technique for studying these interactions.

With the objective of characterizing the bridges and of studying their interactions with their surroundings, we have fabricated Sn and Nb variable thickness microbridges on GaAs substrates and have studied I-V and dV/dI - V for various values of temperature, magnetic field and microwave power. We summarize here some general properties of these bridges.

The I-V characteristics of Sn weak links have a number of properties shared by their Nb counterparts. These include:

1) Hysteretic behavior of the critical current at low temperature.
2) The initial slope of the I-V curve is quite different from the slope of the main curve. We call this the "foot" region of the curve after Octavio *et al*.[4]
3) The bridges show excellent response to externally applied microwave radiation.
4) The curves are rich in subharmonic gap structure.

PROPERTIES OF Sn BRIDGES

The Sn microbridges were fabricated using a simple cross-scribing technique which produced a bridge slightly less than 1 μm square in a Sn film 1500 Å thick formed by evaporation onto a cooled GaAs substrate. The microbridges had a normal state resistance at helium temperature ranging from 0.2 to 35 Ω. The temperature dependence of the I-V curves of two of these bridges is shown in

ISSN: 0094-243X/78/349/$1.50

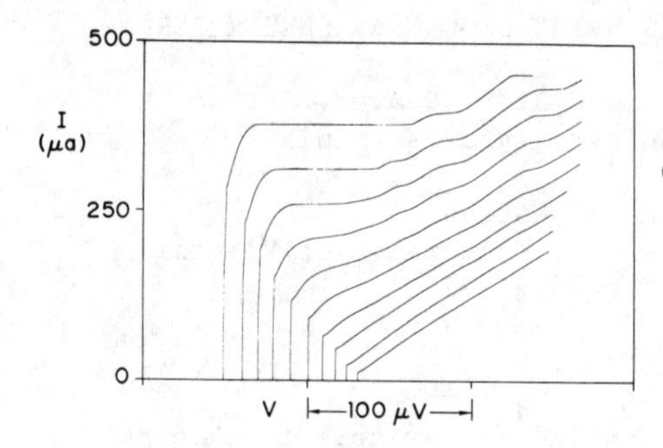

Fig. 1. Temperature dependence of the I-V characteristic of a Sn microbridge for T/T_c from 0.993 to 0.945.

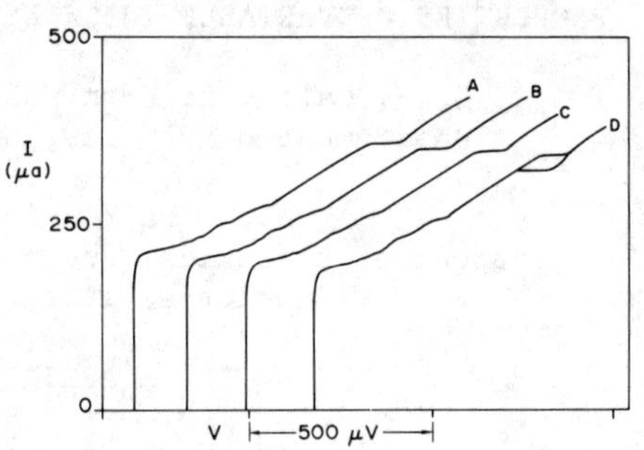

Fig. 2. I-V curve showing a hysteretic switching region at $V = 2\Delta/e$. Temperatures for curves A to D: 3.122 K, 3.144 K, 3.167 K, and 3.192 K respectively.

Figures 1 and 2. These data show the development of the "foot" from a small curving region (small I_c) into a switching instability (larger I_c). Fig. 2 represents I-V data taken over a narrow temperature range. The switching region near V = 650 μV has been identified as the $2\Delta/e$ peak. Curve D was swept in both directions, indicating that this switching region is hysteretic. The first derivative, dV/dI, of these data is shown in Fig. 3.

Derivative data taken near T_c show subharmonic gap structure labeled A through F occurring at voltages $V_n = 2\Delta(T)/ne$ where n = 2,3,...,7. Fig. 4 shows the temperature dependence of these peaks. The experimental data did not fit the gap function very well. However, as shown previously by Octavio et al.[5] a near-perfect fit of the experimental data to the gap function is obtained with a simple

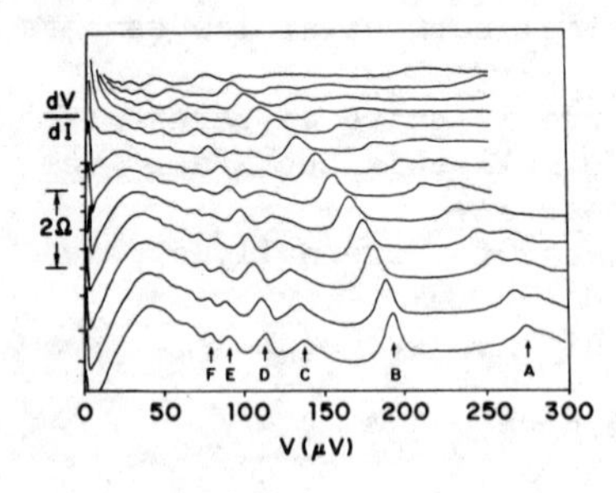

Fig. 3. Dynamic resistance vs voltage taken at various temperatures near T_c. Peak A is the Δ peak; B is $2\Delta/3$; C is $\Delta/2$; D is $2\Delta/5$; E is $\Delta/3$; F is $2\Delta/7$.

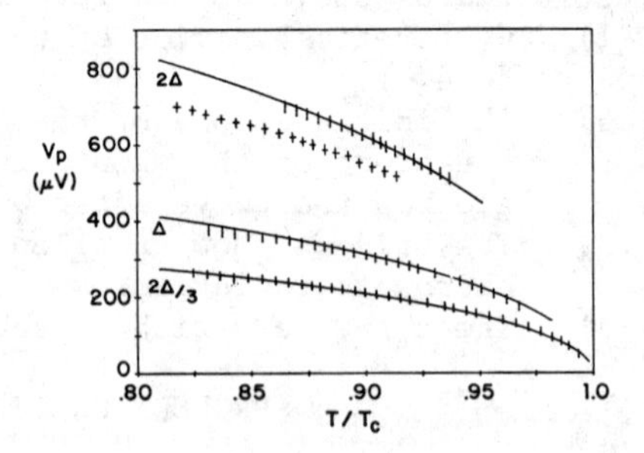

Fig. 4. Temperature variation of three subharmonic gap peaks showing a fit to the B.C.S. gap function after correcting for heating. The points below the 2Δ curve are an example of uncorrected data.

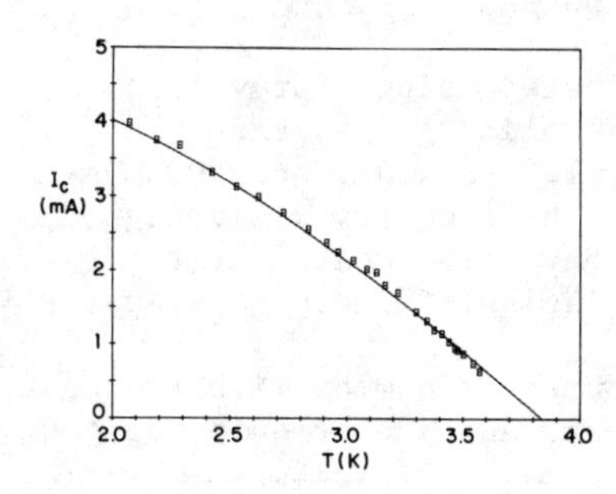

Fig. 5. Critical current as a function of temperature. Data fit to ideal junction form.

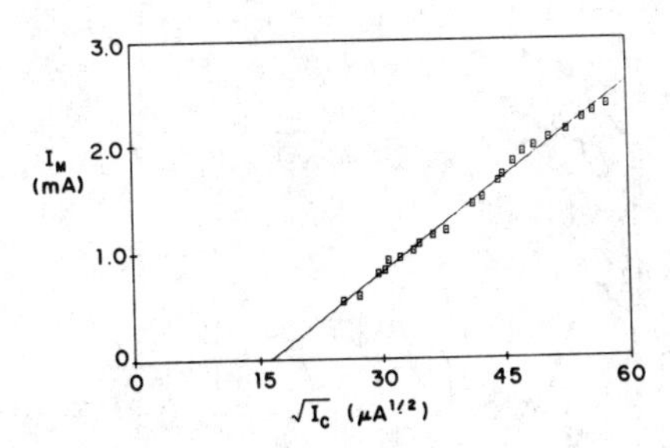

Fig. 6. Hysteretic critical current as a function of I_c.

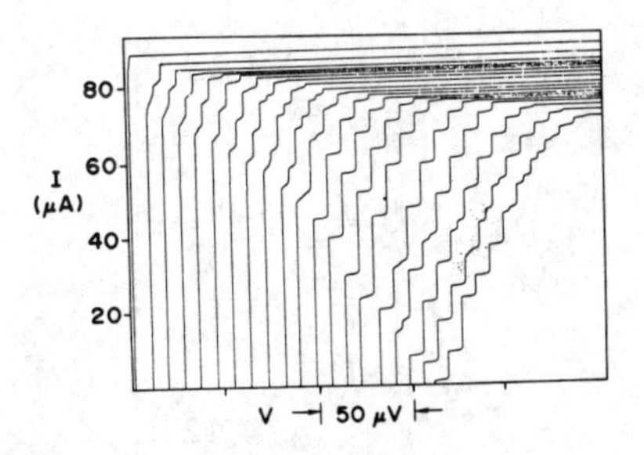

Fig. 7. Microwave response of a high resistance (16.5Ω) Sn microbridge. The applied frequency was 3.95 GHz with the power increasing in 1 dB steps. The left-most curve is taken with no microwave radiation.

correction to the temperature to account for self-heating in the bridge. This correction is:

$$T_{BRIDGE} = T_{BATH} + \alpha \frac{V^2}{R_N}$$

where α is the thermal resistance for heat flow from the microbridge. Determined as a fitting parameter, α was found to be 0.98 K/microwatt. This gives a figure of merit P_o defined by Tinkham et al.[6] of 3.9 μW.

Figure 5 shows the temperature dependence of the critical current over the range T = 2 K to 4 K. These data were fit to the ideal junction relation:

$$I_c = \frac{\pi}{2eR_n} \Delta(t) \tanh \frac{\Delta(t)}{2kT}$$

where t is the reduced temperature T/T_c. The fit for this data with $\Delta(0) = 550$ μeV and $T_c = 3.84$ K was best for $R_n = 0.15$ Ω. The experimental value is $R_{exp} = 0.20$ Ω.

Fig. 6 shows that I_m, the current at which the hysteretic I-V curve returns to zero voltage, is a linear function of $I_c^{1/2}$. This is consistent with various models for hysteresis, but for this data we believe it arises from the voltage dependence of the critical current due to self-heating.[7]

The microwave response of a high resistance Sn microbridge is shown in Fig. 7. At the temperature at which these data were taken, T = 2.236 K, the I-V curve had a large hysteresis (I_m = 68 μA); however, the microwave step response is still observed. In the experiments on Sn bridges, we have not seen any structure characteristic of interaction of the Josephson oscillations with the substrate.

PROPERTIES OF Nb BRIDGES

The Nb microbridges, which were fabricated using a previously described combination of fiber masking, photolithography and chemical etching,[8] were about 0.5 μm long, 1.0 μm wide, and 500 μ Å thick and joined bulk films 5000 Å thick. The geometry was designed for testing them as mixers at 100 GHz so they were located at the center of a stripline choke structure consisting of periodic steps in the width of the line.

The Nb bridges on GaAs have many features in common with the Sn bridges; i.e. hysteresis, microwave response, and subharmonic gap

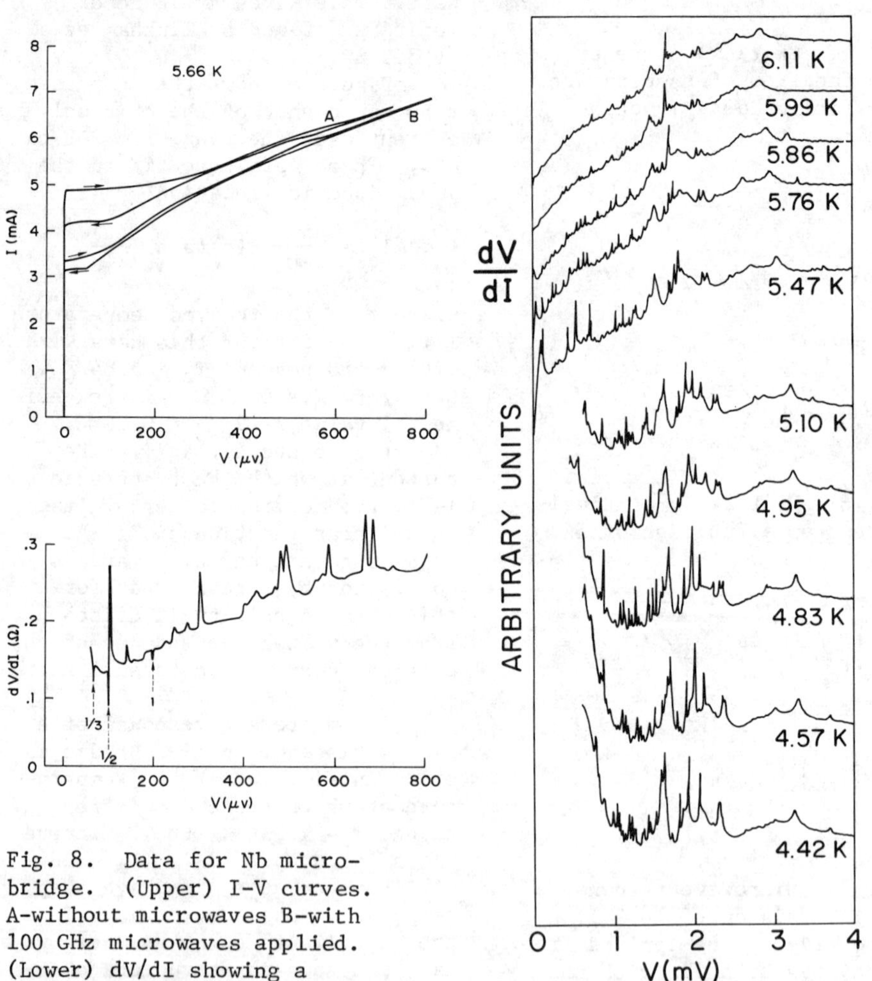

Fig. 8. Data for Nb microbridge. (Upper) I-V curves. A-without microwaves B-with 100 GHz microwaves applied. (Lower) dV/dI showing a microwave-induced step (1) and two subharmonic microwave steps (1/2, 1/3) along with other structure.

Fig. 9. Dynamic resistance of Nb bridges without microwaves showing temperature variation of structure.

structure. However, in contrast to the Sn microbridges, the Nb ones show an abundance of structure on the dV/dI vs V curve which is extremely reproducible and varies systematically with temperature. An example is shown in Fig. 8 along with the I-V curves with and without 100 GHz microwaves. This structure may arise in part from interaction with the adjacent microwave choke. In Fig. 9 we show the dV/dI curve over a wide voltage range and the variation of the structure with temperature.

At present, we are making a systematic study of the structure in the Nb data and determining the effects of a magnetic field on the microbridges. We are assembling an on-line data acquisition system which should improve our ability to study detailed features of the I-V curve and to identify features that may result from interaction with the substrate.

REFERENCES

*Research supported by the Office of Naval Research under Contract N00014-77-C-0440.

1. M. Puma and B. S. Deaver, Jr., Appl. Phys. Lett. 19, 539 (1971).
2. S. N. Artemenko, Rad. Eng.-Elec. Phys. 18, 1466 (1973).
3. B. S. Deaver, Jr., Revue de Physique Appliquée 9, 297 (1974).
4. M. Octavio, W. J. Skocpol, and M. Tinkham, Phys. Rev. B 17, 159 (1978).
5. M. Octavio, W. J. Skocpol and M. Tinkham, IEEE Trans. Magn. MAG-13, 739 (1977).
6. M. Tinkham, M. Octavio, and W. J. Skocpol, J. Appl. Phys. 48, 1311 (1977).
7. A detailed discussion of heating effects in our variable thickness bridges will be published elsewhere by Li-Kong Wang, Dae-Jin Hyun, and B. S. Deaver, Jr.
8. Li-Kong Wang, A. Callegari and B. S. Deaver, Jr., Appl. Phys. Lett. 31, 306 (1977).

EXCESS CURRENT IN Nb AND Nb_3Ga JOSEPHSON JUNCTIONS*

J. E. Nordman and L. L. Houck

University of Wisconsin, Madison, WI 53706

ABSTRACT

A simple density-of-states model is used to show that anomalous behavior observed in the volt-ampere characteristics of Nb_3Ga-Pb tunnel junctions is closely related to that observed in Nb-Pb and Nb-Nb junctions. This model, based on proximity effect considerations, provides an explanation for the following experimental observations: 1) A temperature-dependent negative resistance occurs at a voltage equal to the sum of the half-gaps in some Nb junctions but has not been observed for Nb_3Ga. 2) A current peak at the gap difference voltage is clearly evident only at high temperatures in these same Nb junctions. In Nb_3Ga junctions this peak persists to lower temperatures but its position shifts in the "wrong" direction, toward higher voltages. 3) The current rise near the gap sum voltage is very gradual in most Nb_3Ga junctions and does not correlate at all with the energy gap inferred from a T_c measurement.

INTRODUCTION

Josephson tunnel junctions made with Nb and high T_c Nb compounds commonly exhibit reduced Josephson currents and excess quasiparticle currents for voltages less than the sum of the energy half-gap values for the superconductors on either side of the barrier.[1,2,3] In the case of Nb, a negative resistance region sometimes also appears near the gap sum, its effect becoming enhanced as temperature is lowered. A typical example of this, for a Nb-Nb oxide-Pb junction using RF sputtered Nb[4], is shown in Fig. 1. This behavior has been qualitatively attributed to a thin layer of normal material on the Nb surface which becomes superconducting due to the proximity effect.[1,5]

Junctions made with niobium compounds usually exhibit behavior much worse than Nb junctions in the sense that excess currents are much larger, Josephson currents are small or nonexistent, and apparent gap values are much smaller than predicted by BCS theory. Again, at least part of the cause appears to be a normal surface layer.

In conjunction with a study of sputtered Nb_3Ga films,[6] attempts were made to fabricate usable Josephson junctions and also to correlate tunneling data with T_c measurements. The volt-ampere curves were, however, typically similar to those of Fig. 2. In addition to high excess currents and very small Josephson currents, these junctions exhibit what appears to be a gap difference peak which moves toward higher voltages as temperature decreases, a dependence opposite to that normally seen and predicted by theory. The

* Work supported in part by National Science Foundation Grant ENG 74-24227.

ISSN: 0094-243X/78/354/$1.50

apparent gap is also much smaller than the value inferred from T_c. For the two samples of Fig. 2, the Nb_3Ga half gap should be about 2.25 mV.

Although the characteristics of Figs. 1 and 2 appear to be quite different, it was found that, except for small residual currents at low temperatures, the essential features of both can be described by the same crude density-of-states model described below.

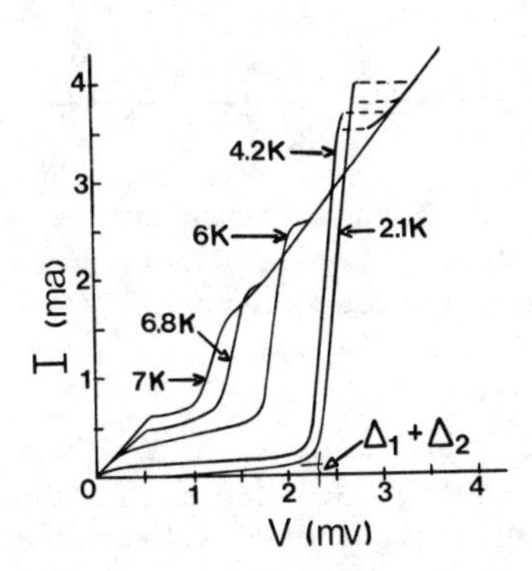

Fig. 1. Volt-ampere curves for a Nb-ox-Pb tunnel junction. Josephson current has been quenched with a small magnetic field.

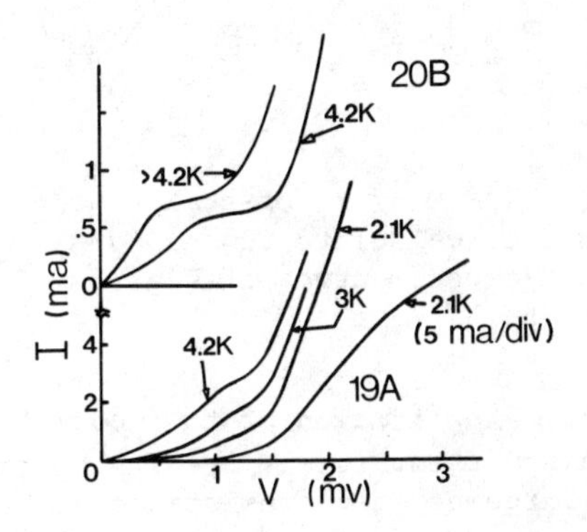

Fig. 2. Volt-ampere curves for two Nb_3Ga-ox-Pb tunnel junctions. $\Delta_2(20B) \approx 0.45$ mV, $\Delta_2(19A) \approx 0.24$ mV. [Note: Scales on >4.2#K curve are uncalibrated]

DENSITY-OF-STATES MODEL

An exact theoretical proximity-effect model is not available to apply to either the Nb or Nb_3Ga devices. However, since there is good evidence that the McMillan theory[7] qualitatively explains the behavior at voltages near $\Delta_1 + \Delta_3$ (see Fig. 3) in Nb,[5] it appears reasonable to use the gross features of that model to construct a simple density of states for both Nb and Nb_3Ga, looked at from the junction (normal) side. Basically what results is a broadened density of states with peaks at both the bulk gap value and at the new gap edge.[7] Similarly to Gilabert, et al.[8] we use a rectangular distribution as shown in Fig. 3.[9] The value of Δ_3 is taken as the half-gap for the material and Δ_2 is obtained from the experimental volt-ampere curve as shown on Fig. 1. Presumably the value of Δ_2 is strongly dependent on the thickness of and/or mean free path in the surface layer.

Volt-ampere curves were calculated numerically using a BCS density of states for Pb and normalizing the rectangular distribution so that the total states remain equal to that of the material when it is normal. The temperature dependencies of Δ_1, Δ_2 and Δ_3 are assumed to be that predicted by BCS except that $\Delta_2(0)$ is not related to T_c. Figure 4 shows these dependencies for the Nb of Fig. 1 and for Pb.

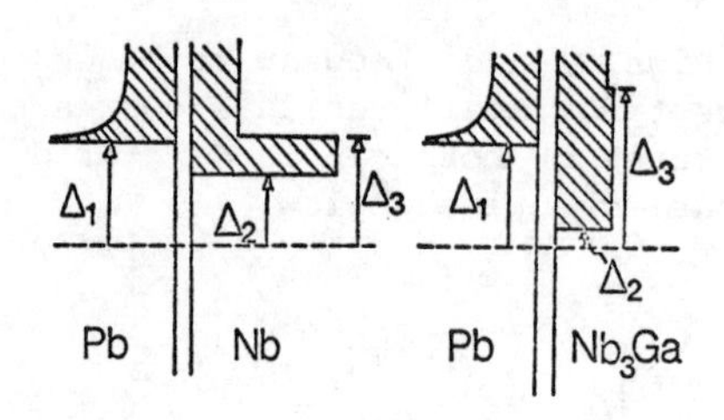

Fig. 3. Density-of-states model used to calculate tunneling volt-ampere curves of Fig. 5 and 6.

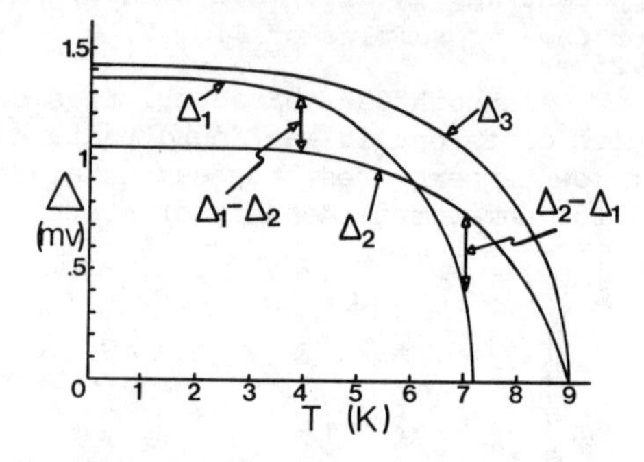

Fig. 4. Assumed temperature dependence of Δ_1, Δ_2 and Δ_3. Numbers correspond to the Nb junction of Fig. 1.

Application of the model to a Nb-Pb junction, with Δ_2 obtained from Fig. 1, results in the curves of Fig. 5. The model shows a large negative resistance at $\Delta_1 + \Delta_3$ which is indeed enhanced at low temperature. It also shows a difference peak at a voltage near $\Delta_2 - \Delta_1$ for $T > 5.5K$ which then disappears and reappears with very small amplitude for low T values. It reappears however at $\Delta_1 - \Delta_2$ and now has the temperature dependence of Δ_1, the Pb gap. This is easily understood from Fig. 4. The current magnitude is very small at these temperatures. Although understandably less sharply defined, the experimental characteristics of Fig. 1 appear to exhibit all of these same features.

Application of the model to Nb_3Ga-Pb junction 19A of Fig. 2 requires using a very small value for Δ_2 of 0.24 mV and a Δ_3 of 2.25 mV. Figure 6 shows the calculated characteristics. For this case it is expected that the difference peak will be at $\Delta_1 - \Delta_2$ for virtually all temperatures below 7 K. Consequently the position of this peak on the voltage axis has the temperature dependence of the Pb gap again and now the current magnitude is quite large. It should be noted that the current peak of the real junction at $\Delta_1 - \Delta_2$ increases by a factor of 2.6 between T = 2.1 K and T = 4.2 K. The increase in the peak current of the model is 2.4 over the same temperature range. Note also that the experimental curve shows a decrease in slope near 2.5 mV which is not evident on the calculated curve. A better fit to the experimental curves for voltages above $\Delta_1 + \Delta_2$ could be obtained by allowing the density of states to bulge more near Δ_2. This might be closer to the predictions of the proximity effect also.[10] However the model is sufficient to show that the current rise is very shallow and that structure at $\Delta_1 + \Delta_3$ should be very difficult to see in the real device.

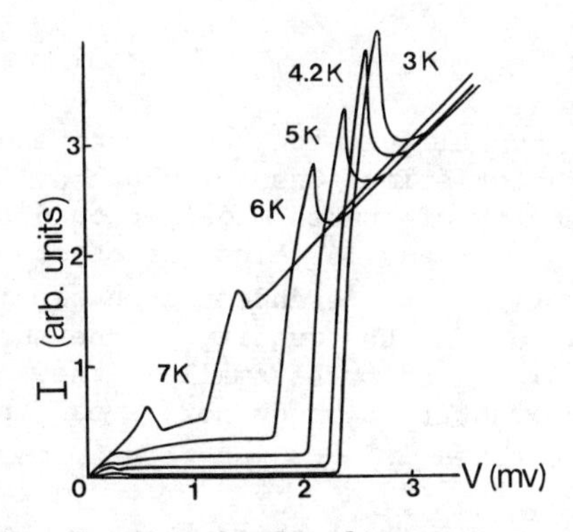

Fig. 5. Computed volt-ampere curves for Nb-Pb junction of Fig. 1, using Δ values of Fig. 4.

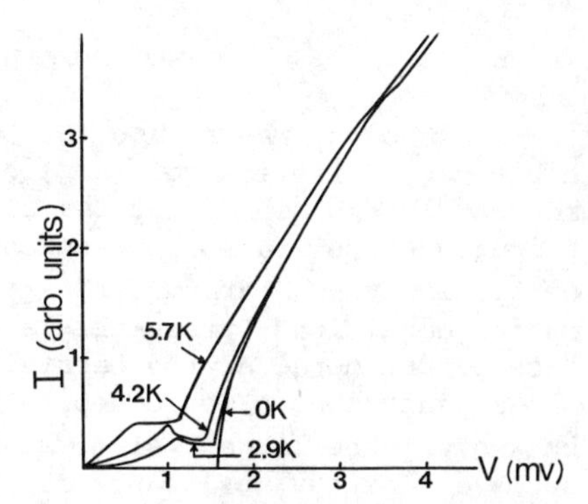

Fig. 6. Computed volt-ampere curves for Nb_3Ga-Pb junction 19A. $\Delta_2(0)$ = 0.24#mV, $\Delta_3(0)$ = 2.25#mV.

OTHER EXPERIMENTAL EVIDENCE

The model appears to be applicable to a large amount of other experimental data obtained by us and by others. For example, Fig. 7 shows the volt-ampere plot for a junction made between a Pb film and the end of a sputter-cleaned Nb wire.[11] For this junction Δ_2 is estimated to be about 0.5#mV. Additional cleaning of such a wire produces a larger Δ_2 as expected, the best VI curve being very similar to Fig. 1.

If one looks closely at published data for Nb - Nb junctions, it is apparent that the model may again be applicable. Figure 1 of Hawkins and Clarke[12] clearly shows evidence of a peak at $\Delta_1 - \Delta_2$ which moves to higher voltages at lower temperature. Broom[2] obtains less excess current but also indicates the observation of a knee near 0.13#mV at 4.2 K. His Fig. 4 can be reinterpreted in terms of our model assuming that the bottom Nb film exhibits a BCS density of states but that the top film has a modified density of states with a Δ_2 value of about 1.3#mV. The corresponding value for the junction of Hawkins and Clarke is about one-half this value.

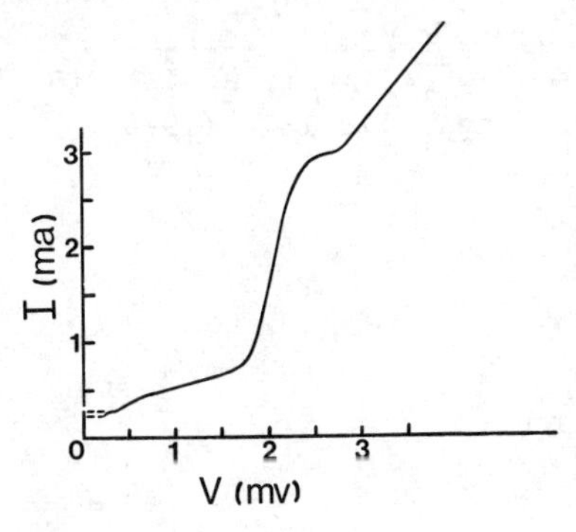

Fig. 7. Volt-ampere curve for a Nb-Pb junction made by deposition of Pb on the end of a sputter-etched and oxidized Nb wire. T = 4.2#K

CONCLUSIONS

Although Nb-Pb and Nb_3Ga-Pb tunnel junctions appear to exhibit very different volt-ampere characteristics, they can both be adequately explained in terms of a density-of-states model which distributes these states between energies Δ_2 and Δ_3 instead of concentrating them near Δ_3. The value of Δ_2 is dependent upon fabrication conditions, presumably associated with the quality of the surface of the metal at the barrier. It is found to be small in the Nb_3Ga junctions. Consequently the characteristics do not reveal the bulk gap. The larger value usually obtained in Nb junctions produces a negative resistance at $\Delta_1 + \Delta_3$.

An interesting result is that a difference peak can move with temperature, in either direction along the voltage axis, depending upon the temperature range and the relative size of Δ_1 and Δ_2.

ACKNOWLEDGMENTS

The authors wish to thank R. J. Burt for providing samples of his RF-sputtered Nb_3Ga films.

REFERENCES

1. L. Y. L. Shen, Superconductivity in d and f Band Metals, ed. by D. H. Douglass (AIP, New York, 1972).
2. R. F. Broom, J. Appl. Phys., 47, 5432 (1976).
3. D. F. Moore, J. M. Rowell, and M. R. Beasley, Sol. St. Com., 20, 305 (1976).
4. L. S. Hoel, W. H. Keller, J. E. Nordman, and A. C. Scott, Sol. St. Elec., 15, 1167 (1972).
5. P. W. Wyatt, R. C. Barker and A. Yelon, Phys. Rev. B, 6, 4169 (1972).
6. R. J. Burt and F. J. Worzala, IEEE Trans., MAG-13, 323 (1977).
7. W. L. McMillan, Phys. Rev., 175, 537 (1968).
8. A. Gilabert, J. P. Romagnan, and E. Guyon, C. R. Acad. Sc. Paris t271, 552 (1970).
9. P. Cardinne, B. Manhes, and J. E. Nordman, Rev. Phys. Appl. 8, 463 (1973).
10. J. Vrba and S. B. Woods, Phys. Rev. B, 3, 2243 (1971).
11. P. Cardinne and J. E. Nordman, Revue Phys. Appl. 8, 467 (1973).
12. G. Hawkins and J. Clarke, J. Appl. Phys. 47, 1616 (1976).

A SUPERCONDUCTING TRANSISTOR*

K. E. Gray
Argonne National Laboratory, Argonne, Illinois 60439

ABSTRACT

A three-film superconducting tunneling device, analogous to a semiconductor transistor, is presented, including a theoretical description and experimental results showing a current gain of four. Much larger current gains are shown to be feasible. Such a development is particularly interesting because of its novelty and the striking analogies with the semiconductor junction transistor.

INTRODUCTION

Although superconductors and semiconductors both have an energy gap in their electronic excitation spectrum, it is the difference in the electrical conductivity of the ground state which leads to the striking difference in their behaviors. The ground state of a superconductor (i.e. Cooper pairs) is a perfect conductor whereas the ground state of a semiconductor is a perfect insulator. Superconductors can, however, duplicate some of the functions of semiconducting devices if the conductivity of the ground state Cooper pairs can be made negligible. This can be accomplished by a thin insulating barrier which blocks Cooper pairs, but through which excitations can quantum-mechanically tunnel. Such a superconducting tunnel junction has a non-linear current-voltage characteristic similar to a semiconducting p-n junction diode except that it is polarity-independent.

Using two such barriers, we have constructed for the first time a superconducting device which is a very close analogy to the semiconductor junction transistor.

The superconducting transistor is a three-terminal device with a current gain. Our device is constructed by stacking three thin films of aluminum on top of each other and oxidizing them to form two insulating barriers between the films. One junction, called the injector, is biased at a voltage greater than $2\Delta/e$, where 2Δ is the Cooper-pair binding energy and e is the electron charge. The resulting current dissociates Cooper pairs and one excited electron (quasiparticle) is created in each of the films, forming this junction for each electron transferred. The additional excitations thus created in the center film can tunnel through the second junction, called the collector. This junction is independently biased at $V < 2\Delta/e$ so that Cooper pairs are not dissociated.

*Work performed under the auspices of the U.S. Department of Energy.

SN: 0094-243X/78/359/$1.50

THEORY

We are interested in the current gain $\Delta I/I^i$, where ΔI is the excess collector current due to the injector current I^i. If the injector junction is biased at $V > 2\Delta$, the current is primarily due to processes which break a Cooper pair and inject one quasiparticle in each film of the injector junction for each electron transfered. Hence, the injection rate per unit volume I_o is $I^i/e\Omega$ where Ω is the volume of the middle film common to the junction. The collector junction is biased below 2Δ so that pairs are not broken, but the current I is proportional to the quasiparticle density N, such that $I = \alpha N$, where α can be calculated using the tunneling Hamiltonian.[1,2] In the presence of injection, the excess collector current is given by $\Delta I = \Delta N/2$, where ΔN is the excess quasiparticle density and the factor of two is because there is direct injection in only one of the films of the collector junction. Combining these expressions, the gain is given by $\Delta I/I^i = \alpha\Delta N/2I_o e\Omega$.

There is nothing in the above which indicates whether or not a current gain, i.e., $\Delta I > I^i$, is possible. Physically a current gain corresponds to the following: A single electron is injected in the middle film creating a quasiparticle excitation which causes more than one electron to flow across the collector junction. This multiplying effect is, however, possible, and it is because of the nature of the two processes contributing to the collector current αN.

To see this, it is necessary to examine the relevant terms in the tunneling Hamiltonian.[1,2] It is convenient to express the normal state electron creation and annihilation operators in terms of quasiparticle operators using the Bogoliubov-Valatin transformation.[3] For tunneling between identical superconductors a and b, at voltages $V_b - V_a < 2\Delta/e$, the only energy-conserving processes are described by the terms $u_{p'}u_p\gamma^+_{p'\uparrow}\gamma_{p\uparrow}$ and $v_{p'}v_p\gamma_{-p'\downarrow}\gamma_{-p\downarrow}$ where u and v are the BCS coherence factors and $\gamma^+(\gamma)$ is the quasiparticle creation (annihilation) operator. These are shown in Figure 1. Integrating over the Fermi function and reduced BCS density of states, one finds in thermal equilibrium the same net current from each term.[2] The first term destroys a quasiparticle in b and creates one in a, while the second term destroys a quasiparticle in a and creates one in b. Each term

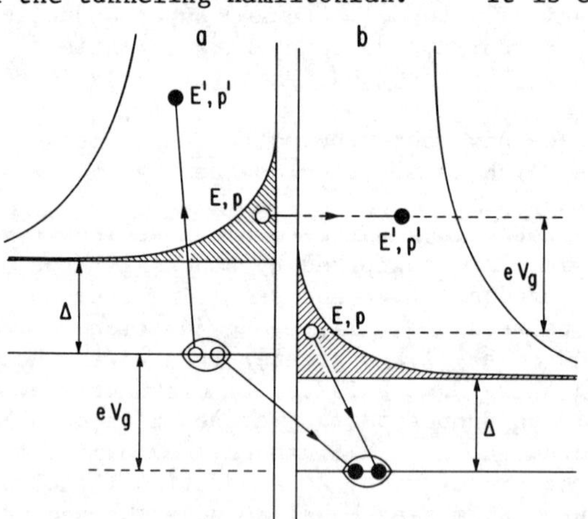

Fig. 1. The two tunneling processes are shown for an initial energy E and a collector bias voltage $V_g < 2\Delta/e$.

transfers one electron across the barrier from a to b.

This clearly indicates that a current gain is possible since a quasiparticle starting in a can go to b and then enter the condensate in a process which returns a quasiparticle to a, etc., each time transfering one electronic charge across the barrier of the collector. In this somewhat more physical picture, the current gain should be the ratio of the tunneling rate Γ_T to the loss rate by recombination Γ_R, since the excess quasiparticles can no longer tunnel back and forth once they recombine into Cooper pairs. Now the tunneling rate of a given quasiparticle[4] is $\alpha/2e\Omega$ and $\Gamma_R = I_o/\Delta N$. Substituting in our previous expression for the gain, we find $\Delta I/I^i = \Gamma_T/\Gamma_R$. Hence if $\Gamma_T >> \Gamma_R$ the injected quasiparticles can tunnel across the collector junction many times before recombining, which implies a current gain.

To maximize the current gain, the smallest value of Γ_R is desirable, indicating that aluminum is a good choice.[5] To evaluate Γ_T, it is a good approximation to set $\alpha = (2N(0)R)^{-1}$ (for collector voltages $\gtrsim \Delta$ and reduced temperatures $T/T_c \lesssim 0.5$). Here N(0) is the single spin density of states and R is the resistance of the collector junction. Hence, $\Gamma_T = (4eN(0)RAd)^{-1}$, so it is desirable to have a small value for the specific junction resistance RA and the film thickness d. Note that RA cannot be too small because the Josephson coupling will interfere with the desired effects. Using reasonable values for aluminum ($\Gamma_R = 10^5$ sec^{-1}, d = 300 Å, RA = 10^{-5} Ωcm^2 and N(0) = 1.7×10^{22} $cm^{-3}eV^{-1}$), we calculate a transfer function $\Delta I/I^i \cong 30$.

EXPERIMENT

In our experimental setup, three aluminum films (thickness 300 Å each) are deposited in a stack on a sapphire substrate to form two tunnel junctions (Fig. 2a). The samples are mounted in a vacuum can on a copper block cooled by helium 3. The collector junction has RA $\cong 6 \times 10^{-6}$ Ωcm^2 and the injector is about 16 Ω. The IV characteristic of the injector in Figure 2b shows the onset of pair breaking at $\cong$ 360 μV. For the collector, IV plots corresponding to various labeled injection currents are shown in Fig. 2c. The current axis is greatly magnified compared with Fig. 2b so that the current due to thermally excited quasiparticles as well as injected quasiparticles can be readily seen. Only the bottom portion of the vertical rising part of the characteristic at $V = 2\Delta/e$ is visible. The negative resistance is due to the non-uniform energy gap in the films which leads to an inhomogeneous distribution of the tunnel current once the voltage corresponding to the smallest gap is reached.[6] For the transistor operation this feature is unimportant since it is the current at $V < 2\Delta/e$ which shows a gain.

In Fig. 2d, the current change in the collector ΔI is seen to increase linearly with injector current, but with a temperature-dependent slope greater than unity. Note that ΔI is always less than I(o), the collector current with $I^i = 0$. Therefore, the saturation effects accompanying overinjection[4] are negligible and the response is linear. The temperature dependence of the observed effect is a convenient check that recombination into Cooper pairs is being observed. The value of $R_{eff} = \Gamma_R/N(T)$ found from these data is in reasonable agreement with earlier measurements[4] if allowance is made for the difference in film thicknesses. At the lowest temperature measured, the current gain is about 4.

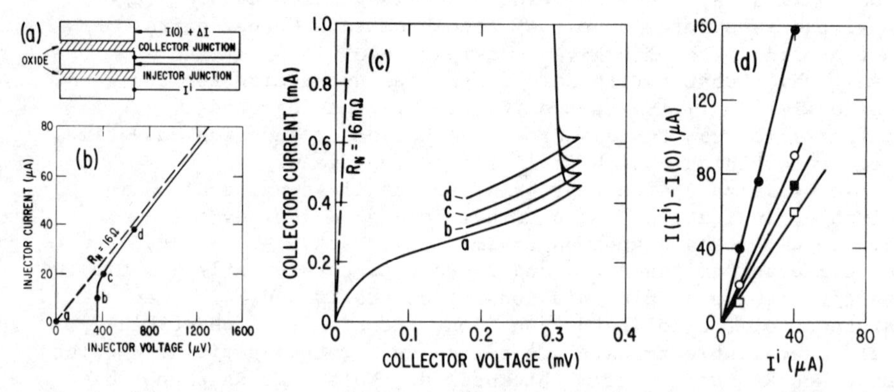

Fig. 2 (a) Schematic experimental configuration. (b) Current-voltage characteristic of the injector junction. The labeled points refer to the injection current for the collector curves shown in (c). (c) The current voltage characteristics of the collector junction. (d) The change in collector current is plotted against injector current showing a temperature-dependent slope $\Delta I/I^i$ greater than unity. The temperature is determined from the quasi-particle current I(o) in the collector with $I^i = 0$: • 0.625 K; o 0.650 K; ■ 0.695 K; ■ 0.703 K.

DISCUSSION

Note that, unlike semiconductor devices, the polarity of both bias voltages is arbitrary--the device works the same for any combination of polarities. For the device shown in Figure 2, the power gain could be as large as twice the current gain, and it is a simple matter to achieve additional current gain by cascading several devices. The voltage gain depends on the bias conditions, but because of the low input impedance the device might first be used on low impedance sources like thermocouples.

Although the device described above is similar to a semiconductor transistor in the common base configuration, it should also be possible to bias in the common emitter configuration. If the collector bias is across the entire stack of films as shown in Fig. 3, the same current I_c performs both functions I^i and I, and a small additional current across the injector junction I_b can control I_c. The current gain, β, can be determined from a calculation of the "base" current I_b applied across the injector junction: $I_b = I_c - I^i$. If I_c is replaced by the equivalent expression $I(o) + \eta I^i$, where

$\eta \equiv \Delta I/I^i$, the result is $I_b = I(o) + (\eta - 1)\, I^i$. Therefore the small signal β is just $(\eta - 1)^{-1}$ which can be made arbitrarily large, if $\eta \simeq 1$. Thus $\eta = \alpha/4e\Omega\Gamma_R \simeq 1$, which can be achieved by using RA $\simeq 10^{-4}$ Ωcm^2 and fine-tuning Γ_R through its dependence on operating temperature and applied magnetic field.[4]

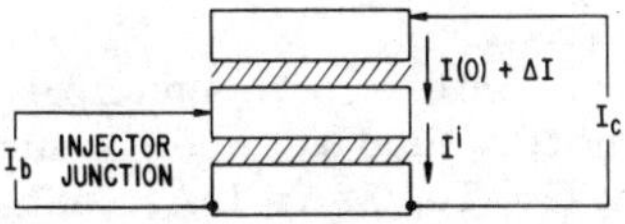

Fig. 3. The common emitter configuration in which the collector and injector are not biased independently.

The analogy with a semiconductor junction transistor can be carried even further. In the semiconductor transistor, minority carriers are injected into the base region from the forward-biased emitter-base diode, and are collected at the back-biased base-collector diode. However, the injected carriers must diffuse across the base region before recombining with the majority carriers which are always present in the base. Thus the diffusion time must be shorter than the recombination time. In the superconducting device, quasiparticles are injected at either polarity when the junction voltage is $2\Delta/e$ or greater. At lower voltages quasiparticles are only collected. Hence, the superconducting transistor uses different magnitudes of the bias voltages instead of the forward- and back-biased diodes in the semiconductor transistor. Because the middle film is so thin, the quasiparticles diffuse across it almost instantaneously, but they have a rather small probability of transmission through the tunnel barrier of the collector junction. Therefore the criterion for current gain is in the comparison of Γ_T with Γ_R indicated above.

REFERENCES

1. M. H. Cohen, L. M. Falicov and J. C. Phillips, Phys. Rev. Letters 8, 316 (1962).
2. B. N. Taylor, Thesis, Univ of Penn. (1963) unpublished.
3. N. N. Bogoliubov, Nuovo cimento 7, 794 (1958), J. G. Valatin, Nuovo cimento 7, 843 (1958).
4. K. E. Gray, J. Phys. F: Metal Phys. 1, 290 (1971).
5. S. B. Kaplan, C. C. Chi, D. N. Langenberg, J. J. Chang, S. Jafarey and D. J. Scalapino, Phys. Rev. B14, 4854 (1976).
6. K. E. Gray and H. Willemsen, J. Low Temp. Phys. 31, 911 (1978).
7. K. E. Gray, A. R. Long, and C. J. Adkins, Phil. Mag. 20, 273 (1969).

SUPERCONDUCTOR-SEMICONDUCTOR DEVICE RESEARCH*

A. H. Silver, A. B. Chase, M. McColl, and M. F. Millea
The Ivan A. Getting Laboratories
The Aerospace Corporation, El Segundo, CA 90045

ABSTRACT

Devices and components which combine both superconductors and semiconductors include Josephson and super-Schottky diodes, superconducting triodes, and hybrids. Semiconductors, such as Si, Ge, Te, GaAs, InAs, InSb, and CdS, are used to form barriers or effective barriers to electron and pair tunneling in the active regions of these devices. Hence such properties as bandgap, barrier height, and mobility become important parameters in tailoring and controlling the device performance. This paper will discuss the motivation for this research, accomplishments to date, and an assessment of the future prospects.

INTRODUCTION

The present emphasis in superconducting electronics traces its origins to the discoveries of tunneling phenomena in superconductors by Giaever[1] and Josephson[2], for which they shared the Nobel prize with Esaki. Subsequently there has been a large expansion of this field on both a fundamental and applications basis. By far the major effort has centered on fabricating and characterizing the active elements, or junctions. Principally at high frequencies and for fast switching times, the detailed nature of the junctions will determine actual device performance. The joint incorporation of superconductors and semiconductors holds promise for the development of greater flexibility, control, reproducibility, yield, and reliability for active junctions.

We discuss the progress, applications, and future prospects for single-particle tunneling (super-Schottky) diodes, Josephson junctions, and FET-like devices. Tunnel junctions are traditionally fabricated with thin-film sandwich technology, where the two conducting (metal) films are separated by a non-conducting barrier through which current flows via tunneling. A principal motivating factor for this research is the substitution of a semiconductor barrier for the native metal-oxide barriers commonly used. Additionally, unique advantages may be derived from the low loss of the superconductors compared to normal (and room temperature) metals in more conventional

*Research supported by ONR and The Aerospace Corporation

ISSN: 0094-243X/78/364/$1.50

devices, and from the specific optical properties of semiconductors. One can further benefit from the selection of materials and processing techniques which have been developed by the semiconductor industry and are not necessarily applicable to metal film technology.

SUPER-SCHOTTKY TUNNELING DIODE

One type of device which has been successfully demonstrated and which has great potential for both video and heterodyne detection at near millimeter wavelengths is the superconducting-semiconducting tunnel diode.[3-6] This device is a variant of the Schottky diode in which the metal contact to the semiconductor is a superconductor. Current flows via electron tunneling across the natural surface barrier in the semiconductor. The principal advantages of these devices are the control and stability of the surface barrier, particularly in the high conductance regime required for high frequency applications, high intrinsic efficiency, and low noise. Disadvantages are parasitic losses in the semiconductor and incomplete control of the surface properties of certain semiconductors.

The conventional Schottky diode is a metal-semiconductor contact for which the conductance is controlled by a surface barrier. It is desirable to have thermionic emission and thermally activated tunneling dominate over pure tunneling in order to realize the greatest nonlinearity. The resulting forward biased exponential electrical behavior[7]

$$I = I_o \exp\left[eV/k(T + T_o)\right] \tag{1}$$

is a result of these thermally stimulated mechanisms. On the other hand, the super-Schottky diode is a pure tunneling-dominated device which obeys Eq. (1) as a result of the thermal distribution of electrons in the heavily doped semiconductor. In order for tunneling to dominate the diode conductance and provide an efficient impedance match to microwave networks, the space charge barrier must be small. The prefactor I_o in Eq. (1) varies as[8]

$$\exp -\left[\frac{2^{3/2} t \phi^{1/2} m^{1/2}}{\hbar}\right]$$

where t is the thickness and ϕ the height of an effective rectangular barrier. In an ideal Schottky barrier, this is replaced by approximately

$$\exp -(2\,\varepsilon^{1/2} m^{1/2} E_B / e\hbar N^{1/2})$$

where N is the donor (acceptor) concentration, E_B is the barrier height, m is the effective mass, and ε is the dielectric constant.[9] The super-Schottky diode is then to be compared with cryogenic devices such as Josephson junctions, superconductor—oxide—normal metal (S-O-M), and M-O-M, and with conventional Schottky diodes.

The performance of the super-Schottky mixer, like a conventional Schottky mixer diode, is directly related to the degree of nonlinearity of its I-V characteristic. The I-V curve of either diode can be represented over the voltage range of interest by Eq. (1), where I_o is determined by the geometrical and material parameters of the diode. The parameter S

$$S = e/k(T + T_o) \tag{2}$$

is a measure of the nonlinearity of the diode and, as such, is of central importance in the determination of the sensitivity of the diode as either a video detector or mixer. Numerically, Eq. (2) becomes

$$S \simeq 11{,}600/(T + T_o) \text{ per volt.} \tag{3}$$

Both experimentally and theoretically for super-Schottky diodes reported here, $T_o \ll 1$ K if $kT \ll e\Delta$ (the superconducting energy gap); therefore, $S = e/kT$ is a reasonable approximation. The value of T_o for Schottky barriers on n-type GaAs is greater than 40 K.[10, 11] Comparing a conventional Schottky mixer with a super-Schottky, both operating at 1 K, implies greater than a 40 to 1 improvement in diode noise temperature. This reduction in noise would imply a corresponding improvement in sensitivity if the conversion loss of the super-Schottky were equivalent to that of the conventional Schottky. As discussed below, conversion loss is a major issue for the super-Schottky. Because of the relatively limited voltage range, the design criteria for obtaining low conversion losses are stringent.

Comparison of the super-Schottky and SOM diodes shows that the SOM device has an impedance to which matching is difficult at high frequencies. This problem arises because the capacitive reactance of the junction is extremely small in comparison with the parallel junction resistance. Matching this highly reactive diode impedance to a typically resistive waveguide impedance of a few hundred ohms represents a formidable problem. However, the impedance level of the super-Schottky diode is inherently lower, less reactive, and more controllable. The decreased impedance level results from the fact that the electron tunneling probability is increased with the super-Schottky because of the exponential dependence on several factors. These are the following:

(1) Schottky barrier heights tend to be smaller than those of the metal-oxide-metal structures.

(2) Tunneling effective masses of carriers in semiconductors tend to be smaller than those in oxide barriers. Lower barrier heights and smaller tunneling masses are particularly encountered with low bandgap semiconductors.

(3) The general shape of the Schottky barrier is more conducive to tunneling. A parabolic as opposed to a rectangular barrier contributes less attenuation to tunneling carriers.

The net result is not only a lower resistive impedance with the super-Schottky barrier but a wider barrier as well, which decreases the junction capacitance, a parasitic element of the diode. This effect is partially compensated by the higher dielectric constant of the semiconductor. However, for materials with low effective masses and low bandgaps, the capacitance per unit area is significantly less with the super-Schottky diode.

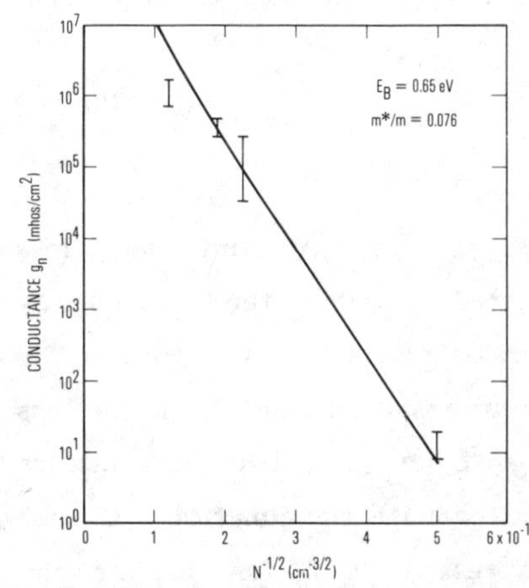

Fig. 1 Normal conductance g_n per unit area of Pb contacts to p-GaAs as a function of hole concentration N. (From Ref. 4,5)

An important advantage of the super-Schottky is the ability to adjust the barrier width (impedance level) by varying the semiconductor doping. Over a large range, $\ln g_n \propto N^{-1/2}$, where g_n is the diode conductance and N the doping concentration. Fig. 1 demonstrates the effect of doping on the conductance of Pb/p-GaAs diodes.

The solid line is the theoretical curve for a barrier height of 0.65 eV and an effective mass ratio of 0.076. One generally strives for high conductance diodes, which also means lowering the parasitic series resistance.

The super-Schottky diode fabricated from Pb/GaAs approaches ideal behavior. Extensive measurements[5,6] at 9 GHz have demonstrated the relevant characteristics and clearly represent a significant advance in low noise sensors. We briefly summarize these properties and discuss future prospects.

MIXER DESIGN CONSIDERATIONS

The noise temperature T_r of a heterodyne receiver referred to the input of the mixer is given by

$$T_r = L_c(T_D + T_{IF}) \tag{4}$$

where L_c is the single-sideband conversion loss of the mixer, T_D is the noise temperature of the mixer diode measured at the IF terminals of the mixer, and T_{IF} is the noise temperature of the IF amplifier. Hence, in fabricating a mixer, utmost attention must be paid to minimizing L_c.

Conversion loss is conveniently expressed as the product of two terms:

$$L_c = L_o L_1 \tag{5}$$

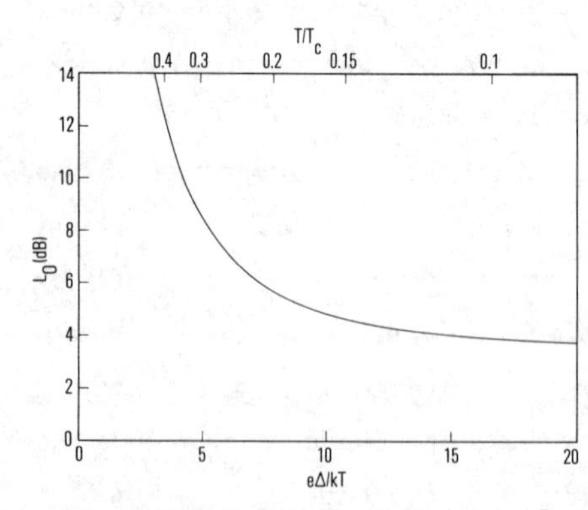

Fig. 2 Intrinsic conversion loss L_o as a function of Δ/kT and T/T_c. (From Ref. 5,6)

where L_o is the intrinsic loss associated with the frequency conversion process that occurs in the nonlinear resistance and L_1 is the loss arising from parasitic impedances associated with the junction. L_o can be expressed as a function of the energy gap voltage Δ, temperature T, and superconducting transition temperature T_c as shown in Fig. 2. Conversion loss of 5 dB is typically expected for S values $\gtrsim 10^4\ V^{-1}$. For Pb/2 x 10^{19} p–GaAs, near theoretical (11,600/T) was measured; however, for 3 x 10^{19} doping, typical values measured from the I-V curves were 5000 V^{-1} near 1 K. If the leakage responsible for this apparent decrease in S was exponential (or non-ohmic in general), significant degradation in the diode noise would have been measured. We will see from the data that no excess noise sources were introduced and this effect on S is a result of "ohmic" parasitic losses.

Fig. 3 is the equivalent circuit of the super-Schottky diode. The loss term L_1 can be expressed as

$$L_1 = 1 + (R_s/R) + \omega^2 C^2 RR_s \tag{6}$$

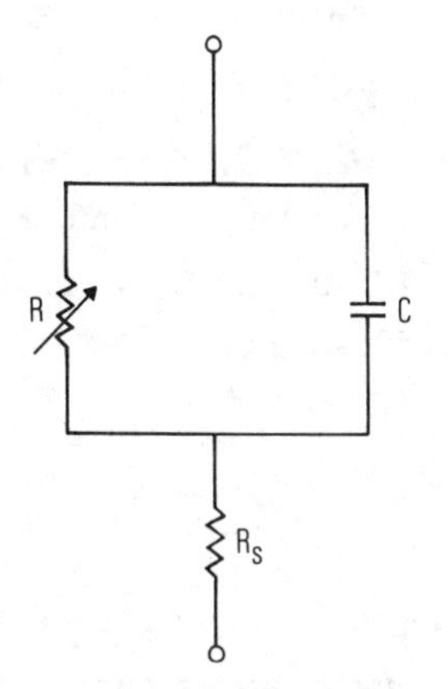

Fig. 3 Equivalent circuit of the super-Schottky diode.

where ω is the signal angular frequency, C is the junction capacitance, and R is the signal input impedance of the LO-pumped nonlinear junction resistance. Several methods are available for minimizing this loss by reducing R_s. Two such methods involve the reduction of the resistivity ρ of the semiconductor. Since $(L_1 - 1)$ is proportional to R_s, which for a small area circular contact is given by $R_s = \rho/2D$ (D is the contact diameter), L_1 can be reduced by choosing semiconductors that have large mobilities and, to a certain degree, are heavily doped. Heavy doping is entirely compatible with the super-Schottky diode as discussed previously. High mobility materials, e.g., n-type InSb, would be the best candidates for millimeter wavelength applications. In this context, why not use n-type GaAs, a very high mobility material and one that is used for conventional high-frequency Schottky mixing rather than p-type GaAs whose mobility is much lower? The obstacle in using n-GaAs lies in the difficulty of obtaining a sufficiently high carrier concentration to obtain a usable impedance level for super-Schottky construction. That is, because of its larger barrier height, n–GaAs would require a larger doping than p-GaAs to obtain the same impedance. The solubility of known donor impurities is smaller than that of acceptor impurities which are presently being used in

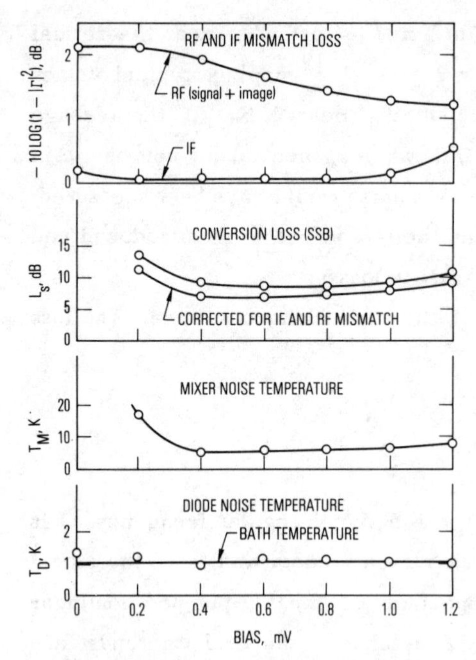

Fig. 4 Measured parameters describing Pb/p-GaAs mixer performance at 9 GHz and 1.06 K. (From Ref. 5,6)

GaAs. A third method to reduce L_1 involves the use of multiple contacts. For the same total area of contact and hence impedance, $(L_1 - 1)$ is inversely proportional to the square root of the number of identical independent contacts connected in parallel.

The performance of Pb/3 x 10^{19} p-GaAs at 9 GHz is the best reported for a video detector and a mixer.[5,6] The latter are summarized in Fig. 4. The conversion loss of 7.5 dB is composed of L_o = 6 dB and L_1 = 1.5 dB. This can be compared with L_o = 4.8 dB and L_1 = 5.1 dB for 2 x 10^{19} p-GaAs.

FUTURE MIXER IMPROVEMENT

The prospects for improvement rely on reducing L_1 without degrading L_o or T_D, essentially minimizing parasitic losses. Several approaches have received attention. Huang and Van Duzer[12, 13] have applied Si technology to fabricating both super-Schottky and Josephson diodes. The principal advantage of their technique is reduction of the series resistance by use of thinned membranes of heavily doped silicon between superconducting electrodes. Satisfactory I-V curves have been measured, but thus far theoretical S values and microwave measurements have not been reported.

We are developing the multiple-contact diode array. The required diode impedance is obtained by making many diodes in parallel with the same total active area. Fig. 5 shows the predicted L_1 for Pb/3 x $10^{19}/cm^3$ p-GaAs based on the fabrication technology used in the measurements reported earlier. Because of the very large peripheral area and the small dimensions involved, new problems and techniques are encountered.

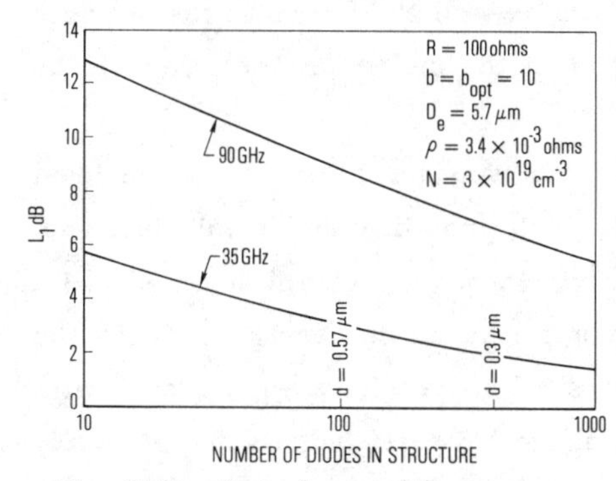

Fig. 5 Predicted parasitic conversion loss L_1 for multicontact diode array of Pb/p-GaAs

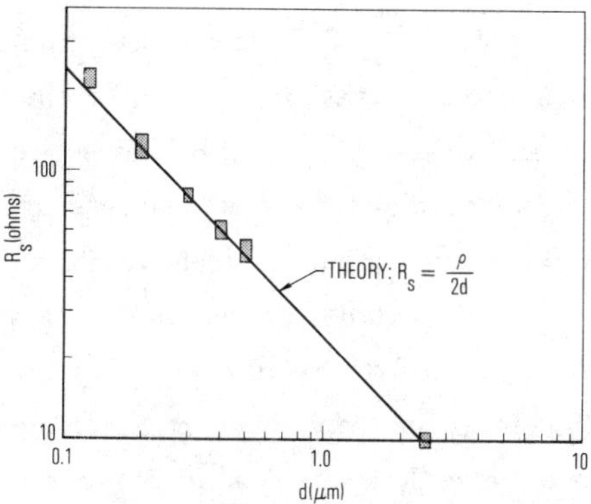

Fig. 6 Measured values of series resistance R_s versus diode diameter d for HFPP.

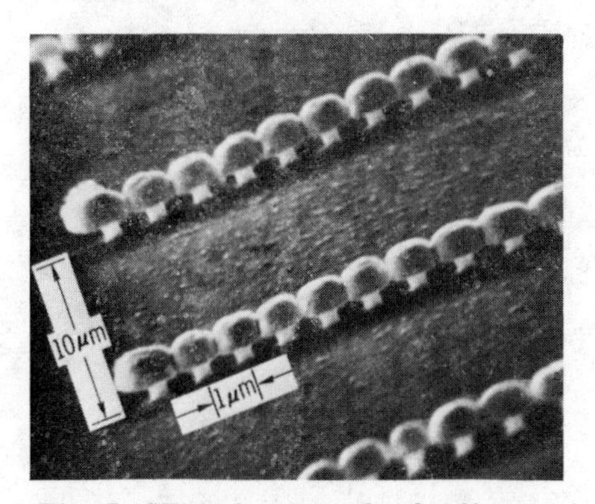

Fig. 7 SEM photograph of a linear multicontact diode fabricated with HFPP. (From Ref. 4,5)

The method of diode formation is by electroplating the small diodes through "windows" in a protective overlayer on the surface.[14] The uniformity of conventional dc plating is poor, and results in values of R_s well above theoretical.[15] We have developed a high field-pulsed plating (HFPP) technique[16] which yields theoretical values of R_s and uniform deposition for contacts as small as 1200 A diameter. Fig. 6 shows measured R_s versus diameter for Pt on n-GaAs with no adjustable parameters in fitting the theoretical curve. A linear diode array under development is shown in Fig. 7, with the insulating layer removed for inspection.[17] The ohmic contact will be made to metal strips plated between the "string-of-pearls" diodes. Contact along the chain is accomplished by overplating as shown.

Parallel efforts to reduce R_s and to modify the tunneling barrier are in progress. We have been able to reduce the barrier on p-GaAs by 150 mV by chemical treatment of the surfaces to lower the Ga/As ratio and increase the As-As surface cross-linking. This treatment is consistent with our electroplating chemistry, thus producing a lower impedance diode. (For n-GaAs this technique raises the barrier 150 mV). Therefore we can expect a single Pb/p–GaAs diode to perform as well at 35 GHz as the previous device did at 9 GHz. Similar improvement in the multi-contact array can be expected at higher frequencies.

For higher frequency applications we consider n-InSb to be an ideal semiconductor substrate because of its high mobility and small bandgap. Predicted L_1 values extend the operating range to several hundred GHz, essentially to the limits imposed by the superconducting energy gap and the quantum limits at cryogenic temperatures. The problems encountered with InSb are barrier formation and surface leakage. Barrier heights are generally of the order of the bandgap. Therefore we expect to utilize this small barrier in InSb to make wide, low impedance junctions. The very high mobility at helium temperatures would essentially eliminate the series resistance.

Most metals make ohmic contacts to n-InSb.[18] Gold is an exception. We have been successful in fabricating a super-Schottky diode using a thin interface of Au between the Pb electrode and n-InSb (Fig. 8). However, we consider this to be a difficult and low-yield technique for obvious reasons. Using chemical surface manipulation along the lines mentioned above for p-GaAs, we have fabricated super-Schottky tunneling diodes on n-InSb. In this case we want an Sb-rich surface and must be careful to avoid destroying the Sb-Sb cross-linking. Fig. 9 shows the Pb energy gap observed by tunneling through a Schottky barrier on n-InSb. We believe that this type of diode will satisfy the requirements for millimeter and submillimeter wave low noise detectors. Fig. 10 projects the parasitic loss term for Pb/n-InSb operating at 1 K.

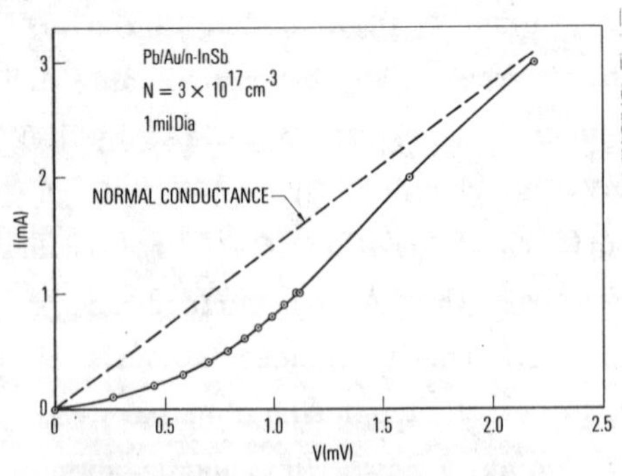

Fig. 8 Superconducting energy gap effect on I-V curve of Pb/Au/ n-InSb junction at 1.4 K.

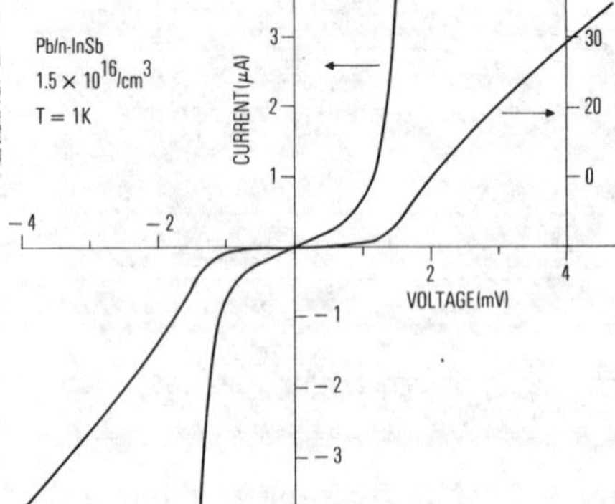

Fig. 9 Super-Schottky I-V characteristic for n-InSb.

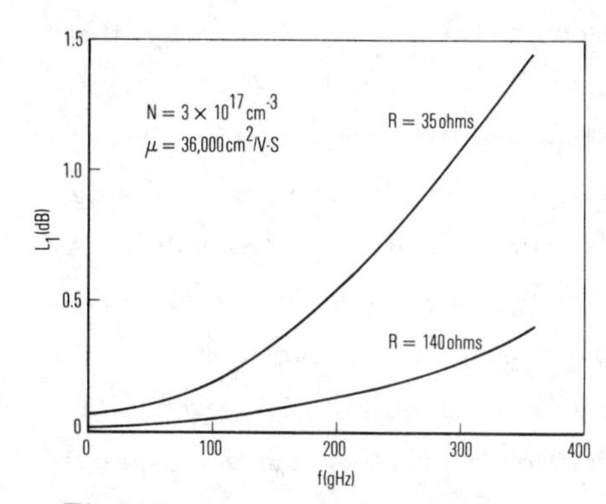

Fig. 10 Predicted parasitic loss L_1 for one Pb/n-InSb super-Schottky mixer diode.

One should not expect mixer noise temperatures of a few degrees K over this entire frequency range even for this ideal diode. Tucker and Millea[19] have analyzed the quantum-mechanical behavior of the super-Schottky diode in the high frequency regime $\hbar\omega > 2kT$, where quantum effects dominate and the performance deviates from the classical predictions due to photon-assisted tunneling. For example, they predict the intrinsic mixer conversion loss to be

$$L_o^{QM} = \left[\frac{(\hbar\omega/2kT)}{\tanh(\hbar\omega/2kT)}\right] L_o^{cl} \qquad (7)$$

where L_o^{cl} is the classical prediction of Fig. 2. At high frequencies the apparent degradation in conversion loss by a factor $\hbar\omega/2kT$ arises because a photon must be absorbed at the signal frequency to drive an electron across the barrier, while only 2kT is required for each electron at the RF frequency. When noise is included, the theory predicts a minimum detectable power which approaches $\hbar\omega B$ at high frequencies for optimal impedance matching and strong local oscillator drive, compared to the classical result of order kTB for $\hbar\omega \ll 2kT$. The super-Schottky is therefore expected to become a photon detector at frequencies $\hbar\omega \gtrsim 2kT$ with mixer performance ultimately limited by photon shot noise.

JOSEPHSON JUNCTION DIODES

Giaever[20] reported one of the first successful experiments with semiconductor barriers in a superconducting device, Pb-CdS-Pb. Here the single-particle tunneling and Josephson pair tunneling were both observed. Furthermore, because of the photo-sensitive properties of CdS, the sensitivity of both effects to light was also observed. Similar semiconductor barrier sandwich devices were studied using CdSe,[21] CdS,[22,23] PbS,[23] Te,[24,25,26]

Se,[25,26] and Ge(Sn)[27] evaporated semiconductor films. Generally these devices exhibit pinholes in the semiconductor unless thick films or successive oxidation is employed. In either case the junction impedance will be high.

Huang and Van Duzer[12,13] developed a unique method for using single crystal Si to form a barrier between the superconducting electrodes. An etching technique is used to produce precisely thinned membranes of heavily doped silicon on which the metal films are deposited. Thicknesses from 400 to 1200 A have been studied. This technique produces useful Josephson (and super-Schottky diodes) for which the barrier properties can be carefully tailored before metal deposition. The physical process is conduction through two very thin Schottky barriers in series with the degenerate semiconductor. A principal problem is the alignment of two counter-electrodes on opposite sides of the membrane.

THREE-TERMINAL DEVICES

Let us consider the possibility of fabricating three-terminal superconducting devices which are analogous to field effect transistors. For this purpose we propose a planar Josephson junction for which the coupling can be controlled by a third, electrically isolated electrode. Furthermore the Josephson coupling should be adjustable with geometry and material electronic properties. We believe a promising approach is to employ a p-type semiconductor which has a naturally inverted n-type surface. Important parameters will be mobility, surface channel thickness, substrate isolation, and ability to contact the surface channel with closely spaced electrodes. By means of the proximity effect the surface channel underneath the superconducting contact will become superconducting and we can achieve Josephson coupling as in S-N-S junctions or proximity bridges.

The inverted channel in conventional FETs is produced by a biased field gate or other fixed surface charges, making it difficult to contact the conducting surface layer. Furthermore, one is always faced with a surface impedance which will reduce or eliminate the proximity-induced superconductivity. To avoid this, we use a semiconductor with a naturally n-type surface, p-InAs. (PbTe also exhibits this property.) We characterize

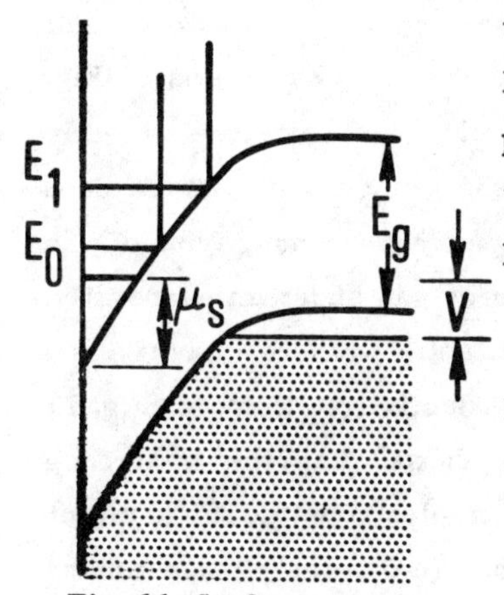

Fig. 11 Surface energy band diagram for a Complete Schottky Barrier under forward bias V. Quantized surface levels are shown for moderately doped p-InAs.

this material as a Complete Schottky Barrier,[28] since the surface Fermi level is pinned above the conduction band even for p-type substrates as shown in Fig. 11. In principle one always makes ohmic contacts to the n-surface of InAs. The "barrier height" to the substrate exceeds the bandgap by the Fermi degeneracy of the surface μ_s. The device would be constructed as shown in Fig. 12, where the superconductor spacing is expected to be $\lesssim 1\ \mu$m.

Experiments to date have shown the problem to be difficult, but the approach promising. They have partly determined the nature of the inverted surface, the metal-semiconductor contact, and how to control each. Both a Josephson FET and a superconducting MESFET are attractive possibilities and are under active investigation. By a superconducting MESFET we mean a non-Josephson device similar to a conventional MESFET except that the normal metal electrodes are replaced by superconductors.

The most important feature is the metal to surface contact. We have established that a direct contact of Pb to the n surface channel can be made with proper plating techniques; vacuum evaporation generally results in an almost atomically thin interfacial barrier.[29] These conclusions were obtained by analyzing the I-V curves from the metal electrode to the substrate on $\approx 10^{18}/cm^3$ doped p-InAs. Evaporated contacts display a prominent negative resistance in the forward direction, similar to a conventional Esaki tunnel diode. Properly plated contacts eliminate the negative resistance feature associated with the interfacial barrier and replace it with a plateau in the forward current.[30] Furthermore, evaporated Pb diodes show only a 50% change in conductance near the energy gap of Pb at 1.4 K; electroplated

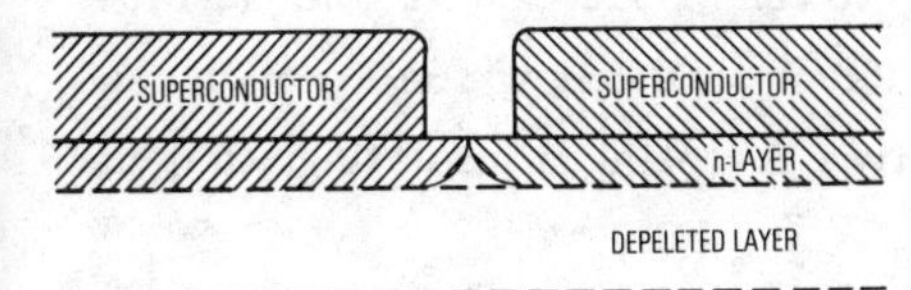

Fig. 12 Schematic diagram of the cross-section of a planar coupled Josephson junction on p-InAs.

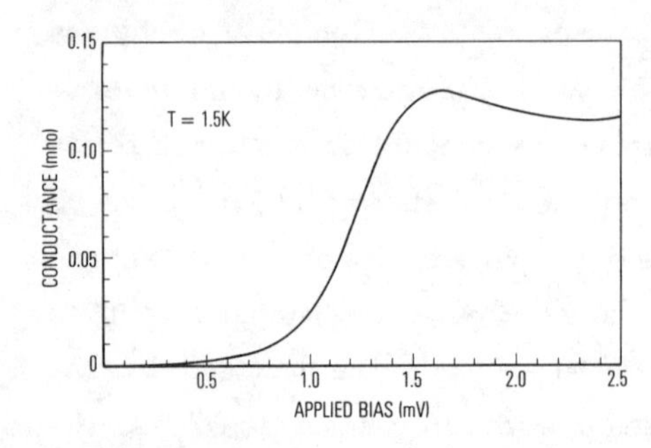

Fig. 13 Conductance of electroplated Pb contacts to heavily doped p-InAs.

contacts show an excellent "super-Schottky" characteristic (Fig. 13). It is therefore clearly established that the n surface does not display superconductivity of the full energy gap under an evaporated Pb contact. (There may be no superconductivity!) However Fig. 13 clearly shows that the n surface under an electroplated Pb contact has the same superconducting energy gap as Pb. We can therefore establish a superconducting contact to the inverted surface channel.

The nature of the free surface of p-InAs and a unique method of control has been demonstrated. This is important for the proposed device since the "controlled" conduction takes place along the "free" surface. "Free" implies only that no electrical contact is made directly on this surface. Earlier studies of the free-surface properties of p-InAs were inconsistent; our initial measurements clearly showed spatial non-uniformity and strong dependence on sample history and aging. We have been able to develop a consistent, qualitative model of the surface electron energy levels and their populations.[28] The etched surface of p-InAs is n-type as stated above, i.e., the surface Fermi level is above the conduction band. However, for moderately doped ($\gtrsim 2 \times 10^{16}/cm^3$) p-InAs the surface channel is non-degenerate and the electron levels are spatially quantized. In our material the surface channel is so thin that the lowest bound state is above the Fermi level and therefore vacant at LHe temperatures. The corresponding energy level diagram is shown in Fig. 11, where E_o and E_1 represent lowest quantized energy bands. However, for lower doped p-InAs the surface channel width increases and the surface levels fall below the Fermi level. Therefore, the free surface of heavily p-type material is insulating; that of light p-type is conducting.

The conductivity of the surface channel can be controlled by bulk doping, substrate voltage, light, and argon ion bombardment. The last mechanism has been studied in detail for 140 V ions and is an effective means of surface tailoring.[31] Almost uniquely, the surface of InAs becomes

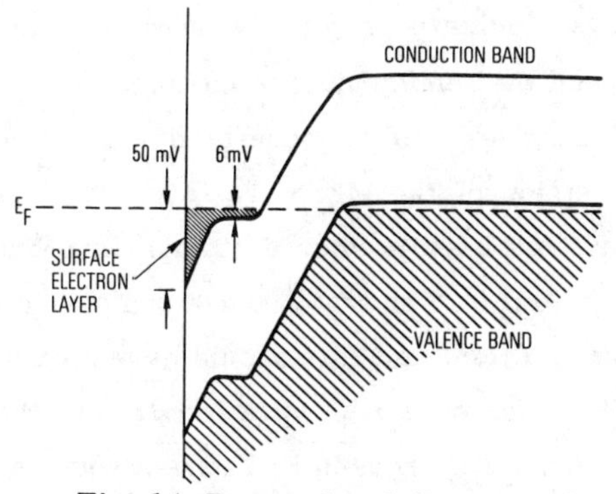

Fig. 14 Energy level diagram of argon ion bombarded p-InAs

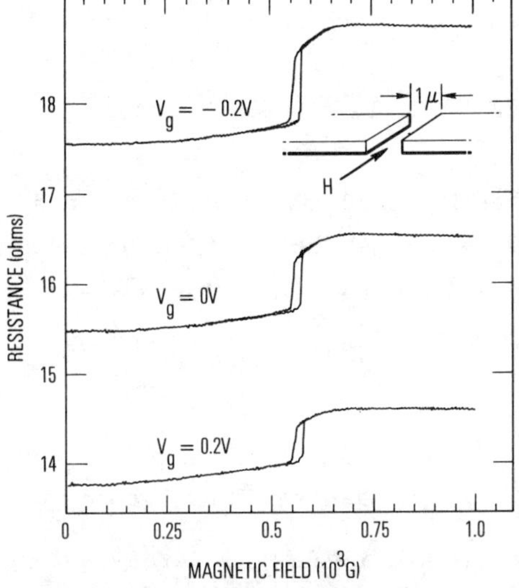

Fig 15 The effect of magnetic field H and substrate voltage V_G on the resistance between two closely spaced electroplated Pb contacts on argon ion bombarded p-InAs (2 x $10^{16}/cm^3$ bulk hole concentration) at 4.2 K.

more conducting (n-type) with such bombardment. From studies of magnetoresistance, Shubnikov-de Haas oscillations, and Hall effect, Flesner, et al.[32] have determined the mobility, surface electron density, and channel width as a function of bulk doping, argon flux, and substrate voltage. Channel width can be controlled from several hundred A to 1.5 μm with mobilities reaching 3.3 x 10^4 cm^2/ V-S for the wider channels. Electron volume densities appear to be 2 x $10^{16}/cm^3$, with the surface density reaching 2 x $10^{12}/cm^2$. Fig. 14 shows the deduced band diagram for argon-bombarded surfaces, although the detailed physical processes which lead to its formation are not understood.

The surface resistance of closely spaced superconducting contacts on p-InAs (Fig. 12) are being investigated[32] for micron-dimension spacing. Fig. 15 shows the resistance between 1 μm- spaced Pb contacts 100 μm wide on argon bombarded 2 X $10^{16}/cm^3$ p-InAs. The surface channel is $\approx$ 1 μm deep and the substrate is insulated from the surface by the depletion layer. The three curves represent different bias voltages applied to the substrate. The magnetic field dependence was measured for H parallel to the surface but perpendicular to the current flow as shown in the inset. In addition to the large change at H_c(Pb), there are significant resistance changes when the Pb is both superconducting and normal. Furthermore, all these variations depend on substrate bias. These experiments indicate strong superconducting interactions between the Pb electrodes and the semiconductor surface.

The superconducting InAs FET may lead to two useful devices, the Josephson FET and a superconducting version of the MESFET. Properties of the planar-coupled Josephson device have not been predicted at this time, but one expects gate controlled performance well into the microwave region. An InAs MESFET with superconducting source, drain, and gate should have a gain-bandwidth product approaching 150 GHz based on 1 μm gate length and the increased mobility measured. Since most noise sources in FETs are associated with electron temperatures and parasitic resistances, we can hope to achieve much lower noise figures for these superconductor-semiconductor FET devices than have been reported for GaAs FETs.

ACKNOWLEDGEMENTS

We wish to acknowledge W. Garber, R. J. Pedersen, J. R. Tucker, and F. L. Vernon, Jr. of Aerospace and L. Flesner and H. Suhl of UCSD for helpful discussions and the use of their data and analyses. We also acknowledge experimental support from M. F. Bottjer, R. Gowin, R. Robertson, and V. Schnepp.

REFERENCES

1. I. Giaever, Phys. Rev. Lett. 5, 147 (1960); Phys. Rev. Lett. 5, 464 (1960).
2. B. D. Josephson, Phys. Lett. 1, 251 (1962); Advan. Phys. 14, 419 (1965).
3. M. McColl, M. F. Millea, and A. H. Silver, Appl. Phys. Lett. 23, 263 (1973).
4. M. McColl, R. J. Pedersen, M. F. Bottjer, M. F. Millea, A. H. Silver, and F. L. Vernon, Jr., Appl. Phys. Lett. 28, 159 (1976).
5. M. McColl, M. F. Millea, A. H. Silver, M. F. Bottjer, R. J. Pedersen, and F. L. Vernon, Jr., IEEE Trans. Magn. MAG-13, 221 (1977).
6. F. L. Vernon, Jr., M. F. Millea, M. F. Bottjer, A. H. Silver, R. J. Pedersen, and M. McColl, IEEE Trans. Microwave Theory Tech. MTT-25, 286 (1977).
7. F. A. Padovani and G. G. Sumner, J. Appl. Phys. 36, 3744 (1965).
8. R. Stratton, J. Phys. Chem. Solids 23, 1177 (1962).
9. F. A. Padovani and R. Stratton, Solid-State Electron. 9, 695 (1966).
10. F. A. Padovani, J. App. Phys. 37, 921 (1966).
11. T. J. Viola, Jr. and R. J. Mattauch, Proc. IEEE 61, 393 (1973).
12. C. L. Huang and T. Van Duzer, Appl. Phys. Lett. 25, 753 (1974).

13. C. L. Huang and T. Van Duzer, IEEE Trans. Electron. Devices ED-23, 579 (1976).
14. M. McColl, W. A. Garber, and M. F. Millea, Proc. IEEE 60, 1446 (1972).
15. D. T. Hodges and M. McColl, Appl. Phys. Lett. 30, 5 (1977).
16. M. McColl, A. B. Chase, and W. Garber, J. Appl. Phys. (in press)
17. M. McColl, D. T. Hodges, and W. A. Garber, IEEE Trans. Microwave Theory Tech. MTT-25, 463 (1977).
18. M. McColl and M. F. Millea, J. Electronic Materials 5, 191 (1976).
19. J. R. Tucker and M. F. Millea, to be published.
20. I. Giaever, Phys. Rev. Lett. 20, 1286 (1968).
21. G. Lubberts and S. Shapiro, J. Appl. Phys. 43, 3958 (1972).
22. M. Mehbod, W. Thijs, and Y. Bruynseraede, Phys. Stat. Sol. a32, 203 (1975).
23. I. Giaever and H. R. Zeller, J. Vac. Sci. Technol. 6, 502 (1969).
24. J. Seto and T. Van Duzer, Appl. Phys. Lett. 19, 488 (1971).
25. T. Tsuboi, Phys. Lett. 56A, 472 (1976).
26. P. Cardinne, B. Manhes, and M. Renard, Proc. 1972 Appl. Supercond. Conf., IEEE Pub. No. 72CH0682-5-TABSC, 565 (1972).
27. E. L. Hu, L. D. Jackel, and R. W.Epworth, Bull. Am. Phys. Soc. 23, 94 (1978).
28. M. F. Millea and A. H. Silver, J. Vac. Sci. Tech., (in press) (1978).
29. M. F. Millea, M. McColl, and A. H. Silver, J. Electronic Materials 5, 321 (1976).
30. M. F. Millea and A. H. Silver, Bull. Amer. Phys. Soc. 22, 407 (1977).
31. M. F. Millea, A. H. Silver, and L. D. Flesner, Thin Solid Films (to be published).
32. L. D. Flesner, A. H. Silver, H. Suhl, and M. F. Millea, (to be published).

JOSEPHSON TUNNELING THROUGH Ge-Sn BARRIERS

E. L. Hu, L. D. Jackel, and R. W. Epworth
Bell Laboratories, Holmdel, N. J. 07733

ABSTRACT

Josephson tunneling has been observed through thick (∿ 600Å) barriers of coevaporated Ge-Sn mixtures. The barrier transparency can be adjusted by altering the ratio of the mixture components: for a given barrier thickness critical current densities have been varied over six orders of magnitude. In addition the critical current density can be controlled by varying barrier thickness. For thick barriers, or barriers with low Sn content, high tunneling resistances and no supercurrent result. Barriers of intermediate thickness or moderate Sn content yield junctions with electrical properties very similar to conventional hysteretic oxide-barrier junctions. Thin or high Sn-content barriers give junctions with non-hysteretic I-V characteristics similar to those of a superconducting microbridge.

INTRODUCTION

Josephson tunnel junctions surpass other weakly-coupled superconducting systems in the ease with which they are reproducibly fabricated and in their high RI_c products, the normal state resistance times the critical current. However, there have been two major obstacles to wider applications of tunnel junctions, both resulting from the thin metal oxides (20Å to 50Å thick) used to form the tunneling barriers. First, the oxides are often mechanically fragile. Second, the sandwich geometry junctions, in which the thin barrier separates the junction electrodes, are intrinsically high in parasitic capacitance. This capacitance causes hysteresis in the junction I-V characteristics, complicating Josephson junction logic circuit design and making the junctions generally unsuitable for magnetometry and radiation detection. Such considerations motivate research into artificial barriers which may be made thicker than oxide barriers, yet allow comparable tunneling transmission. Ge-Sn barriers of several hundred angstroms thickness have been fabricated; variation of the Ge-Sn relative composition as well as of film thickness provides the degrees of freedom necessary to span the range of useful weak-link characteristics. For example, tunnel junctions with Ge-Sn barriers can be fabricated with non-hysteretic I-V characteristics similar to those of microbridges, but with much higher normal state resistances.

HYSTERESIS OF I-V CHARACTERISTICS

A distinguishing feature in a superconducting weak link I-V characteristic is the amount of hysteresis. For tunnel junctions, hysteresis is determined by the parameter β_C:

ISSN: 0094-243X/78/380/$1.50

$$\beta_C = 2\pi I_c R^2 C/\phi_o \qquad (1)$$

where I_c is the critical current, R is the normal state resistance, C is the junction capacitance, and ϕ_o is the flux quantum (2.07×10^{-15} Wb). Hysteresis occurs for $\beta_c \gtrsim 0.7$. In a tunnel junction, $RI_c \sim 1.5\Delta$, where Δ is the superconducting energy gap of the electrode. For tin, $\Delta \sim 0.6$ mV, so that (1) becomes

$$\beta_c = 2.5 \times 10^9 \text{ (Amps/Farad) } C/I_c \qquad (2)$$

Thus the amount of hysteresis is determined by the ratio of capacitance per unit area to the critical current density, j_c. In an oxide-barrier junction, the capacitance per unit area is $\sim 10^{-5}$ Farad/cm^2, requiring $j_c \sim 3 \times 10^4$ A/cm^2 in order to produce a non-hysteretic I-V characteristic. With such a high critical current density, it is not possible to obtain a normal resistance greater than $\sim$ 1 ohm, or critical currents less than 1 mA, using conventional photolithography to define the junction area. In contrast, Ge-Sn barrier junctions have been fabricated with barrier layers $\sim$ 30 times thicker than high current density oxide barriers. The reduction in capacitance per unit area allows non-hysteretic I-V's to be observed for critical current densities of $\sim 10^3$ A/cm^2. By varying the critical current density one can obtain a variety of I-V characteristics. The critical current density of the Ge-Sn barrier can be controlled by varying the barrier thickness or by changing the relative concentrations of Sn and Ge. To achieve this variation, the techniques of Molecular Beam Epitaxy are employed in the junction fabrication.

FABRICATION

The junctions are sandwich structures of superconducting electrodes separated by a Ge-Sn barrier. A (typically) Sn base electrode pattern, 2000 Å thick, is evaporated through a photoresist mask onto a sapphire substrate. The Dolan[2]-Dunkleberger[3] oblique evaporation technique is used to give the Sn film a tapered edge: evaporation is made through an undercut mask onto a rotating substrate. The sample is placed into the MBE apparatus and the electrode surfaces are sputtered clean. The barrier film is co-evaporated from separate Sn and Ge ovens at a combined growth rate of $\sim$ 100 Å/minute. Evaporations have also been made using a single oven containing both Ge and Sn in particular proportions. The background vacuum maintained during the growth of the barrier is typically 10^{-7} Torr. The sample is then removed from the MBE chamber, replaced into a conventional evaporator, and a counter electrode of Sn or Pb (2000Å thick) is deposited through a metal mask onto the underlying film.

EXPERIMENTAL JUNCTION I-V CHARACTERISTICS

The effect of variations in barrier composition is shown in

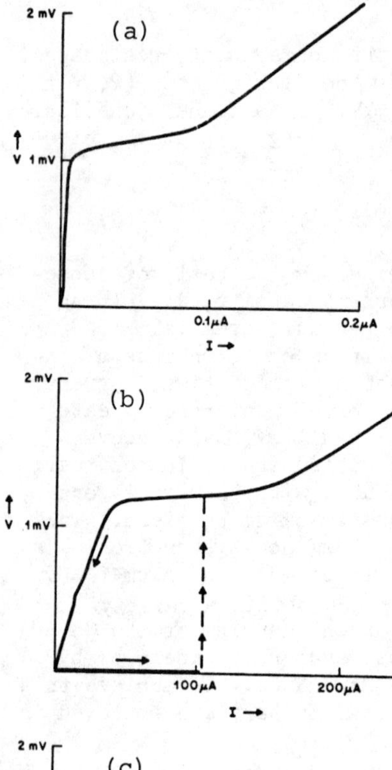

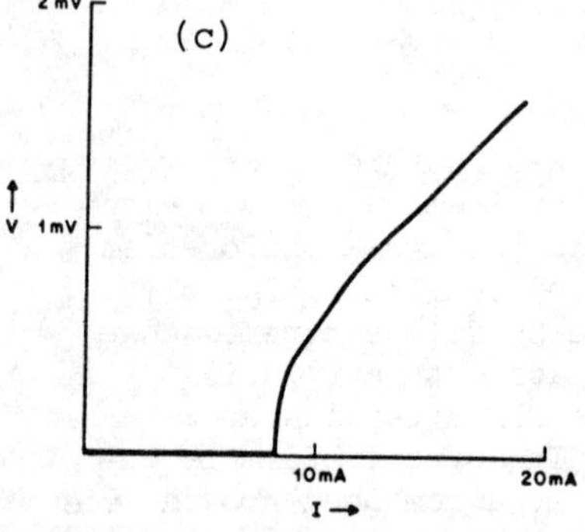

Fig. 1. The effect of barrier composition on junction I-V characteristics. In a) the Sn/Ge ratio is 1:4, in b) it is 1:2 and in c) it is 1:1. In all three samples the barrier thickness was ∿ 600 Å. The junction areas for a) and b) were 10^{-4} cm^2, while that of c) was 10^{-5} cm^2. All electrodes were Sn.

Figure 1. The junction of Figure 1a has a barrier with a Sn to Ge composition of ratio 1:4. The normal tunneling resistance is so high that the expected critical current of 0.1 microamps, corresponding to $j_c \sim 10^{-3}$ A/cm^2, is suppressed by thermal fluctuations. For the junction of Figure 1b, the proportion of Sn in the barrier film was increased to obtain a Sn to Ge ratio of 1:2. The I-V shows standard oxide-barrier characteristics, with a critical current density of ∿ 1 A/cm^2. For Figure 1c, the proportion of Sn to Ge was increased still further to a value of 1:1. The critical current density of 10^3 A/cm^2 is sufficiently high that the critical current is slightly suppressed because the junction size is now $\gtrsim \lambda_J$, the Josephson penetration depth. The I-V characteristic of this junction is very similar to that of a variable thickness microbridge[4]. The range of I-V's shown in Figure 1 can also be obtained by a variation of the barrier thickness for a fixed Ge:Sn ratio.

Figure 2 describes a junction with a Pb base electrode and Sn counter electrode, with a Ge:Sn barrier of composition Sn:Ge = 2.3. The thickness of the barrier material here is ∿ 400 Å. The use of the Pb base electrode has apparently reduced the amount of leakage current observed.

The non-hysteretic junctions should be a good choice for microwave and millimeter wave radiation detectors. Microwave-induced steps have been observed up to 1.5 mV with 10 GHz radiation, indicating that heating is not a problem in these junctions. Steps were also observed with 100 GHz radiation. A typical rf response curve is shown in Figure 3.

CONCLUSION

We have employed a new technique

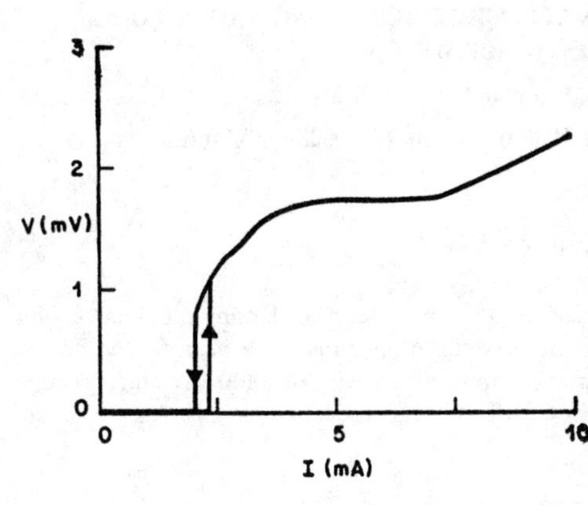

Fig. 2. I-V characteristic for junction having Pb base electrode and Sn counter-electrode. Barrier composition of Sn/Ge = 2:3, 400 Å thick. Junction area was ∿ 2.6 x 10^{-6} cm^2.

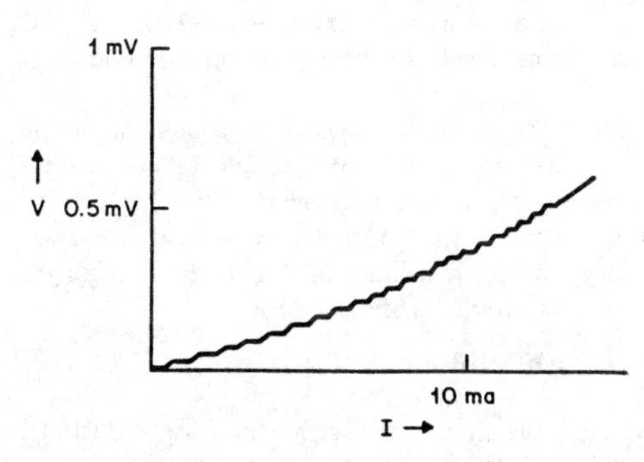

Fig. 3. The effect of 10 GHz radiation upon a junction with the same barrier composition as the one shown in Figure 1c.

of fabricating Josephson tunnel junctions where we can tailor the form of the junction I-V characteristic by varying the composition and thickness of the barrier material. Their electrical characteristics and ease of fabrication make these junctions natural choices for use in non-latching Josephson junction logic circuits[5]. Because of their high resistance, these junctions should also allow nearly two orders of magnitude improvement in the energy resolution of d.c. SQUID magnetometers[6].

REFERENCES

1. D. E. McCumber, J. Appl. Phys., 39, 3113 (1968) and W. C. Stewart, Appl. Phys. Lett., 12, 277 (1968).
2. G. J. Dolan, Appl. Phys. Lett., 31, 337 (1977).
3. L. N. Dunkleberger, to be published, J. Vac. Sc. Tech., Jan-Feb (1978).
4. M. Octavio, W. J. Skocpol and M. Tinkham, IEEE Trans. on Magn., MAG-13, 739 (1977).
5. Ge-Sn Barrier junctions with moderate hysteresis could be used in the non-latching schemes of H. W. K. Chan and T. Van Duzer, IEEE J. Solid-State Circuits, SC-12, 73 (1977), while non- hysteretic junctions could be used in non-latching interferometer logic, H. H. Zappe, IEEE MAG-13, 41 (1977).
6. Claudia D. Tesche and John Clarke, IEEE Trans. on Magn., MAG-13, 41 (1977).

SUPER-SCHOTTKY AND JOSEPHSON-EFFECT DEVICES USING NIOBIUM ON THIN SILICON MEMBRANES

Lynette B. Roth, John A. Roth, and Paul M. Schwartz

Hughes Research Laboratories, 3011 Malibu Canyon Rd., Malibu, California

ABSTRACT

Super-Schottky and Josephson-effect devices have been fabricated using niobium on thin silicon membranes; and their dc (I, V) characteristics have been measured. The rugged, reproducible Nb/Si super-Schottky devices exhibit at 1.9K a non-optimized sensitivity of 3800 V^{-1} corresponding to a noise temperature of 3K. The transition from Giaever to Josephson tunneling is observed as the silicon barrier thickness is reduced.

INTRODUCTION

Recent work[1,2,3,4] has indicated that superconductor-semiconductor devices may achieve a level of performance in high frequency mixer and logic applications competitive with conventional superconducting devices, e.g. oxide barrier Josephson junctions. Specifically a mixer input noise temperature of 6K at X-band has established the Pb on GaAs super-Schottky as the most sensitive microwave detector[1,2]. Super-Schottky devices have been made using thin silicon membranes also[3]. Thinning of the semiconductor should reduce the parasitic series resistance hence should increase the high frequency cut-off. Josephson tunneling has been observed in superconductor-semiconductor devices with thinned, single crystal silicon[4] or evaporated tellurium barriers[5]. Semiconductor barrier thicknesses lie in the range of 100 Å to 1000 Å while those of oxide barriers are on the order of 10 Å. The thicker semiconductor barriers should exhibit better thermal cycling behavior and more reproducible device characteristics.

We have fabricated both super-Schottky and Josephson-effect devices using niobium on thin silicon membranes. Niobium has been chosen over soft superconductors, i.e., Pb, for its superior chemical and mechanical stability as well as its somewhat larger critical temperatures (T_c) and superconducting gap. In addition results from these devices may indicate the feasibility of our next generation devices using niobium-based, high T_c superconductors. In the remainder of this article we highlight fabrication techniques and describe the current-voltage characteristics of these devices.

EXPERIMENTAL PROCEDURE

We have fabricated our first superconducting devices in a vertical geometry. Fig. 1a illustrates this structure: A small area, ultra-thin silicon membrane sandwiched between a superconducting niobium top contact and a normal metal, e.g., Al, or superconducting back contact for the super-Schottky or the Josephson effect devices, respectively. We have adopted the double etch pit method, developed by C.L. Huang and T. Van Duzer[3] to fabricate the thin silicon membrances (Fig. 1b and 1c). This technique has produced high yields (95%) of thin membranes, viz, 1000 Å to 1800 Å thick. These membranes

ISSN: 0094-243X/78/384/$1.50

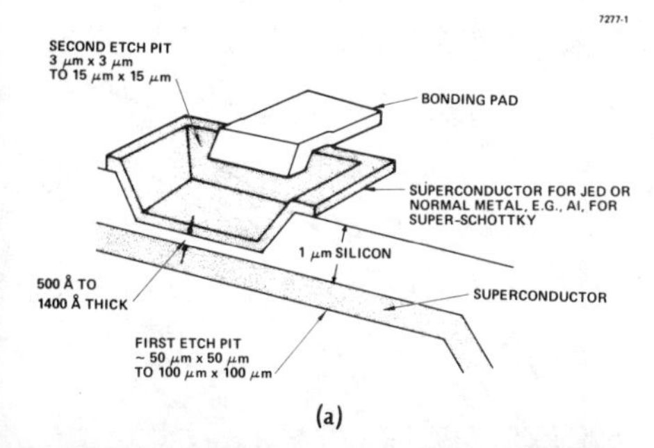

(a)

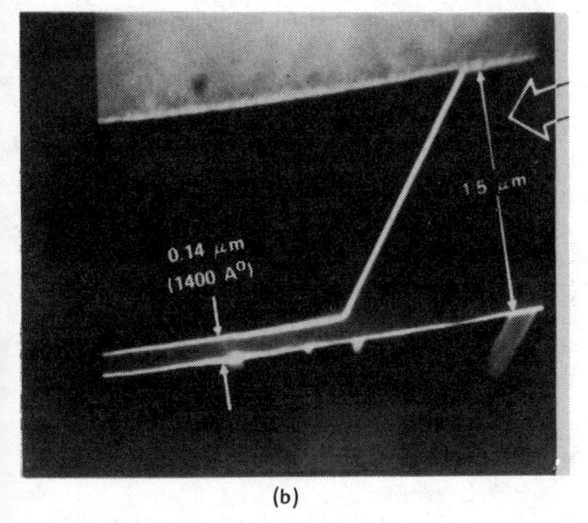

(b)

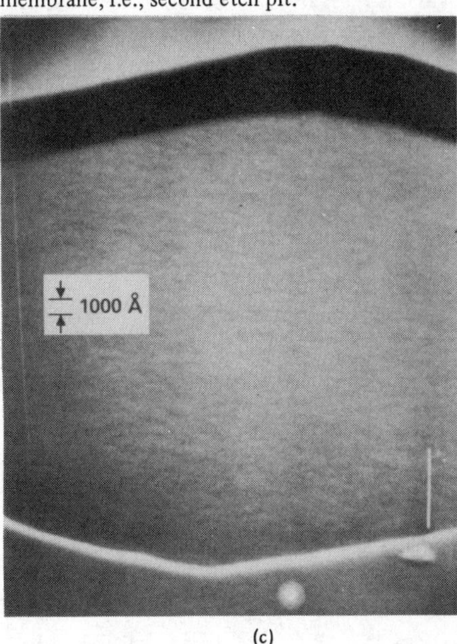

(c)

Fig. 1: Structure of superconducting devices: (a) complete structure of super-Schottky or Josephson–effect device; (b) cross section of 1400 Å silicon membrane; (c) surface of silicon membrane, i.e., second etch pit.

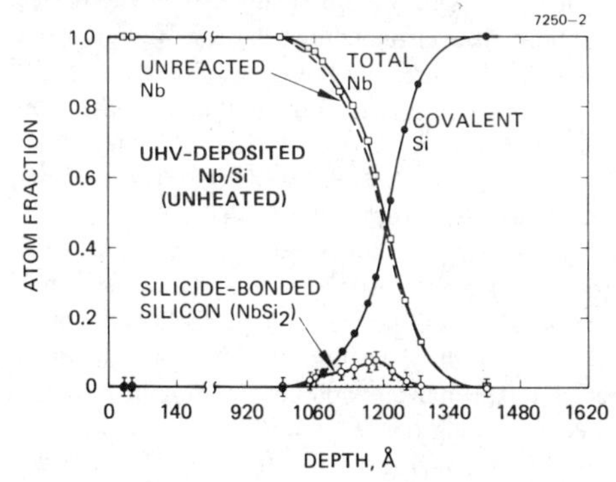

Fig. 2: Depth variation of reacted and un-reacted atomic species in a Nb film deposited in UHV onto sputter-cleaned and annealed Si.

show remarkable uniformity over large areas, e.g., less than a 100 Å difference in thickness between devices in a 250 mil x 500 mil area. Thickness variations within an individual membrane are typically 50 Å.

Surface contaminants are removed from the silicon membrane by sputter cleaning and annealing, in-situ, prior to electron beam deposition of niobium in ultra-high vacuum (10^{-11} torr base pressure). The surfaces are Auger-analyzed for contaminants, i.e., O, C, Ar, and examined for crystallinity by low energy electron diffraction (LEED). We find none of those contaminants above the limits of detectability of our instruments, e.g., 0.1% for oxygen, and observe a fair LEED pattern after annealing. Preliminary work on Nb/Si super-Schottky devices fabricated on chemically cleaned silicon demonstrated a nearly linear (I, V) for these devices, suggesting the presence of a gapless layer at the Nb/Si interface. This result provided the motivation for employing the sputter cleaning and annealing technique.

We have performed Auger spectroscopy depth profiles to determine the nature of the Nb/Si interface. Figure 2 shows the depth variation of unreacted (covalent) Si, reacted (silicided) Si, and total Nb. These curves were obtained by analyzing peak shapes as well as amplitudes[6] in order to separate the atom fractions into reacted and unreacted species.

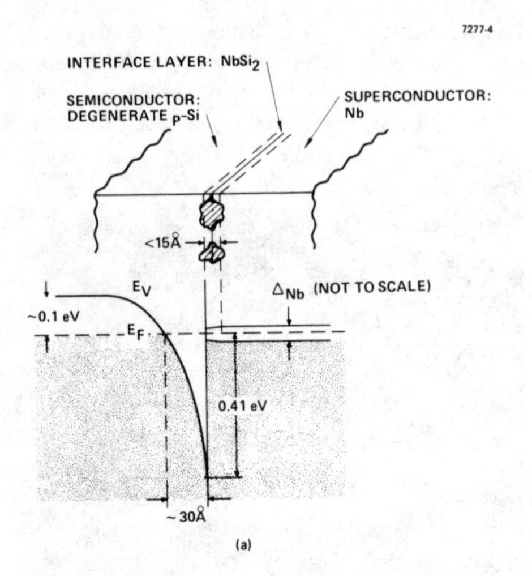

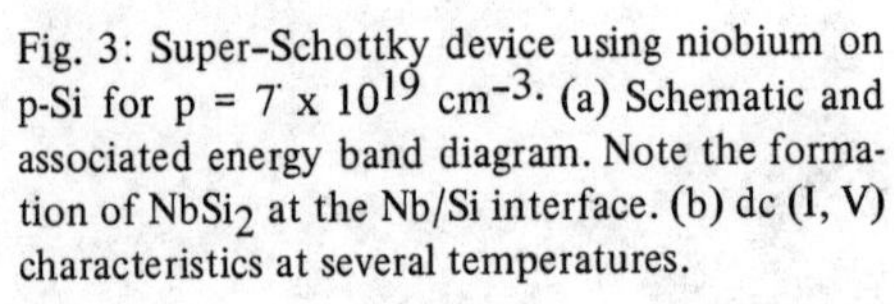

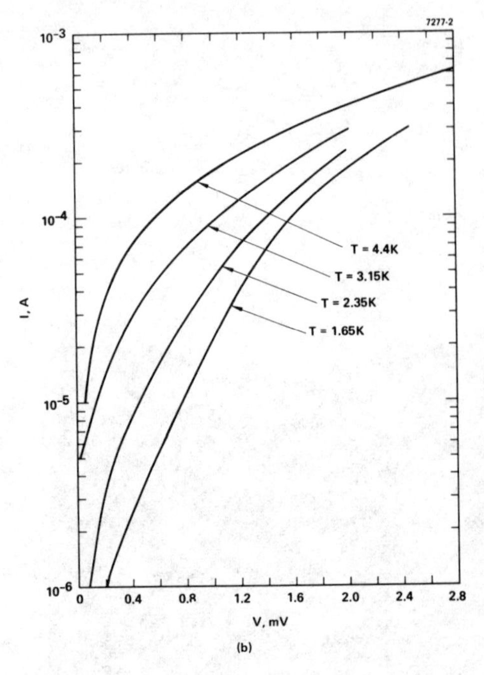

Fig. 3: Super-Schottky device using niobium on p-Si for $p = 7 \times 10^{19}\ cm^{-3}$. (a) Schematic and associated energy band diagram. Note the formation of $NbSi_2$ at the Nb/Si interface. (b) dc (I, V) characteristics at several temperatures.

RESULTS AND DISCUSSION

Figure 3b shows a set of (I, V) characteristics of a typical Nb/p-Si super-Schottky device. These devices, similar to Pb/p-GaAs[1,2] and Pb/p-Si[3], exhibit Giaever tunneling behavior. We have fitted the measured (I, V) of our devices to an expression for a circuit containing a resistance R_S in series with the super-Schottky shunted by a second resistor R_H. The tunneling current through the super-Schottky is given by $I = I_S \sinh (S V)$[2,7] where I_S is the saturation current and $S = q/kT$. The following parameters gave a best-fit to one of our devices operating at $T_a = 1.9K$: $S = 3800V^{-1}$; $I_S = 1.4 \times 10^{-5}A$; $R_H = 140\ \Omega$; and $R_S = 0.9\ \Omega$. This sensitivity corresponds to a temperature $T_d = q/kT$ of 3K.

The doping profile of the silicon membrane for this device, as all devices reported in this paper, is shown Fig. 4a. The dopant level increases monotonically from $7 \times 10^{19}\ cm^{-3}$ to $1.5 \times 10^{20}\ cm^{-3}$ from one surface to the other of the membrane. We are seeking to optimize the super-Schottky performance by tailoring the doping profile. For example, we can vary the hole concentration at the dominant Schottky-barrier, thus optimizing junction resistance and capacitance, while maintaining high dopant levels throughout the remainder of the membrane, hence keeping the series resistance to a low value.

Device performance may be improved further by minimizing the amount of $NbSi_2$ formed at the interface e.g., by cooling the Si substrates during Nb deposition. Our analysis of the Auger depth profiles, shown in Fig. 2, indicates a non-uniform distribution of interfacial silicide. Figure 3a depicts our model of the Nb/Si interface: pockets of $NbSi_2$ randomly nucleated along the boundary and large regions where niobium makes direct contact with the silicon. It has been proposed that metal silicides initially form in an amorphous layer[8]. If interfacial $NbSi_2$ is likewise amorphous, it may resemble amorphous niobium-silicon mixtures which superconduct at 2K[9]. In this case niobium can enhance the superconducting gap in the silicide by the proximity effect thereby reducing the effect of the interfacial layer characteristics.

Fig. 4 shows the appearance of pair tunneling current as the silicon membrane thickness decreases. The device structure, illustrated in Fig. 4a, uses niobium on both sides of the membrane. We have observed that the use of niobium rather than aluminum as a back contact has little effect on our super-Schottky characteristics. The Nb/Si/Nb devices with a 1800 Å thick membrane displays a typical

Giaever tunneling characteristic (Fig. 4b). Pair tunneling is first observed in 1700 Å membranes as an extra current at low voltages (Fig. 4b). The pair current contribution increases rapidly at low temperatures; note that the thinner membrane exhibits a stronger non-linearity, i.e., Giaever tunneling behavior, at 4.4K. Fig. 4c shows that the (I, V) characteristic of a 1400 Å membrane is dominated by pair tunneling. The current increases at all voltages as the temperature decreases, in contrast to Giaever tunneling behavior. For this silicon barrier thickness and doping level, the pair coupling through the junction is weak, hence circuit parasitics and noise strongly influence the shape of the (I, V) characteristic[10].

Josephson tunneling current for a 1000 Å thick silicon barrier device is shown in Fig. 5. The arrows along the current scale indicate the zero-voltage current. We have observed weak, periodic dependence of the supercurrent with magnetic field where the period is in the neighborhood of one flux quantum. Screening of the magnetic field in the niobium sheet covering our device array or wide junction behavior may have reduced the modulation amplitude. The best I_C R product, measured for these devices, is $I_C R = 2.25$ mV at T = 1.9K where the normal state resistance R = 0.9 Ω and the critical current $I_C = 2.5$ mA. This value is 90% of the theoretical value for niobium tunnel junctions at low temperatures.

Clearly the (I, V) characteristic at finite voltage is complex. Currently we are studying the (I, V) characteristics as a function of doping density and membrane thickness in order to understand the behavior of these silicon barrier Josephson devices. We have observed that the Josephson coupling significantly decreases with the degree of degeneracy in the silicon. Specifically, Si barriers with doping level of 3×10^{19} cm^{-3} at one Nb/Si interface exhibit weak coupling through 800 Å barriers similar to that seen for more heavily doped 1400 Å membranes.

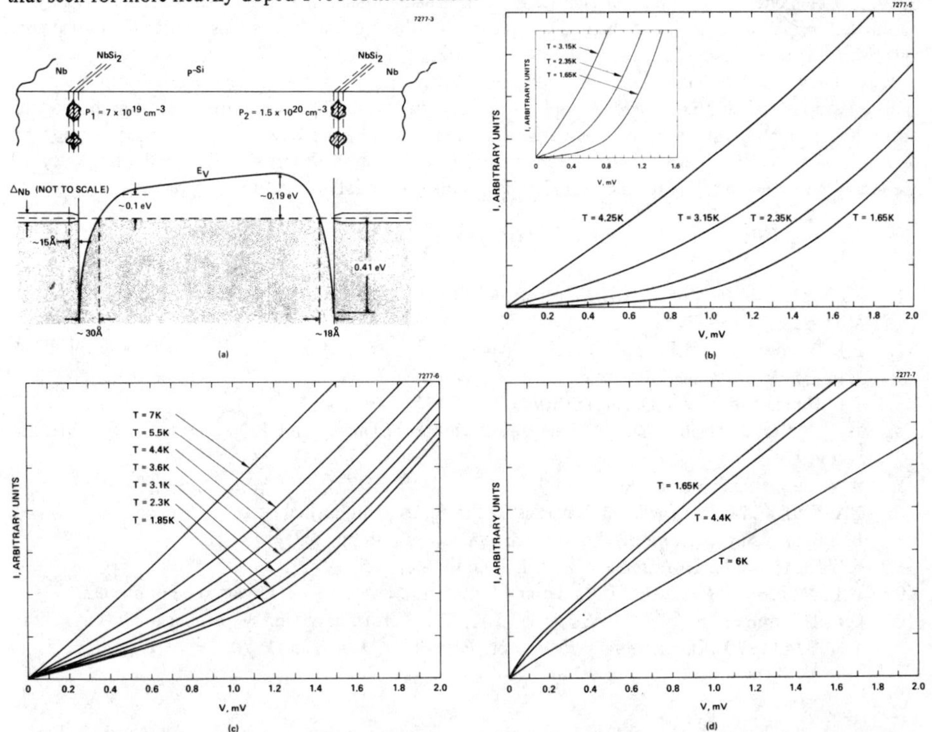

Fig. 4: dc (I, V) characteristics of the Nb/Si/Nb superconducting device, shown in (a), for the following sequence of membrane thicknesses: (b) 1800 Å, (c) 1600 Å and (d) 1400 Å. This sequence along with Fig. 5 shows the transition from Giaever to Josephson tunneling for silicon barrier devices with the indicated dopant profile.

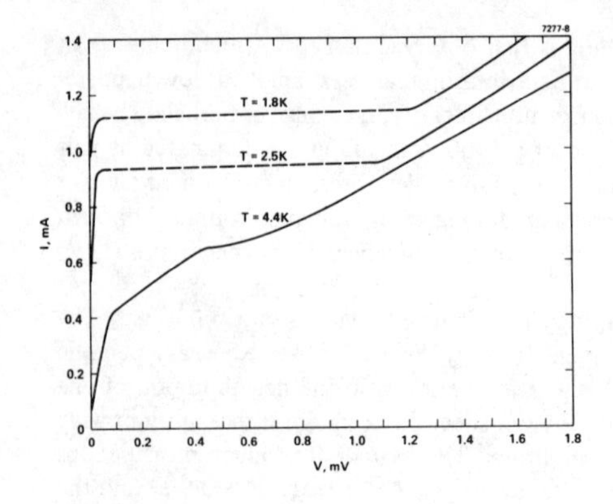

Fig. 5: (I, V) characteristic of Nb/Si/Nb device, shown in Fig. 4a, with membrane approximately 1000 Å thick. Zero-voltage current indicated by arrows along current scale.

CONCLUSIONS

We have demonstrated both super-Schottky and Josephson-effect devices using niobium on ultra-thin silicon. We have found the super-Schottky devices to be rugged and reliable: its non-linear (I, V) characteristic reproducible after many thermal cycles. Assuming the actual rf characteristics follow those predicted from the dc characteristics, the Nb/Si super-Schottky device appears suitable for mixer elements in low-noise receivers. Our results also indicate that super-Schottky diodes may be fabricated using niobium-based high T_c superconductors for use in systems constrained to higher operating temperatures.

On the other hand the semiconductor barrier Josephson-effect device requires more study before its potential in mixer or logic applications can be assessed. We have produced rugged and uniform single crystal, silicon barriers with high yields: and we have developed techniques to control Josephson device characteristics (to be reported in a future publication). In view of the large pay-off we feel it is worth the effort to understand and optimize this relatively complex device.

REFERENCES

1. M. McColl, M.F. Millea, A.H. Silver, M.F. Bottjer, R.J. Pedersen, and F.L. Vernon, Jr., IEEE *MAG-13,* 221 (1977).
2. F.L. Vernon, Jr., M.F. Millea, M.F. Bottjer, A.H. Silver, R.J. Pedersen, and M. McColl, IEEE *MTT-25,* 286 (1977).
3. C.L. Huang and T. Van Duzer, IEEE *ED-23,* 579 (1976).
4. M. Schyfter, J. Maah-Sango, N. Raley, R. Ruby, B.T. Ulrich, and T. Van Duzer, IEEE *MAG-13,* 862 (1977).
5. W.Y. Lum and T. Van Duzer, J. Appl. Phys. *48,* 1693 (1977).
6. J.A. Roth and C.R. Crowell, J. Vac. Sci. Technol., to be published.
7. L. Solymar, *Superconductive Tunneling Applications,* 38 pp (Wiley, 1972).
8. R.W. Bene and R.M. Walser, J. Vac. Sci. Technol. *14,* 925 (1977).
9. W.L. Johnson, T.C. Tsuei, S.I. Raider and R.L. Laibowitz, J. Appl. Phys., to be published.
10. D.E. McCumber, J. Appl. Phys. *39,* 3113 (1968); J. Kurkijarvi and V. Ambegaokar, Phys. Lett. *31A,* 314 (1970); H. Kanter and F. Vernon, Phys. Lett. *32A,* 155 (1970).

IMPROVED MATERIALS FOR SUPERCONDUCTING ELECTRONICS

M. R. Beasley
Departments of Applied Physics and Electrical Engineering
Stanford University
Stanford, California 94305

ABSTRACT

The potential advantages of hard and/or high transition temperature superconductors in superconducting electronics are discussed and recent progress in fabricating actual devices from these materials is reviewed.

I. INTRODUCTION

In any emerging technology there are always seemingly countless materials problems. Indeed it is frequently just these materials problems that ultimately determine the success or failure of a given device or technology in practice, never mind the promise in principle. At the same time, certain classes of materials often have general advantages that make their exploitation particularly attractive, never mind the materials problems. This would appear to be the case in superconducting electronics where to date we have not used the best superconducting materials available. In this paper the present status of superconducting electronic technology is reviewed and those general opportunities and problems that might be exploited or solved by use of improved superconducting materials are identified. We shall see that the interesting materials are the so-called hard (i.e. physically and chemically robust) and the high transition temperature (high-T_c) superconductors. Although the advantages of these materials are easily recognized, it has long been felt - and borne out by much experience in the past - that their use was too difficult. Recent work on these materials and devices fabricated from them indicate, however, that this may no longer be the case. Indeed, the initial successes have been quite encouraging.

II. PRESENT STATUS OF SUPERCONDUCTING ELECTRONIC TECHNOLOGY

To better define what improvements are desirable in superconducting electronics let us briefly review the present status of the technology. Basically there are two extremes. On the one hand there is the integrated, thin-film circuit technology based on Pb-alloy tunnel junctions developed by IBM. On the other hand there are a variety of commercially available discrete junction devices based on Nb or Nb-alloy point contacts. (Much research has also been carried out on thin-film weak-link Josephson junction devices, but they have yet to find widespread application.) Thus, where the technology is quite advanced, it is based on mechanically soft

SSN: 0094-243X/78/389/$1.50

materials, the failure mechanism of which under thermal cycling is known to be related to their soft nature. By contrast, where hard materials have been used, the technology is basically at a cat's whisker level, which is not to deny some very impressive electrical performance. In any event, neither of these technologies incorporate the high-T_c materials, which have the best potential performance. Clearly what is desired is an all thin film, mechanically hard, and, if possible, high-T_c circuit technology.

The impact of such a technology would be substantial. The improved materials would in all probability allow: (1) robust, long-lived, thermally cyclable circuits; (2) high temperature (10-20K) operation (at some penalty in reduced device performance) using small, cheap, closed-cycle cryogenic refrigerators; and (3) the ultimate performance (speed, sensitivity, etc.) at low temperatures. Any of these could materially improve the prospects for widespread application of superconducting electronic devices.

III. AVAILABLE MATERIALS

In Table I we show a list of superconducting materials commonly considered of practical interest for superconducting electronics, along with some of their properties and methods of fabrication in thin film form. In brief the table shows that the high-T_c materials are also hard materials, that the energy gap (which is an overall good figure of merit for the material in applications) increases with T_c as expected from BCS theory ($\Delta(0) \simeq 1.78\ k_B T_c$), and that their fabrication in thin film form is admittedly more difficult than for the low-T_c materials.

Because the high-T_c materials are harder to work with than the more popular Sn, In, and Pb, or even the hard superconductor Nb, they have received less attention by the superconducting device and device physics communities. At the same time, the superconducting device community, which is familiar with these materials, has been more concerned with the high-field, high-current applications. Fortunately, the situation is changing. Very good thin films of these materials can now be routinely made and a number of promising devices using these films have been reported.

The various techniques required to deposit thin films of these materials are indicated in the table. In most cases they involve multi-component physical vapor deposition, usually dual electron beam deposition[1] or sputtering from a composite target.[2] When the constituent materials are not all available in solid form (e.g. NbN), reactive deposition in the appropriate gas has been used.[3] The growth morphologies and textures of these films can be complicated, but are generally well-behaved for sufficiently thin films ($\lesssim$ 5000 Å). We note, however, that the superconducting transition temperature often decreases in very thin films ($\lesssim$ 300 Å) and is also usually a very sensitive function of stoichiometry and disorder. The material science and detailed superconducting properties of these films are far from fully understood, but in several cases the situation is sufficiently under control empirically[4] that device work can proceed

TABLE I

POTENTIAL MATERIALS FOR SUPERCONDUCTING DEVICES

	Material	T_c (K)	$\Delta(0)$* (meV)	$\xi(0)$ (Å)	Mechanical Properties		Thin Film Deposition Technique
Low-T_c	In	3.4	0.51	$\lesssim$ 3600	Soft		Therm. Evap.
	Sn	3.7	0.56	$\lesssim$ 2300	Soft		Therm. Evap.
	Ta	4.5	0.68	450	Hard		e-beam
	Pb/Pb alloys	7.2	1.35	$\lesssim$ 800	Soft		Therm. Evap.
	Nb	9.3	1.5	$\lesssim$ 400	Hard		sput; e-beam
	Nb-Ti	7-10	(1.5)	$\sim$ 30-50	Hard		e-beam
High-T_c	Mo-Re	15	(2.3)	$\sim$ 50	Hard	Heated Substrate $T_s > 500$ C	sput.
	NbN	16	(2.4)	$\sim$ 40	Hard		reactive sput. reactive e-beam
	V_3Si	16.9	2.5	$\sim$ 30	Hard		e-beam
	Nb_3Sn	18.3	3.3	$\sim$ 30	Hard		e-beam; sput.
	Nb_3Ge	23.3	3.9	$\sim$ 30	Hard		sput; e-beam

*Values given in parentheses are calculated from BCS theory. Other values have been experimentally measured.

and, indeed, frequently provides valuable diagnostic information about the materials themselves.

IV. IMPROVED DEVICES

A. Hard Materials

The desirability of harder thin film superconducting devices has been recognized for a long time and has stimulated, for example, considerable work on Nb/Pb and Nb/Nb tunneling Josephson junctions.[5] By using the hard superconductor Nb as the base electrode, very rugged devices can be made. The electrical characteristics of these devices are not found to be ideal, however. The I-V curves are usually distorted and the dielectric constant of the oxide tunneling barrier (Nb_2O_5) has been found to have an unfortunately large dielectric constant ($\varepsilon \simeq 30$), leading to an excessive parasitic capacitance. The Cal-Tech group[6] has developed a hard thin-film weak-link Josephson-junction circuitry based on hard but low T_c materials (e.g. Ta). The low transition temperatures and circuit impedances of these devices appears to have discouraged wider use, however. These various studies on low-T_c hard superconductors have nevertheless motivated and guided the more recent work using higher T_c materials described below.

B. Hard, High-T_c Materials

The initial attempts to make Josephson junction devices with high-T_c materials involved weak-link structures. The first such reported attempt was the work of Janocko, et al.[7] who observed weak ac Josephson steps in 3 μm × 3 μm Mo-Re microbridges at ∿ 12K. However, it now appears that these steps reflect synchronized flux flow and not classical Josephson behavior. Fujita, Kosaka, Ohtsuka, and Onodera[8] have made weak-link Josephson junctions from NbN films epitaxially grown on MgO substrates by reactive sputtering. While it is not entirely clear, these bridges appear to be of the SS'S ($T_c' < T_c$) proximity type where the T_c of the bridge region was reduced (due to thinning or radiation damage) during the sputter-etching process used to form the bridge. These bridges have been successfully used in RF SQUIDS's operating up to ∿ 15K and with a field sensitivity of 7×10^{-9} G. Palmer, Notarys, and Mercereau[9] have made SS'S proximity bridges out of polycrystalline NbN and dual electron beam codeposited Nb_3Sn in which the T_c of the bridge region was lowered by reducing its thickness. (Although, because sputter-etching was used, radiation damage may also have been involved.) Using these Nb_3Sn bridges Palmer, et al., also fabricated dc SQUID's that operated at ∿ 17K. Their SQUID's had very small sensing areas, however, (typically 20 μm × 50 μm), implying field periodicities on the order of a few mOe. Falco[10] has also successfully fabricated a Nb_3Sn SQUID. He used sputtered Nb_3Sn and reduced the T_c of the bridge by means of radiation damage. He reports RF SQUID operation at $T \simeq 14.5$K. His most recent results are summarized in these proceedings.[11] Finally we note that weak RF-induced Josephson steps have been observed in Nb_3Ge bridges.[12] However, these bridges are as yet poorly characterized and the observed steps are almost certainly due to synchronized vortex flow.

These preliminary results with high-T_c weak-links are uneven at best, but SQUID operation has been demonstrated at high temperatures, which certainly is significant. Considerably more work will be required to fully characterize these devices and understand their detailed operation. It is clear on general grounds, however, that high-T_c weak-link Josephson devices must be of the SS'S or SNS type because of the very small Ginzburg-Landau coherence lengths of the high-T_c materials (See Table I). Conventional microbridges would have to be too small ($\simeq$ 50-100 Å) to be feasible with present microfabrication technology. Very little experimental work on SNS bridges has been reported, although some work in this direction is in progess. Such bridges are very attractive and could in principle approach nearly ideal Josephson behavior and impedance levels $\simeq 1\ \Omega$ over the entire temperature range form $0 < T < T_c$ using current state-of-the-art microlithography.

The first thin film tunneling junctions incorporating high-T_c superconductors were fabricated by Moore, Rowell, and Beasley[13] on Nb_3Sn films produced using dual electron beam codeposition techniques of Hammond. A Pb counterelectrode was used, so these are hard but not all high-T_c devices. Subsequently, good tunneling has also been obtained on V_3Si[14] and Nb_3Ge[15] films. In the case of V_3Si it was found necessary to use an artificial Si barrier to get good tunneling. Some typical I-V curves showing Josephson tunneling are shown in Fig. 1. Note that $\Delta_1 + \Delta_2$ for these junctions are very large. The Josephson tunneling properties of these junctions have recently been studied in detail by Howard, Rudman, Moore, and Beasley.[16] In the case of Nb_3Sn/Pb junctions, for which considerable data are now available, the results are gratifying indeed. Essentially ideal Josephson tunneling is observed and from a study of the Fiske modes, the dielectric constant of the oxides is found to be $\sim$ 8 rather than $\sim$ 30 as found with Nb-based Josephson junctions. Thus these junctions are electrically superior to Nb/Pb junctions and also easier to make. Techniques to incorporate such junctions into practical superconducting circuits have not yet been developed, however. Since Nb/Nb junctions have been successfully made, high-T_c, A15 superconducting tunnel junctions with a hard Nb counterelectrode seem entirely feasible, and since artificial barriers seem to work with these materials, all high-T_c tunnel junctions cannot be ruled out. These are developments for the future, however.

As a final point we mention, in passing, that high-T_c Nb_3Sn rf cavities have recently been successfully produced with very good performance, particularly considering the limited amount of effort expended to date.[17] At 4.2K the Nb_3Sn cavities had a surface resistance two orders of magnitude lower than Nb, as would be expected by virtue of the much larger energy gap of Nb_3Sn. This advantage obviously will increase at high temperatures. Also, the maximum surface field measured was 1000 Oe making these very high

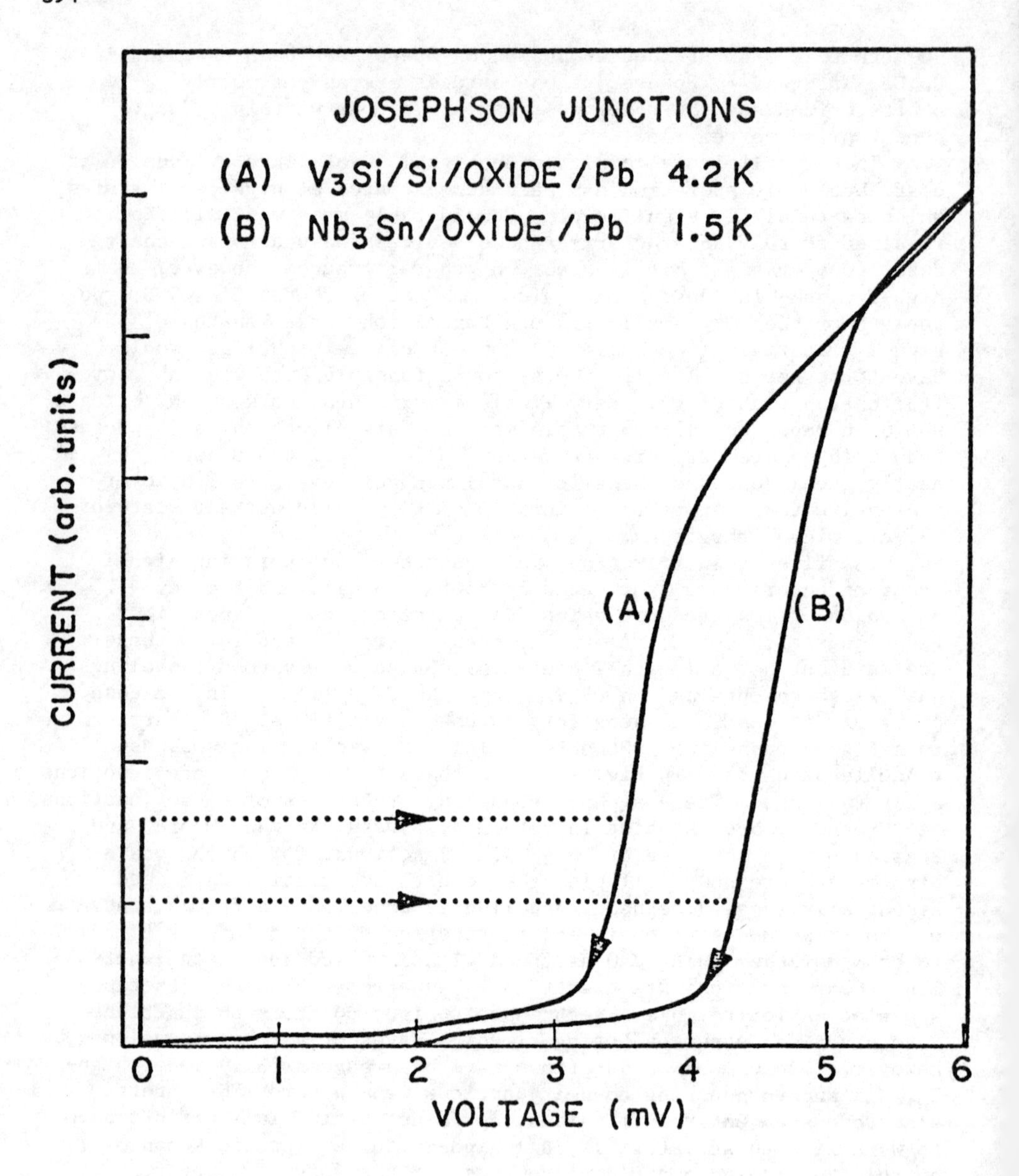

FIG. 1--Josephson tunnel junctions on Nb_3Sn and V_3Si. The counter-electrode for these junctions is Pb[3].

power cavities as well. At 1.5K, Nb cavities are better; however, due to excessive residual losses of the Nb_3Sn in its present state of development. In principle, Nb_3Sn is capable of losses four orders of magnitude lower than Nb at 1.5K, but better materials will be required to reach this level of performance.

To conclude, it is increasingly evident that the superior superconducting properties of the hard and high-T_c superconductors can be successfully exploited in superconducting electronic devices now that good thin films of these materials are available. Some optimism seems warranted, therefore, but only time will tell how fully these gains can be translated into a complete circuit technology.

ACKNOWLEDGEMENTS

Since the author is hardly a "materials man" he gratefully acknowledges instruction from and discussions with T. H. Geballe, R. Hammond, J. M. Rowell, R. E. Howard, and D. F. Moore. The support of the Office of Naval Research and the Joint Services Electronics Program is also gratefully acknowledged.

REFERENCES

1. R. H. Hammond, IEEE Trans. Mag-11, 201 (1975); R. H. Hammond, J. Vac. Sci. Technol. 15, No. 2 (1978).

2. J. R. Gavalar, Appl. Phys. Lett. 23, 480 (1973); R. T. Kampwirth, J. W. Hafstrom and C. T. Wu, IEEE Trans. MAG-13, 315 (1977).

3. M. A. Janocko, J. R. Gavalar, C. K. Jones, and R. Blaugher, J. Appl. Phys. 42, 182 (1971); see also Ref. 8.

4. B. E. Jacobson, S. Mahajan, R. H. Hammond, and J. Salem, to be published.

5. See for example, R. F. Broom, R. Jaggi, R. B. Laibowitz, Th. O. Mohr, and W. Walker, Proceedings of the 14th International Conference on Low Temperature Physics, Helsinki, paper T.044; R. F. Broom, J. Appl. Phys. 47, 5432 (1976); and references therein.

6. H. A. Notarys and J. E. Mercereau, J. Appl. Phys. 44, 1821 (1973); R. B. Kirschman, J. A. Hutchby, J. W. Burgers, R. P. McNamara, and H. A. Notarys, IEEE Trans. MAG-13, 731 (1977).

7. M. A. Janocko, J. R. Gavaler, and C. K. Jones, Proceedings of the 1972 Applied Superconductivity Conference, Annapolis, Maryland, IEEE Pub. No. 72CM0682-5-TABSC.

8. T. Fujita, S. Kosaka, T. Ohtsuka, and V. Onodera, IEEE Trans. Mag-11, 739 (1975); see also S. A. Wolf, F. J. Rachford, and M. Niesenoff, J. Vac. Sci. Technol. 15, No. 2 (1978).

9. D. W. Palmer, H. A. Notarys, and J. E. Mercereau, Appl. Phys. Letters 25, 527 (1974).

10. C. T. Wu and C. M. Falco, Appl. Phys. Letters 30, 609 (1977); C. T. Wu and C. M. Falco, J. Appl. Phys. 49, 361 (1978).

11. Paper by C. M. Falco these proceedings.

12. M. A. Janocko, J. R. Gavaler, and C. K. Jones, IEEE Trans. MAG-11, 880 (1975); R. B. Laibowitz, C. C. Tsuei, J. C. Cuomo, J. F. Ziegler, and M. Hatzakis, IEEE Trans. MAG-11, 883 (1975).

13. D. F. Moore, J. M. Rowell, and M. R. Beasley, Solid State Commun. 20, 305 (1976).

14. D. F. Moore, R. B. Zubeck, and M. R. Beasley, Bull. Am. Phys. Soc. 22, 289 (1977).

15. J. M. Rowell and P. H. Schmidt, Appl. Phys. Lett. 29, 622 (1976).

16. R. Howard, D. Rudman, D. F. Moore, and M. R. Beasley, Bull. Am. Phys. Soc., March Meeting, Washington, D.C., 1978; and to be published.

17. B. Hillenbrand, H. Martens, H. Pfister, K. Schnitzke, and Y. Uzel, IEEE Trans. MAG-13, 491 (1977).

GRANULAR SUPERCONDUCTORS FOR SQUIDS*

G. Deutscher
Tel Aviv University, Ramat Aviv, Tel Aviv, Israel

ABSTRACT

The low critical current density of granular superconductors makes them attractive for thin film SQUID applications. A survey of the properties of granular superconductors such as granular aluminium, lead and niobium is presented, and the actual performance of some granular SQUIDS is discussed. In addition to their convenient low critical current density, granular SQUIDS also appear to have low intrinsic noise levels.

INTRODUCTION

In the past, thin film R. F. SQUIDS have usually been devices working rather close to their critical temperature. The reason is that in order to optimize signal amplitude the SQUID should have a critical current i_c of order ϕ_o/L where ϕ_o is the flux quantum and L is the SQUID inductance[1]: This requirement leads to i_c values of the order of 1 to 10 μA, which for usual bridge dimensions (thickness 1000 Å, width one micron) and typical superconductors (low temperature critical current j_c density 10^6 A/cm^2) can be achieved only near T_c. This is not a very practical way of operation.

There is, therefore, an interest in low critical current density superconductors for SQUID applications since they would allow operation in a wider temperature range without resorting to submicron bridges. For the typical bridge dimensions referred to above, low temperature j_c values of a few 1000 A/cm^2 would be quite satisfactory. Such values can be obtained in granular superconductors[2]. Another requirement is a reasonably strong Meissner effect: This may be a problem in granular superconductors because a low critical current density implies a large penetration depth. Finally we would like the granular bridge to have a low intrinsic noise level. This also poses a problem because in order to have a sinusoidal current-phase relationship it is generally accepted that the bridge length should not be much larger than the coherence length, and we know that granular superconductors have short coherence lengths[3].

In section I we review the methods of fabrication and structure of granular materials and the nature of their superconducting transition. Section II is devoted to critical current densities and

*Research supported by the ISRAEL NATIONAL COUNCIL for RESEARCH and DEVELOPMENT and the KARLSRUHE NUCLEAR RESEARCH CENTER.

ISSN: 0094-243X/78/397/$1.50

critical temperatures. In section III we discuss actual SQUID operation of granular superconductors with emphasis on intrinsic noise level.

I. FABRICATION, STRUCTURE AND SUPERCONDUCTING TRANSITION OF GRANULAR MATERIALS.

Granular superconductors are composed of two elements: A superconducting metal, often in the form of grains, and a non-superconductor with a low normal state conductivity such as a semiconductor or oxide. Both elements are essentially mutually insoluble. The normal state conductivity σ_n of the mixture depends on the relative concentrations and nature of its constituents. But properties characteristic of granular superconductivity are obtained for values of σ_n in the range of 10^{+4} $(\Omega cm)^{-1}$ to 10^2 $(\Omega cm)^{-1}$, irrespective of the nature of the constituents and detailed structure of the mixture.

Granular materials can be prepared by vacuum deposition in a number of ways[3]: Evaporation of the metal in the presence of oxygen resulting in co-deposition of metal and oxide ($A\ell$-$A\ell_2O_3$, Sn-Sn_xO_y), co-evaporation or co-sputtering of the constituents (Aℓ-Ge, Pb-Ge, Aℓ-SiO_2). The metal-oxide combination can also be prepared by anodization of the pure metal film (NbN-Nb_xO_y)[4], and by oxidation of the individual metallic particles[5].

The structure of the granular materials is often such that the metal is in the form of grains surrounded by the oxide or semiconductor. A typical example is Aℓ-Ge shown in Fig.1a. This is not, however, always the case: In other systems such as Pb-Ge (Fig.1b), the structure is much more random [6].

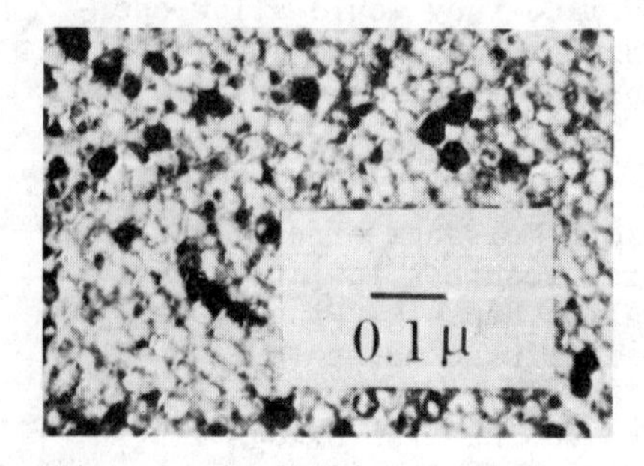

Fig.1a. Regular granular structure of Aℓ-Ge.

Fig.1b. Random structure of Pb-Ge.

The normal state conductivity decreases strongly when the metal volume fraction X approaches a critical value X_c, below which superconductivity is quenched. The value of X_c varies a great deal. It is typically of the order of 50% or even more in regular granular structures such as shown in Fig.1a[3], and much lower (15%) in more random structures (Fig.1b)[6]. Properties characteristic of granular superconductivity are observed for both kinds of structures in the same range of σ_n values, although they occur at very different concentrations (around 60% for Aℓ-Ge and 20% for Pb-Ge for instance).

When X_c is approached, the superconducting building blocks (or clusters of building blocks) become weakly coupled. If the metallic crystallites are large enough to exhibit a superconducting transition when they are isolated (grain size d $\gtrsim$ 50 to 100 Å), weak coupling effects can be directly observed in the shape of the superconducting transition: It splits in two[7]. The resistance drops by a certain amount at the critical temperature of the grains, T_{co}, then decreases slowly until it eventually disappears at a lower critical temperature, T_{cp}, where superconducting coherence is established through the weak intergrain links. Similarly, two critical currents are observed below T_{cp}: A low critical current i_{cp} where a fraction of the resistance is restored and a higher current i_{co} where the full normal state resistance is recovered. This double transition has been observed in anodized NbN[4] and in Aℓ-Ge[8] and is believed to be a general property of large-grain granular superconductors[7]. In the small grain case only the lower critical temperature T_{cp} can be observed because thermodynamic fluctuation effects tend to wash out the superconducting transition in the small grains[7]. In any event weak intergrain coupling is at the origin of granular superconductivity, and both large-grain and small-grain systems are suitable for SQUID applications.

II. CRITICAL CURRENT DENSITY AND CRITICAL TEMPERATURE.

Results for the critical current density of $A\ell\text{-}A\ell_2O_3$ films[9] are shown in Fig.2. It can be seen that increasing values of the normal state resistivity ρ_n lead to much-reduced values of j_c, especially when ρ_n reaches the range of 1000 μΩcm. Low values of j_c are then observed in a significant temperature range below T_c. Reduction of j_c from the bulk Aℓ value ($\sim 10^6$A/cm^2) is several orders of magnitude.

Results for Aℓ-Ge are shown in Fig.3 as a function of composition. They show a close correlation between σ_n and j_c[8]. In practice this is very useful because the j_c value of a particular sample can now be accurately predicted from a simple room temperature measurement of σ_n. Another observation is that useful critical current densities (a few 1000 A/cm^2 or less) are achieved in a significant concentration range. This means that thin film alloys appropriate for SQUID application can be prepared with conventional control methods of alloy concentration. Similar results have been obtained for

Pb Ge[8].

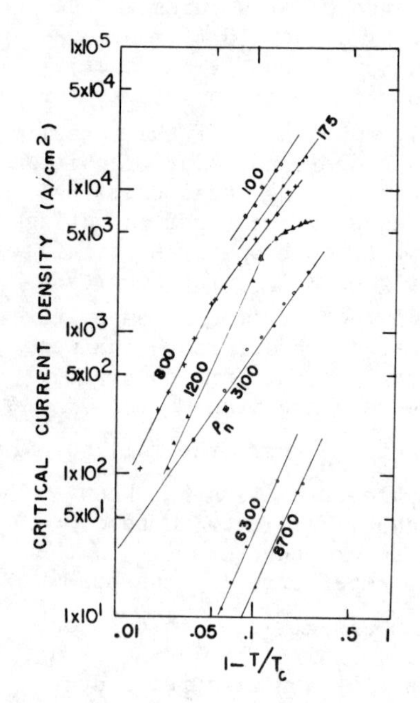

Fig.2. Critical currents of $A\ell$-$A\ell_2O_3$ for various values of the normal state resistivity (in μΩcm).

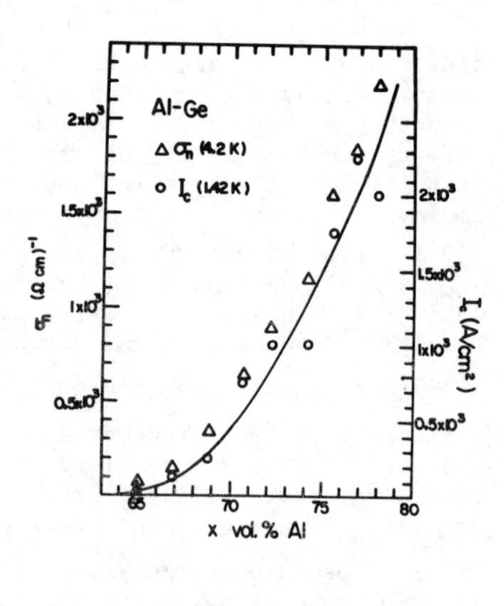

Fig.3. Composition dependence of the critical current I_c and the normal state conductivity σ_n of $A\ell$ Ge.

An important point is that a large reduction of j_c can be achieved without lowering the critical temperature T_c. Typically, a three-orders-of-magnitude reduction of j_c is accompanied by a 10% decrease of T_c. Eventually, of course, both T_c and j_c go to zero as the critical concentration X_c is reached. But the concentration dependence of T_c and that of j_c are fortunately completely different. From Fig.3 one sees that j_c approaches 0 with zero slope as $X \to X_c$, while from Fig.4a and 4b, T_c approaches 0 with infinite slope when $X \to X_c$. The difference between the functional dependences $j_c(X)$ and $T_c(X)$ actually provides the basis for SQUID applications of granular superconductors.

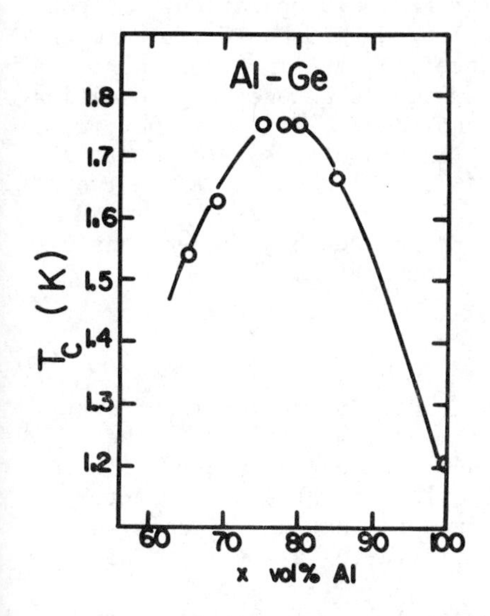

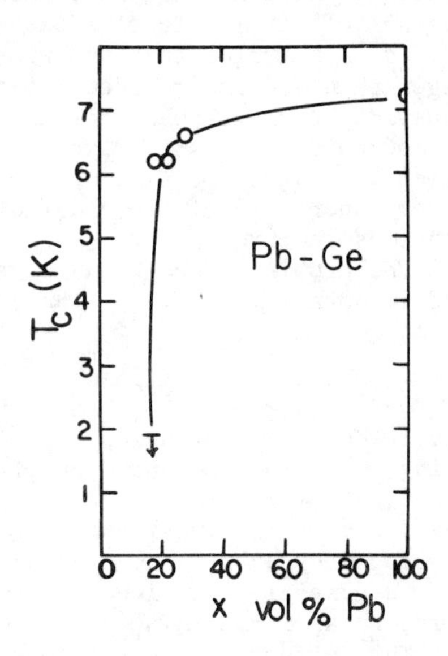

Fig.4a. Composition dependence of the critical temperature of Aℓ-Ge.

Fig.4b. Composition dependence of the critical temperature of Pb-Ge.

In the range $X_c \ll X < 1$, T_c may be essentially composition-independent like in Pb Ge (Fig.4b) or be a decreasing function of X like in Aℓ Ge (Fig.4a). In the latter case $T_c(X)$ exhibits actually a maximum. The microscopic mechanisms responsible for the T_c enhancement observed in many systems are not well understood (for a review, the reader is referred to Ref.3). Both disorder and surface effects at grain boundaries have been invoked to explain enhancement effects. What is important for the application of granular superconductors is that granularity does not have an adverse effect on T_c (except of course very close to X_c), and may even sometimes enhance it.

III. SQUID OPERATION OF GRANULAR SUPERCONDUCTORS.

SQUID operation of granular superconductors has been reported for $A\ell\text{-}A\ell_2O_3$ [2,9] and anodized Nb[10] and NbN[4]. In all cases SQUID operation was observed in wide bridges (10 μm or more) thanks to the low critical current densities, and low noise levels have been reported[9,10].

For reasons mentioned in the introduction, low noise levels in granular SQUIDS seem surprising. First, a small critical current density implies a large penetration depth and the existence of a Meissner effect is clearly necessary for RF SQUID operation. Second, it is common belief that the weak link length should not be much longer than the superconducting coherence length ξ in order to have a sinusoidal current-phase relationship, which is a necessary condition for low noise levels[1]: However, the condition $L < 10\ \xi$ is not fulfilled in granular weak links because ξ in granular superconductors is very short and of the order of 100 Å [3]. We consider these two questions in turn.

The problem posed by a large penetration depth has been considered in detail in Ref.9. The relevant parameter is

$$x = \frac{\lambda^2}{r\ d} \quad ,$$

where λ is the penetration depth, r the radius of the cylinder supporting the thin film and d the film thickness. For $x \ll 1$ the magnetic flux is almost entirely excluded from the cylinder, while for $x \gg 1$ there is almost complete penetration. The skew parameter γ introduced by Jackel and Buhrman[1] is found to increase linearly with x. In general, the signal-to-noise ratio varies as γ^{-1} [1], and therefore as x^{-1}.

This effect becomes pronounced when $x \gtrsim 1$. Therefore, for practical values of the parameters $r = .1$ cm and $d = 1000$ Å, λ should not exceed 10 μm for good SQUID performance. Such high values of λ are actually achieved in granular materials for $\rho_n \sim 5000$ μΩcm (Fig.5). Higher values of ρ_n are therefore not favorable, at least when the superconducting ring is entirely made of the granular superconductor. Should higher values of ρ_n be desired, then only the weak link itself should be granular and the rest of the ring made of a regular superconducting film.

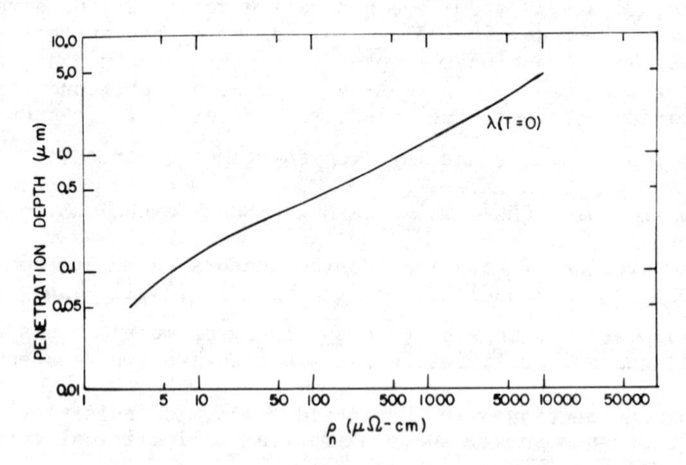

Fig.5. Penetration depth of the magnetic field as a function of the normal state resistivity ρ_n for $Al\text{-}Al_2O_3$.

The strong Meissner effect observed in the granular SQUIDS of ref. 9 and 10 is in good agreement with the reported low noise level (close to intrinsic).

On the other hand, the problem posed by the short coherence length $\xi << L$ is not so well understood. It is clear from the experimental results[9,10] that the condition for low noise level $L < 10\ \xi$ is relaxed in granular weak links where low noise levels have been observed while $L \sim 10^2$ to $10^3\ \xi$. The reason must probably be that the phase of the order parameter in these links is actually a discontinuous function of position, most of the variation occurring at the grain boundaries. Indeed, near the concentration X_c the behaviour of the weak link is controlled by the few remaining superconducting intergrain junctions. So maybe it is not so surprising that the granular weak link behaves like a point contact or Josephson tunnelling junction rather than as a conventional weak link bridge. We note in this connection that λ is of the order of L, which may allow coherent behaviour of the individual junctions.

CONCLUSION

Granular superconductors seem appropriate for RF SQUIDS. The relatively large bridge dimensions allowed by their low critical current densities make granular SQUIDS easy to produce. They have been found to have a low noise level. Operation at 4.2 K is possible by using for instance Pb- based or Nb-based granular films.

ACKNOWLEDGEMENTS

This review was prepared in close collaboration with Ralph Rosenbaum, Michael Rappaport and Dov Abraham at Tel Aviv University. Stimulating conversations with Stuart Wolf are gratefully acknowledged. The helium gas for this research was supplied by the ONR.

REFERENCES

1. L. D. Jackel and R. A. Buhrman, J. Low Temp. Phys. 19, 201 (1975).
2. G. Deutscher and R. Rosenbaum, Appl. Phys. Lett. 27, 366 (1975).
3. B. Abeles, in Applied Solid State Science, Vol.VI, R. Wolfe, Ed. (Academic Press, 1976).
4. S. Wolf and W. H. Lowrey, Phys. Rev. Lett. 39, 1038 (1977).
5. P. Hansma and J. Kirtley, J. Appl. Phys. 45, 4016 (1974).
6. G. Deutscher, M. Rappaport and Z. Ovadyahu, to be published.
7. G. Deutscher, Y. Imry and L. Gunther, Phys. Rev. B10, 4598 (1974).
8. G. Deutscher and M. Rappaport, to be published.
9. D. Avraham, G. Deutscher, R. Rosenbaum and S. Wolf, submitted to J. Low Temp. Phys.
10. F. J. Rachford, S. A. Wolf, J. K. Hirvonen, J. Kennedy and M. Nisenoff, I.E.E.E. Transactions on Magnetics, Vol.MAG-13, 875 (1977).

THE SURFACES OF HIGH-T_c "Nb_3Ge" FILMS AS STUDIED BY ELECTRON TUNNELING AND AUGER ELECTRON SPECTROSCOPY†

R. Buitrago*, L. E. Toth, A. M. Goldman and M. Dayan
School of Chemical Engineering and Materials Science and
School of Physics and Astronomy, University of Minnesota
Minneapolis, Minnesota 55455

ABSTRACT

Auger electron spectroscopy (AES) has been used to understand the nature of the surfaces of the top atomic layers of high-T_c sputter-deposited films of Nb_3Ge. Pb-Nb_3Ge tunneling junctions of high quality have been fabricated with oxidized thin aluminum barriers.

For superconducting tunneling to be used to determine microscopic parameters such as λ and μ^*, it is critical that the chemical composition and structure of the layers within a coherence volume of the tunneling barrier have the same chemical composition and structure as the part of the system which determines its macroscopic properties. In the study of high-T_c intermetallic compounds such as Nb_3Ge the problem is exacerbated by the short coherence length which is on the order of 50 Å, by the great difficulties in preparing single-phase high-T_c films which are homogeneous across their thickness,and by the fact that the layers of material adjacent to a tunneling barrier prepared by the conventional thermal oxidation are rich in germanium.[1] These problems have led us to develop a procedure based on proximity-effect tunneling.[2]

In the case of Nb_3Ge it is relatively easy to deposit films by sputtering, CVD or electron beam deposition and achieve T_c's greater than 20 K. The primary problem is that "Nb_3Ge" is a metastable phase. We have found that Nb_3Ge films with a 20 K transition temperature may contain up to five Nb-Ge compounds. These are the A15, Nb_5Ge_3(H), Nb_5Ge_3(T), a quasiamorphous phase characterized by well defined diffraction peaks at d = 2.66 Å and d = 1.33 Å, and the amorphous phase. Figure 1 shows the phase relationships we have

†Supported in part by the National Science Foundation under Grant DMR 76-01370-A01 and by the Central Administration of U. of M.

*Fellow of El Consejo Naçional de Investigaciones Científicas of Technological de la República Argentina.

ISSN: 0094-243X/78/404/$1.50

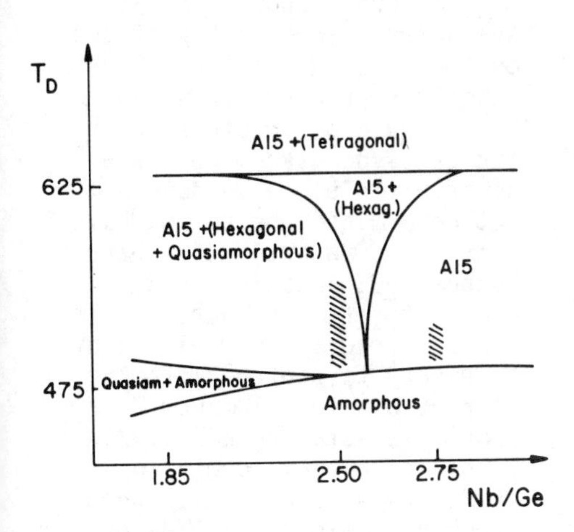

Fig. 1. Phase relationships of sputter-deposited Nb_3Ge films. T_D is the substrate temperature during deposition in °C.

found as a function of substrate deposition temperature and Nb/Ge ratio of the sputtering target. Domains where transition temperatures in excess of 21 K occur are also shown. Maximum T_c's occur in regions very close to boundaries between two phases; in fact, films with T_c's > 21 K usually contain a small amount of a second phase which is not an A15. As one moves away from these boundaries, T_c decreases slightly and the amount of non-A15 phase increases. It is possible to produce films with T_c > 20 K that are (1) single phase A15 or (2) about 1/3 A15, 2/3 non-A15. Thus knowing that T_c is high is not necessarily an indication that tunneling data will accurately reflect the properties of the A15.

A second problem is illustrated in Fig. 2 which shows the localized chemistry at the top surface of a sputter-deposited Nb_3Ge film with a T_c > 21 K. A pronounced Ge-rich layer exists in the first 50-70 A of the original surface. We have shown that this Ge-rich layer develops as the film is formed.

Thermal oxidation usually used to produce barriers is not necessarily appropriate when the electrodes are compounds. In general, preferential oxidation of one element occurs. For the specific case of Nb_3Ge, the top surface of the electrode beneath the oxide no longer has the same composition as the bulk of the film, assuming that it was homogeneous in the first place, because as the oxide forms the remaining elements get concentrated directly beneath the oxide layer. In the case of Nb_3Ge the oxidation process results in an enhancement of the Ge-rich layer. It is highly probable that such a layer was responsible for the lower energy gap observed by Rowell and Schmidt[3] in their tunneling studies of Nb_3Ge.

The procedure that we have developed to circumvent the above difficulties is an adaptation of the proximity-effect tunneling configuration to Nb_3Ge. First, single or nearly single phase Nb_3Ge films with high T_c's are sputter-deposited. A post-anneal of four hours at 550°C reduces the amounts of non-A15 phases to the detection limit of X-rays or below it without increasing T_c. Next, the Ge-rich surface layer is removed by a low-voltage glow-discharge or by ion-milling in the Auger system. A thin layer of Al (50 ± 20 A) is next deposited in a vacuum of 10^{-7} Torr. This layer is then oxidized in air for three minutes. The Al layer is masked with collodion at the same time. Finally a counterelectrode of Pb is deposited under UHV conditions.

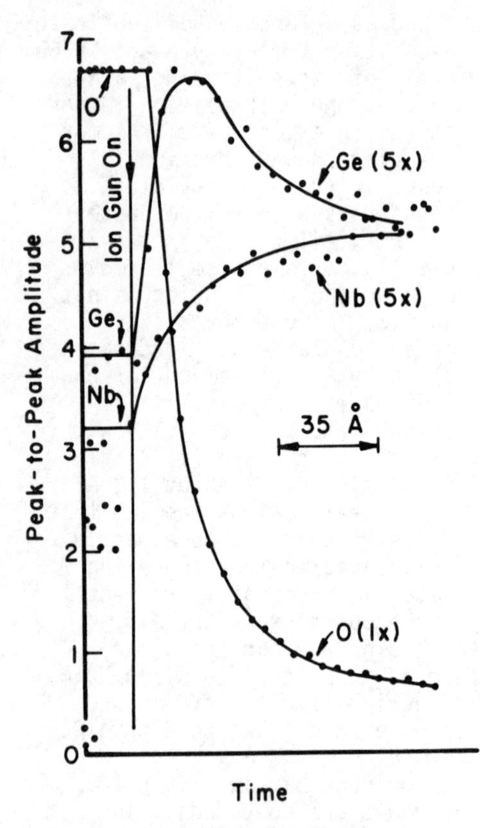

Fig. 2. Auger electron depth profile of an "Nb_3Ge" film with a Ge-rich layer of 70 A. T_c was 21.8 K.

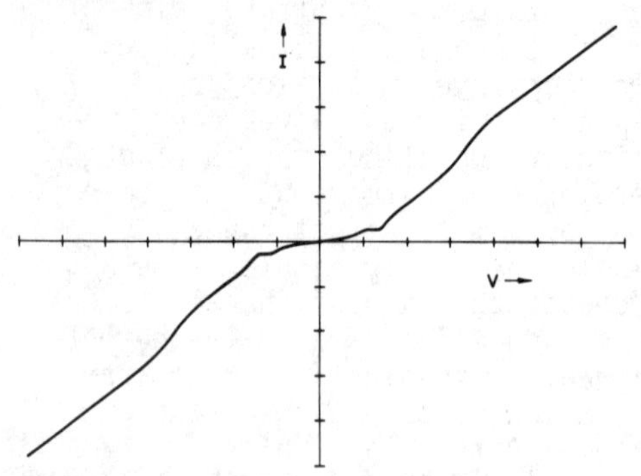

Fig. 3. Current-voltage characteristic of a nominal Nb_3Ge-Al-Al_2O_3-Pb junction. The scales are 10^{-3} V/division and 10^{-5}A/division. T = 4.2 K.

Figure 3 shows the I-V characteristic at 4.2 K of one of the first Nb_3Ge-Al-Al_2O_3-Pb junctions prepared this way. An energy gap structure which can be associated with Nb_3Ge is clearly visible. We have not yet determined whether we have fabricated an artificial barrier Nb_3Ge-Al_2O_3 junction or a proximity-effect junction. In the former case it would be unnecessary to contend with the corrections for aluminum density-of-states in analyzing tunneling data in order to study the strong-coupled phonon structure of Nb_3Ge.[4] The presence of two gaps in the I-V characteristic suggests that the surface layer is not single-phase Nb_3Ge. It remains to be demonstrated that enriched surface layers can be removed and the surface appropriately annealed so that a single-gap characteristic of a homogeneous material can be obtained. However, this technique of junction fabrication appears to be sufficiently reproducible that a systematic study of the dependence of I-V characteristics on such treatment seems possible.

REFERENCES

1. R. Buitrago, L. Toth and A. M. Goldman, submitted to J. Appl. Physics.
2. E. L. Wolf and J. Zasadzinsky, Physics Lett. 62A, 165 (1977).
3. J. M. Rowell and P. H. Schmidt, Appl. Phys. Lett. 29. 622 (1976).
4. Gerald B. Arnold, Bull. Am. Phys. Soc. 23, 263 (1978) and to be published.

RF SPUTTER ANODISATION OF LEAD-INDIUM FILMS STUDIED BY ELLIPSOMETRY

G.B. Donaldson, Department of Applied Physics, University of Strathclyde, Glasgow, Scotland.

H. Faghihi-Nejad*, Department of Physics, University of Lancaster, England.

ABSTRACT

We have used simple ellipsometric equipment to study the RF sputter oxidation of a series of lead-indium alloy films containing from 0-46 atomic % In. Our results show monotonic oxide growth for most films, with saturation thicknesses (3.0 - 9.0 nm) being reached within ~ 10 min for small In concentrations, and ~ 40 min for higher In concentrations. For films of 26-36 atomic % nominal In concentration, the oxide thickness rose to an intermediate maximum before decreasing to the equilibrium value. The In_2O_3 content of the oxide layer was disproportionately large and, in particular, was 100% for all alloys containing more than 28 atomic % of indium.

INTRODUCTION

Lead-indium alloy films have shown considerable promise as materials on which to grow Josephson-junction oxide barriers. Eldridge et al[1] have used ellipsometry to study the developments of oxide layers grown thermally in atmospheres of oxygen over periods of up to five hours. We have looked ellipsometrically at similar films on which the barrier has been grown using the technique described by Greiner[2], in which a low power RF discharge in oxygen is used to produce a dynamic equilibrium between the rate at which oxide is removed by sputtering and that at which new oxide is being formed by activated oxygen atoms. The work was undertaken to further elucidate a technique of ours[3] which is known to produce junctions with excellent critical current characteristics.

APPARATUS AND METHOD

We used Joule-heated boats in a standard evaporator (Fig.1) to deposit films on Corning 7059 glass substrates which were mounted on a RF target suspended from the bell jar top plate.

*Present address: College of Science, Jundi-Shapur University, Ahvaz, Iran.

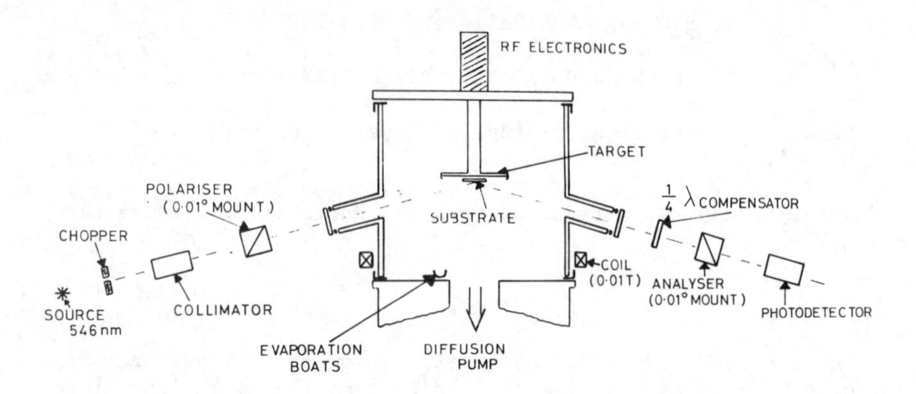

APPARATUS LAYOUT

Figure 1

Polarized green (546 nm) light entered the vacuum chamber through a strain-free window mounted on a side arm, was incident on the substrate at 70°. The reflected beam passed through another strain-free window on a second side arm, and was detected with a photomultiplier after passing through a quarter-wave compensator and an analyser. Extinction settings of the polariser and analyser could be made to 0.04°, and from them we determined Δ and ψ, the parameters describing the change which reflection produces in the ellipticity of the polarisation of the incident beam. For bare metal surfaces we used Δ and ψ in standard formulae[4] to determine the real (n) and imaginary (k) parts of the refractive index; in the case of oxide layers we determined their refractive index (n) and and thickness (d) by assuming that their absorption was zero and applying the approximate formulae of Archer[5].

Our films were evaporated at 5×10^{-6} torr as described by Emmanuel et al.,[3] with indium being deposited first and lead overlaying it. The total thickness was about 200 nm and the relative thicknesses of the two layers represented the required alloy ratio. We made ellipsometric measurements of n and k on the films immediately after evaporation and for pure lead and pure indium found good agreement with other workers[1,2]. On alloys, however, we found that Δ and ψ varied in a manner consistent with the diffusion of

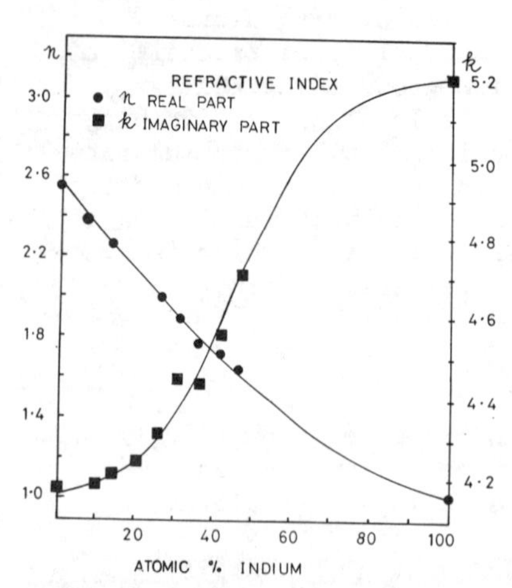

MEASURED OPTICAL CONSTANTS OF LEAD–INDIUM ALLOYS

Figure 2

indium into the lead layer, reaching equilibrium after about 30-90 minutes at room temperature. The limiting n and k values fit a roughly linear interpolation, based on nominal concentration, between lead and indium. (Fig.2) This suggests that the diffusion process homogenises the film completely.

To prepare insulating layers, we admitted 10-100 millitorr of pure oxygen. After 5 minutes, during which a few angstroms of oxide were observed to form, we started an RF discharge between the target (as anode) and ground, using a 10^{-2}T axial magnetic field to concentrate the discharge. A simple single valve 13.5 MHz oscillator circuit was used, and the power incident on the films was of order 0.6-4 mW-cm^{-2}: the substrate temperature rise was negligible. The discharge was stopped occasionally to make ellipsometric measurements of n and d. No oxide growth could be detected during these intermediate intervals.

OXIDATION RESULTS AND DISCUSSION

Thickness of oxide layer

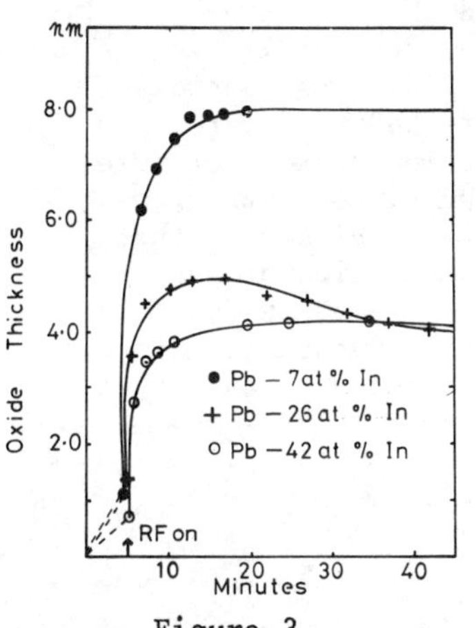

Figure 3

Figure 3 shows the growth of oxide for three films of different indium concentrations under typical pressure and power conditions. The top curve is characteristic of films containing 0-20 atomic % of indium, with initial growth rates of about 3.5 nm-min^{-1}, decreasing monotonically until a saturation thickness X_s is reached after about 15 minutes. Simple behaviour is also observed with concentrations of indium greater than 35%, and with pure indium films: initial growth rates are slower and the time constants are longer than in the case of lead-like films, with 40 minutes being required to reach saturation in the case of pure In. The most striking behaviour, however, is observed in the range 26-36 atomic % In (middle curve). Here, for power levels greater than 2 mW-cm^{-2}, the oxide thickness rises with a short time-constant and reaches an intermediate maximum before decreasing to its equilibrium value. We do not fully understand the mechanism for this behaviour, but believe that it may be due to preferential sputtering of islands of lead oxide such as are postulated by Eldridge et al.[1] in their thermal oxidation work. After initial rapid removal of material from lead-rich islands the controlling time constant could become the rate at which lead can diffuse through the lead-indium alloy or indium oxide. It is not clear, however, how this can account for a decrease in the measured barrier thickness.

The saturation thickness X_s was found to increase with increasing oxygen pressure: for example in a specimen of 8 atomic % In subjected to 5 $mW\text{-}cm^{-2}$ power, the first thickness was 74 nm in a 10-millitorr discharge, but this increased to 84 nm when the pressure was raised to 50 millitorr. When the pressure was returned to 10 millitorr, X decreased again over about 20 minutes to the 74 nm value, demonstrating the true dynamic equilibrium of the sputter-etch procedure. The dependence of X_s on RF power was not clear-cut however: in general for lead-rich alloys as in Fig.3 (top), X_s was decreased by increasing the power, but for alloys with more than 30 atomic % In. X_s was a weakly increasing function of RF power under our operating conditions.

Oxide composition

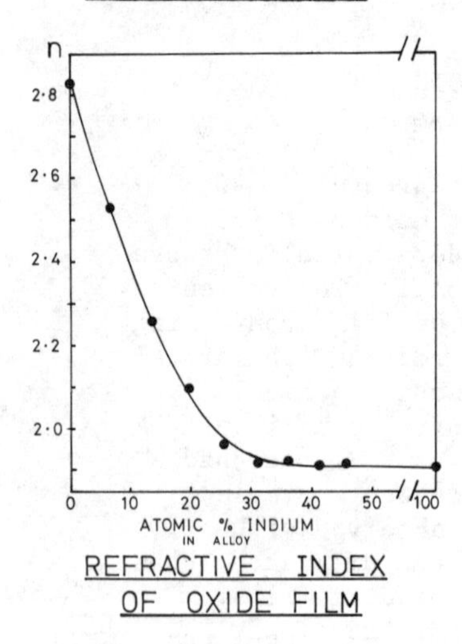

REFRACTIVE INDEX OF OXIDE FILM

Figure 4

The refractive index (n) of the anodised layer was found to vary very little during oxidation, and in Fig. 4 we show how it varied with indium concentration. The data agree well with those of Eldridge et al.[1], obtained in their natural oxidation work at 760 torr. In particular the oxide index decreases from that of pure PbO at a zero indium concentration to that of pure In_2O_3 at all indium concentrations above 30%. This conclusion - that the effective composition of the barrier is much enriched In_2O_3 - is in agreement with evidence from ion scattering spectroscopy [3] and Auger spectroscopy. [6]

CONCLUSION

RF sputter-etching is an effective and rapid way of producing oxide barriers of controlled thickness on lead-indium alloy films. Only inexpensive equipment is needed. The composition of the oxide films is similar to that obtained when the growth takes place more slowly by natural oxidation. The mechanism of oxidation is not simple, however, as is shown by the intermediate maximum which is seen as oxide grows on films of intermediate composition (26-36 atomic % In), and which possibly arises by preferential sputtering from small islands of enriched Pb composition with consequent inter-grain Pb diffusion within the film.

ACKNOWLEDGMENTS

The work was supported by the Science Research Council, and one of us (HFN) acknowledges a studentship from the Iran Ministry of Science and Higher Education. We are both grateful to W.T. Band and F.M. Livingstone for advice and assistance.

REFERENCES

1. J.M. Eldridge, D.W. Dong and K.L. Komarek, J. Electron. Mater. 4, 1191 (1975).

2. J.H. Greiner, J. Appl. Phys. 45, 32, (1974) and 42, 5151 (1971).

3. A. Emmanuel, G.B. Donaldson, W.T. Band and D. Dew-Hughes, I.E.E.E. Trans. Magnetics, MAG-11, 763 (1975).

4. See for example M. Born and E. Wolf, Principles of Optics, 2nd ed., Pergamon, London (1964).

5. R.J. Archer, J. Electrochemical Soc. 104, 619 (1957).

6. N.J. Chou, S.K. Lahiri, R. Hammer and K.L. Komarek, J. Chem. Phys. 63, 2758 (1975).

CRYOCOOLERS FOR SUPERCONDUCTIVE ELECTRONICS*

J. E. Zimmerman
Cryogenics Division
National Bureau of Standards
Boulder, Colorado 80303

ABSTRACT

Superconductive electronic devices are generally operated and will continue to be operated in liquid-helium cryostats. Appropriate small closed-cycle cryocoolers for these devices do not exist. Development of cryocoolers, including efficient use of conventional gas refrigeration cycles and innovative thoughts on new and unconventional methods, could greatly enhance the usefulness of small-scale superconductive electronics. Large cryocoolers are already well-developed and can be adapted to the requirements of large-scale systems, such as computers, requiring a watt or more of refrigeration.

INTRODUCTION

In this paper I will refer repeatedly to two sources[1,2]. Ref. 1 is the proceedings of a 1977 conference on "Applications of Closed-cycle Cryocoolers to Small Superconducting Devices," and is publicly available. Ref. 2 is the proceedings of a 1973 conference on "Closed-Cycle Cryocoolers," and is available for "limited distribution." A third significant source of information and speculation is the proceedings of the 1975 Navy Study on Superconductive Electronics.[3] Other useful and readily available references are the proceedings of the 1976 SQUID conference[4] and the proceedings of the 1976 NATO study on small superconducting devices[5].

This author's demonstrated qualifications as an authority on cryocoolers consist of his having built (1) in 1949, a Simon single-expansion helium liquefier, a mechanism which at the same time was made obsolete by the Collins Helium Cryostat, (2) a few years later, a cyclic magnetic refrigerator for temperatures down to 0.1K, a machine which was immediately replaced by the helium dilution refrigerator, and (3) recently, a laboratory-model Stirling cryocooler specially adapted to operate a SQUID magnetometer. The trend established by the earlier work probably has been reversed by the recent achievement, since Stirling machines are now being extensively developed both for pumping heat and for doing work, and this conference is evidence that the SQUID and other superconductive devices are seen as having great future potential. In fact, it should also be pointed

---

*Contribution of the National Bureau of Standards

Support in part by the Office of Naval Research and the Naval Air Systems Command

out that cyclic magnetic cooling systems are currently being developed by Steyert and others for temperatures between 1 and 300K, and it is suggested below that the single-expansion process of liquefying helium may still be of some use in certain systems (p.81 and p.49, ref. 1).

Very few closed-cycle cryocoolers are now in use, or have ever been used, to operate superconducting devices. The reason for this became evident in 1947 or shortly thereafter, when the Collins Helium Cryostat became generally available. This machine was advertised and sold as a "cryostat", a device for maintaining a constant low temperature. Very quickly it became obvious that the machine was more cost-effective if used not as a cryostat but as a liquefier, for servicing any number of separate small liquid-helium evaporative cryostats.

ESSENTIAL HEAT INPUTS INTO CRYOSTATS

With only a few isolated exceptions, (e.g. Hartwig. p.135, ref. 1) liquid-helium cryostats have been used for all research in superconductivity and superconductive devices up to the present. Now, however, the beginning of change is apparent. As superconducting devices are found to be useful for a variety of purposes unrelated to low-temperature physics, users are searching for ways of reducing the cost and inconvenience of liquid-helium cryogenics. One approach is to make dewars larger and more efficient, with hold times of the order of 10^3 to 10^4 hours (e.g. the 180-liter VLF SQUID cryostat of Dinger _et al_., with a hold time of 100 days[6], the 800-liter Infrared Astronomy Satellite (IRAS), with a projected hold time of one year[7], and a similar 800-liter satellite for the Stanford relativity experiment, projected to operate one year in orbit at 2K[8]. More indicative of the nature of things to come, we hope, are the systems in which a cryocooler is incorporated into a medium-sized liquid-helium dewar to cool one or more radiation shields and greatly reduce the evaporation rate. Temperatures well below 20K can be maintained with available small commercial coolers. Black-body radiation at 20K is only 9.1 mW/m^2, and if this figure can be reduced by a factor of 100 by the use of highly reflecting surfaces, giving $91\mu W/m^2$, the corresponding evaporation rate due to radiated heat is 0.0031 $\ell/day\text{-}m^2$. A spherical container with a surface area of $1m^2$ would hold 95 liters, giving a time of 3×10^4 days (80 years) for complete evaporation. In most practical cases, with a radiation shield at 20K and a helium bath at 4K, the main heat input will be conduction along mechanical supports and electrical leads, but even so, the heat leak can in principle be made very small. If a 95-liter container is supported by spun-glass cords, 10 cm long, calculated to withstand 100 G loading in three orthogonal directions (six cords), and assuming a total mass twice that of the helium itself, the heat leak is calculated to be 2mW, giving a hold time somewhat less than 4 years (40 years for 10G loading). The enthalpy change of helium between 4 and 20K is about 4 times the heat of vaporization, so if the vapor can be used to cool the supports, this evaporation rate might be reduced considerably, at the expense of geometric complexity. Size of electrical leads depends upon the particular devices in the cryostat. Let us consider (as we often do) an rf SQUID and suppose that the rf line to it is a thin-film stripline with a parallel dc resistance of 0.1Ω between 4 and 20K. Invoking the Wiedemann-Franz law leads to a heat leak of about 45 μW, which is smaller than any of the other heat leaks calculated above.

This exercise in arithmetic is highly idealized, but it does indicate the possibility of greatly reducing the evaporation rate of liquid helium by using a cryocooler in a carefully-designed system. A 70-liter cryostat incorporating a cryocooler to maintain the inner radiation shield at 15K has been built by a commercial firm (W. Goree, private communiation). This system demonstrated a hold time greater than one year (0.19ℓ/day evaporation rate). The heat leak into the helium, about 5.5 mW, was calculated to be due almost entirely to the 75 mm-diameter stainless steel neck tube, which provided easy access to the helium bath and also served as mechanical support for the reservoir.

Conventional liquid-helium cryostats, such as those mentioned above, without liquid-nitrogen jackets, typically evaporate from one to several liters per day. One liter per day indicates a heat input into the liquid of 30 mW. The total enthalpy change of the vapor between 4 and 300K is roughly 70 times the heat of vaporization. The cooling capacity of the vapor, proportional to the specific heat Cp, is nearly independent of temperature over most of the range. This is not, in general, the optimum distribution of cooling capacity vs. temperature. A closed-cycle cryocooler with the optimum distribution is therefore, in principle, more efficient than an evaporative cryostat. Related to this is the important point that, for many small low-power superconducting devices, the lion's share of the cooling capacity is required at higher temperatures to intercept heat radiated and conducted from room temperature. To specify the performance of a cryocooler only in terms of its cooling capacity at the cold end, as is commonly done (e.g. p.iv, ref. 3), is therefore misleading.

CLOSED-CYCLE CRYOCOOLERS VS EVAPORATIVE CRYOSTATS

The relative merits of closed-cycle cryocoolers and evaporative cryostats have been debated many times. In 1966, Winters and Snow showed that an evaporative cryostat is economically preferable to a cryocooler for a heat load of 100 W (3300 ℓ/day evaporation rate) if the duty cycle is less than 95 hours per year for 10 years[9]. If one may extrapolate from this by 4 or 5 orders of magnitude, it might be concluded that a cryocooler could not possibly be economical at milliwatt refrigeration levels. Nevertheless, economic analyses are imperfect at best, and may change with time as the data base changes. One such change now occurring is that superconductive electronics is becoming a widely used technology with large duty cycles. Another change which one may expect in the future is that helium gas will become more expensive as the supply runs out. There is also a possibility that inventive genius may change the economics of small cryocoolers by discovering better and simpler refrigeration methods (see for example W. A. Little, this conference, also R. Radebaugh, p.7, W. A. Little, p. 74, W. A. Steyert, p.81, R. Radebaugh, p.93, and J. E. Zimmerman and R. Radebaugh, p.49, ref. 1). A factor which might significantly affect the economics of cryocoolers is the development of high-T_c devices (M. Beasley, this conference and p. 167, ref. 1).

COST AND RELIABILITY OF CRYOCOOLERS

Data on cost and physical characteristics of practically all cryocoolers built up to 1973 have been collected by Strobridge[10]. More data on small cryocoolers, primarily for infrared sensors and space applications at temperatures of 10 to 100K, are given in reference 2 (W.S. Sims, p. 1, V. C. Johnson, p. 11, and W. D. Clarke III, p. 17). This reference provides interesting information (p. 15) such as that the Air Force had expended over 9 x 10^6 dollars, as of 1973, on two objectives (1) "decrease the cost of ownership for airborne cryogenic coolers, and (2) "increase the unattended life of cryogenic coolers for spacecraft to something in excess of three years." The second objective had not, and probably has not yet, been achieved. The reference contains only one paper relating to cryocooler technology for elemental superconductors (W. Lawless, p 417, ref. 2, see also R. Radebaugh, p.7 and p.93, ref. 1) but several papers (e.g. B. Renyer, p 147, ref. 2) describe cryocoolers capable of cooling high-T_c devices. Reference 2 also contains some speculation on cost of cryocoolers. S. Horn _et al._, p. 73, and W. Sims, p. 1, ref. 2, have estimated prices of 1000 to 2000 dollars for the very simplest (Vuilleumier) single-stage miniature 77 K closed-cycle cryocoolers, in lots of 1000. Miniature multistage Stirling and Vuilleumier machines for low temperatures, 6 to 8K, should not necessarily cost much more than this, since, by using the stepped-displacer principle of Fokker and Kohler[11], the number of essential moving parts (approximately 5) is precisely the same as for a single-stage machine, and only the displacer geometry is more complicated. Other information of general interest from reference 2 are the comments on reliability of closed-cycle cryocoolers. W.Clarke III, p.17, stated that "failures occur between 120 and 300 hours of cooler operating time on all Air Force airborne systems," which includes Joule-Thomson, Stirling, and Gifford-McMahon cryocoolers. Chellis, p.111, ref. 2, reports a mean time between failure of 20000 hours for several hundred Gifford-McMahon cryocoolers used (on the ground) to cool parametric amplifiers for satellite communications and also for military purposes. Clarke and Chellis were reporting on different systems with different reliability requirements, but the contrast is nevertheless impressive. More recent information on reliability is given by Chellis (p.109, ref. 1) on 20K machines and by Higa and Wiebe (p.99, ref. 1) on 4K machines. About 30 or 40 of the latter are in use to cool masers for deep-space probes.

LARGE AND SMALL MACHINES AND APPLICATIONS

In the literature and in private conversation, cryocoolers with one to several watts of refrigeration at the cold end are often called "small" or "miniature." In the context of this paper, such machines will be referred to as "large." This paper will deal primarily with concepts of cryocoolers for devices such as single Josephson junctions or small arrays of junctions whose inherent refrigeration requirements are in the sub-milliwatt level. At least one author (W. A. Little, p. 74, ref. 1) has pointed out that present commercial cryocoolers are grossly mismatched to such devices. With regard to large superconducting systems such as computers, which may actually require a watt or or more of refrigeration, present cryocooler technology (F. Chellis, p.109, R. Longsworth, p.45, W. Higa, p.99, ref. 1) should be readily adaptable, perhaps with some effort to reduce vibration levels (see for example, J. Cox and S. Wolf, p.123, ref.1). The computer application of cryocoolers was considered in detail by van de Hoeven and Anacker (p.221, ref.1), who indicate that refrigeration requirements may fall in the range of 0.1 to 10W.

A few "large" machines have been used for superconducting devices and research. The first such use was by W. Hartwig and his colleagues to cool a superconducting Pb cavity (for a review of this and other uses see p.135, ref. 1). Operation of a SQUID susceptometer is described by D. Vincent (p.131,ref. 1), and a cryocooler has been used or several years for low-temperature research by W. Little (private communication). Both of the latter two machines were originally provided by the Office of Naval Research (E. Edelsack) for operational evaluation with superconducting devices.

It is arguable that there is a need for small cryocoolers for such applications as SQUID magnetometers for geothermal prospecting (W.D. Stanley, p.205, ref. 1, and J. Clarke, ref. 5), gradiometers for magnetic anomaly detection[12] and for biomagnetic research (reviewed by S. Williamson and L. Kaufmann, p. 177, ref. 1),

Josephson voltage standards (L. Holdeman and C. Chang, p.243, ref. 1), microwave and millimeter-wave detectors (J. Edrich, p. 159 and M. Nisenoff, p.221, ref. 1) and a variety of other laboratory research instruments. There have been virtually no studies of small or miniature cryocoolers other than the recent one by W. A. Little on microminiaturizing the Joule-Thomson process (p.74, ref. 1).

As far as I am aware, the only experimental small (milliwatt) cryocooler work currently going on is that by me and my colleagues on a low-speed low-power 4-stage split Stirling machine for operating a Nb SQUID at 8.5K (p.59, ref. 1). This machine is unique in several respects, (1) it requires an order-of-magnitude-lower input power than comparable machines that have been built for the same temperature range, (2) it employs a gap regenerator, making a displacer and heat exchange mechanism of ultimate simplicity, and most important (3) nylon and epoxy-glass composite materials are used for the displacer and cylinder to minimize magnetic interference caused by moving parts near the low-temperature end.

The requirements for small cryocoolers were estimated by M. Simmonds (p. 207, ref. 1). Except for the susceptometer, his estimates of refrigeration requirements for various devices are all of the order of a few mW. Even these estimates are considerably higher than the above calculations would indicate, but it may be assumed that he is biased more toward commercial realism than academic idealism. In any case, even if extremely low refrigeration power satisfies the steady-state requirements, cooldown times may be too long for some applications. Estimates of acceptable cooldown times and other parameters vary considerably (M. Simmonds, p.207 and M. Nisenoff, p.221, ref. 1).

FUTURE TRENDS

With regard to the future of superconductive electronics, there is a great need for intelligent but uninhibited analysis and experimentation on cooling methods for small superconducting devices and systems. Classical gas-cooling systems like Stirling and Gifford-McMahon machines are highly developed and effective for temperatures as low as 6 to 8K. This is tantalizingly close to the temperature range needed for most present devices. It is quite adequate for high-T_c devices, and although the latter will undoubtedly serve some of the needs of superconductive electronics, lower-temperature cryocoolers will be essential for lower noise devices, for devices like the super-Schottky diode[13] which must operate well below the transition temperature, and to make use of elemental superconductors with various special properties. To span the temperature region from about 8K down to 1 or 2K, the development of solid-state processes as alternatives to Joule-Thomson or other processes requiring compressors would be highly desirable. On the other hand, perhaps a low-temperature low-pressure stage using helium (or helium 3) gas in a Stirling cycle would be practical. For very low refrigeration power, a few cm^3 of liquid helium produced by a single expansion, with a starting and shield temperature of 8K, could last practically forever. Long-term reliability is probably the most difficult thing to achieve with mechanical systems, so we should emulate Steyert by considering other refrigeration processes for higher temperatures as well. Even though conventional gas refrigerators may be hard to beat, there are numerous ways to improve and simplify these systems--gas bearings, thermal compressors, absorption-desorption compressors, resonant spring-mounted or free pistons and displacers driven by mechanical reactive coupling rather than by crankshafts and connecting rods, and (no doubt) hundreds of other good ideas.

Radebaugh (p.7, ref. 1) has written an unusually informative and stimulating review of refrigeration fundamentals. In addition to the commonly-used gas and gas-liquid systems, he describes a broad range of other systems, tells what kind of forces (electric, magnetic, tensile, centrifugal, etc.) are required to change the entropy, and estimates the minimum useful entropy change. An example of the many less-well-known systems is a rubber-band refrigerator, with which a modest amount of refrigeration might be obtained in the neighborhood of room temperature.

Space applications of superconductive electronics pose special problems for cryocoolers in terms of weight, power and reliability. Work on large cryocoolers (mostly classified) under the sponsorship of Wright-Patterson Air Force Base is continuing, with a multi-stage Vuilleumier cryocooler under development for launching in a spacecraft within a few years. For certain farout missions, cooling of superconducting devices by radiation into space has been suggested. I have estimated that a black aluminum sheet 1 mm thick, carefully shielded from the sun and not too close to any of the planets or their satellites, and attached to the spacecraft by a practical low-conductivity suspension, should reach a temperature of 5 to 10K in about 2 days. The estimate is based upon evidence that the total energy flux in interplanetary space, including zodiacal light, starlight, and cosmic background, but not including sunlight, adds up to something like 10 to 100 $\mu W/m^2$. Solar energy flux at the earth's orbit is 1.3 kW/m^2. A number of applications of superconductive and other devices in space were reviewed by E. Tward and P. Mason and by J. Vorreiter and C. McCreight (p.227, and p.153, ref. 1).

As for specific suggestions, there is a well-defined need for a cryocooler for a biomagnetic SQUID gradiometer. This instrument would operate in a stationary, well-controlled environment, where self-generated vibration and magnetic interference could be reduced by careful mounting and shielding techniques. If an economical and reliable cryocooler could be produced for this application, it would make magnetocardiography and pneumomagnetic measurements acceptable for general clinical use. Experience gained with such systems could lead naturally to development of portable instruments for geomagnetism and magnetic anomaly detection. Another, less demanding, application for a small cryocooler was suggested by L. Holdeman and C. Chang (p.243, ref. 1), a simplified portable Josephson voltage standard. Systems such as these could serve as indicators of the viability of small cryocoolers for many other applications, and should not require large amounts of money for their development.

REFERENCES

1) Applications of Closed-Cycle Cryocoolers to Small Superconducting Devices, edited by J. E. Zimmerman and T. M. Flynn. National Bureau of Standards, Special Publication 508 (April, 1978) (in press).

2) Closed-Cycle Cryogenic Cooler Technology Applications, Vol. 1, Tech. Report AFFDL-TR-73-149. Wright-Patterson Air Force Base (Dec. 1973).

3) 1976 Navy Study on Superconductive Electronics, edited by A. H. Silver, Office of Naval Research Report No. NR319-110 (Aug. 1976).

4) Superconducting Quantum Interference Devices and their Applications, edited by H. D. Hahlbohm and H. Lubbig. Walter de Gruyter, Berlin (1977).

5) Superconducting Machines and Devices, Vol. B1, edited by S. Foner and B. Schwartz, Plenum Press, New York (1974). Primarily on very large systems.

6) R. J. Dinger, J. R. Davis, and M. Nisenoff, NRL Memorandum Report 3256. Naval Research Laboratory. (Mar. 1976).

7) V. Vorreiter, private communication. See also ref. 8.

8) C. W. F. Everitt, *et al.*, To Perform a Gyro Test of General Relativity. Final Report on NASA Grant 05-020-019. High Energy Physics Laboratory, Stanford University (July 1977).

9) Technology of Liquid Helium, edited by R. H. Kropschot, B. W. Birmingham and D. B. Mann. NBS Monograph 111. (Oct. 1968) p. 85.

10) T. R. Strobridge. Cryogenic Refrigerators - an Updated Survey. NBS Tech. Note 655 (1973).

11) G. Prast, Philips Tech. Rev. 26, 1 (1965). J. Kohler, The Stirling Refrigeration Cycle. Scientific American, 242, 119 (1965).

12) W. M. Wynn, C. P. Frahm, P. J. Carroll, R. H. Clark, J. Wellhoner, and J. M. Wynn, IEEE Trans. MAG-11, 701 (1975).

13) M. McColl, M. F. Millea and A. H. Silver, IEEE Trans. MAG-13, 221 (1977).

DESIGN AND CONSTRUCTION OF MICROMINIATURE CRYOGENIC REFRIGERATORS

W. A. Little
RAI, 15 Crescent Dr., Palo Alto, CA, 94301

ABSTRACT

The small size and small refrigeration requirements of superconducting sensors make these devices a poor match to conventional cryogenic refrigeration systems. We have considered the design and construction problems of building microminiature refrigerators for the cooling of these devices which would match their cooling requirements. Preliminary work is described on the construction of a first stage 77 K refrigerator built on a Si-wafer using photolithography.

INTRODUCTION

During the past decade new electronic devices have been developed which are based on the Josephson effect in superconductors. These include supersensitive magnetometers, gradiometers, voltage standards, bolometers and logic elements. The power dissipated by these devices in the cryogenic environment is typically of the order of microwatts. Traditionally the low-temperature environment has been provided by a bath of liquid helium. This is not particularly convenient and an effort has thus been made in recent years to use small closed-cycle refrigerators for this purpose. However, these refrigerators typically have a capacity on the order of watts and are thus poorly matched to the minute refrigeration requirements of the devices. In addition the refrigerators are noisy, they vibrate and their moving parts modulate the local magnetic field. For these reasons we have considered the problem of building a microminiature refrigerator which would match more closely the refrigeration requirements of the devices and, in addition, would be quiet in operation.

DISCUSSION

We chose to study the Joule Thomson system because it is simple and has no moving parts even though it is less efficient than one using the Stirling or Gifford-McMahon cycles. Earlier[1] we had developed a set of scaling laws based on the operation of a counterflow heat exchanger. If in such an exchanger d is the diameter of the gas lines, ℓ their length and $\dot{m}$ the mass flow per unit time, then one can show that for refrigerators of the same efficiency, $d \approx (\dot{m})^{0.5}$, $\ell \approx (\dot{m})^{0.6}$ and the cooldown time $\approx (\dot{m})^{0.6}$. The refrigeration capacity is proportional to $\dot{m}$ and thus for a large reduction in the refrigeration capacity the gas lines must be of very small diameter. For example, for a refrigerator of a

ISSN: 0094-243X/78/421/$1.50

milliwatt capacity d $\approx$ 25 microns and ℓ a few cms. Some of the advantages which accrue though with this reduction of scale are the following: A cooldown time of a few seconds is anticipated and an operating time of many thousands of hours on a standard cylinder of gas. The fine gas lines of the heat exchangers also introduce another feature. In all conventional counter-flow heat exchangers the gas flow must be sufficient to maintain turbulent flow through the exchanger; otherwise the poor mixing of the gas will result in a large temperature difference between the two gas streams. With a great reduction in the size of the gas lines the entire gas flow can lie within the boundary layer and turbulent flow is then unnecessary for an efficient exchange of heat. One may thus operate under streamline conditions, free of vibration and turbulence noise. This should have some value for SQUID devices.

The problem remains as to how to construct such a refrigerator whose components are 10 to 100 times smaller than their conventional counterparts. The use of finned tubing and soldered or welded construction becomes difficult and expensive if not impossible. For these reasons we have been exploring the possibility of using photolithography for the construction of the refrigerator. I would like to present some details of progress in this area.

Several different approaches have been considered and the most advanced makes use of the extensive work done in the semiconductor industry on single crystal silicon wafers. We have constructed the major part of a J-T refrigerator designed to produce 100 milliwatts of refrigeration at 77°K on a 2" Si-wafer. The procedure we use is the following:[2] A 2" wafer of p-type Boron-doped Si oriented in the (100) direction and 300 microns in thickness is used. This is oxidized using a wet oxygen treatment to produce a 9000 Å oxide layer on the polished face of the wafer. Photoresist is spun on, prebaked, exposed under a mask illustrated in Figure 1, developed and post-baked. Where the photoresist is removed, the exposed oxide is etched away with a buffered HF etch exposing the underlying silicon. The silicon oxide then acts as an inert mask to allow the gas lines to be etched in the Si. This is done using an anisotropic etchant, ethylene diamine, resulting in V-grooves provided the grooves are oriented in one of the (110) directions. Upon the completion of the etch the photoresist and oxide layers are stripped from the Si and the wafer is cleaned. An optically flat pyrex plate is then anodically bonded to the Si surface.[3] Input and output lines are then drilled in the reverse side of the wafer and a second etch used to remove excess Si. Stainless steel hypodermic tubing gas lines are then bonded to the reverse side with epoxy. The refrigerator is then complete.

The advantage of the photolithographic technique is that it allows one to construct the entire refrigerator - gas manifold, particulate filter, heat exchanger, Joule-Thomson expansion nozzle and liquid collector - in one step. The refrigerator illustrated in Figure 1 is designed to operate from room temperature to 77 K using N_2. It could also be used as an add-on to a closed cycle

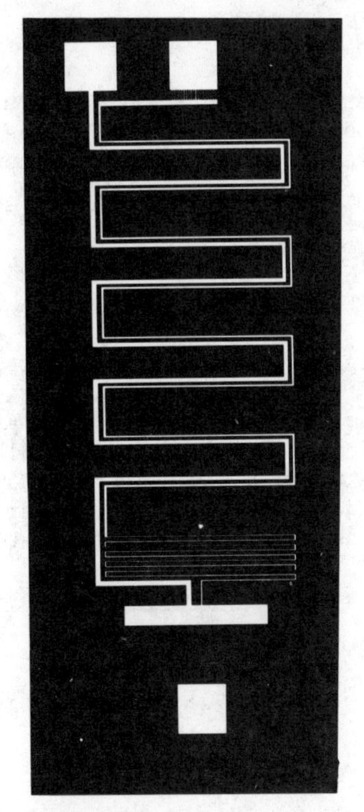

Figure 1. Photoresist mask used for etching microminiature refrigerator showing particulate filter, heat exchanger capillary expansion tube and liquid collector.

20 K cooler to produce refrigeration at 4 K using He.

Using a different mask one could equally well build on the same size wafer a refrigerator using three gas cycles; N_2, H_2 and He. This would refrigerate to 4 K.

The problems one faces in implementing this are related to the high thermal conductivity of Si, which is comparable to that of Cu at room temperature. This requires one to use very thin sections of Si, which are fragile, and to remove all excess Si near the gas lines. Secondly, very thin ($\approx$ 200 micron) optically polished pyrex plates must be used to seal the gas lines. For these reasons we have considered an alternative approach using instead of the Si a Cu-covered circuit board as the substrate or stainless steel wafers. This work has been done by Richard Hollman, a graduate student at Stanford. The same design has been etched into the surface of the copper or stainless steel films but an improved surface finish is needed to obtain a reliable seal to the cover plate. Work is in progress to develop a reliable sealing technique.

ACKNOWLEDGMENTS

I am indebted to Jim Mercereau and Harry Notarys of Cal Tech for advice and suggestions on several aspects of the work, in particular, on photolithography and adhesives. I would also like to thank Jacque Beaudoin, Stephen Terry, and Zora Norris, and the members of the Integrated Circuits Laboratory, Stanford University, for their advice, enthusiasm, and help.

REFERENCES

1. W. A. Little, Proceedings of the NBS Cryocooler Conference, Boulder, Colorado, 1977 (to be published).
2. Stephen C. Terry, "A Gas Chromatography System Fabricated on a Silicon Wafer Using Integrated Circuit Technology", Technical Report No. 4603-1, NASA Grant NGR-05-020-690, May 1975, Stanford Electronics Laboratory, Stanford University, Stanford, California; this contains a description of the techniques used here on Si.
3. G. Wallis and D. I. Pomerantz, J. of Applied Phys. Vol. 40 No. 10, 1969, p. 3946.

DIGITAL JOSEPHSON TECHNOLOGY - PRESENT AND FUTURE

Nancy K. Welker
Fernand D. Bedard
Laboratory for Physical Sciences
College Park, MD 20740

ABSTRACT

The area in which superconductive electronics is likely to have its broadest ranging impact is in the use of Josephson-junction devices for large-scale digital processing. This prediction arises from the fact that no semiconductor devices simultaneously possess the short switching delays (∿10 picoseconds) and low power dissipation (<1 microwatt/device; speed-power product $<10^{-16}$ joules/gate) exhibited by today's Josephson devices. Even projections for the five-to ten-year time frame indicate an advantage in speed-power product for Josephson devices of two to four orders of magnitude. The advent of Large Scale Integration (LSI; $\sim 10^3$ gates/chip) necessitates such low power devices since the best conventional cooling techniques will handle power densities of only tens of watts/cm^2. However, packing densities must be high for any high-speed technology to prevent the propagation delays from limiting the clock speed. Josephson devices appear ideally suited as a solution to the LSI problem. Naturally, a number of obstacles must be overcome before this can become a reality. Elemental or alloy superconductors must be found with which one can make reliable junctions which can be cycled between room temperature and cryogenic temperatures a sufficient number of times to permit servicing and normal operating routines. Techniques must be developed for fabricating Josephson junctions which are sufficiently well controlled to provide reproducible critical currents from device to device and from chip to chip. Systems problems such as how to provide power and clocking to circuits must be solved. Packaging is perhaps the most critical problem facing this technology. It has been demonstrated that on-chip speeds are very fast, but unless these speeds can be essentially maintained off-chip, the technology will lose its competitive edge over semiconductor techniques. During the past several years a great deal has been accomplished in all of these areas. This paper will attempt to present the current status of digital Josephson technology and to indicate some of the directions it will take in the future.

INTRODUCTION

Superconductivity has been proposed and tried with mixed success for a number of application areas. A major example was the development of cryotrons as computer switching elements which received great attention and funding, but whose performance did not exceed that of room temperature competitors, and, therefore, did not justify the exotic cooling required. With the development of

ISSN: 0094-243X/78/425/$1.50

the Josephson device, however, and with the use of readily available semiconductor fabrication techniques, we find that the greatest impact of superconductivity upon the applications world may well be in high-performance digital electronics.

The reason that the Josephson junction as a computer switching element is regarded as exceptionally promising in spite of the reluctance to go to low temperatures is shown in Figure 1. It gives

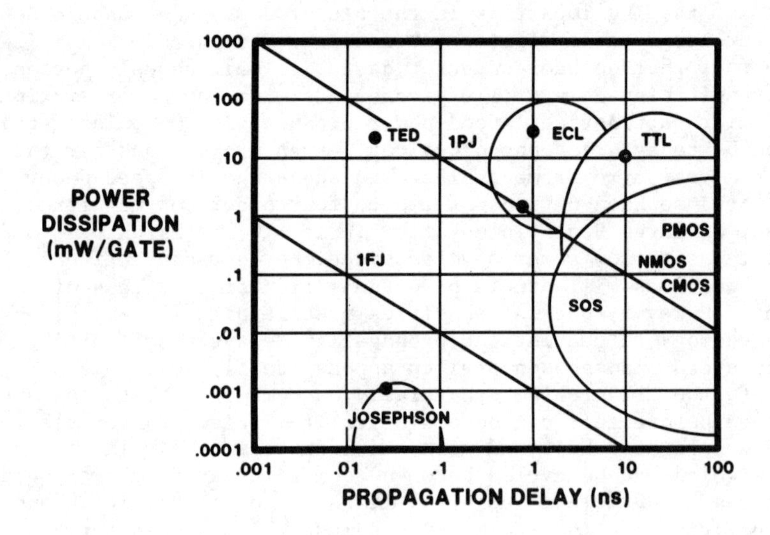

Figure 1. Power dissipation per logic gate versus propagation delay for representative semiconductor and Josephson devices. Lines of constant power-delay products of 1 picojoule and 1 femtojoule are shown.

the power dissipation per logic gate versus propagation delay for Josephson devices and for a broad sampling of semiconductor devices. It is immediately obvious that Josephson devices are very fast with switching times on the order of 5-30 picoseconds[1,2] and fully loaded logic delays on the order of 100 ps. There are, however, semiconductor devices which are just as fast, e.g., GaAs transferred electron devices (TED's). The big difference in performance potential arises when we look at the power dissipation axis. The Josephson devices dissipate 1 μwatt/gate or less[1] while other high-speed devices dissipate milliwatts/gate. Another way to look at this is in terms of power-delay product. The semiconductors cluster around the 1 picojoule line while the Josephson devices are below the .1 femtojoule line. Power density becomes critically important with the advent of Large Scale Integration (LSI; $\geq 10^3$ gates/chip). Present large computers typically have on the order of 10^5 gates total. For reasons of improved performance, reliability and cost, LSI is clearly mandated for future systems. One can envision 10^5 or more Josephson devices on a cm^2 chip with no heating problems while that density of fast semiconductors is not feasible since the best conventional cooling techniques will handle power densities of only tens of watts/cm^2. Table I indicates the power that can be transferred from a chip in various cooling media.[3] For the

TABLE I - Heat Transfer from planar surfaces.

Medium	Q_m(W/CM^2/DEG)	Q_m(W/CM^2) for ΔT = 20K
Free Air	$\sim 10^{-3}$	∿.02
Forced Air	$\sim 5 \times 10^{-2}$	∿1
Boiling Fluorocarbon	∿1	∿20
Boiling Helium	∿.05	--

semiconductor devices the only alternative is to go to lower packing densities; but, since on-chip signal propagation rates are $\sim 10^{10}$ cm/sec, this delay, rather than the switching speed, will become the limiting factor on the clock rate. Since in one logic delay (∿100 ps) a signal will propagate only ∿1 cm, one would ideally like to have the largest dimension in his entire system to be on the order of 1 cm. It is only with the very low-power Josephson devices that one can even approach this goal.

MATERIALS AND FABRICATION

Many types and geometries of Josephson junctions have been discussed at this meeting, but the type used nearly always in digital applications is the thin film tunnel junction with its hysteretic current-voltage characteristics shown in Figure 2. This type is preferred because the high resistance below the gap voltage, V_g, allows the transfer of most of the junction current into the external load. Devices can be either latching, in which case the load resistor R_1 causes the junction to be set to the gap voltage where it will remain "latched" until the gate current is lowered; or non-latching, in which the load resistor R_2 causes the junction to be set at or below V_{min} from which it will spontaneously reset to zero. In addition to not requiring a reset cycle, non-latching logic has the advantage of power levels which are down by several orders of magnitude from latching logic. Most often the tunneling barrier is a native oxide of the base electrode grown either in air or by dc or rf glow discharge. This oxide must be extremely thin (∿30Å) in order for the Josephson phenomena to exist, thus putting severe demands on the control of the fabrication process. One way around this problem is to use a deposited semiconductor barrier[4] so that the barrier thickness can be on the order of 10^2-10^3Å. The

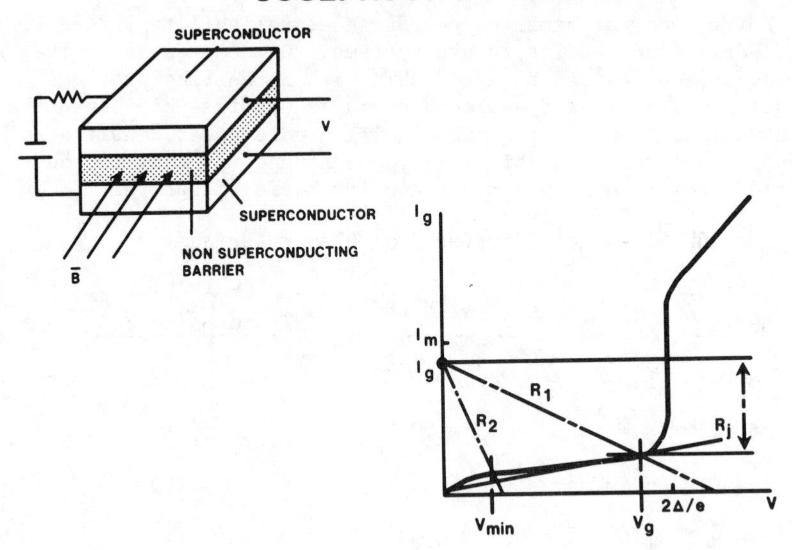

Figure 2. Josephson tunnel junction and its gate current (Ig) versus voltage (V) characteristic.

high resistive damping of the semiconductor barrier also lowers the Q of the device and thus suppresses the cavity resonances which can be a problem in oxide barrier junctions. Similarly, devices have been fabricated by selectively thinning a silicon substrate in the proper locations and depositing the base electrodes and counter-electrodes on opposite sides of the substrate which then forms the tunneling barriers.[5] An interesting variation on this idea is to place the base electrode and the counterelectrode in close proximity on a semiconducting substrate and allow tunneling to occur laterally through the substrate.[6,7]

There is really one fundamental requirement on the fabrication techniques used in digital technology, i.e., that they be reproducible. When one is building an instrument such as a magnetometer containing a single junction, one can select from a large number of samples or, alternatively, tailor the rest of the circuit to match the junction characteristics. However, if we are to build digital systems containing hundreds of thousands of Josephson devices, we must be able to obtain the same characteristics, e.g., critical current, from device to device and from chip to chip. A big step in that direction was taken with the development by Greiner of an rf oxidation process for forming the tunneling barrier on Pb.[8] In this process during one half-cycle the sample is sputter-cleaned, and during the other half-cycle oxidation takes place. As shown in Figure 3, the rate of change of oxide thickness is the difference between the oxidation rate and the sputtering rate. The oxidation

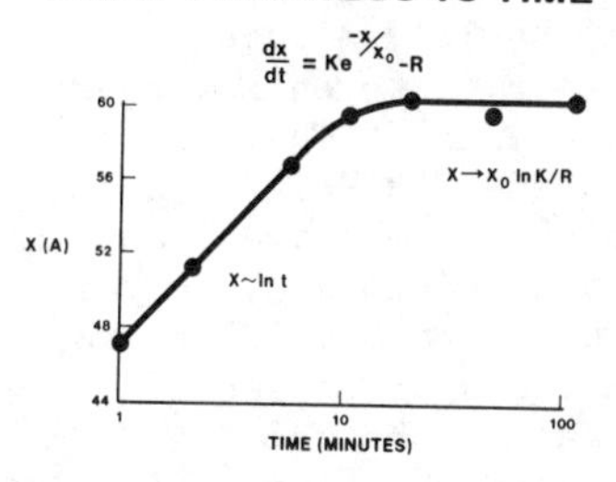

Figure 3. Barrier thickness versus duration of RF oxidation time for Pb tunnel junctions. R is the sputtering rate, and K and X_0 are oxidation parameters.

rate decreases with increasing oxide thickness and depends on the oxidation parameters K and X_0 while the sputtering rate is a constant R. Thus, when the two terms are equal, an equilibrium thickness $X_L = X_0 \ln (K/R)$ is reached. By adjusting such parameters as partial pressure of oxygen and sputtering voltage, one can obtain the desired oxide thickness and, hence, critical currents, and repeat them from run to run. The essential feature is that, after a certain thickness is achieved, the process shows little or no time dependence so that it is easily repeatable.

Another concern in digital devices is the choice of superconducting materials. Most early work was done on the soft elemental superconductors such as tin[9] and lead[10] since they are easy to deposit and oxidize. However, tin should be avoided if possible since its low transition temperature would necessitate operation below the boiling point of liquid helium. The problem with lead is even more severe. It has been repeatedly observed that lead films experience mechanical stress during thermal cycling because of differential thermal contraction with the most desirable substrate materials. This causes dislocation flow and/or grain boundary deformation and leads to the growth of hillocks and whiskers.[11] Thus, junctions formed from pure lead films on glass or silicon substrates will develop shorts upon returning from 4.2 K to room temperature. One of the means investigated to improve this situation is to alloy the lead with other materials such as indium and gold to form intermetallic compounds in order to improve mechanical stability.[12] Figure 4 shows the number of cycles between room temperature and 4.2 K withstood by various film compositions before the first hillock was observed in 1 cm^2. To test this method on actual Josephson tunnel junctions, strings of 19 gates having 25 x 25 μm junctions were fabricated on silicon wafers.[13] The base electrodes were a Pb-In-Au alloy while the counterelectrodes were a Pb-Au alloy. Typically one failure was observed after 50 cycles with some gates withstanding 200 cycles.

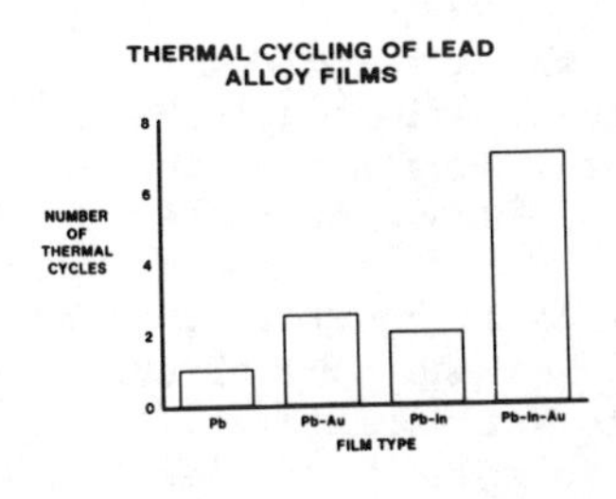

Figure 4. Number of cycles between room temperature and 4.2 K withstood by various lead alloy films before the first hillock was formed in 1 cm^2.

Another approach to the cyclability problem is to move away from the lead alloys to harder materials such as niobium.[14,15] They have much higher tensile strength than the lead alloys and are, therefore, much more stable under conditions of differential thermal contraction encountered during temperature cycling. They do, however, have the disadvantages of less than ideal BCS I-V characteristics, more difficult fabrication and higher dielectric constants of the native oxides which lead to higher capacitance and, therefore, slower switching speeds of the devices. One possible solution to this problem is to form Nb_3 Sn-Oxide-Pb junctions which appear to have a dielectric constant approximately equal to that of the Pb alloys.[16] It is clear from all of these results that the cyclability problem has not been solved but that good progress has been made in that direction. Once again, a solution to this problem is much more critical in digital technology than in other applications. If large-scale Josephson-device computers are to become a reality, individual devices must be able to withstand a significant number of temperature cycles with an extremely low probability of failure. Otherwise, every warm-up of the machine for repair will induce more faults than it will correct. Fortunately, once a machine is operating, few warm-ups should be necessary since there is very little chemical activity at 4.2 K, and few material failures should occur.

JOSEPHSON DEVICES AND CIRCUITS

When using a Josephson junction as a switching device, some means must be provided for switching the junction out of the zero voltage state. One way is to apply a magnetic field whose effect is to reduce the Josephson critical current. Thus, increasing applied flux can be made to decrease the maximum allowed Josephson current to a point below the gate current level, and the junction will switch. The field is created by a control line running over the junction. An alternative to magnetic field switching is current switching. In this case a bias current is combined with the input current to exceed the critical current and switch the junction.

Figure 5 shows a logic gate employing magnetic field switching.[17] It consists of a single junction across a terminated superconducting transmission line. From the threshold curve (I (gate) versus I (control)), it can be seen that when the control currents E, F and G are antiparallel to the gate current, it becomes a three input AND gate, i.e., for one unit of

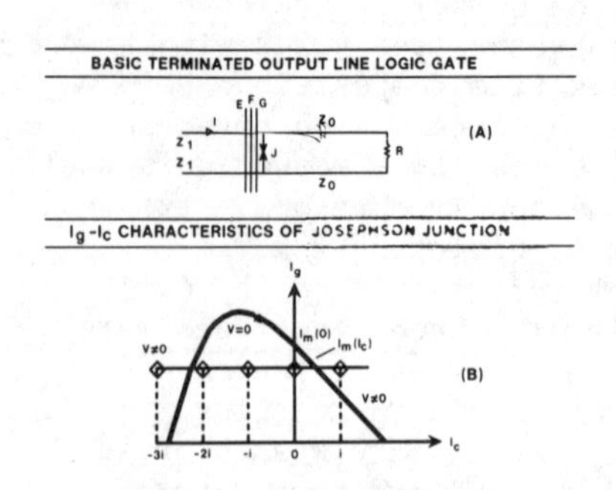

Figure 5. (A) Circuit for a three input magnetically switched in-line Josephson junction logic gate. (B) Threshold curve for an in-line gate over a ground plane.

gate current, three units of control current are needed to switch to the $V \neq 0$ region. For control currents parallel to the gate current, it becomes a three input OR gate. One unit of control current will switch the junction to the $V \neq 0$ state, diverting the gate current into the load resistor. Fanout is achieved by placing other junctions under the transmission line. The asymmetric shape of the threshold curve arises from having the controls in line with the junction over a ground plane.

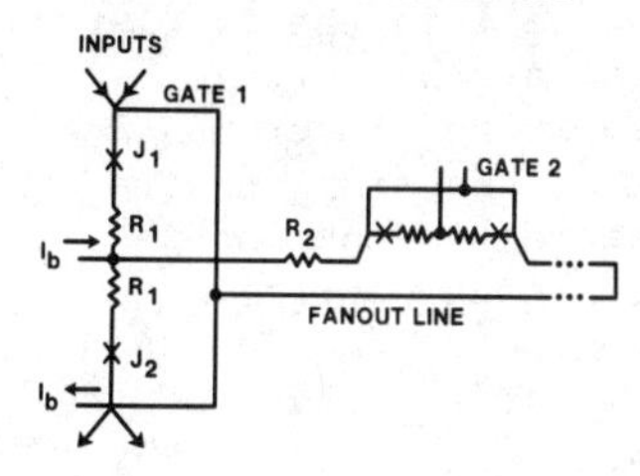

Figure 6. Current-switched Josephson logic gate with fanout gate.

Figure 6 is an example of current-switched or direct-coupled logic.[18] It consists of an input loop and a fanout loop into which the bias current, I_b, is diverted if the junctions are switched. Initially, the junctions conduct most of the current since $R_1 << R_2$. When the inputs are applied, most of their current also goes through the junctions because of inductive splitting. Therefore, the inputs and the bias add in J_2, and it will switch if the critical current is exceeded. Its current is then diverted to J_1; and, when this junction also switches, the current is diverted into the fanout line. By adjusting the magnitude of the inputs, AND and OR functions can be implemented. Care must be taken in direct coupled logic to provide isolation between circuits, but an advantage is gained by eliminating the last layer of metallization. Ultimately, speed considerations may be the deciding factor on which form of logic is chosen.

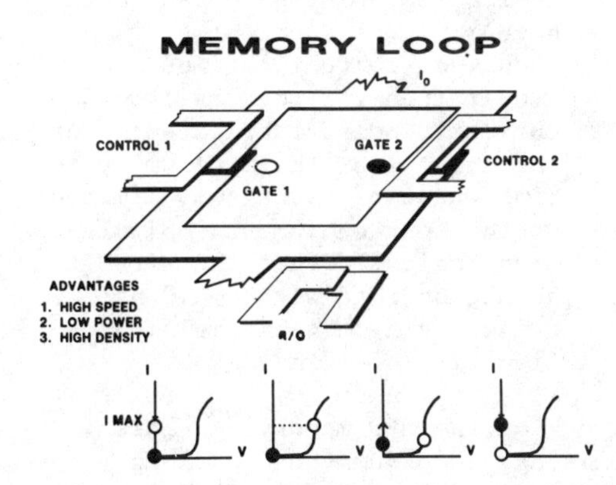

Of course, it is also necessary to be able to perform memory functions, and a simple memory cell is shown in Figure 7. Here current can be stored in either clockwise or counterclockwise direction to indicate a "0" or a "1", and the loop itself acts as a control for the read-out junction.

A variation on this memory cell is the single

Figure 7. Josephson memory loop with read-out gate. I-V curves trace loop through a state change (a) all current through gate 1; (b) gate 1 switched; (c) current decaying in gate 1 and increasing in gate 2; (d) all current through gate 2.

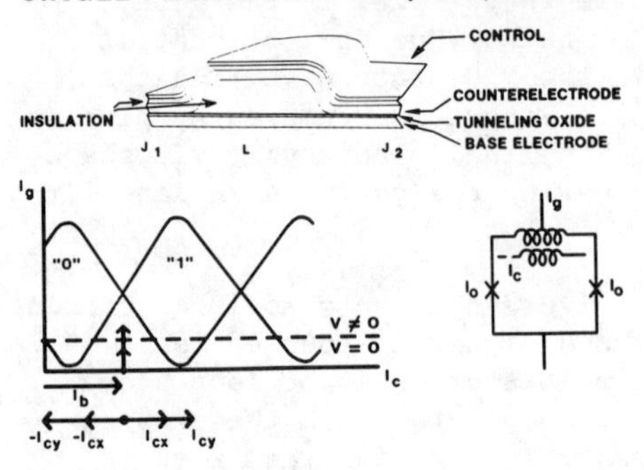

Figure 8. Single flux quantum (SFQ) Memory cell with threshold curve and equivalent circuit.

flux quantum or SFQ cell[19] shown in Figure 8. In this case the loop is not in the plane of the substrate but is raised out of the plane by a layer of thick insulation forming a very compact cell for mass memory applications. In operation the bias current, I_b, forces the cell into the overlap region of the vortex modes where either no flux or one flux quantum is stored in the loop. Appropriate control currents can then be used to select the state of the loop.

All of these types of Josephson devices have been incorporated into relatively complex circuits during the past several years. Figure 9 shows an 8 x 8 bit, non-destructive read-out, random access Josephson memory, fully decoded on a 6.35 x 6.35 mm chip.[20] Figure 10 shows a 4-bit multiplier consisting of a 4-bit adder with ripple carry and an 8-bit accumulator shift register.[21] Both of these circuits employ in-line latching gates and both have minimum line widths of 25 μm, so switching times are slower than for the small devices currently being fabricated. For the multiplier the minimum cycle time was 6.67 ns but was limited by the test equipment. Simulation indicated that the circuit should operate with a 3.0 ns cycle giving a 4-bit multiplication time of 12 ns.

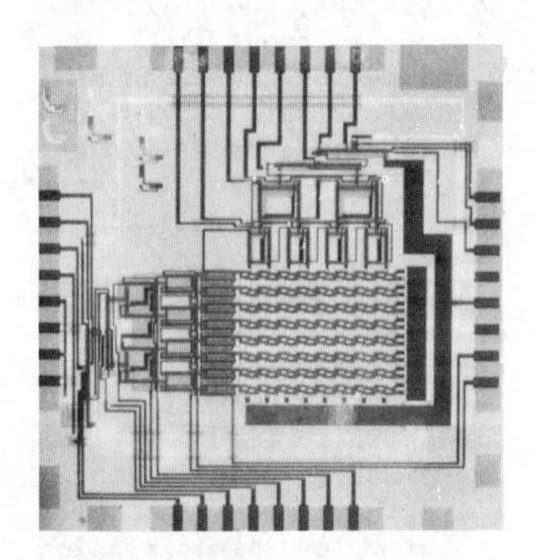

Figure 9. 64 bit, non-destructive read-out, random access Josephson device memory.

Figure 11 shows the latest work on the SFQ memory.[22] This chip contains 2048 SFQ memory cells and 2048 dummy cells along with the drivers and decoders needed to address and read out the memory array. Here read-out is destructive so the information must be rewritten after each read cycle. If fully populated, this chip would contain 16 Kbits of SFQ memory. The circuit is made up of lead alloy films with a minimum line width of 2.5 μm. Testing indicates that an access time of 15 ns and a cycle time of 30 ns should be achievable for the 16 Kbit array.

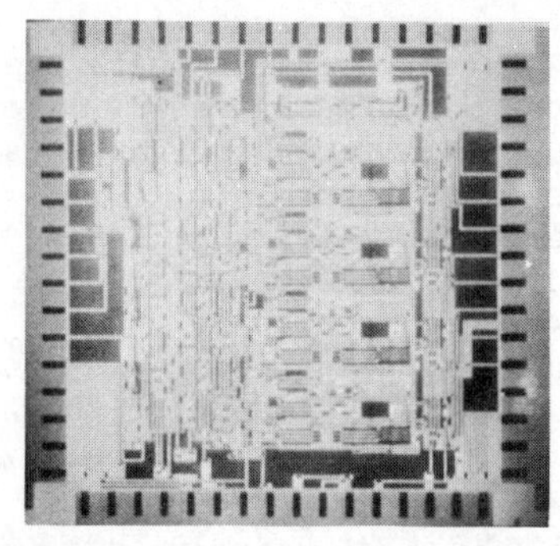

Figure 10. 4-bit Josephson device multiplier.

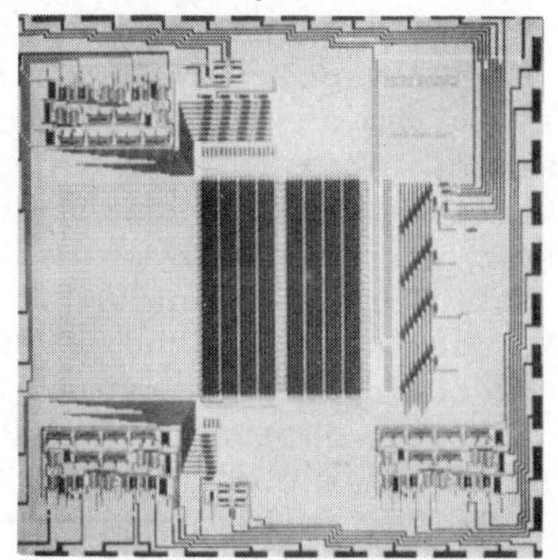

Figure 11. Single flux quantum (SFQ) memory cross-section of a 16 Kbit chip.

A structure similar to the SFQ cell has been used in logic applications and is known as an interferometer.[23] Figure 12 shows the threshold characteristic for a three-junction interferometer where the areas and, therefore, the critical currents of the junctions are in the ratio of 1, 2, 1. The greatest advantage in the use of interferometers for field-switched devices is that the shape of the threshold curve is entirely determined by the LI_0 product with larger LI_0 products producing steeper, narrower lobes. Therefore, the current levels can be reduced to increase speed and packing density. Also the capacitance can be reduced and, for a given metallurgy, is determined only by the lithographic limit on junction size. This leads to very fast, low-power devices. It has recently been reported[24] that for OR and AND gates constructed from three-junction interferometers, the delays were respectively 42 and 95 ps. This was achieved in a technology where transmission line widths and junction diameters were 5 μm.

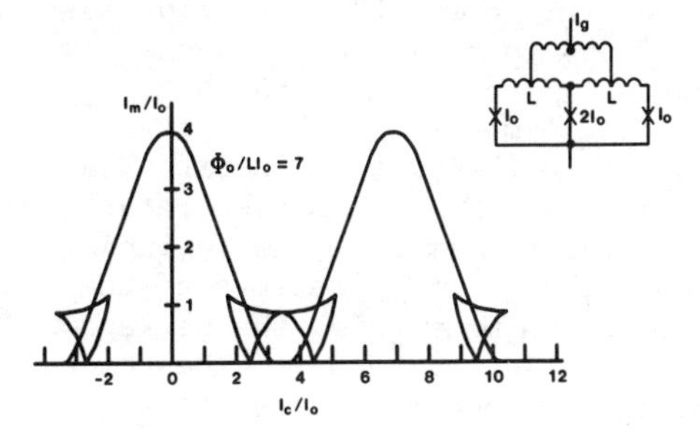

Figure 12. Threshold curve of three-junction interferometer with equivalent circuit.

SYSTEM CONSIDERATIONS

Most of the subjects under the heading of system considerations fall into the category of "things to do in the future," and they are extremely important since Josephson technology should be considered a total system technology. The extremely high performance will be fully realized only if the entire system is at cryogenic temperatures with a minimum amount of communication with room temperature components. This is another case where Josephson technology appears to offer some unique advantages. All interconnections in a Josephson system should be made using superconducting transmission lines which are practically lossless and distortion free. For example, calculations and measurements have been made on 25 μm superconducting transmission lines over a ground plane whose characteristic impedance is 5Ω.[25] For a 50 ps rise time pulse the additional delay due to 1000 transmission line crossovers was only 3 ps while the total propagation delay was ∿300 ps. Crosstalk between an x and y transmission line was found to be 0.2% in the worst case. These problems seem very tractable on-chip, but a great deal of work needs to be done to assure that these transmission line characteristics can be maintained through higher order packaging levels.

Many other system problems remain to be solved. Ways must be found to provide power to a Josephson system and to distribute it to large numbers of both memory and logic circuits. Clocking must also be provided and ways must be found to minimize clock skew from one section of a processor to another. A problem which is not unique to Josephson technology but is common to all LSI systems is that of testing and debugging. Since no chip with 1000 circuits can be probed directly whether it is a 4.2 K or room temperature, test circuits must be designed into the system from the beginning. Finally, the mechanical as well as the electrical aspects of the total package must be addressed since, in many ways, the package is the key to the eventual success of this technology. The unique feature of the technology is that the low power dissipation at high speed allows very high packing density. If this is not achieved for other reasons, the great promise of this technology will not be realized.

CONCLUSIONS

Remarkable and rapid progress has been made in digital Josephson technology in the past decade. Much of this progress is due to the ability to fabricate circuits which match design parameters. Since the Josephson equations are well understood, these circuits can be accurately simulated before fabrication. This progress has also been heavily dependent upon the advances in fabrication developed for the semiconductor industry. As this industry continues to improve its photolithography and electron-beam lithography, these advances will be readily available to further improve the speed, power and performance which Josephson technology promises.

If we consider the status of this technology with regard to fundamental limits we find that there is still much room for improvement. For present devices, the lowest energy stored is on the order of 10^{-18} joules,[2,26] while kT at 4.2 K is $\sim 5 \times 10^{-23}$ joules, leaving much room for improvement. From an uncertainty-principle point of view, $\Delta E \, \Delta t \simeq \hbar$, that amount of of energy should allow us to establish the system state in a time on the order of 10^{-16} sec. Finally, following Keyes[3], if we assume that the bit information energy generated by a logic device must be $\sim$ kT and will be sent via a transmission line of $Z_0 \approx 10\Omega$, then the minimum power becomes:

$$p = V^2/Z_0 = \frac{(kT/q)^2}{Z_0}$$

$$\simeq 10 \text{ nW/gate} \quad .$$

Clearly these assessments are not completely independent of one another, but they do provide bounds from different points of view. Nevertheless, with such refinements as non-latching logic and smaller devices, considerable improvement can be expected, and packing densities will be limited only by the lithographic and packaging techniques available.

Nearly all of the problem areas in this technology have been investigated to some extent. For some, acceptable solutions have been found; for others more work needs to be done.

REFERENCES

1. W. Anacker, Proc. IEEE Int. Solid State Circuits Conf., 162 (1975).
2. J. H. Magerlein and T. A. Fulton, Bull. Am. Phys. Soc., 22, 374 (1977).
3. R. W. Keyes, Proc. IEEE, 63, 740 (1975).
4. W. Y. Lum, H. W. Chan and T. Van Duzer, IEEE Trans. Magn., MAG-13, 48 (1977).
5. C. L. Huang and T. Van Duzer, IEEE Trans. Magn., MAG-11, 766 (1975).
6. T. Van Duzer, Private Communication.
7. A. H. Silver, Private Communication.
8. J. H. Greiner, J. Appl. Phys., 45, 32 (1974).
9. J. Matisoo, Proc. Symp. Physics of Superconducting Devices, N-1 (1967).
10. P. Pritchard and W. Schroen, IEEE Trans. Magn., MAG-4, 320 (1968).
11. S. K. Lahiri, J. Appl. Phys., 46, 2791 (1975).
12. S. K. Lahiri, J. Vac. Sci. Technol., 13 148 (1976).
13. S. Basavaiah and J. H. Greiner, J. Appl. Phys., 48, 463 (1977).
14. K. Schwidtal and R. D. Finnegan, Proc. Appl. Superconductivity Conf., 562 (1972).
15. R. F. Broom, R. Jaggi, R. B. Laibowitz, Th. O. Mohr and W. Walter, Proc. 14th Int. Conf. on Low Temp. Phys, 4, 172 (1975).
16. R. E. Howard, D. Moore, D. Rudman, and M. R. Beasley, Bull. Am. Phys. Soc, 23, 261 (1978).

17. D. J. Herrell, IEEE Trans. Magn, MAG-10, 864 (1974).
18. J. H. Magerlein and L. N. Dunkelberger, Proc. Appl. Superconductivity Conf. (1976).
19. P. Gueret, Th. O. Mohr and P. Wolf, IEEE Trans. Magn., MAG-13, 52 (1977).
20. W. H. Henkels and H. H. Zappe, (To be published IEEE J. Solid State Circuits).
21. D. J. Herrell, IEEE J. Solid State Circuits, SC-10, 360 (1975).
22. R. F. Broom, P. Gueret, W. Kotyczka, Th. O. Mohr, A. Moser, A. Oosenbrug and P. Wolf, Proc. IEEE Int. Solid State Circuits Conf. (1978).
23. H. H. Zappe, IEEE Trans. Magn., MAG-13, 41 (1977).
24. M. Klein, D. J. Herrell and A. Davidson, Proc. IEEE Int. Solid State Circuits Conf., 62 (1978).
25. Y. L. Yao, IEEE Trans. Parts, Hybrids and Packaging, PHP-12, 236 (1976).
26. T. A. Fulton and L. N. Dunkelberger, Appl. Phys. Lett., 22, 232 (1973).

SQUID DIGITAL ELECTRONICS*

J. P. Hurrell and A. H. Silver
The Ivan A. Getting Laboratories
The Aerospace Corporation
El Segundo, CA 90245

ABSTRACT

The use of SQUIDs enables the ultimate switching speeds and low power dissipation of Josephson devices to be achieved in digital logic systems. In addition, unique circuit capabilities become apparent which can simplify the logic circuits. These concepts will be applied to fast analog to digital conversion. A continuous signal is applied to a single-junction SQUID and converted to a pulse train of flux quanta. These quanta are counted in a scalar consisting of a linear array of coupled DC SQUIDs. Scaling is achieved by utilizing the bistable nature of a double-junction SQUID biased at the half-flux point. In this way fast, single-flux quantum logic can be achieved.

INTRODUCTION

Digital superconducting electronics involves the performance of Josephson junctions in superconducting circuits. Logic is performed by driving these junctions through superconducting to normal transitions, enabling current steering to occur. These transitions may be field-driven or current driven. Only in a SQUID configuration,[1,2] however, is the superconducting phase θ across the junction directly controlled. The SQUID combines the properties of the junction with those of fluxoid quantization and provides both quantum (or digital) states which have no dissipation and a proper impedance match for the junction at this quantum limit. As a result, this configuration will ultimately provide the fastest and lowest energy logic schemes.

Three types of SQUID encompass most of the phenomena appropriate to these issues. The single-junction SQUID (SJ-SQUID) or rf SQUID consists of a single junction in a superconducting ring of inductance L and damped by a load resistance R. In the resistive SQUID (R-SQUID), L and R are placed in series across the junction. The double-junction SQUID (DJ-SQUID) or dc SQUID consists of two junctions in the superconducting ring. Gross properties of these rings are controlled by the parameter $\beta = 2\pi L I_c / \phi_o$ where I_c represents the critical current of a junction. We restrict the use of the name SQUID to devices with $\beta \sim 1$.

*This work was supported by The Aerospace Corporation.

ISSN: 0094-243X/78/437/$1.50

This paper addresses the subject of analog-to-digital conversion. A SQUID design is presented which possesses the capability for high speed and precision, large dynamic range, and sensitivity limited only by fundamental noise contributions.

GAIN-BANDWIDTH CONSIDERATIONS

When the junction phase θ advances through 2π a junction energy $\sim \phi_o I_c$ becomes available for the circuit. Some of this energy is stored in L and the junction capacitance C. The rest is available for dissipation in R. A time constant (L/R) determines the rate at which this transfer to R is made. In both SQUIDS and R-SQUIDS with $\beta \sim 1$, the switching time for θ matches (L/R) and efficient transfer of energy occurs. For large β, the situation is different. In SQUIDS the switching time remains comparable to (L/R) and hence fast switching can be maintained with large R, until R becomes larger than the junction resistance. At the gap frequency, this restricts β to unity once again. In R-SQUIDS with large β, the switching rate for θ approaches the gap frequency, though the voltage across R increases only as (L/R). Consequently this transfer can be effected at the gap frequency only when $\beta = 1$ as before. Hence, in the fastest circuits, we should anticipate $\beta \sim 1$.

In a single fluxoid transition, the energy transferred to R will be $\sim \phi_o I_c$ in a time greater than or equal to $\phi_o / V_g = \nu_g^{-1}$ where V_g is the gap voltage and ν_g the corresponding frequency. Consequently, in a latched mode of operation in which the junction is cycling continuously, power can be transferred at a maximum rate $I_c V_g$ as expected. To induce the switching event, a certain amount of energy $\alpha \phi_o I_c$ must be supplied (which is stored in the capacitance in the latched mode). The minimum value of α is usually determined by margins for successful operation. This leads to the gain-bandwidth product = $V_g I_c / \alpha \phi_o I_c = \nu_g / \alpha$. Then a latching time T can be associated with a gain $(\nu_g T/\alpha)$. A natural lower limit on T is $1/\nu_g$, corresponding to an unlatched mode of operation at the highest frequency of operation for which the gain is only $1/\alpha$. In logic circuits gain is required for two important functions. First, fanout or bifurcation is necessary to perform arithmetrical operations; this can only be achieved with gain. Second, gain is usually required to generate logic gates where some degree of isolation between input and output circuits is necessary. As the speed of operation is increased, the gain margins are reduced and the circuits become more difficult to design.

Ultimately, circuits utilizing the full-speed capability of Josephson junctions will of necessity involve single fluxoid events in SQUIDs operating with small gain margins. At this level a number of subtle phenomena appear which can be used to advantage in these circuits. Some of these have already been exploited in the fabrication of single-flux quantum memory cells.[3] We propose a scheme for fast analog-to-digital conversion which also exploits in a unique way some of these phenomena to simplify the process and yet retains high speed operation.

ANALOG TO DIGITAL CONVERSION

Digitizing an analog signal can be accomplished in many ways. Those involving counters, such as dual slope integration and voltage-to-frequency conversion, tend to possess small signal bandwidths limited by the counting speed. More direct approaches utilize comparators; either a single comparator is used with successive approximations or a system of fully parallel comparators is used for high-speed operation. The latter system requires approximately 2^N comparators to generate a data stream corresponding to a resolution of N bits. In addition a stable resistive ladder is needed to provide the comparator levels. An aperture time $\tau_a \leqslant (2^N 2\pi f_s)^{-1}$ is required to convert accurately a signal of bandwidth $< f_s/2$ where f_s represents the sampling frequency.

Details of a superconducting A/D converter have been published[4] which utilized standard logic gates to generate a sample and hold circuit followed by parallel comparators. Four-bit resolution was achieved at a signal bandwidth of 30 Mhz. This modest result was not meant to represent state-of-the art performance, merely conceptual verification. McDonald[5] has proposed a different scheme using the Josephson effect to perform direct voltage-to-frequency conversion. Our approach utilizes a SJ-SQUID to quantize the signal directly into a pulse stream which is fed into a counter composed of a sequence of DJ-SQUIDs operating as scalars. The bit count is

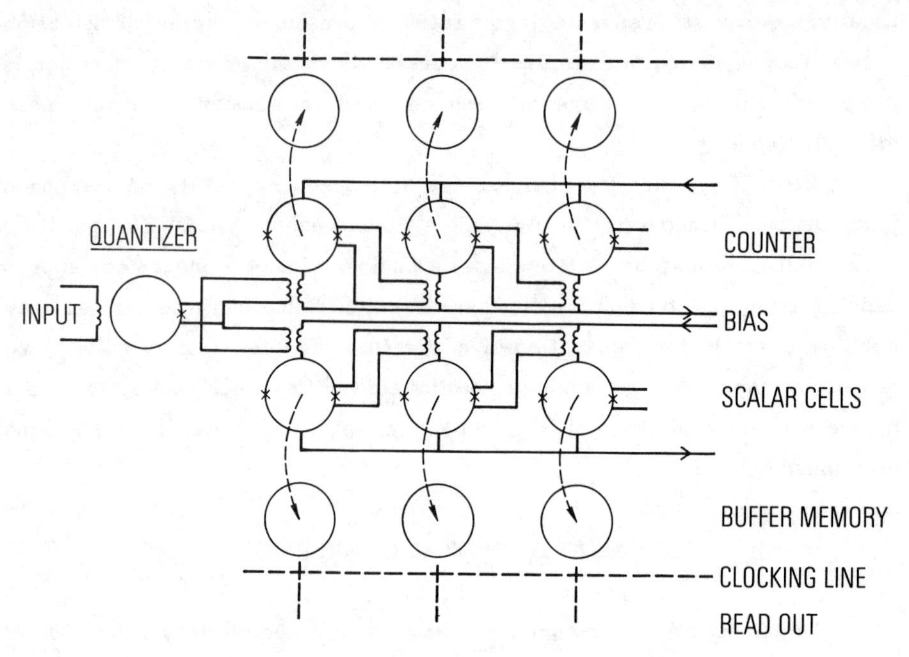

Fig. 1 Conceptual design for A/D conversion.

gated out of the scalar memory into a buffer memory at the sampling frequency. The conceptual design is shown in figure 1. The design of this circuit directly addresses the issues of ultimate aperture times and fluxoid electronics.

QUANTIZER

The stationary states[1] of a SJ-SQUID provide an ideal measure with which to quantize a signal in the form of applied flux because every fluxoid state is identical. Unfortunately this also means that no one state is unique. Hence changes in the signal are quantized but its absolute value must be determined by other means. During a sample time τ_s, the flux in the SQUID will increase or decrease in direct proportion to the applied flux changes in units of ϕ_o. The maximum number of transitions which can be detected in a sampling interval (τ_s / τ_a) is a measure of the bit resolution (N). By counting the increasing and decreasing pulses, a digital conversion of the signal is performed. When the β-value of the SQUID is close to unity, switching hysteresis is small and the bit count will be accurate. There exists ample evidence[6] that SJ-SQUIDs can be made to switch quickly, but there are complications in the design of microwave SQUIDs which will be discussed later.

If the SQUID is heavily loaded by a transmission line in order to provide critical damping or overdamping, the switching transient will launch a pulse of energy $\lesssim \phi_o I_c$ onto the line without ringing. The polarity of the pulse indicates whether the flux in the SQUID is increasing or decreasing. Every time a pulse is launched the signal flux will have changed by ϕ_o. When these pulses are collected and counted, A/D conversion is performed. The pulses will appear in a random fashion and not be quite identical because they depend upon the rate of change of input flux to a certain extent. This variation in pulse shape should remain small until $\dot{\phi}$ exceeds $\phi_o(R/L)$. A pair of counters separately scales the pulses of opposite polarity.

SCALAR CELL

The process of scaling requires storage of information and sequential processing. In superconducting memory cells, ones and zeroes can be stored as circulating currents in SQUIDs. In particular, SQUIDs biased at the half-integral flux point possess states of opposite circulating current with equal energies. In a SJ-SQUID for which $1 < \beta < 7.9$ there exist just two such states. During transitions between them, flux flows in or out through the Josephson junction, creating voltage pulses of opposite polarity. In the DJ-SQUID, a similar situation exists, namely, the possibility of two states of equal energy, but counter-circulating currents, at the half-integral flux point. Figure 2 exhibits the stationary flux states of a DJ-SQUID for two different values of bias current. When I = 0, the two stationary states at the half-

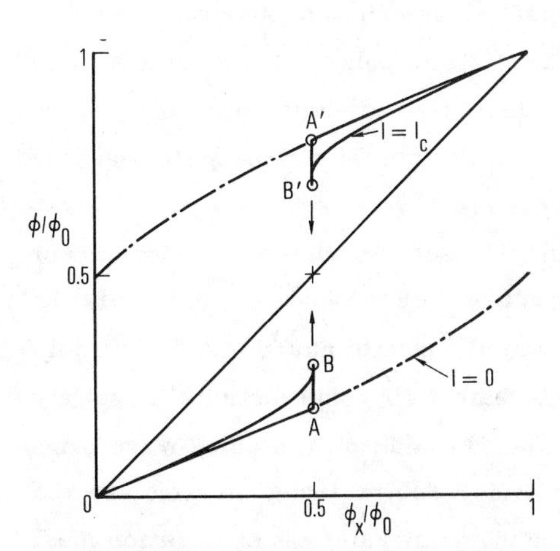

Fig. 2 Magnetization curve for DC SQUID with $\beta = \pi$.

integral flux point are represented by points A, A′. As the current is increased to I_c, the critical current at this flux bias for $\beta = \pi$, these points move to B, B′, respectively. When I exceeds I_c, the SQUID moves into the voltage state. The switching behavior is different from that of the SJ-SQUID because the flux is not required to flow in and out through the same junction. In fact, by applying a bias current to the SQUID it becomes energetically more favorable for flux to flow in and out of alternate junctions reminiscent of flux shuttling.[7]

When the bias current exceeds the critical value, this flux shuttling occurs continuously, giving rise to an oscillating current with a frequency half that of the accompanying voltage pulses across the SQUID.[8] If this current is used to drive an LC circuit resonance, steps are induced on the I-V characteristic.[9] By keeping the bias current below critical, this circulating current becomes stationary with the direction of circulation dependent upon past events.

If the bias current were momentarily raised above critical and then returned to sub-critical, the direction of this circulating current may be reversed. Whether this reversal actually occurs or not depends upon many factors such as pulse amplitude and width, circuit damping, β-value and bias. Assuming it does occur, it provides an opportunity for voltage pulses to appear across each junction at half the rate that the current pulses are being applied to the bias. This is only a partial description because a weaker fundamental response does occur at the current pulse rate as well as this strong sub-harmonic response. Nevertheless a sub-harmonic response is equivalent to dividing by two. Consequently the DJ-SQUID possesses the rudimentary logic operation required in scaling.

Whether the SQUID alone can be used as the logic circuit for successive scaling operations depends upon the usual requirements of adequate isolation between input and output circuits and sufficient gain to drive the next element. Both these requirements become difficult to satisfy in circuits involving single fluxoid events. In addition the pulse waveforms must be appropriate. We may approach this issue by trying to simulate the operation of a pair of these cells and finding what gain/loss or isolation must be inserted between the cells for successful operation. To illustrate this problem, we show the results of a lumped circuit analysis for the circuit

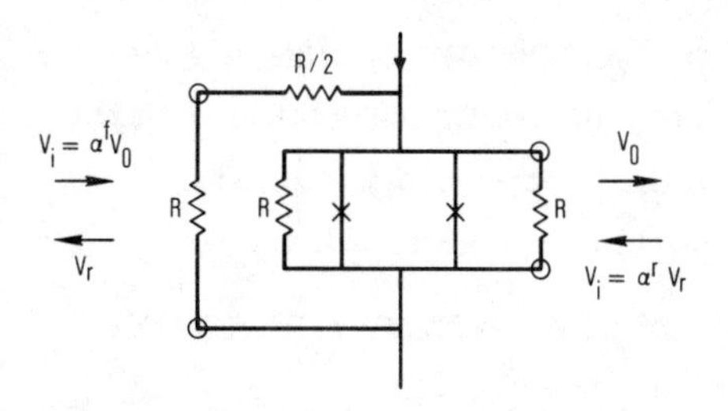

Fig. 3 Simulation for forward/reverse operation.

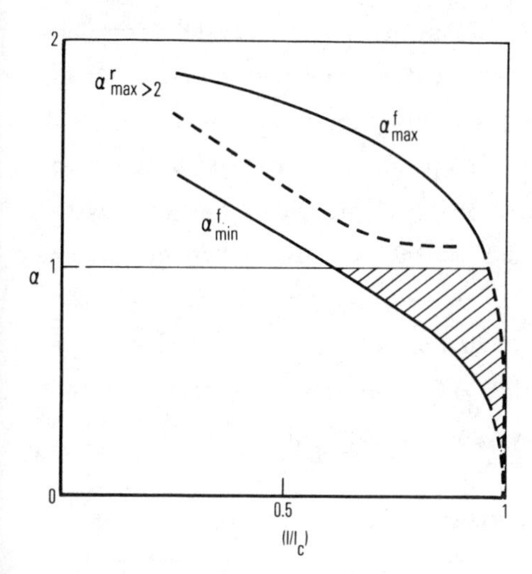

Fig. 4 Operating range of α for $\beta = \pi$.

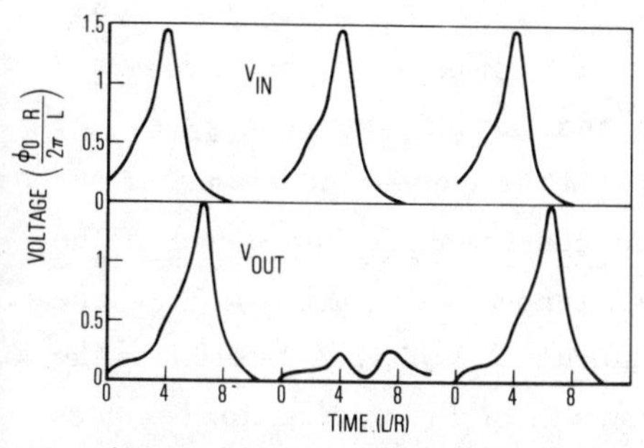

Fig. 5 Pulse shape illustrating scaling operation for $\beta = \pi$, $I = 0.7\ I_c$.

shown in figure 3.

The circuit simulates both the forward and reverse operation in a matched system and allows evaluation of α^f_{max}, α^f_{min} to determine an operating range in which scaling can occur. α^r represents the isolation required in the reverse direction for safe operation.

If a common range of α can be found, then scaling may occur without further amplification/ isolation between scaling cells. The results are summarized in figure 4. The shaded portion represents a useful operating range. If this operating margin turns out to be too small, then amplifiers or isolation diodes will have to be incorporated. Nevertheless these calculations do illustrate the fact that this DJ-SQUID scalar cell does possess some degree of appropriate isolation and amplification. This isolation may be visualized as a consequence of the partial orthogonality between the antisymmetric circulating current mode and the symmetric bias current mode of operation.

A typical pulse shape is shown in figure 5 and represents a soliton-like solution for the pulse

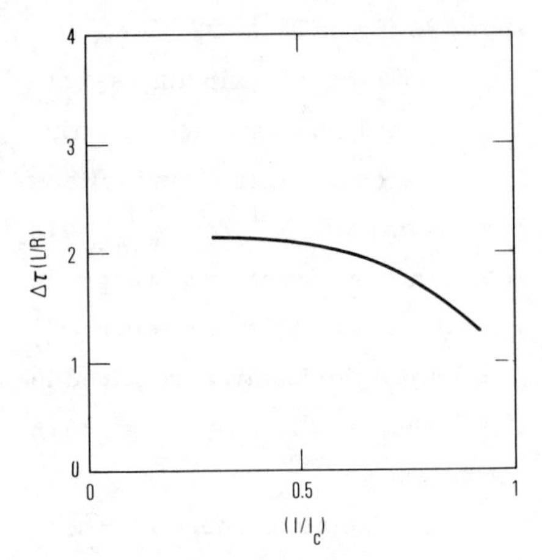

Fig. 6 Pulse width along dotted trajectory for $\beta = \pi$.

which is propagated during the scaling operations. Figure 6 shows the pulse width associated with the $\alpha(I/I_c)$ trajectory drawn as a dotted curve in figure 4.

BUFFER MEMORY/READ OUT

Binary information is continuously recorded in the scalar cells. After every sample time τ_s, this information must be recorded in a buffer memory and read out in parallel fashion at a rate $(1/\tau_s)$. This process can be performed with latched logic and clocked operations. Probably logic gates designed with superconducting interferometers will be suitable. If, however, their sensitivity proves to be inadequate, a sensing scheme similar to the parametric quantron[10] may be required. In that logic gate, information is written at low β_l but transmitted at high β_h. The device operates at the single flux quantum level with an effective gain of (β_h/β_l).

DESIGN CONSIDERATIONS

Heavy damping of the DJ-SQUID is necessary to prevent latching into the voltage state below the critical current mentioned earlier. Consequently the scalar cell and quantizer will be constructed from SQUIDs with $\beta \sim 1$ and critically damped. We envisage microstrip technology with matched impedances and Josephson tunnel junctions. Design parameters are governed by essentially the (L/R) time constant τ and flux quantum energy E. At critical damping $\tau = L/2R = 2RC$ where L is the lumped inductance of the SQUID, R is the circuit loading equal to the characteristic impedance of the microstrip Z, and C the junction capacitance. E equals $\phi_o^2/2L$ and determines the SQUID inductance. The condition $\beta = \pi$ then determines the critical current of the junction. W, the width of the microstrip, becomes proportional to $E\tau$. For a uniform geometry where the junction width equals W, the junction length δ is proportional to τ. Figure 7 illustrates these

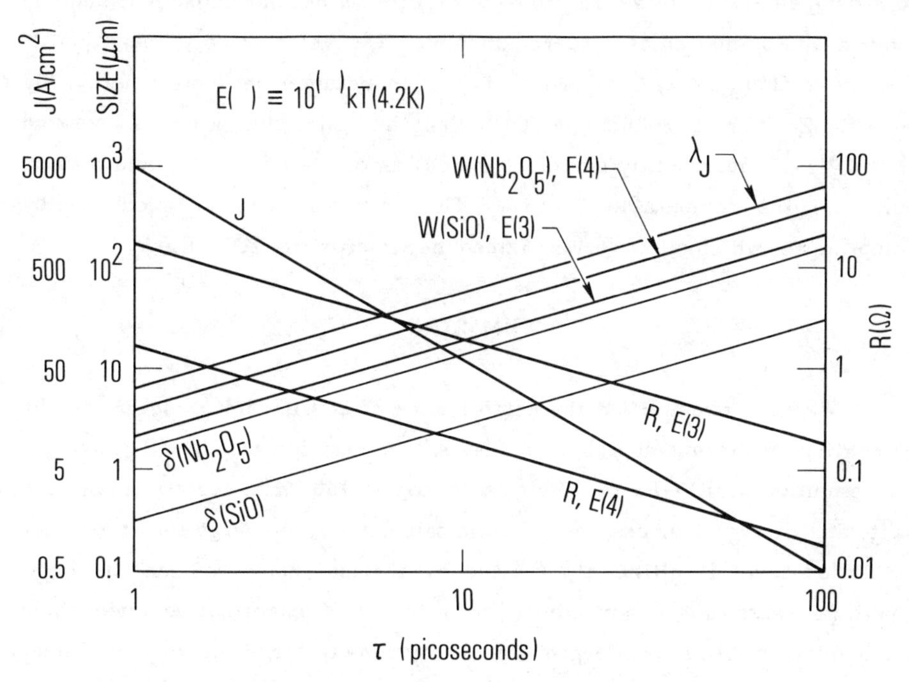

Fig. 7 Parameters for critically damped SQUID with $\beta = \pi$.

relationships. Two microstrip-dielectric-spacer choices were considered: 1500 Å SiO and 500 Å Nb_2O_5. These choices determine W and δ. If the system is to be overdamped, the plasma frequency is raised by increasing the current density and reducing δ. By specifying E in units of kT(4.2K), the immunity to thermal noise is specified. Spontaneous switching of the SQUIDs will occur when E is too small. The working margin (α) which governs the gain ($1/\alpha$) must be raised as E is reduced. Practical limitations associated with penetration lengths restrict the maximum value of E.

The choice of critical damping is consistent with the usual requirement for inhibiting junction hysteresis, namely $\beta_c = 2\pi I_c R^2 C/\phi_o < 1$. Critical damping results in $\beta_c = \beta/4$ and consequently satisfies this condition for small β-values.

Two points should be raised concerning geometry. First, a junction of non-uniform width couples more efficiently to a microstrip where current flows preferentially along the edges. IBM have worked with sinusoidal or "bow-tie" shaped junctions.[11] Second, a uniform microstrip geometry leads to a damped SQUID in which propagation effects become important and the lumped approximation is rendered invalid. The velocity of propagation v may be written as $Z/\mathcal{L}$ where $\mathcal{L}$ is the inductance per unit length. By equating Z= R and introducing a SQUID length ℓ such that $\mathcal{L}\ell = L$, we find $\ell = 2v\tau$. Consequently switching speeds associated with τ would lead to pulse lengths comparable to ℓ. This situation can be modified by compromises which allow varying impedance microstrips to be used.

SUMMARY

We have exemplified the performance of SQUID digital electronics by presenting an elegant approach to fast A/D conversion which will provide 4 bit resolution at 10 GHz or 10 bit resolution at 150 MHz, corresponding to a switching time of 1 picosecond. At the same time the energy sensitivity per resolution level is ultimately limited by thermal noise. Since the input quantizer represents essentially an infinite set of quantization levels in a single element, this device provides an extremely large bit number (linear dynamic range) with a small number of active elements. This device should find application not only in wideband signal processing, but also in signal detection and processing where high precision and large dynamic range are required at more modest bandwidths. One example of the latter is digital magnetometry where the proposed A/D converter should outperform both rf and dc SQUIDs. This assertion is based on the fact that the A/D converter introduces only fundamental thermal noise at liquid helium temperatures and quantum noise, while providing the same responsivity to the signal as the more conventional SQUID magnetometers.

ACKNOWLEDGEMENT

We gratefully acknowledge useful discussions with H. Kanter, Y. Song, H. Suhl, F. L. Vernon, Jr. and J. E. Zimmerman.

REFERENCES

1. A. H. Silver and J. E. Zimmerman, Applied Superconductivity 1, ed. by V. L. Newhouse (Academic Press, N.Y., 1975), p. 2.

2. J.Clarke, Superconductor Applications: SQUIDs and Machines, ed. by B. B. Schwartz and S. Foner (Plenum Press, N.Y., 1977), p. 67.

3. H. H. Zappe, Appl. Phys. Lett. 25, 424 (1974).

4. M. Klein, IEEE ISSCC, Digest of Technical Papers XX, 202 (1977).

5. D. G. McDonald, private communication in 1976 Navy Study on Superconductive Electronics, ed. by A. H. Silver (ONR Report No. NR 319-110), p. 43.

6. H. Kanter and F. L. Vernon, Jr., IEEE Trans. on Magnetics MAG-13, 389 (1977).

7. T. A. Fulton, R. C. Dynes and P. W. Anderson, Proc. IEEE 61, 28 (1973).

8. C. D. Tesche and J. Clarke, J. of Low Temp. Phys. 29, 301 (1977).

9. J. E. Zimmerman and D. B. Sullivan, Appl. Phys. Lett. 31, 360 (1977).

10. K. K. Likharev, IEEE Trans. on Magnetics MAG-13, 242 (1977).

11. A. Moser, to be published.

FAST SUPERCONDUCTING INSTRUMENTS*†

Richard E. Harris and C. A. Hamilton
National Bureau of Standards
Electromagnetic Technology Division
Boulder, CO 80303

ABSTRACT

The emerging technology for fabricating superconducting integrated circuits offers the possibility of remarkable new instruments. The advantages of this form of electronics are high device speed and low dissipation, combined with lossless, dispersionless, properly terminated transmission lines. A number of possible new instruments are presented. It is shown that a small group can successfully fabricate the superconducting integrated circuits required for these new instruments.

INTRODUCTION

Scientific experimentation, military devices, and the potential arrival of very fast computers all require new fast measurement capabilities. Josephson tunneling logic is a compelling candidate for this new class of instruments. Its known high speed and low power dissipation, its past successes in this field (some Josephson instruments are now available commercially), and its newly developed fabrication technology make it most attractive. In the following we shall discuss some of the requirements of fast instruments, indicate how superconducting electronics meets those requirements, and give a few examples of the kind of instruments that might result. Finally, we discuss progress toward such instruments at the National Bureau of Standards, as an example of what can be accomplished by a small group.

I. REQUIREMENTS

Speed

The speed of a Josephson junction is at least as great as the fastest semiconductor devices. Its intrinsic response time [1] has been shown to be determined by the superconducting energy gap 2Δ as

$$\tau_J = \hbar/2\Delta \quad .$$

For lead, which has an energy gap of 2.5×10^{-3} mV, $\tau_J = 0.27$ ps.

†Supported in part by the Office of Naval Research under contracts N00014-77-F-0048 and N00014-77-F-0019.

A more serious limitation is due to the intrinsic capacitance C of a typical Josephson device which consists of two planar electrodes separated by an exceptionally thin insulating layer. An $R_N C$ time constant (R_N is the junction normal state resistance) provides a measure of the response time of the device [2]. Although data on this time constant are scarce for the alloys from which junctions are usually made, R_N can be estimated from a patent [3] for an alloy of Pb with 6.5% In. The capacitance C can be approximated from other data. The resulting dependence of $R_N C$ on the current density j_c and the barrier thickness is shown by the solid line in Fig. 1. The left ordinate gives $R_N C$ normalized to the intrinsic response time. The right ordinate gives the time in ps for a lead alloy junction. The numbers along the curve give the oxide thickness in Angstroms. The x's are actual values from several papers. The figure is probably correct to half an order of magnitude. Reliable junctions have been commonly fabricated with critical current densities up to at least 2×10^3 A/cm^2. This corresponds to $R_N C/\tau_J$ = 13. Thus for lead, $R_N C$ = 3.5 ps, an extremely low value when compared to conventional electronic devices.

The fastest junction reported in an operational logic device had j_c = 2.8×10^4 A/cm^2 ($R_N C$ = 0.87 ps). It may also be possible to fabricate reliable devices having higher critical current densities. Thus $R_N C$ times approaching the intrinsic device limit may be possible. Other parameters serve to make it difficult to achieve this performance in a circuit, but at least one calculation [4] suggests switching times shorter than 1 ps will be possible.

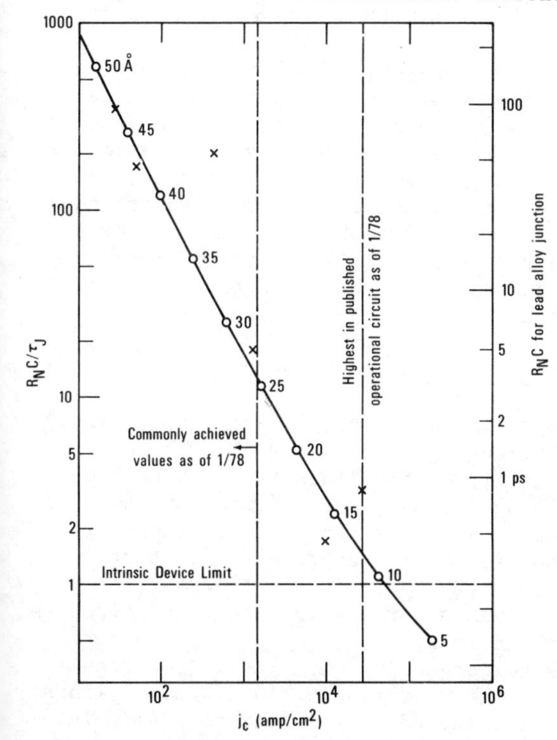

Fig. 1. Response time of a Josephson tunnel junction, when limited by $R_N C$, as a function of critical current density j.

Dissipation

High-speed devices necessarily require small size. Recall that signals propagating at the speed of light travel at most 0.3 mm (0.012 in.) in 1 ps. If an instrument consists of many interconnected elements which must interact with each other within the response time of the instrument, then its maximum size will be determined by the desired speed of operation. We must thus consider the possibility of 1 ps instruments which have a maximum dimension of only a few tenths of a mm.

If the elements in a circuit are located close together, the total dissipation must be sufficiently small that the operating temperature can be maintained. The heat produced by very fast conventional devices already places a severe limit on the number which can be located on the same integrated circuit chip.

Heat dissipation has been shown to be of fundamental origin by Keyes [5]. It varies as the square of the operating temperature. Thus cryogenic devices should offer significant reductions in dissipation.

In particular, Josephson junctions offer 1000 to 10,000 times lower dissipation than semiconductor devices. Thus while the density of fast semiconductor devices is seriously limited, it should be possible to fabricate rather complex superconducting integrated circuits without serious dissipation problems. For this reason, low dissipation is probably a more significant advantage of superconducting electronics than the high speed of the Josephson junction. For example, the Transferred Electron Device (TED) exhibits a response time which is comparable to that of a Josephson junction, but the heat dissipation is much greater.

Terminated Superconducting Transmission Lines

Superconducting technology offers not only a remarkably fast, low-dissipation active element, but also makes possible the use of nearly lossless, dispersionless microstrip lines [6]. These lines are composed of a superconducting ground plane, covered by an insulator, with a narrow superconducting strip on top. Their high quality persists from dc up to frequencies approaching that of the superconducting energy gap, about 700 GHz [7]. In contrast, the miniaturization of normal state lines for use in integrated circuits produces excessive dispersion and loss.

The usual technique in superconducting circuits is to transmit signals from one logic level to the next via overlaid control lines. These lines can be accurately terminated with a resistor [6]. This is another important advantage since it reduces interference from reflections and maintains the highest possible speed.

II. EXAMPLES OF POSSIBLE SUPERCONDUCTING INSTRUMENTS

The use of new lithographic techniques for fabricating superconducting circuits makes possible not only fast new instruments, but also improvements in superconducting instruments of existing design.

Extensions of Existing Techniques

The ac Josephson effect now provides a reference voltage with which the U. S. legal volt is maintained. This reference is of the order of 5 to 10 mV. As we shall see below, lithographic thin film fabrication techniques may make possible a substantial increase in this reference level.

The use of arrays of Josephson junctions as microwave detectors offers promise. Such arrays would have higher impedance and lower saturation levels than single junctions. Their fabrication is carried out conveniently only by using lithographic techniques. Additionally, the use of the quasiparticle current as the nonlinear element in a detector [8] may be a significant competitor for existing devices.

Present SQUIDs are rf-biased primarily because of the difficulty of fabricating more than one junction in a device. Dc-biased SQUIDs, consisting of at least two Josephson junctions fabricated lithographically, would not require the complicated room temperature rf-biasing circuitry, but would still provide the same or better sensitivity and signal-to-noise ratio. The new fabrication techniques might also allow a low-noise superconducting amplifier to be placed on the same chip as the SQUID.

Finally another paper at this conference suggests the replacement of the whisker-type contacts normally applied to frequency synthesis with a Josephson thin-film device [9].

Important Elements of New Instruments

There are a variety of possible new instruments. Those to be discussed here make use of two simple components, a current comparator and a sample-and-hold circuit. Some instruments may also make use of digital logic.

Current Comparator

A Josephson gate is essentially a current comparator. In digital circuits it must determine only whether a signal represents a "0" or a "1". However, it can also be used in an analog sense [10]. If the gate is biased in the zero voltage state with some current flowing through it, then the gate will switch at some well-defined current flowing through the control line over the gate. One can vary the switching point by providing a bias current through a second control line over the gate. Thus an adjustable current comparator results. It should operate as fast as any Josephson gate -- that is, ultimately in less than 1 ps.

Sample-and-Hold

Sample-and-hold circuits are used to sample a signal in a very short time, and maintain it until it can be characterized by a slower device. One possible kind of superconducting sample-and-hold device would store a persistent current in a superconducting loop. The current would be switched in and out of the loop using a Josephson junction. Since the maximum flux in the loop, LI_o, is quantized in units of the flux quantum, 2.07×10^{-15} Wb, the inductance of the loop would have to be large enough to permit the loop to contain enough flux quanta to achieve the desired resolution. The speed of response of such a simple circuit would be determined by the time required for the Josephson junction to go from the normal state to the superconducting state. Times as short as 1 ps might be achieved.

Logic Circuits

Logic operations may also be essential in some instruments. A counter for example must perform logical operations. Such elements of an instrument might be quite similar to those in superconducting computers.

A Few Possible Josephson Instruments

Analog Sampling Device

Published [10] and unpublished reports of sampling devices suggest that such instruments have already been constructed having a time resolution of a few ps. Such sampling devices are constructed from current comparators of the type described above. Present conventional sampling oscilloscopes have a time resolution of about 25 ps.

Fast Counters

It should be possible to construct fast counters using superconducting technology. The speed of response would be determined by the speed of the circuitry for the least significant bit. More significant bits would change state more slowly and would therefore not be as important. Careful engineering of the first stage might bring the response time down to a few picoseconds, corresponding to a counting rate of order 500 GHz. Present maximum counting rates are about 500 MHz.

A/D Converters

In conventional electronics the fastest analog-to-digital converters are of the parallel type. That is, the converter contains one current (or voltage) comparator for every possible level of the input analog signal. Thus, a 4-bit converter might have 16 comparators. Subsequent circuitry then condenses the 16 bits of level information into a 4-bit binary word. A similar design in superconducting technology would also provide the highest possible speed.

A/D conversion is subject to limitations. One cannot accurately digitize a signal which is changing faster than the resolution of the converter. Thus, sample-and-hold circuits are used to maintain a signal until it can be digitized. The aperture time t_{ap} of a sample-and-hold circuit may determine the fastest signal which can be digitized. During this period the signal is allowed to change no more than one part in 2^n for an n-bit converter. Assuming a sine wave at frequency f, one finds

$$f_{max} = (\pi 2^n t_{ap})^{-1} .$$

Thus the achievement of the minimum aperture time represents a high priority task if extremely high-speed A/D converters are to be realized. For example, sampling with 4-bit accuracy at a rate of about 10 GHz would be possible with aperture times below about 15 ps. Present A/D converters having this accuracy operate in the range of hundreds of MHz. A long range goal might be to reduce t_{ap} below a few ps.

Digital Instruments for Real-Time Signal Processing

Given a high-speed digital technology, it becomes possible to conceive of real-time applications of computational techniques which previously were done off-line, or which could be applied only to very slow signals. One such example involves Fourier analysis of signals. Such signals might pass through a superconducting A/D converter of the type discussed above, and then be digitally analyzed using a special-purpose fast Fourier transform processor designed to perform these calculations at the highest possible speed. Such an FFT processor could produce a variety of instruments. Digital filters could be produced which might have rapidly tunable frequency, lineshape, etc., possibly controlled by the incoming signal. In addition a multichannel spectrum analyzer having high speed and a large number of channels should be feasible. Such analyzers are now being developed, using conventional technology. This device could also be coupled with processing, possibly using special purpose pattern recognition circuits, to examine the data in real time.

III. PROGRESS TOWARD SUPERCONDUCTING THIN-FILM INSTRUMENTS AT THE NATIONAL BUREAU OF STANDARDS

While the preceeding discussion suggests a number of possible areas in which superconducting electronics can make significant contributions, there remains the question of the difficulty of fabricating them. Our experience shows that considerable success can be achieved by a group of three people. The following discussion of our problems and successes is offered for the consideration of others who might enter the field or attempt to assess its future impact.

The effort is about two years old. The first year was largely concerned with the procurement and assembly of the required equipment. This includes a complete mask-making facility, three vacuum evaporators, and the equipment required to do photoresist lift-off processing. The work to date has been largely directed towards reproducing techniques reported in the literature.

Considerable time was devoted to the production of photoresist layers with sufficient undercutting so that tearing would not occur in the lift-off process. If the deposited metal layer is connected where it crosses the edge of the photoresist, it will tear during lift-off and ragged edges result. The method finally adopted involves soaking the photoresist in chlorobenzene [11] prior to development. Figure 2 shows such a photoresist layer with 4000 Å of lead alloy deposited on top. The chlorobenzene treatment results in an undercut of about 2000 Å which is sufficient to make the deposited film discontinuous at the edge.

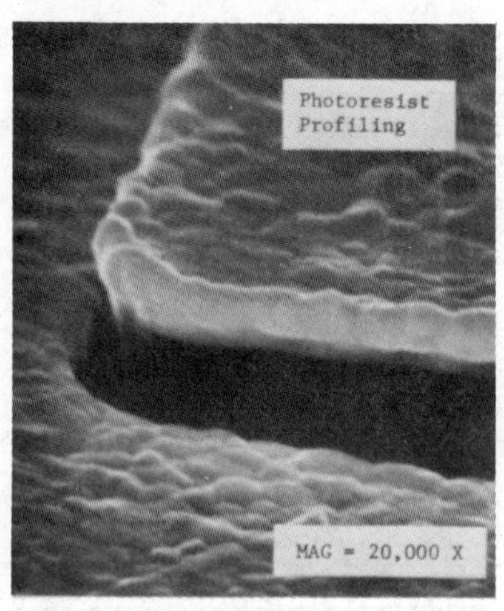

Fig. 2. Lead alloy film evaporated on a 2 μm thick photoresist layer. The photoresist is undercut due to treatment with chlorobenzene.

Control of edges is essential for the production of junctions with good I-V curves and predictable critical currents. Our measure of the quality of a junction is the ratio of the instantaneous resistance at 2 mV to that at 3 mV (i.e., 2 mV/I(2mV) and 3 mV/I(3mV)). Typical values of this ratio for our good junctions are five to six. While it is possible to fabricate good junctions by simply overlapping the electrodes, we have found that junctions made through windows in a layer of silicon monoxide are more consistently of high quality.

The oxide barriers are formed by rf sputter oxidation [12]. Figure 3 gives an indication of the reproducibility and transferability of the process. The figure plots the critical current density as a function of oxygen pressure for fixed rf power and an oxidation time of ten minutes. The bottom curve is derived from the paper which first reported this technique [12]. We do not

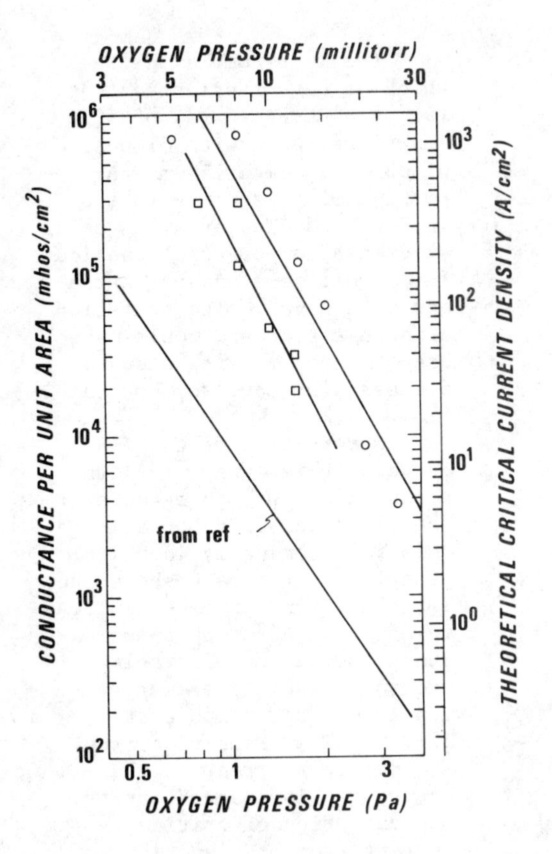

Fig. 3. Conductance per unit area of rf sputter oxidized tunnel junctions as a function of oxygen pressure. Cathode self bias was regulated at 45 volts and rf power was approximately 10 W during oxidation. The lower line is from reference 12. The square and circle points represent our data taken under somewhat different conditions. See text.

characterize our rf conditions in the same way, but a comparison of this curve with ours illustrates the great sensitivity of the process to the rf conditions. The top curve has been reported previously by us [13]. Since then we have made slight changes in the vacuum system and have begun the use of SiO windowed junctions. We are now obtaining the results indicated by the middle curve. The difference suggests that there are some parameters which are not sufficiently well understood and therefore not adequately controlled.

Another important property of these junctions is their ability to cycle repeatedly between room temperature and helium temperature. We have performed automated cycling tests using a low-helium-consumption apparatus [14]. Figure 4 is a plot of normalized junction resistance versus the number of thermal cycles. Most of the devices tested changed very little for the first 10 to 15 cycles and degraded rapidly after that. Lahiri and Basavaiah [15] have recently reported that the use of thinner films can substantially improve cyclability. Nevertheless the junctions we have already made are sufficiently stable to demonstrate prototype instruments.

Storage of these devices is also an important consideration. Data thus far indicate that storage at room temperature can produce significant changes in a period of a few days while junctions stored at -15°C remain essentially unchanged for up to eight months.

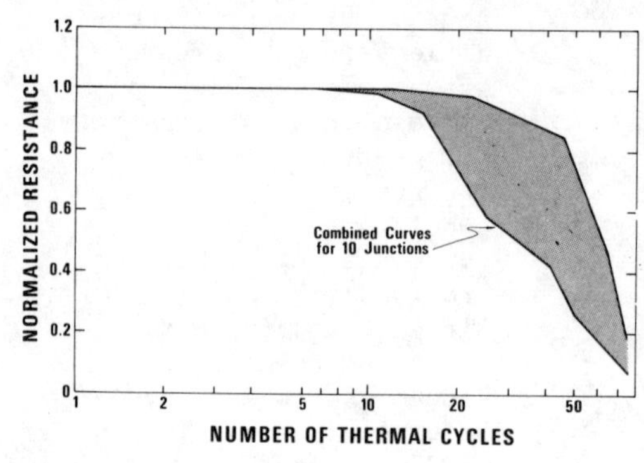

Fig. 4. Normalized junction resistance as a function of the number of thermal cycles.

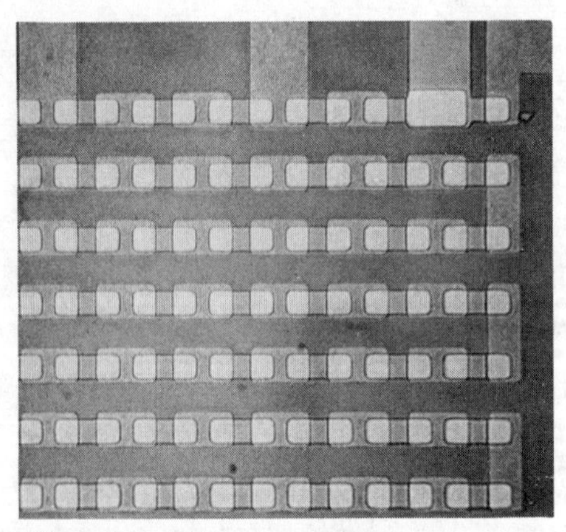

Fig. 5. Photograph of a portion of a series array of 5276 junctions.

The most complex device which we have fabricated thus far is a series array of 5276 junctions, each with dimensions of about 15 μm x 15 μm. A portion of this array is shown in Figure 5. The array is being studied as a potential candidate for a voltage standard at the one volt level. The operation of such a standard would make use of zero-crossing steps as discussed by Levinson, et al. [16].

Observation of the I-V curve of this array provides a convenient way to measure the reproducibility of a large number of nominally identical junctions. Figure 6 shows the I-V curve for our first successful array. Although these junctions had considerable excess current, the energy gap is clearly visible at 12.5 V. The transition from the dc supercurrent to the energy gap is smeared out by the variation of critical current from one junction to the next. Viewed with higher sensitivity, this portion of the curve is actually a staircase (nominally 5276 steps). Assuming that each junction switches to the energy gap voltage, a plot of (dV/dI) versus I yields a probability density function for the critical current. For the example shown in the Fig., the mean critical current is 0.9 mA with a standard deviation of ± 12%. In this example, trapped flux is certainly responsible for a large part of the spread.

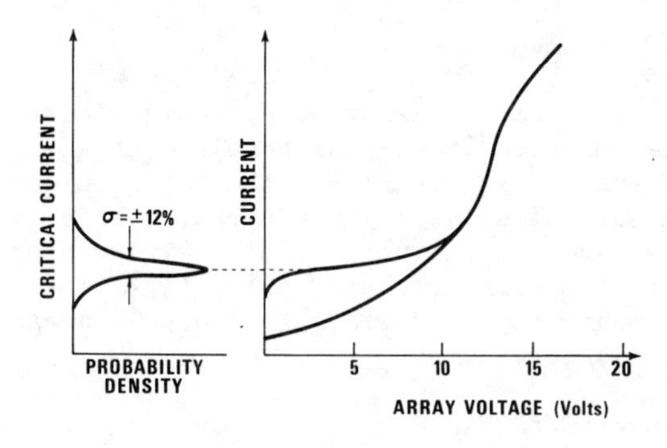

Fig. 6. The I-V curve for a series array of 5276 junctions. The slope of the upper branch of the curve gives a probability density function for the critical current of individual junctions.

Our effort is currently moving toward the incorporation of junctions in high-speed circuits. This requires a mounting which is capable of transmission of high-speed signals to and from the chip [17]. Figure 7 is a cross-sectional view of our approach to signal coupling. Twenty-four 50 Ω coax lines are terminated at a 2.5 cm diameter printed circuit board. Striplines (50 Ω) on the printed circuit board are laid out so as to contact the lead pads of a chip pressed against the board. By fabricating a long superconducting stripline on the chip, a time domain reflectometer can be used to measure the risetime achievable with the chip mount. These measurements indicate that the risetime for transmission of a signal at room temperature to a superconducting stripline on the chip is less than 70 ps. Through use of both niobium oxide and silicon monoxide as the dielectric we have managed to fabricate superconducting striplines with predictable impedances in the range of 0.1 to 10 Ω. This is a useful range for matching to Josephson junctions.

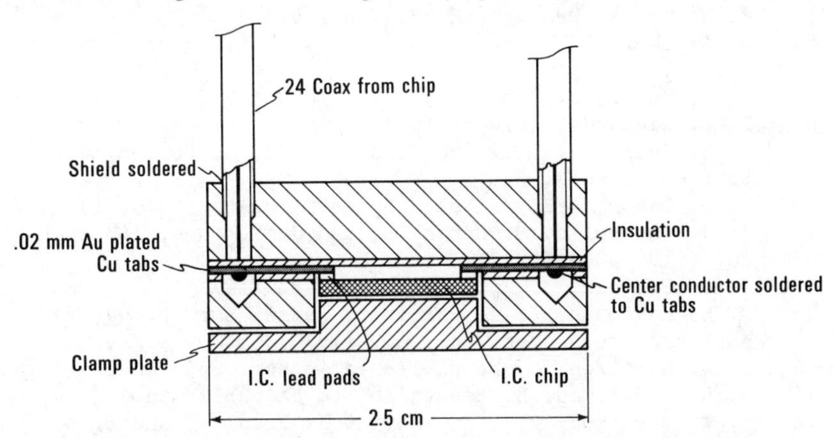

Fig. 7. Cross-sectional view of a chip mount for making 24 high frequency contacts. Signals can be transmitted from room temperature onto the chip with a bandwidth of about 5 GHz.

CONCLUSION

Superconducting integrated circuits utilizing Josephson devices and lossless, dispersionless transmission lines make possible the construction of a variety of useful new instruments. The fabrication technology required is becoming sufficiently well-known that even a small group can use it. Success in constructing portable refrigerators for these instruments would make them even more attractive [18]. Over the next few decades such instruments may find widespread use in areas where the highest possible speeds are required.

ACKNOWLEDGEMENTS

We are grateful to a number of people for stimulating and useful discussions: L. H. Holdeman, W. Henkels, R. L. Kautz, C. Kircher, F. L. Lloyd, D. G. McDonald, D. B. Sullivan, and H. H. Zappe.

REFERENCES

1. R. E. Harris, Phys. Rev. B13, 3818 (1976).
2. The actual response time may deviate from R_NC in a way which will depend on the circuit to which the junction is attached.
3. J. M. Eldridge and J. Matisoo, U. S. Patent Number 3,816,173, June 11, 1974.
4. D. G. McDonald and R. L. Peterson, J. Appl. Phys. 48, 5366, (1977); NB, τ_g in this reference is 2π times bigger than our τ_J.
5. R. W. Keyes, Proc. IEEE 63, 740 (1975).
6. W. Anacker AFIPS Conf. Proceedings: Fall Joint Computer Conf. 41, 1269 (1972).
7. R. L. Kautz, J. Appl. Phys. 49, 308, (1978).
8. P. L. Richards, this conference.
9. D. B. Sullivan, L. B. Holdeman, and R. J. Soulen, Jr., this conference.
10. H. H. Zappe, IEEE J. Solid State Circuits SC-10, 12 (1975).
11. B. J. Canavello, M. Hatzakis, and J. M. Show, IBM Tech. Disc. Bull. 19, 4048 (1977).
12. J. H. Greiner, J. Appl. Phys. 42, 32 (1971).
13. R. H. Havemann, C. A. Hamilton, and Richard E. Harris, to be published in J. Vac. Soc. Am. (1978).
14. C. A. Hamilton, to be published in Rev. Sci. Inst. (1978).
15. S. K. Lahiri and S. Basavaiah, to be published.
16. M. T. Levinson, R. Y. Chiao, M. J. Feldman, and B. A. Tucker, Appl. Phys. Lett. 31, 776 (1977).
17. This mount is quite similar to one described to us by W. Henkels and H. H. Zappe.
18. J. E. Zimmerman, this conference.

A CURRENT-SWITCHED FULL ADDER FABRICATED BY PHOTOLITHOGRAPHIC TECHNIQUES

J. H. Magerlein*, L. N. Dunkleberger and T. A. Fulton
Bell Laboratories, Murray Hill, New Jersey 07974

ABSTRACT

We describe a one-bit full-adder based on Josephson tunnel junctions and fabricated by photolithographic techniques. The design employs current-switched gates with latching characteristics, direct-coupled control currents and parallel fanout. The linewidths are 25 μm. Simulated and indirectly measured delays/gate are approximately 100 psec.

INTRODUCTION

This report describes a one-bit full adder employing Josephson tunnel junctions in a current-switched arrangement. The circuit is similar to that described by Magerlein and Dunkleberger[1] with some differences in design, and with the important practical difference that the fabrication is by photolithographic techniques. The results reported are for three particular circuits selected from a larger total number of circuits which were carried through fabrication and testing procedures. Two of these circuits worked as full adders in all data conditions, albeit they required individual adjustment of the bias currents of the various gates. The third, which conformed more closely to the design values for critical currents and resistors, operated as a half-adder with the various gates having a common bias current. Simulated and indirectly measured delays/gate for the half-adder were approximately 150 psec and for the full adders approximately 70 psec.

LOGIC DESIGN

The circuit contained nine gates, arranged as in Fig. 1. Gates A, B, C, and Ã provided the input bits. The 2/3 gate, a

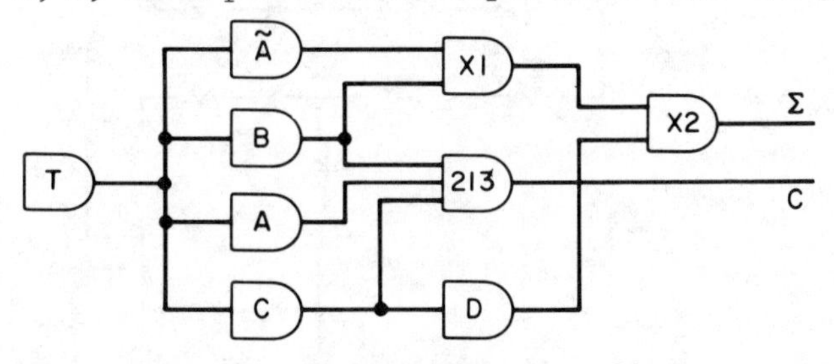

Fig. 1. Logic design for adder

*Present address: IBM Research Center, Yorktown Heights, New York

SSN: 0094-243X/78/459/$1.50

three-input AND gate, supplied the CARRY function while the two "exclusive OR" gates X1 and X2 supplied the SUM function. Gate D provided one stage of delay between C and X2 and gate T provided the overall timing necessary to this design. The arrangement is similar to that of ref. 1 except for the presence of the $\tilde{A}$ which contained the same logical information as the A gate but with an opposite sense of current bias, for reasons described below.

CIRCUIT DESIGN

Figure 2 shows a typical portion of the circuit layout. Here also the design is similar to that of ref. 1 with differences as noted. "Goalpost"[2] switching elements were used throughout. The circuits are latching, requiring a reduction in bias current to provide reset. The fanout current diverted from a gate switched to $V \neq 0$ acts as a control current for subsequent gates by "direct-coupling" as in ref. 1, with the current splitting up around the minor loop in proportion to the inductances L_1' and L_2'. The fraction which flows through L_1' adds to the bias current of the junction in that arm and causes switching. The "direct-coupling" approach requires that the junctions be biased in parallel. Unlike the scheme in ref. 1, in the present design the fanout current proceeds immediately to ground from this point (the ground return occurs on an overlaid ground plane).

Control of two different gates by a single gate is achieved where required by splitting the fanout line and the current into two separate equal parts as in Fig. 2 each of which controls a separate gate. In situations where a gate controls only one subsequent gate, such as D acting on X2, half the fanout current is diverted to

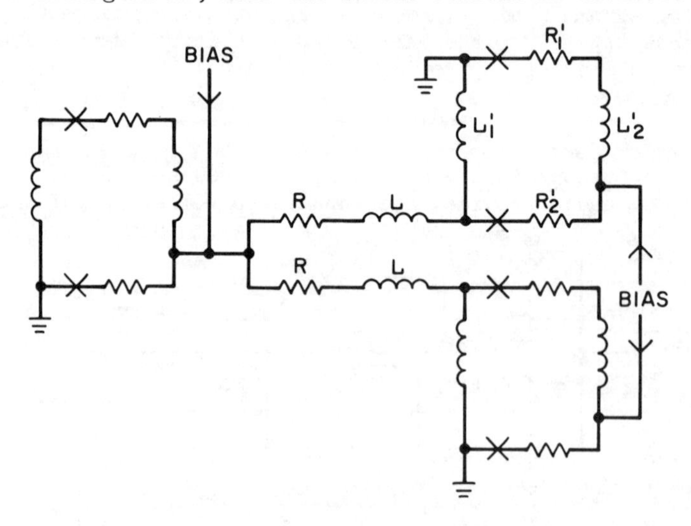

Fig. 2. Parallel-fanout scheme

ground in order to equalize the influence of each gate on its down-stream neighbors. This approach, "parallel fanout", has reduced crosstalk compared to the patterned ground-plane approach of ref. 1. Also the fact that the fanout line need not visit several gates in series, the use of the ground plane for current return and the elimination of the ground plane slots all serve to reduce the fanout line inductance compared to that in ref. 1, providing improved speed. Parallel fanout does however suffer a concomitant reduction in gain due to the reduced fanout current levels. Since improvements in gain go roughly as the one-half power of the fanout current level in this design[2], the reduction in margins is less than a factor of two. One could increase the fanout current by adding further junctions in parallel to the goalposts which would be switched by the diverted goalpost current and add their bias currents to the total. This approach would allow parallel fanout with reasonable gain to more than two gates.

The use of direct coupling and parallel fanout in this circuit required that the two control inputs for the "exclusive OR" gates have current biases of the opposite sign in order to provide cancellation of the induced loop currents when both inputs are present. For this reason an additional input gate $\tilde{A}$ was used to provide the same information as A to X1 but with a current bias of the opposite sign. Similarly the current in D was given the opposite polarity to that of X1.

The relative sizes of the individual junctions in the A, B, C, D, and $\tilde{A}$ gates were adjusted to the ratio of 1.51:1 with their bias resistors in the same proportions. This was done to maximize the switching margins for these gates given an anticipated induced loop current of 0.26I where I is the bias current of the gates. Similarly for the 2/3 gate, the junction areas and bias resistors were in the ratios of 1.20:1 and 2.20:1. The two junctions and bias resistors of gates X1 and X2 were necessarily symmetric.

The design values of the circuit parameters were 65 pH and 0.5 Ω for the fanout inductances and resistances respectively (the net values for the two parallel paths), $L_1' = 5$ pH and $L_2' = 21$ pH for the minor loop inductances and $R_1' + R_2' = 0.075\ \Omega$ and $(1/R_1' + 1/R_2')^{-1} = 0.0175\ \Omega$ for the minor loop resistors. The design critical current for X1 was 0.62 mA per junction with a corresponding 0.80 mA bias current, with similar values for the other gates. The expected margins were 1.66 for the OR gates, 1.45 for the 2/3 gate and 1.51 for the XOR gates, where these numbers refer to ratio of the maximum and minimum values of critical current for which these gates would still work correctly for the given bias current. The expected maximum delay per gate was about 150 psec. This corresponds to the time at which the induced current through L_1' reaches its maximum value measured from the onset of voltage in the gate which provides the control current. The average gate speeds are less than this and depend on how close the bias current is to the critical current.

FABRICATION

The circuits occupied an area of 2.5 x 3.7 mm in the center of a 76 mm-by-254 mm Si chip. The superconducting layers were Pb-In-Au alloys,[3] the resistors were Ag, the insulation for crossovers a Ge-SiO sandwich and the ground plane insulation was a negative photoresist. The pattern definition for the first four layers, the two electrodes, the resistors and the crossover insulation, utilized a photoresist stencil technique.[4] Via openings in the photoresist ground plane insulation was developed in the usual way, and the final ground plane was defined by a metal shadow mask. Cleaning of layers at various stages was achieved by ion milling or by glow discharge. Formation of the oxide tunneling barrier was carried out in a dc glow discharge in oxygen. Film thicknesses were about 0.2-0.3 μm for the superconductors and crossover insulation, 0.05 μm for the resistors and 0.8 μm for the ground-plane insulation. Linewidths were 25 μm in all cases except where the junction area adjustments called for somewhat larger or smaller values. Figure 3 shows a microphotograph of the circuit.

OPERATION

The samples were tested at 4.2 K in a He exchange gas environment. Common failure modes involved shorts through the ground-plane insulation and mismatch of the critical current levels with the resistors. Testing generally involved first examining and recording the I-V curves of all gates. If the I-V curves were free from obvious problems, measurements were then made of the reduction of the critical current of a gate by repeated rapid firing of the other

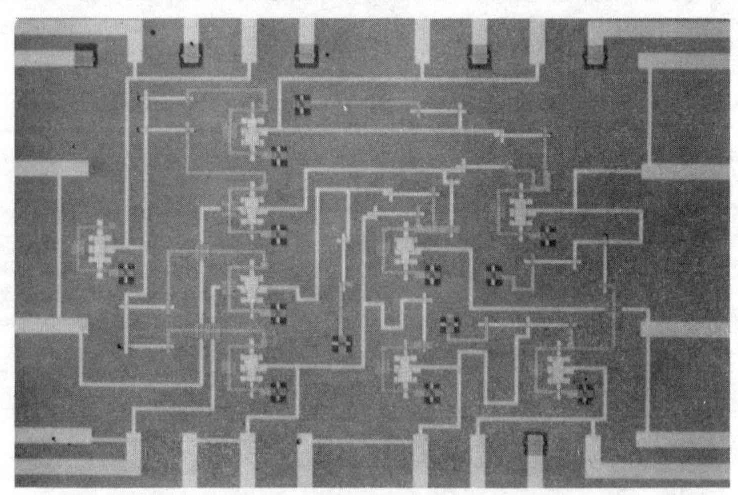

Fig. 3. Photomicrograph of circuit layout. The ground plane is not present.

gates. Depressions of the I_c by $\lesssim$ 30% were typical with minimal crosstalk. This value is smaller than that of ref. 1 due mainly to reduction of the minor loop time constant by contact resistance in R_1' and R_2'.

For the two working full adders the fanout resistors were larger than desired for the critical currents of 2.3 and 4.7 mA per gate, causing further reductions in the diverted current and hence the gain. Some problems were also created by the non-negligible magnetic-field effects on the critical currents caused by the flow of the control currents in the underlying electrodes. In principle these could be used to aid the switching process, but in our particular design they acted in opposition to the current-induced switching for some of the gates and aided in others, further reducing the margins of operation. The combination of these effects made it necessary to adjust the bias currents of the gates individually to achieve correct operation in all data conditions.

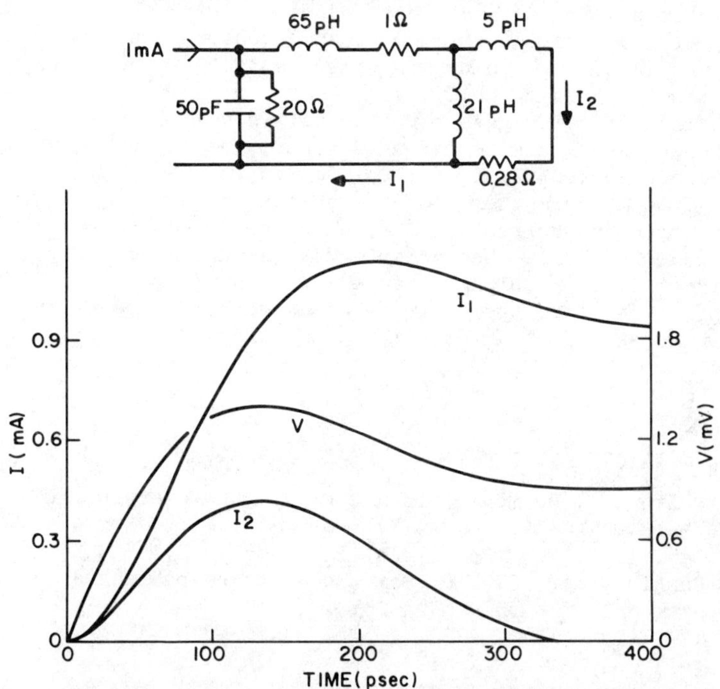

Fig. 4. Simulation of switching behavior. The inset shows the equivalent circuit used for simulation. Fanout current I, the current I_3 through L_1', and the voltage V on the junction are shown. The magnitude of the current I_3 would be two times smaller if parallel fanout were taken into account. The parameters are appropriate for the half-adder circuit. The junction is treated as a 10 Ω resistor for $V < 2.5$ mV and as having zero differential resistance at $V = 2.5$ mV.

Numerical simulation of the gate interaction for the particular circuit parameters of the full adders gave a maximum delay/gate of 60 to 70 psec. [The simulation used lumped circuit elements and treated the junction as though the dc I-V curve was appropriate (see Fig. 4).] Direct measurements of the delay were not attempted. However, the size of the I_c reduction by the fan-in currents is determined rather directly by the switching time through the intermediation of the time constant of the minor loop, which for these samples was about 40 psec. The observed I_c reductions were in rough accord with the 60 to 70 psec switching time.

For a third, somewhat ill-starred circuit having a shorted C gate (thanks to a poorly-placed tweezer mark), the critical currents were nearly at their design values of 0.5 mA per junction and the resistors were such that I_cR was about 1 mV, a reasonable match. This sample worked as a half-adder (with A, B, and Ā inputs) with the same bias current on each gate over an approximately 10% range of bias currents. Critical current reductions were about 2/3 that of the expected values, due mainly to some contact resistance in the minor loop resistors, reducing its time constant (to about 100 psec) and hence its integrating ability.

The simulated delay/gate for this sample was 135 psec. Figure 4 shows the simulated fanout current, the junction voltage and the induced current in L_1' for this circuit. Measurements of the delay from T to Ā, or actually of the change in the delay at the two extremes of the bias current of Ā at which proper operation occurred, gave 200 ± 100 psec delay. More accurate measurements were precluded by the cabling arrangements. Here again, however, the I_c depressions from the upstream gates were in approximate accord with those expected for a 135 psec delay.

REFERENCES

1. J. H. Magerlein and L. N. Dunkleberger, IEEE Trans. Magn., MAG-13, 585 (1977).
2. T. A. Fulton, J. H. Magerlein and L. N. Dunkleberger, IEEE Trans. Magn., MAG-13, 56 (1977).
3. S. K. Lahiri, J. Vac. Sci. Tech. 13, 148 (1976).
4. L. N. Dunkleberger, J. Vac. Sci. Tech (to be published Jan. 1978).

A MEMORY DEVICE UTILIZING THE STORAGE OF ABRIKOSOV VORTICES AT AN ARRAY OF PINNING SITES IN A SUPERCONDUCTING FILM

A. F. Hebard and A. T. Fiory
Bell Laboratories, Murray Hill, N.J. 07974

ABSTRACT

A non-destructive readout (NDRO) random access memory based on the use of a two-dimensional array of strong pinning sites in a film in the mixed state is proposed. Information is stored as the off-center displacement of vortices in overlay films. The overlay films are electrodes of a Josephson junction, which detects the position of the vortex by means of the critical current suppression.

This paper considers the design of devices which can exploit the high packing density of elements in two dimensions implied by the use of vortices in the superconducting mixed state. The methods investigated here are based upon the fabrication of patterned films that locally manipulate the vortex positions. Developments in recent years point towards a manner in which this may be accomplished in multiply-layered superconducting films.

In the mixed state of a homogeneous type II superconductor, the equilibrium configuration of the fluxoids is a triangular lattice. Recent studies [1,2] of the pinning of vortices in holes 1 μm in diameter in aluminum films have shown that a hole is a deep potential trap. Oxygen-doped aluminum was used for the host material because the fine scale of the microstructure gives a low background of random pinning. Extension to high H_{c2} materials has not been made yet, although it appears that holes may generally produce much stronger pinning than that usually encountered in deposited films.

We are proposing that a film in the mixed state, with single quantum vortices being created by an externally applied transverse magnetic field and stored at localized trapping sites, can be used as a substrate for a high-density NDRO random access memory. The plan view of one possible design is shown in Fig. 1 and a cross-sectional view at one of the pinning sites is shown in Fig. 2. The thick ground plane film G contains a triangular array of strong pinning sites. The area A of the memory cell determines the magnitude of the applied field through $B = \phi_o/A$. A key feature of the design is in the use of a Josephson junction with electrodes J1 and J2 overlaying each pinning site. The active junction areas (shaded in Fig. 2) result from overlapping film strips and are laid out so as to form linear arrays of series-connected junctions. On top of each junction are a pair of insulated control or select lines, S1 and S2, which pass over the cell as shown in the figures. These lines are made thicker than the junction electrodes, so that there is partial screening of the upper J2 electrode and the quantized flux of the trapped vortex threads the junction asymmetrically. Consequently, the vortex at the edge of the J2 electrode is displaced with respect to the pinning site in J1 and the ground plane. Josephson

SN: 243X/78/465/$1.50

junction detection of the vortex is possible because the misaligned vortex produces a component of magnetic field in the plane of the tunnel barrier, which gives rise to a critical current suppression [3].

In order to write a bit in a particular memory cell, it is necessary to reversably switch the vortex position between region A (logical "0") and region B (logical "1"). This is done by applying half-select currents to both S1 and S2, which induce screening currents to flow in J2. The current at the edge of J2 initiates vortex motion towards the center of the junction. Screening currents in J2 are able to do this because the junction electrodes are wider than the S1 and S2 lines. The flux then also cuts the S1 and S2 lines and the vortex is switched to the other off-center position. The net current flowing in the J1 electrode is relatively weaker, due to screening.

Non-destructive readout of the contents of a particular cell is achieved by simultaneously applying a bias current, less than the junction critical current, to the selected column of junctions, together with a half-select current pulse to S1. The polarities of these currents are chosen so that the net force on the vortex tends to be cancelled. The shielding currents flowing on the surface of J2 in response to the current in S1 give rise to a component of field in the barrier region which for reading logical "1" adds to the field of the misaligned vortex, switching the selected junction to a finite voltage state, or for reading logical "0" subtracts, with the

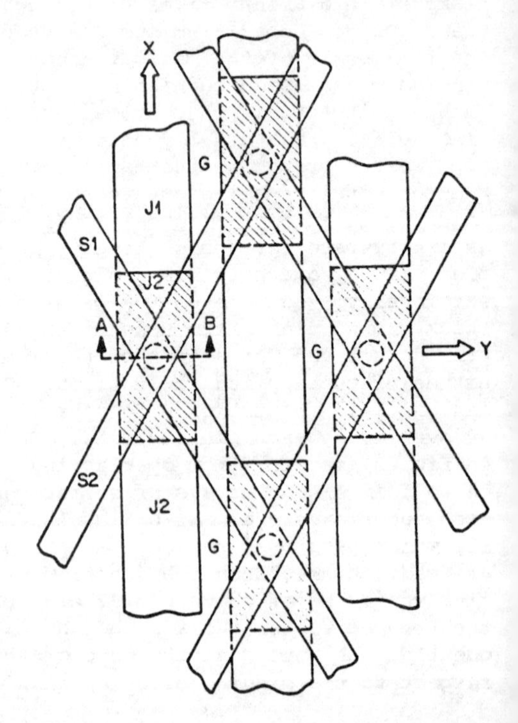

Figure 1. Mixed state memory array. The ground plane G and lower Josephson junction electrode J1 contain common pinning sites (dashed circles) spaced on a triangular lattice. Overlaying current select lines S1 and S2 intersect directly over the pinning site. The shaded regions represent those areas in which the electrodes J1 and J2 overlap to form series-connected Josephson tunnel barriers.

selected junction remaining at zero voltage. The expected switching times of about 1 ps for the small capacitance junctions are fast compared to the propagation delay expected in the transmission line formed by the series-connected junctions and the ground plane G [4].

We show in Fig. 3 the results of a calculation of the critical current suppression for the relative displacement of a vortex pair threading a dual-film junction. The calculation was carried out for an idealized model where one of the vortices is displaced a distance s in a circular junction of radius R. To facilitate computation the film thickness $d_F \gg \lambda$ and $\lambda \gg \xi$, where λ is the penetration depth and ξ the coherence length. Image vortices of opposite polarity, shown as I_1 and I_2 in the figure, are positioned at distances R^2/s so that the radial component of the surface current density on the inner surfaces of J1 and J2 vanishes at the circumference, thereby minimizing the external magnetic field energy. Using this calculation, Fulton et al.[3] estimated that the typical lateral misalignment of a vortex in their 5 μm-wide junction is about 0.1 μm, which is also the thickness of their films. Recently Hebard and Eick [5] used a similar technique for larger area junctions and again found that the mean displacement was comparable to the film thickness. These interesting results show that the displacement actually exceeds the range of the maximum vortex-pair coupling force, which is between λ and 2λ, depending upon the film thickness [6]. Thus the pinning force dominates the coupling force in the lead films studied. Since these films had a microtexture of size comparable to the film thickness, it appears that the pinning displacements to be expected will be on

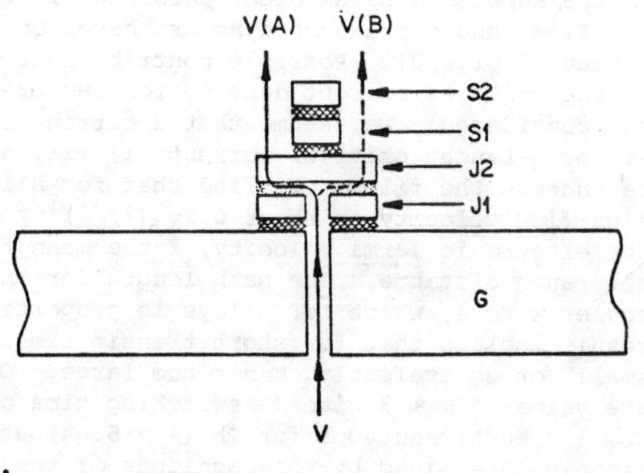

Figure 2. Cross sectional view of A-B of Fig. 1, showing a trapped vortex line V-V at two equivalent sites, V(A) and V(B) corresponding to logical "0" and "1", respectively. The tunnel barrier region between electrodes J1 and J2 is shaded and the insulating regions between S1 and S2, S1 and J2, and J1 and G are cross-hatched.

the order of the size of the microtexture. It is reasonable to expect progress in making films of finer scale texture than those previously studied.

The design specifications of a memory array such as is proposed here can be described in terms of the length λ, ξ and the dimensions of the films. Type II superconductors with $\kappa = \lambda/\xi > 1/\sqrt{2}$ are required in order to assure sufficiently strong pinning at the holes. The vortex trapping potential is estimated from a London model calculation for a vortex in a cylindrical cavity of radius r_h [7] and is given as $(\phi_o/4\pi\lambda)^2 d_F \ln(2r_h/\xi)$. A hole of radius $r_h = \lambda/2$ is therefore an effective vortex trap. The thickness of the junction films must be at least about λ to maintain adequate detection sensitivity. Also, the change in the flux threading the junction during the read cycle must be an appreciable fraction of ϕ_o, which is satisfied if κ is about 3, or greater. Using the coupling potential between a pair of vortices in two films [6], we estimate that for our geometry the vortex displacement in J2 can be on the order of 5λ without depinning the vortex trapped in J1.

The control strips S1 and S2 are several times λ in thickness, so that they provide a barrier against vortex alignment in J1 and J2 and yet are thin enough to permit nucleation of vortices in these strips during the write cycle. A typical width for S1 and S2 might be 5λ and for J1 and J2 about 10λ. The maximum packing density is therefore limited by the magnitude of λ. The base film G should be several times λ in thickness, greater than the total thickness of the films overlaying the holes.

Our estimates of the switching time and energy dissipation during the write cycle are based on an extrapolation of measured properties of vortex motion in the mixed state. The relaxation time of the superconducting order parameter is expected to be the limiting time, and our assumptions are based on time-dependent Ginzburg-Landau theory [8]. Possible contributions from non-linear effects at the critical current density for temperatures well below T_c were not considered. We assume that a current of about 20% of the Ginzburg-Landau critical current [1] will nucleate and drive a vortex across the films. We find that for alloy films the maximum flux-flow velocity is about $0.2v_F(\ell/\xi_o)^{1/2}$ for $T = T_c/2$, where v_F is the electronic Fermi velocity, ℓ the mean free path, and ξ_o the BCS coherence distance. The path length for the vortex displacement scales with λ, which for alloys is proportional to $\ell^{-1/2}$. This result implies that for short transit times ℓ should not be too small, or equivalently, κ not too large. Our estimates for a moderate value of $\kappa \approx 3$ gives a switching time of about 25ps for Nb ($\lambda = 900\text{Å}$) and 40ps for Pb ($\lambda = 500\text{Å}$) at 4.2K. Energy dissipation is determined by the magnitude of the critical current, and is about 5×10^{-10} erg. With new sub-micron patterning techniques currently under development [9] we feel that a 3 μm separation between the holes is an achievable goal, which would give a bit density of about $10^7 cm^{-2}$.

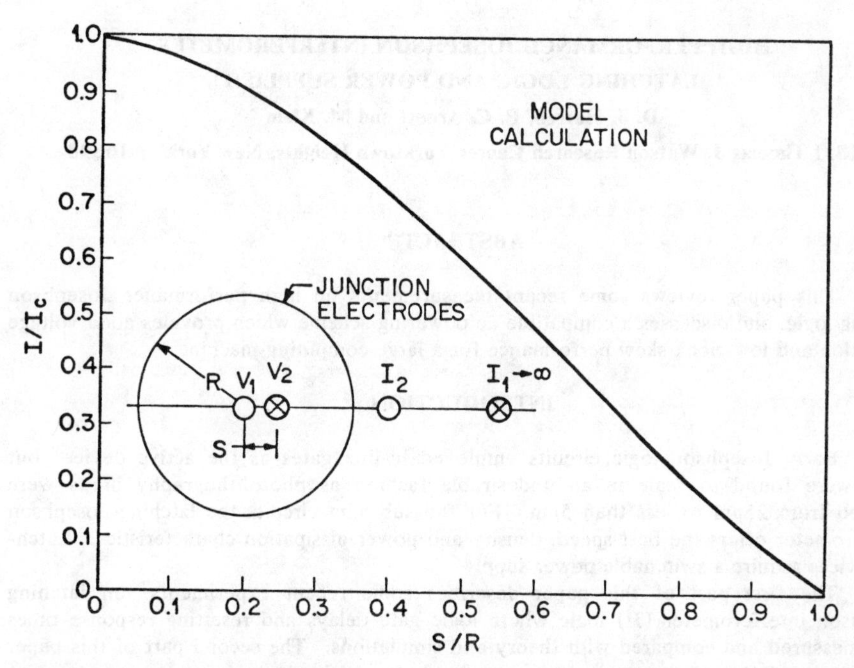

Figure 3. Calculated critical current of a circular junction as a function of the misalignment separation of a single flux line which threads the center of the bottom electrode and emerges at a radial distance s in the top electrode. The flux line sources S_1 and I_2 and sinks S_2 and I_1 are discussed in the text.

REFERENCES

1. A. F. Hebard, A. T. Fiory and S. Somekh, IEEE Trans. Magn. MAG-13, 589 (1977).
2. A. T. Fiory, A. F. Hebard and S. Somekh, Appl. Phys. Lett. 32, 73 (1978).
3. T. A. Fulton, A. F. Hebard, L. N. Dunkleberger and R. H. Eick, Solid State Commun. 22, 493 (1977).
4. H. H. Zappe, IEEE Trans. Magn. MAG-13, 41 (1977).
5. A. F. Hebard and R. H. Eick, J. Appl. Phys. 49, 338 (1978).
6. J. R. Clem, Phys. Rev. B 9, 898 (1974).
7. G. S. Mkrtchyan and V. V. Schmidt, Zh. Eksp. Teor. Fiz. 61, 367 (1971) [Soviet Phys. - JETP 34, 195 (1972).
8. M. Yu. Kupriyanov and K. K. Likharev, Fiz. Tverd. Tela 18, 1947 (1976) [Soviet Phys. Solid State 18, 1133 (1976)].
9. A. N. Broers and R. B. Laibowitz, this conference.

HIGH-PERFORMANCE JOSEPHSON INTERFEROMETER LATCHING LOGIC AND POWER SUPPLIES

D. J. Herrell, P. C. Arnett and M. Klein

IBM Thomas J. Watson Research Center Yorktown Heights, New York 10598

ABSTRACT

This paper reviews some recent measurements on high-performance Josephson latching logic, and discusses a compatible ac powering scheme which provides good voltage regulation and low clock skew performance for a large computing machine.

INTRODUCTION

Early Josephson logic circuits employed in-line gates as the active device[1] but these were found to scale in an undesirable fashion as photolithography limits were changed from 25μm to less than 5μm. For the sub-5μm circuits the latching Josephson interferometer offers the best speed, density and power dissipation characteristics. Latching devices require a switchable power supply.

The first part of this paper describes some recent experiments[2] on latching Josephson interferometer (JI) logic where logic gate delays and resetting response times were measured and compared with theory and simulations. The second part of this paper describes the design of a powering scheme suitable for driving these latching logic circuits.[3]

LATCHING JI LOGIC CIRCUITS

The basic three junction 1:2:1 interferometer originally described by Zappe[4] has optimal operating margins when used as a simple binary switch (Fig. 1). Optimum threshold curve margins are realized with $LI_o \sim \Phi_o/4$, where L is the loop inductance and I_o the Josephson critical current of the smaller junction in the loop. $\Phi_o = 2.07$ mVpS is the flux quantum.

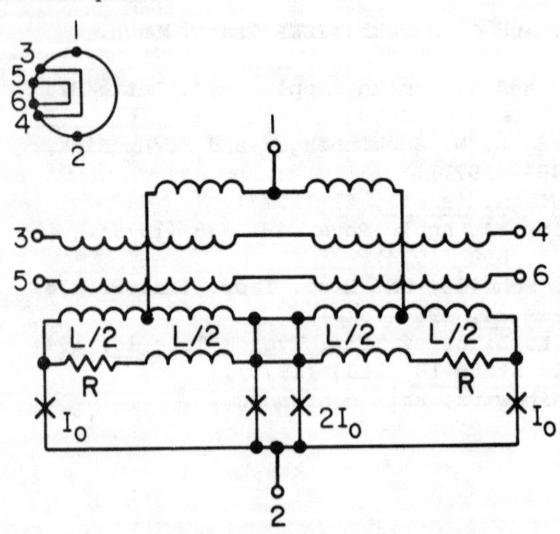

Fig. 1: Symmetrically fed 1:2:1 Josephson interferometer with two control lines I_{C1} (3-4) and I_{C2} (5-6). Optimum $LI_o = 0.25\Phi_o$. The damping resistors are shown as R. Circuit symbol shown inset, top left.

ISSN: 0094-243X/78/470/$1.50

In order to avoid switching into device resonances it is necessary to damp the interferometer inductances with small resistors.[5] The requirement that the series inductance of this damping resistor be small, together with the inability to conveniently attach resistors to the Josephson junction counterelectrode metallurgy without interdiffusion problems, forces the use of the base electrode structure for the interferometer loops (Fig. 2).

Such JI's, fabricated[6] with 5μm diameter junctions and 2.5μm overlaying control lines, form the basis of the latching JI logic gate family shown schematically in Fig. 3. The OR function is obtained by either of the overlaying control lines being activated by a logic 'one' current, i of the order of 0.2 mA for an $I_m(0) = 4I_o = 0.4$ mA and a nominal supply current level, I of 0.32 mA. The AND gate is formed following an original suggestion by Zappe by two JI's in parallel. Both JI's require input control currents before a logical output signal is generated. It should be noted that if the individual JI's of the AND gate have each two controls then the logic function (A+B) • (C+D) can be generated.

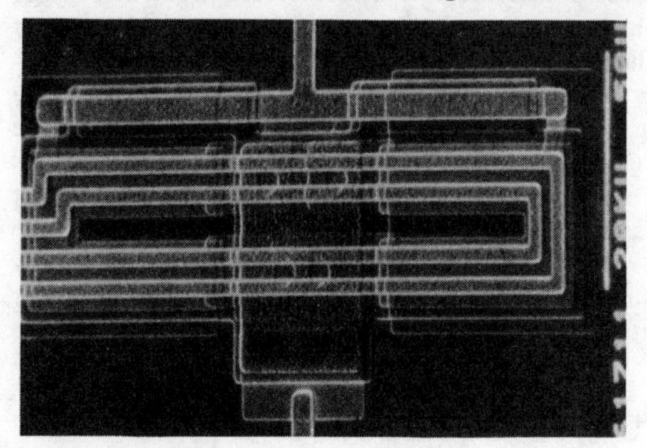

Fig. 2: SEM of Josephson interferometer. Junction diameters are 5μm. Connection 1 is at the top of the photograph, and connection 2 at the bottom. The control lines enter and leave the left-hand side and run completely over the interferometer loops.

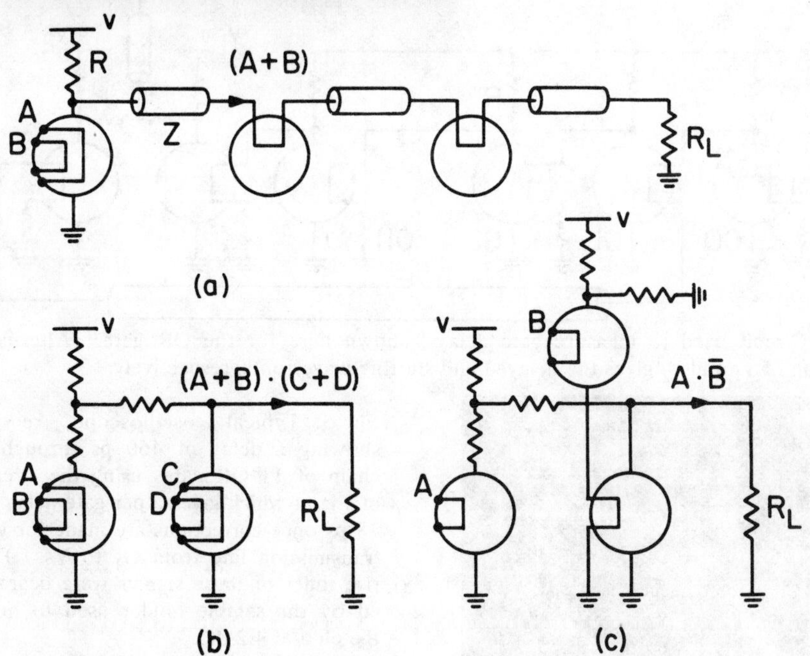

Fig. 3: JI logic family of (a) OR, (b) AND and (c) INVERTER gates. The OR gate is shown with a serial fan-out of two JI's.

The output from either of these logic gates can communicate in a serial fashion to similar logic gates with, in principle, no limit to fan-out, provided a matched transmission line and matched control crossing of fan-out JI's is employed.

The circuits were interconnected with superconducting transmission lines of 5μm nominal width. Allowing for penetration–depth effects in calculating characteristic impedance and desiring a fast transmission line not seriously slowed by such effects determines that Z should be in the range of 5 to 10 ohms for a width of 5 μm.[7] A value of 7 ohms was chosen for these experiments. The individual JI's have a total junction area of 25π μm^2 which gives a device capacitance of the order of 3 to 4 pF at a Josephson current density of 500 Acm^{-2}. This capacitance and impedance results in a latching JI (as opposed to a non-latching JI). To effect non-latching operation with these circuits would require a very low characteristic impedance and matching resistor (< 1 ohm) which leads to impractically wide lines in logic circuits for which there is a further requirement of high areal density. For example, $Z < 1$ ohm requires $w > 60$ μm with an oxide thickness of 0.18μm and penetration depths, λ_i into the superconductors of 0.086 and 0.154 μm.

To operate these latching circuits all must start out in the zero state (V = 0) at the beginning of a logic cycle. As the data front ripples through the combinatorial logic the state of particular logic gates will change to "one" ($V \neq 0$). Once the data front has passed and latched in registers then the entire combinatorial network logic has to be reset to the zero state by momentarily interrupting the power supply. Since the JI's have symmetrical characteristics the circuits operate equally well irrespective of the supply polarity and thus it is sufficient to supply an alternating polarity trapezoidal power waveform. The latched circuits will automatically reset to the zero voltage state as the power supply waveform traverses zero.

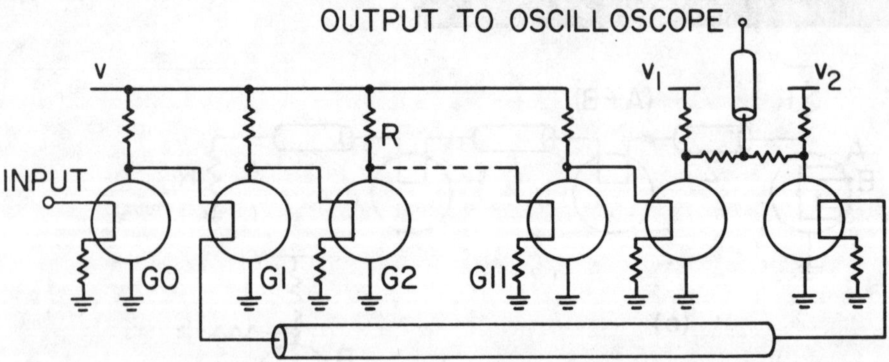

Fig. 4: Circuit used to measure gate delays, shown here for the OR gate. Alternate excitation of v_1 and v_2 gives the delayed and starting wavefronts, respectively.

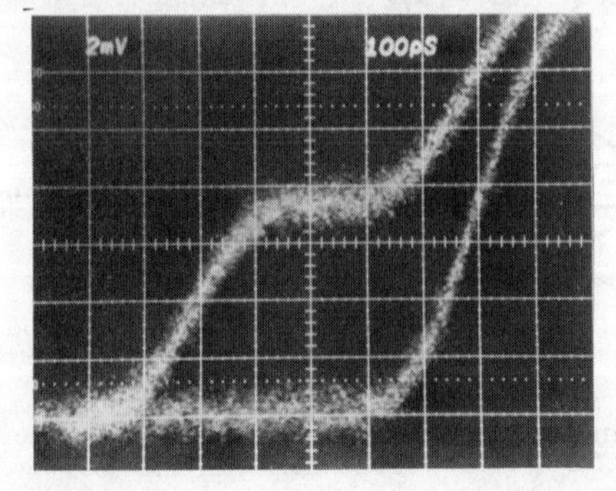

Fig. 5: Typical oscilloscope record showing a delay of 460 ps through a chain of 11 OR gates using the circuit of Fig. 4 which gives a per gate delay of 48 ps once corrections are made for the transmission line from G_1 to V_2. The rise times of these signals were degraded by the sample holder used to hold the chip at 4.2°K.

An experimental chip was designed to measure the OR and AND gate delays by making a serially connected chain of these circuits. The time a signal entered the chain was compared to the time the logic signal exited the chain through the use of JI logic circuits alternately activated to give two wavefronts to be recorded on a conventional sampling oscilloscope system (Figs. 4 and 5). The measured JI logic gate delays were found to depend, as expected, on the level that the individual JI supply current level (I) was set relative to the maximum Josephson supercurrent of the JI [$I_m(0)$]. Delay figures for $I/I_m(0)$ set at approximately 0.80 were taken as being representative of anticipated operation level once allowance is made of tolerances and noise margins.

The average OR gate delay, loaded with a fan-out of one was measured at 43 pS whereas the average AND gate delay was measured at 105 pS. Only the AND delay of B activated before A was measured as this is the slower mode of this gate operation (refer to Fig. 3) since upon the arrival of the last input signal to the gate the supply current has to be first diverted into the second JI before it can switch. These measurements were found to be in very good agreement with detailed ASTAP simulations. Based on these simulations the faster AND gate mode is expected to give a delay of 40 pS. Thus the average AND delay is 73 pS. The overall logic family delay with a fan-out delay of one is thus 58 pS.

In a similar experiment where a single OR gate was loaded with a fan-out of 22 OR gates the fan-out delay was measured at 14 pS/fan-out in excellent agreement with theory.

Simply an AND gate together with an OR gate does not constitute a logic family unless complementary logic is performed throughout the combinatorial network. In general, an INVERTER of some form is desirable. It has to be timed because of obvious error problems that would arise by placing an ordinary (i.e., non-timed) inverter in the middle of a combinatorial network composed of latching logic circuits. An inverter, which consisted of an OR gate followed by an AND gate (refer to Fig. 3), was successfully tested.

All circuits were found to reset satisfactorily at the maximum gate current slewing rate available from the external equipment (2 $mAnS^{-1}$) whereas punchthrough limits were estimated to lie between 2 and 3 $mAnS^{-1}$ for these circuits. Thus assuming a conservatively set operation level of 0.5 $mAnS^{-1}$ at a gate supply current of 80% of $I_m(0)$ it appears quite feasible that these circuits can be operated with a machine cycle time less than 5 nS. This would allow 4 nS for the powered-up portion of the trapezoidal power waveform and 1 nS for the switch-over to opposite power polarity. External equipment limitations in the experiment limited the cycle time to 4 nS (3 nS powered and 1 nS reset) (Fig. 6).

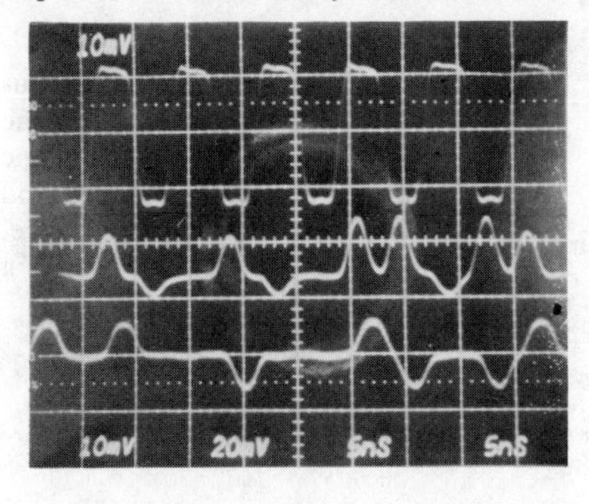

Fig. 6: Bipolar power operation of OR gate with a 4 ns logic cycle time. The top trace shows the bipolar power supply waveform. The middle trace is the input monopolar data. The bottom trace corresponds to the output voltage developed across the OR gate terminating resistor, and shows data polarity following power polarity.

POWER SUPPLIES

The JI circuits require a very fast switchable power supply. It should be capable of being distributed to all logic circuits in a machine, have good regulation qualities when the logic gates are operated, and not dissipate too much power into the LHe environment. The scheme[3] reviewed here has these qualities as well as some further desirable and very necessary properties.

Starting at the logic chip level it was recognized as being very important to regulate the power supply very close to the actual JI logic circuits as indicated by the following discussion. Current to the individual JI circuits is defined by a dropping resistor from a voltage bus. As a circuit switches, the current drawn from the bus changes slightly, (Fig. 7) inducing a disturb voltage on the bus which in turn generates a disturb current in an adjacent logic circuit. The longer the bus length from the circuit to the regulating device then the larger will be the power bus disturb. Typically it is the bus inductance, L_B, that is important and this should be minimized. Thus, referring to Fig. 7,

$$\Delta v = L_B \frac{\Delta I}{\Delta t}$$

where $\delta I = \Delta v/R$, and $I = v/R$.

We have chosen to regulate the power supply voltage directly on the chip by using a series chain of n Josephson junctions biased at the gap characteristic ($v = n2\Delta/e$). Typically either two or four gap voltages will suffice, giving a power supply voltage level of between 5 and 12 mV. These Josephson junctions have very low slope resistance at the gap and thus act as excellent regulators (Fig. 8).

The bipolar trapezoidal power waveform (Fig. 8b) can be conveniently derived by clipping an incoming sinusoidal current waveform by these regulator Josephson junctions. The fine structure in the resultant trapezoidal waveform close to $V = 0$ due to the Josephson pair tunneling characteristic does not perturb the JI logic circuits that are powered by this waveform. In certain cases this detail is absent as the Josephson junction regulators may, depending on their size, exhibit punchthrough. It is clear that the Josephson junction regulator size has to be chosen to ensure that the supply current does not exceed the top of the current step at the gap voltage. That is, referring to Fig. 8(b),

$$I_1 < I_{step}$$

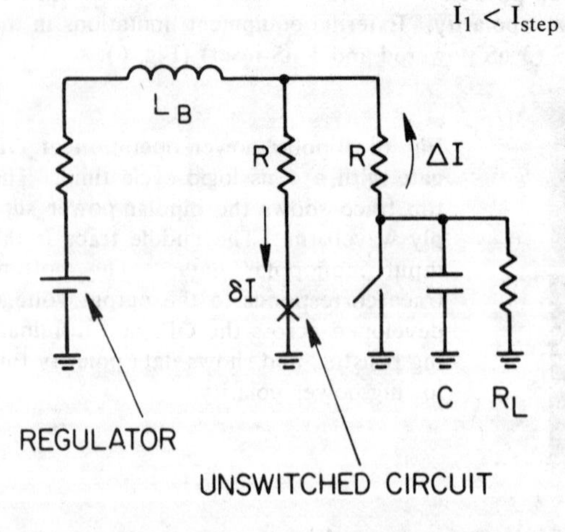

Fig. 7: Equivalent circuit of the regulated power distribution network. The switch represents the Josephson logic gate. C is the device capacitance and R_L the load (terminating) impedance. The signal rises exponentially with a time constant of approximately $R_L C$.

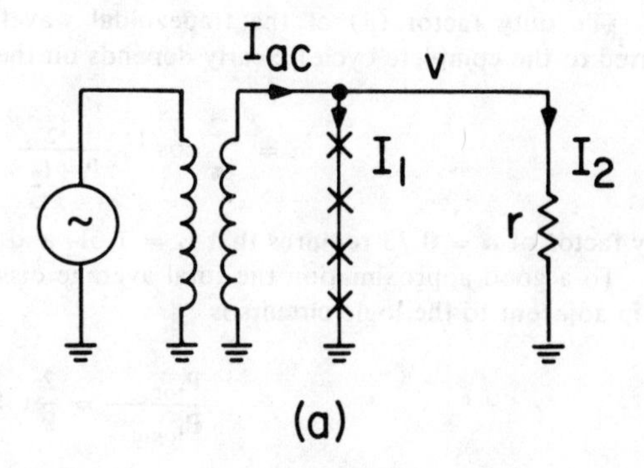

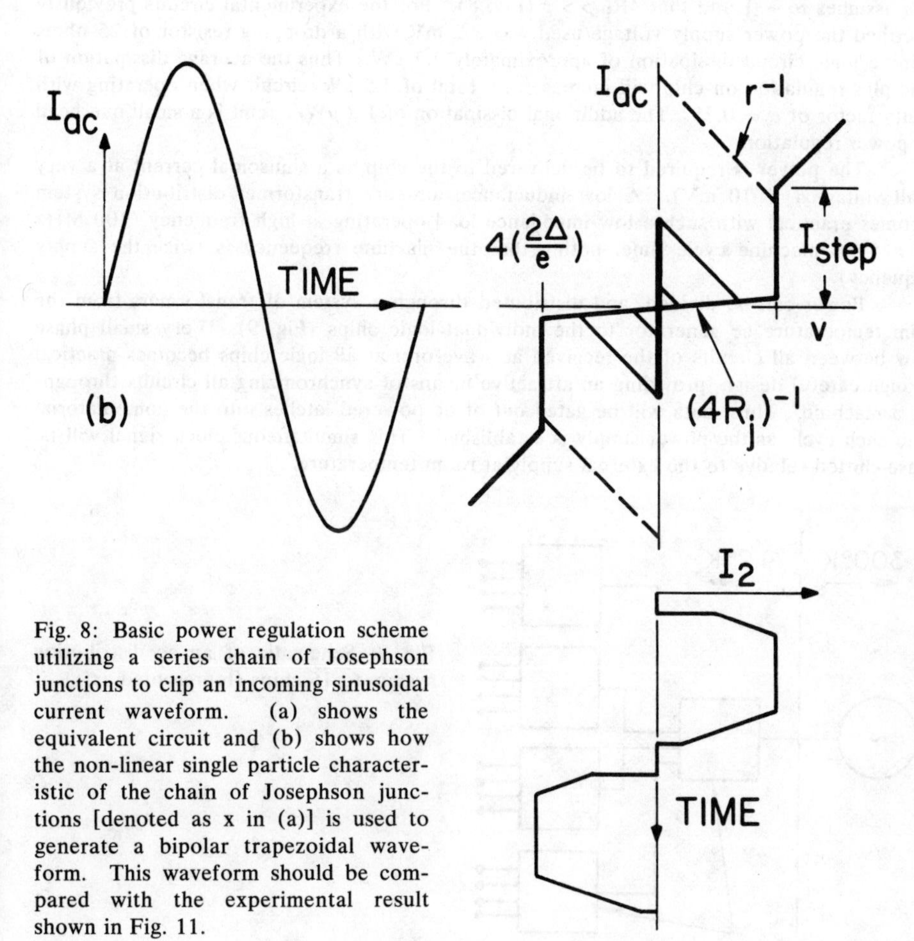

Fig. 8: Basic power regulation scheme utilizing a series chain of Josephson junctions to clip an incoming sinusoidal current waveform. (a) shows the equivalent circuit and (b) shows how the non-linear single particle characteristic of the chain of Josephson junctions [denoted as x in (a)] is used to generate a bipolar trapezoidal waveform. This waveform should be compared with the experimental result shown in Fig. 11.

The duty factor (α) of the trapezoidal waveform (powered–up portion duration compared to the complete cycle) clearly depends on the relative magnitudes of I_1 and I_2.

$$\alpha = \frac{2}{\pi} \cos^{-1} \frac{I_2}{I_1+I_2}$$

A duty factor of $\alpha = 0.75$ requires that $I_1 = 1.6I_2$ and hence $I_{ac} = 2.6I_2$.

To a good approximation the total average dissipation with regulation of power on the chip adjacent to the logic circuits is

$$\frac{P_{tot}}{P_{logic}} = \frac{2}{\pi} \left(1 + \frac{I_1}{I_2} \right)$$

This assumes $\alpha \to 1$ and that $4R_j >> r$ (Fig. 8). For the experimental circuits previously described the power supply voltage used was 8.2 mV with a dropping resistor of 25 ohms, giving a logic circuit dissipation of approximately 2.7 μW. Thus the average dissipation of logic plus regulation on-chip will increase to a total of 4.5 μW/circuit when operating with a duty factor of $\alpha = 0.75$. The additional dissipation of 1.8 μW/circuit is a small overhead for power regulation.

The power is required to be delivered to the chip as a sinusoidal current at a very small voltage (~ 10 mV). A low–inductance, air–core transformer distribution system becomes practical with such a low impedance load operating at high frequency (100 MHz for a 5 nS machine cycle time, noting that the machine frequency is twice the supply frequency).

Power can be divided and distributed through a system of transformers from the room temperature ac generator to the individual logic chips (Fig. 9). Very small phase skew between all circuits of the received ac waveform at all logic chips becomes practical through careful design, providing an attractive means of synchronizing all circuits throughout a machine. Thus, data will be gated out of dc powered latches into the combinatorial logic each cycle as the power supply is established.[2] This simultaneous clock signal will be phase-shifted relative to the external supply at room temperature.

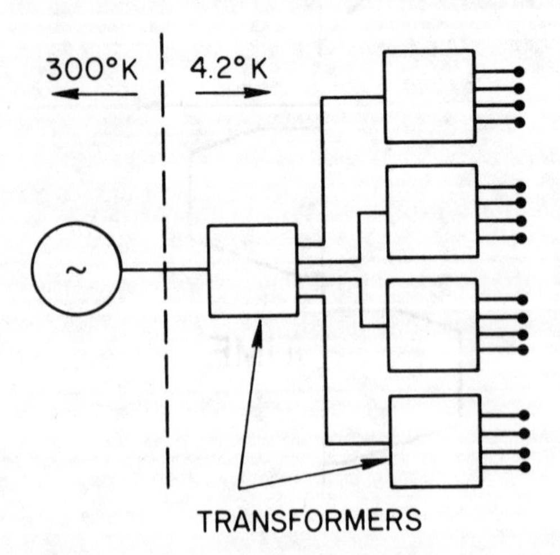

Fig. 9: Schematic of an ac distribution system to 16 chips (represented as •).

The performance of the components of such a power system were experimentally evaluated[3] with a transformer formed by two overlaying superconducting lines crossing a hole in a superconducting groundplane (Fig. 10). Large regulating Josephson junctions were found to behave as expected, giving regulated power supply waveforms identical to detailed ASTAP simulations (Fig. 11). The simple 1:1 transformers were found to have a mutual inductance of 325 pH with a coupling coefficient of 0.93. Experimental operating frequencies were limited by present test equipment.

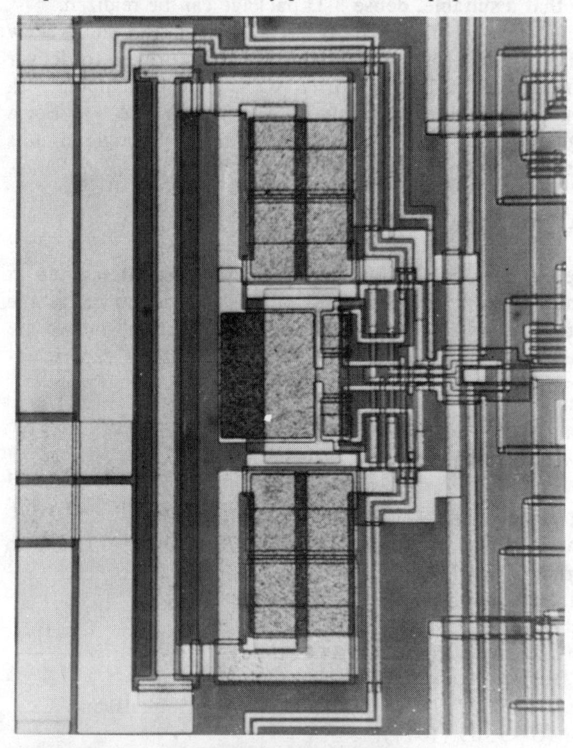

Fig. 10: Micrograph of an experimental circuit showing transformer and regulators. The transformer, which is to the left of the photograph, consists of a single primary overlying a single secondary conductor, both of which lie over a hole in the groundplane. The transformer length was 700μm with a width of 80μm.

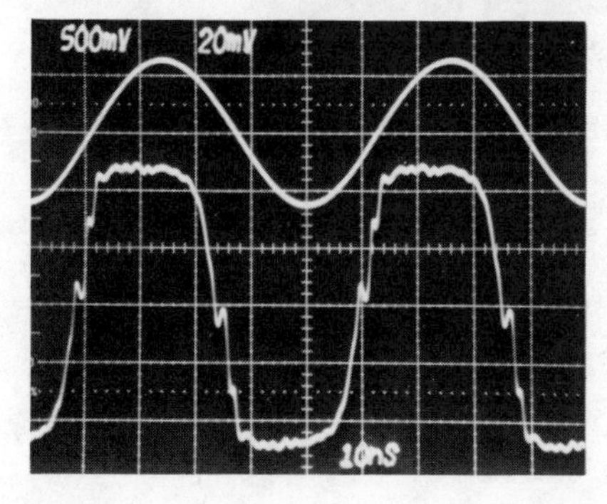

Fig. 11: Measured power waveform of the circuit of Fig. 10. The signal has an amplification of 5. This measurement was made at low repetition rates to avoid undesired sample holder resonances.

CONCLUSIONS

Latching Josephson Interferometer logic circuits have been experimentally verified at 5μm as having an average delay with a fan-out of one of 58 pS and a delay per serial fan-out of 14 pS. These circuits operated with an average dissipation level of 2.7 μW/gate and an equipment limited successful resetting rate better than 2 $mAnS^{-1}$. Such performance characteristics promise machine cycle times less than 5 nS with an active cycle of 4 nS (duty factor of 75%) providing that a suitable, dense 3-D package can be realized.

These performance characteristics are matched by an ac power and small skew clock distribution system which relies on thin–film superconducting transformers to deliver sinusoidal current to the chips. On the logic chips the power is regulated by clipping with a series string of two or more Josephson junctions connected in series ($v = n2\Delta/e$). For a duty cycle of 75% this power regulation and subsequent distribution to the logic circuits raises the average power dissipation level to approximately 4.5 μW/circuit.

ACKNOWLEDGEMENTS

The contributions of numerous members of the Exploratory Cryogenics group at IBM Research are gratefully acknowledged. In particular, we wish to acknowledge the excellent work of S. Baliozian in mask design and sample holder construction and we are also particularly indebted to J. Greiner and his group for the fabrication of the experimental circuits.

REFERENCES

1. D. J. Herrell, IEEE Journal of Solid-State Circuits, *SC-10*, 360 (1975).
2. M. Klein, D. J. Herrell, and A. Davidson, presented at ISSCC 78 (Feb. 15-17) and to be published in the IEEE Journal of Solid-State Circuits.
3. P. C. Arnett and D. J. Herrell, to be published.
4. H. H. Zappe, Applied Physics Letters, *27*, 432 (1975).
5. H. H. Zappe and B. S. Landman, J. Appl. Phys. *49*, 344 (1978).
6. L. M. Geppert, J. Greiner, D. J. Herrell and S. Klepner, to be published.
7. M. Klein, IEEE Trans. Mag., *MAG-13*, 59 (1977) and W. H. Henkels and C. J. Kircher, IEEE Trans. Mag., *MAG-13*, 63 (1977).

SUPERCONDUCTIVE ELECTRONICS IN FUTURE COMPUTER DESIGNS

H. H. Zappe

IBM Thomas J. Watson Research Center Yorktown Heights, New York 10598

ABSTRACT

Recent work and trends in experimentally verified designs of Josephson logic and memory devices and circuits are discussed.

INTRODUCTION

The in-line gate, with which as an active device, rather complex logic[1] and memory[2] circuits have been built, has by now been almost entirely replaced by the three junction interferometer.[3,4] The interferometer composed of three small quasi-point junctions, interconnected by superconducting inductances, has the advantage of being a very sensitive low-current device in which the junctions and hence the capacitance can be made as small as technology allows.

The high-speed potential of interferometer logic circuits has recently been demonstrated[5], and, although the logic functions realized are not yet as complex as those demonstrated with in-line gates,[1] much progress has been made. Basically, the dynamics of the device is now so well understood that it can be modeled with great accuracy, and new ultra-high performance logic and memory circuits are emerging. The results of these efforts, which bring Josephson computer applications to a certain maturity, are presently being prepared for publication. It is the aim of this paper to discuss these new trends and to indicate in the references where the various authors intend to submit their work.

INTERFEROMETER DYNAMICS

Statically, the interferometer is well characterized particularly as long as the junctions are small as compared to the size of the inductances.[6,7,8] Corrections for short structures in which the junctions themselves contribute in part to the total device inductance have recently also been evaluated.[9,10] A major problem surfaced in device dynamics. As in in-line gates self-induced resonance current steps appear in the I-V characteristic of interferometers. The only difference is that there are only as many current steps as meshes in the device, e.g., two in a three-junction interferometer. These steps if intersected by the load line prevent the device from switching to full voltage.

Initial experiments[3,4] suggested that the amplitudes of interferometer resonance steps are small; in fact, less than 20% of the zero field threshold current I_{mo}. However, subsequent investigations indicated a rather substantial dependence on internal damping and on asymmetrically connected external loads.

To reach a basic understanding of resonance effects a theory was derived[11] based on partial linearizations of this intrinsically non-linear problem. The main outcome was that the dc step amplitude I_p of a resonance occuring at a time-averaged voltage V requires an additional dissipation I_pV which has to be equal to the ac power dissipated by all resistors that carry ac currents. It was further shown that, throughout the interferometer design range of interest, the ac voltage components can be well approximated by a sinusoidal perturbation. This leads to a very simple formulation of the dependence of resonance amplitudes on damping.

In essence, resonance amplitudes are small in low Q interferometers in which the oscillations between the device inductances and the junction capacitances are largely damped out. As the Q is increased, resonance amplitudes increase, up to a maximum of I_p

ISSN: 0094-243X/78/479/$1.50

$\simeq 0.6\ I_{mo}$. If the Q is increased further, I_p decreases again, largely because the junction current becomes so non-linear that it ceases to drive the resonance circuit at the fundamental frequency. A similar behavior had been obtained by Werthamer and Shapiro[12] using analog simulations of a junction coupled to a resonantor; however, numerically the results are different.

Unless one can design with $I_p = 0.6\ I_{mo}$, high–Q interferometers are difficult to use since any additional load will decrease Q and hence increase the resonance amplitudes. In addition the step amplitudes in the high–Q case can dynamically increase depending on how fast the device current is swept through the resonance. For this reason, planar interferometers in which the resonance amplitudes are adjusted through resistors connected across the device inductances were proposed and were used to verify the resonance theory.[13] In this investigation excellent agreement was found between theory and experiment, verifying not only the basic assumptions, but also the Stewart-McCumber[14,15] junction model on which the theory is based.

As a result of these changes in device design, the turn-on delay had to be re-evaluated. It is a delay between the time of crossing of the switching threshold by a control current and the effective onset of the voltage transition. In high Q devices the turn-on delay is very small.[16] In contrast, Klein[17] and Harris[18] found independently that in very low–Q devices the turn-on delay can in extreme cases be as large as tens of picoseconds. Harris[18] derived an analysis of turn-on for the two-junction interferometer. The delay is shown to be a strong inverse function of control current amplitude so that it can be made very small if the device inputs are sufficiently overdriven. Furthermore, it scales with $\sqrt{C}$ where C is the junction capacitance. This indicates that this delay can be further decreased as the devices are miniaturized.

There are now four delay components in a Josephson logic circuit. These are the turn-on delay, the device risetime which is proportional to CZ_o where Z_o is the output line impedance, the propagation delay in the transmission lines and the fan-out delay associated with the inductive discontinuity in the output line at the crossing of each fan-out gate.[4] These loaded logic delays in circuits designed with damped planar interferometers have been experimentally verified.[5]

LOGIC AND MEMORY DESIGNS

Logic circuit designs have recently been extended to approaches in which for instance the AND function is realized with more than one device.[5,19] In this gate one of the inputs cannot be arbitrarily overdriven and as a result the control currents have a lower and an upper bound. This in turn causes a relatively large turn-on delay.

A logic family composed of gates with high gain, that accept arbitrarily large control currents, has recently been developed by Gheewala.[20] In this multi-device approach the gate inputs are, as before, connected to control lines; however, the devices within a gate operate through non-linear addition of directly injected currents. Non-linear current injection is best visualized by considering a two-junction interferometer having one current injected into the device center and the other directly into a junction. With properly chosen inductance and junction current ratios, such a device has the characteristic of a near–ideal AND gate, in that either current can be fairly large without switching the device whereas relatively small simultaneous inputs cause switching. Simulations indicate that this logic family will have better performance than that recently investigated.[5,19]

Memory designs are also undergoing changes as one can judge from recent work on DRO arrays.[21] Presently the most complex NDRO memory circuit is to our knowledge an experimental 64-bit fully decoded RAM;[2] still designed with in-line gates both in the cells and in the decoders. In present memory designs these gates are, as in logic, replaced by interferometers. The reason is that with interferometers the array current levels can be lowered. Lower current levels, in turn, decrease the time required to transfer current into the superconducting array loops[22] since this time is to first order proportional to IL/V, where I is the array current, L is the loop inductance and V the voltage across the driving gate.

Starting with such considerations, Henkels[22] has recently optimized and experimentally investigated the design of 5μm technology, NDRO RAM ring-cells in which both the write and the sense gate were replaced by interferometers. These cells which stored less than ten flux quanta did operate at low current levels in excellent agreement with design expectations.

The decoder of the 64-bit array[2] was distinctly a restrictive design. First, since the array loops are part of the decoder, the access time is relatively large and can be improved only if a large number of decoder gates are stacked in series in every decoder branch. Secondly, and more importantly, the crosstalk between the selected and unselected array lines is intrinsically large. To circumvent these problems Faris[23] proposed, designed and experimentally investigated a loop decoder. It consists of a modular arrangement of interconnected superconducting loops. A set of so-called address loops into which, depending on the address, current is or is not transferred, act as address register. Other loops which interrelate the address loops perform the decoder function proper. In a 5μm technology design, current transfer time into the address loops was measured to be 200 ps and the decoding delay was 30 ps per stage. In both cases excellent agreement was found between detailed measurements and simulations.

To prevent relaxation oscillations the decoder must be triggered by a short pulse. For this reason, Faris[24] designed and experimentally tested a pulse generator which is able to produce 26-ps-wide pulses. It was found to properly operate and was successfully used to drive his decoder.

CONCLUSION

The most striking feature of Josephson digital device and circuit design work is that it is based on very simple yet very effective and surprisingly accurate models. In all the investigations discussed here, the small interferometer junctions were modeled using the Stewart-McCumber[14,15] model except that the junction resistance was made voltage dependent so as to reflect the single-particle tunneling I-V characteristic. This is physically not entirely correct but yields very good results. Short sections of superconductors were modeled as inductances, long sections as transmission lines and in both cases losses in the penetration region have so far been neglected. If we add resistors to these few circuit elements, complex circuits can easily be simulated using for instance a design program such as ASTAP.[25]

More work is being done. Kautz[26] has recently evaluated losses in superconducting lines and Alsop et al.[27] have developed computational methods to calculate the magnetic fields fringing around miniaturized superconducting structures situated over a superconducting groundplane. This leads directly to an area of investigation which needs more attention, namely the investigation of ultra-miniaturized devices and circuits. Without a doubt the speed and performance potential of Josephson computer circuits is impressive, but if, as we believe at present, the Josephson technology will one day be the basis of a new computer species, then it is time now to secure its future.

ACKNOWLEDGEMENTS

The encouragement by the various authors to present a short preview of their work is appreciated. Many thanks go to Dr. B. Deaver for providing a transcript of my talk, which served as the basis for this paper.

REFERENCES

1. D. J. Herrell, IEEE Journal of Solid-State Circuits, *SC-10*, 360 (1975).
2. W. H. Henkels and H. H. Zappe, submitted to IEEE Journal of Solid-State Circuits.
3. H. H. Zappe, Appl. Phys. Lett. *27*, 432 (1975).
4. H. H. Zappe, IEEE Trans. Magn. *MAG-13*, 41 (1977).
5. M. Klein, D. J. Herrell and A. Davidson, presented at ISSCC 78 and to be published in IEEE Journal of Solid-State Circuits.
6. B. S. Landman, IEEE Trans. Magn. *MAG-13*, 871 (1977).
7. W. H. Henkels and C. J. Kircher, IEEE Trans. Magn. *MAG-13*, 63 (1977).
8. M. Klein, IEEE Trans. Magn. *MAG-13*, 59 (1977).
9. P. Gueret, unpublished.
10. W. H. Henkels, partly included in reference 22.
11. H. H. Zappe and B. S. Landman, J. Appl. Phys. *49*, 344 (1978).
12. N. R. Werthamer and S. Shapiro, Phys. Rev. *164*, 523 (1967).
13. H. H. Zappe and B. S. Landman, J. Appl. Phys. *49*, May (1978).
14. W. C. Stewart, Appl. Phys. Lett. *12*, 277(1968).
15. D. E. McCumber, J. Appl. Phys. *39* 3113 (1968).
16. H. H. Zappe, unpublished talk presented at the Conference on Transient Effects in Josephson Junctions, San Diego, CA, March, 1977.
17. M. Klein, unpublished.
18. E. P. Harris, to be submitted to J. Appl. Phys.
19. D. J. Herrell, P. C. Arnett and M. Klein, this issue.
20. T. Gheewala, to be submitted to Appl. Phys. Lett.
21. R. F. Broom, P. Gueret, W. Kotyczka, T. O. Mohr, A. Moser, A. Oosenbrug and P. Wolf, presented at ISSCC 78 and to be published in IEEE Journal of Solid-State Circuits.
22. W. H. Henkels, to be submitted to IEEE Journal of Solid-State Circuits.
23. S. M. Faris, to be submitted to IEEE Journal of Solid-State Circuits.
24. S. M. Faris, to be submitted to Appl. Phys. Lett.
25. Advanced Statistical Analysis Program, Program Reference Manual (IBM Corp. White Plains, NY, 1973).
26. R. L. Kautz, J. Appl. Phys. *49*, 308 (1978).
27. L. E. Alsop, A. S. Goodman, F. G. Gustavson and W. L. Miranker, submitted to J. Comp. Phys.

LIST OF PARTICIPANTS

Robert Adde
Institut d'Electronique
Bât. 220
Université Paris-Sud, 91605
Orsay, France

James C. Aller
National Science Foundation
1800 G Street, N.W.
Washington, D. C. 20550

Bruce Andeen
CTI-Cryogenics
Kelvin Park
Waltham, Massachusetts 02154

John T. Anderson
Hansen Laboratories of Physics
Stanford University
Stanford, California 94305

Allan Aqualino
Department of Physics
University of Virginia
Charlottesville, Virginia 22901

Tom Aton
Department of Physics
University of Virginia
Charlottesville, Virginia 22901

Ted E. Batchman
Electrical Engineering Department
University of Virginia
Charlottesville, Virginia 22903

A. Barone
Laboratorio di Cibernetica del CNR
Via Toiano 2
80072 Arco Felice
Napoli, Italy

Daniel Barr
Semiconductor Device Laboratory
Electrical Engineering Department
University of Virginia
Charlottesville, Virginia 22903

Malcom R. Beasley
Department of Applied Physics
Stanford University
Stanford, California 94305

Fernand D. Bedard
Laboratory for Physical Sciences
4928 College Avenue
College Park, Maryland 20740

William C. Black
S.H.E. Corporation
4174 Sorrento Valley Boulevard
San Diego, California 92121

T. G. Blaney
Division of Electrical Science
National Physical Laboratory
Teddington, United Kingdom

Truman G. Blocker
Texas Instruments Incorporated
MS 118
P. O. Box 225936
Dallas, Texas 75265

Charles Boghosian
Army Research Office
P. O. Box 12211
Research Triangle Park
Durham, North Carolina 27709

Bradley G. Boone
Applied Physics Laboratory
The Johns Hopkins University
Johns Hopkins Road
Laurel, Maryland 20810

J. W. Boring
Department of Engineering Physics
A & M Building
University of Virginia
Charlottesville, Virginia 22903

A. I. Braginski
Westinghouse Research &
Development Center
1310 Beulah Road, 401-3A15
Pittsburgh, Pennsylvania 15235

Richard G. Brandt
Office of Naval Research
1030 East Green Street
Pasadena, California 91106

Alec N. Broers
IBM Watson Research Center
Yorktown Heights, New York 10598

Thomas J. Bucelot
Department of Physics
University of Virginia
Charlottesville, Virginia 22901

Spencer A. Buckner
Texas Instruments, Inc.
Mail Station 332
P. O. Box 6015
Dallas, Texas 75222

Robert Buhrman
Applied Physics
Clark Hall
Cornell University
Ithaca, New York 14853

Max Burbank
CTF Systems, Inc. No. 15
1750 McLean Avenue
Port Coquitlam
British Columbia, Canada V3C 1M9

Blas Cabrera
Physics Department
Stanford University
Stanford, California 94305

Alessandro Callegari
Department of Physics
University of Virginia
Charlottesville, Virginia 22901

Isidoro E. Campisi
Department of Physics
Louisiana State University
Baton Rouge, Louisiana 70803

George A. Candela
National Bureau of Standards
BH 223, Room A 259
Washington, D. C. 20234

Mark S. Chang
Moore School of Electrical Engineering
University of Pennsylvania
Philadelphia, Pennsylvania 19104

Raymond Chiao
Physics Department
University of California
Berkeley, California 94720

John Claassen
Naval Research Laboratory
Washington, D. C. 20375

John Clarke
Department of Physics
University of California
Berkeley, California 94720

William L. Clinton
Division of Materials Science
Department of Energy
Washington, D. C. 20545

F. D. Colegrove
Texas Instruments, Inc.
Mail Stop 332
P. O. Box 6015
Dallas, Texas 75222

R. V. Coleman
Department of Physics
University of Virginia
Charlottesville, Virginia 22901

Michael Cromar
Department of Physics
University of Rochester
Rochester, New York 14627

Edward J. Cukauskas
Code 5280
Naval Research Laboratory
Washington, D. C. 20375

Michael R. Daniel
Westinghouse Research &
Development Center
1310 Beulah Road
Pittsburgh, Pennsylvania 15235

Roger A. Davidheiser
TRW Systems
1 Space Park
Redondo Beach, California 90278

Kenneth L. Davis
Code 5253
Naval Research Laboratory
Washington, D. C. 20375

Bascom Deaver, Jr.
Department of Physics
University of Virginia
Charlottesville, Virginia 22901

D. R. Decker
National Radio Astronomy Observatory
2015 Ivy Road
Charlottesville, Virginia 22903

Guy Deutscher
Department of Physics and Astronomy
Tel Aviv University
Ramat-Aviv, Tel Aviv, Israel

G. J. Dolan
Bell Labs
Room 1D304
600 Mountain Avenue
Murray Hill, New Jersey 07974

Gordon B. Donaldson
Department of Applied Physics
John Anderson Building
University of Strathclyde
Glasgow, Scotland G4 ONG

D. H. Douglass
Department of Physics
University of Rochester
Rochester, New York 14627

E. A. Edelsack
Code 427
Office of Naval Research
800 North Quincy Street
Arlington, Virginia 22217

Ole Eg
Physics Laboratory I
H. C. Oersted Institute
Universitets Parken 5
2100 Copenhagen Ø, Denmark

William M. Fairbank
Physics Department
Stanford University
Stanford, California 94305

Charles M. Falco
Solid State Science Division
Argonne National Laboratory
Argonne, Illinois 60439

Sadeg M. Faris
IBM Corporation
P. O. Box 218
Yorktown Heights, New York 10598

James J. Finley
Department of Physics
University of Virginia
Charlottesville, Virginia 22901

Nicolaos Flytzanis
Department of Physics
University of Virginia
Charlottesville, Virginia 22901

Carl Franck
Department of Physics
University of Virginia
Charlottesville, Virginia 22901

Chuck Friedberg
Department of Physics
University of Virginia
Charlottesville, Virginia 22901

Theodore Fulton
Bell Labs
Room 1D461
600 Mountain Avenue
Murray Hill, New Jersey 07974

N. Garcia
Department of Physics
University of Virginia
Charlottesville, Virginia 22901

Chris H. Galfo
Department of Physics
University of Virginia
Charlottesville, Virginia 22901

A. K. Ganguly
Naval Research Laboratory
Washington, D. C. 20375

J. C. Garland
Department of Physics
Ohio State University
Columbus, Ohio 43210

Robert Gayley
Physics Department
State University of New York at Buffalo
Amherst, New York 14260

Linda Geppert
IBM Watson Research Center
Yorktown Heights, New York 10598

Meir Gershenson
Department of Physics
Serin Physics Laboratory
Rutgers University
Frelinghuysen Road
Piscataway, New Jersey 08854

Robin P. Giffard
Department of Physics
Stanford University
Stanford, California 94305

James F. Goff
NSWC/White Oak Laboratory
Silver Spring, Maryland 20910

Allen M. Goldman
School of Physics
University of Minnesota
Minneapolis, Minnesota 55455

William L. Goodman
Applied Physics Systems
3274 Ramona Street
Palo Alto, California 92665

Roy G. Goodrich
Physics Department
Louisiana State University
Baton Rouge, Louisiana 70803

Kenneth E. Gray
Building 223
Argonne National Laboratory
Argonne, Illinois 60439

Donald Gubser
Naval Research Laboratory
Washington, D. C. 20375

Clark A. Hamilton
National Bureau of Standards
325 Broadway, 276.04
Boulder, Colorado 80303

William O. Hamilton
Department of Physics and Astronomy
University of Rochester
Rochester, New York 14627

Erik P. Harris
IBM Watson Research Center
P. O. Box 218
Yorktown Heights, New York 10598

Jay H. Harris
Engineering Division
National Science Foundation
1800 G Street, N.W.
Washington, D. C. 20550

Richard E. Harris
Cryogenics Division, 275.08
National Bureau of Standards
Boulder, Colorado 80302

John T. Harvell
CTI-Cryogenics
Kelvin Park
Waltham, Massachusetts 02154

Arthur F. Hebard
Bell Labs
Room 1D460
600 Mountain Avenue
Murray Hill, New Jersey 07974

R. A. Hein
Code 5280
Naval Research Laboratory
Washington, D. C. 20375

John B. Hendricks
Physics Department
University of Alabama
Huntsville, Alabama 35807

William Henshaw
Department of Physics
University of Virginia
Charlottesville, Virginia 22901

Dennis Herrell
IBM Watson Research Center
P. O. Box 218
Yorktown Heights, New York 10598

George B. Hess
Department of Physics
University of Virginia
Charlottesville, Virginia 22901

Ross Hill
Department of Physics
University of Virginia
Charlottesville, Virginia 22901

Louis B. Holdeman
National Bureau of Standards
Room A247, Building 220
Washington, D. C. 20234

Richard E. Howard
Ginzton Labs
Stanford University
Stanford, California 94305

Evelyn L. Hu
Bell Labs
4D-423
Holmdel, New Jersey 07733

John J. Hudak
DIRNSA
Department of Defense
5245 Hayledge Court
Columbia, Maryland 21045

Robert R. Humphris
RLES - Thornton Hall
University of Virginia
Charlottesville, Virginia 22903

John P. Hurrell
The Aerospace Corporation
P. O. Box 92957
Los Angeles, California 90009

D. J. Hyun
Department of Physics
University of Virginia
Charlottesville, Virginia 22901

L. D. Jackel
Bell Labs
4D-335
Holmdel, New Jersey 07733

Michael Janocko
Westinghouse Research Labs
Pittsburgh, Pennsylvania 15235

Don Jillie
Sperry Research Center
100 North Road
Sudbury, Massachusetts 01776

Leonard M. Kahn
Department of Physics
University of Virginia
Charlottesville, Virginia 22901

Robert T. Kampwirth
Argonne National Laboratory
M.S. 223-A109
9700 South Cass Avenue
Argonne, Illinois 60439

Y. H. Kao
Physics Department
State University of New York
Stony Brook, New York 11794

Mark B. Ketchen
IBM Watson Research Center
P. O. Box 218
Yorktown Heights, New York 10598

John Ketterson
Northwestern University
628 Garret
Evanston, Illinois 60201

C. A. King
Physics Department
Loyola University
New Orleans, Louisiana 70118

Randall K. Kirschman
Jet Propulsion Laboratory
183-401
4800 Oak Grove Drive
Pasadena, California 91103

Teun M. Klapwijk
Department of Applied Physics
Delft University of Technology
Lorentzweg 1, Delft, The Netherlands

Charles S. Korman
Department of Physics
University of Virginia
Charlottesville, Virginia 22901

Harry Kroger
Sperry Research Center
100 North Road
Sudbury, Massachusetts 01776

Robert Laibowitz
IBM Watson Research Center
P. O. Box 218
Yorktown Heights, New York 10598

Donald N. Langenberg
Department of Physics
University of Pennsylvania
Philadelphia, Pennsylvania 19104

Kei-Fung Lau
TRW Defense and Space Systems Group
Building R1, Room 1070
One Space Park
Redondo Beach, California 90278

James A. Leaverton
TRW Systems
One Space Park
Redondo Beach, California 90278

I. Lefkowitz
Army Research Office
P. O. Box 12211
Research Triangle Park
Durham, North Carolina 27709

C. H. Li
Honeywell, Inc.
Corporate Technology Center
10701 Lyndale Avenue, South
Bloomington, Minnesota 54420

P. E. Lindelof
Physics Laboratory I
H. C. Oersted Institute
Universitets Parken 5
2100 Copenhagen Ø, Denmark

William A. Little
Department of Physics
Stanford University
Stanford, California 94305

Frances L. Lloyd
National Bureau of Standards
Cryogenics Division 275.08
Boulder, Colorado 80303

J. E. Lokken
Defense Research Establishment
Pacific F.M.O.
Victoria, British Columbia
Canada VOS 1BO

H. Lübbig
Physikalisch-Technische Bundesanstalt
7 Berlin 10, Abbestrabe 2-12
Berlin, West Germany

James Lukens
Department of Physics
State University of New York
Stony Brook, New York 11794

Jean Maah-Sango
Electronics Research Laboratory
University of California
Berkeley, California 93720

Juri Matisoo
IBM Watson Research Center
Yorktown Heights, New York 10598

Robert Mattauch
Electrical Engineering Department
University of Virginia
Charlottesville, Virginia 22903

David B. McElvein
CTI-Cryogenics
Kelvin Park
Waltham, Massachusetts 02154

James Mercereau
Physics Department
California Institute of Technology
Pasadena, California 91125

George Miller
U.S. Naval Oceanographic Office
NSTL Station
Bay St. Louis, Mississippi 39522

H. J. Mueller
Naval Air Systems Command (O3C)
Washington, D. C. 20361

Akira Nakamura
Electrotechnical Laboratory
5-4-1, Mukodai-machi, Tanashi
Tokyo, Japan

Martin Nisenoff
Code 5285
Naval Research Laboratory
Washington, D. C. 20375

James E. Nordman
Department of Electrical & Computer Engineering
University of Wisconsin
Madison, Wisconsin 53706

William Oelfke
Florida Technological University
Department of Physics
Orlando, Florida 32816

James Opfer
Hewlett-Packard
3500 Deer Creek Road
Palo Alto, California 94304

Ho Jung Paik
Physics Department
Stanford University
Stanford, California 94305

Daniel Pascal
Bât. 220, Salle 268
Université d'Orsay 91405
Orsay, France

James L. Paterson
Department of Physics
University of Virginia
Charlottesville, Virginia 22901

Niels F. Pedersen
Physics Lab I
The Technical University of Denmark
Lyngby, Denmark DK 2800

Robert A. Peters
Physics Department
Catholic University
Washington, D. C. 20064

John Philo
United Scientific Corporation
1400 Stierlin Road
Mountain View, California 94043

Walter Podney
Physical Dynamics, Inc.
P. O. Box 556
La Jolla, California 92038

Daniel Prober
Becton Center
Yale University
New Haven, Connecticut 06520

Marcello Puma
Departamento De Fisica
Universidad Simon Bolivar
Apartado 80659
Caracas, Venezuela

Stanley A. Reible
MIT Lincoln Laboratory
Lexington, Massachusetts 01824

Paul L. Richards
Department of Physics
University of California
Berkeley, California 94720

Rogers Ritter
Department of Physics
University of Virginia
Charlottesville, Virginia 22901

Horst Rogalla
Institut für Angewandte Physik der
Justus Liebig-Universität Giessen
Heinrich-Buff-Ring 16
6300 Giessen, Germany

Lynette B. Roth
Hughes Research Laboratories
3011 Malibu Canyon Road
Malibu, California 90265

John Rowell
Bell Laboratories, 1D-367
Murray Hill, New Jersey 07974

John Ruvalds
Department of Physics
University of Virginia
Charlottesville, Virginia 22901

Ronald Sager
Physical Dynamics, Inc.
P. O. Box 556
La Jolla, California 92038

Robert D. Sandell
Physics Department
State University of New York
Stony Brook, New York 11794

S. E. Schnatterly
Department of Physics
University of Virginia
Charlottesville, Virginia 22901

Ivan Schuller
Physics Department
UCLA
Los Angeles, California 90024

Sidney Shapiro
Department of Electrical Engineering
University of Rochester
Rochester, New York 14627

Arnold H. Silver
The Aerospace Corporation
P. O. Box 92957
Los Angeles, California 90009

W. J. Skocpol
Department of Physics
Harvard University
Cambridge, Massachusetts 02138

Joanne Snare
Physics Department
University of North Carolina
Chapel Hill, North Carolina 27514

J. C. Solinsky
ARACOR
1223 East Argues Avenue
Sunnyvale, California 94086

Gordon L. Spencer
U.S. Army Foreign Science and
Technology Center
220 Seventh Street, N.E.
Charlottesville, Virginia 22901

Charles V. Stancampiano
Department of Electrical Engineering
University of Rochester
Rochester, New York 14627

Samuel R. Stein
Time and Frequency Division
National Bureau of Standards
Boulder, Colorado 80303

D. B. Sullivan
National Bureau of Standards
Cryogenic Division
Boulder, Colorado 80303

Yuan Taur
NASA/GISS
2880 Broadway
New York, New York 10025

Barry N. Taylor
National Bureau of Standards
Building 220, Room B258
Washington, D. C. 20234

Eric D. Thompson
Electrical Engineering & Applied
Physics Department
Case Western Reserve University
Cleveland, Ohio 44106

Michael Tinkham
Physics Department
Harvard University
Cambridge, Massachusetts 02138

Bryan Troutman
IBM
18100 Frederick Pike
Gaithersburg, Maryland 20760

Bruce T. Ulrich
Department of Physics
University of Geneva
32 Boulevard d'Yvoy
CH-1211 Geneva 4, Switzerland

John I. Upshur
Department of Physics
University of Virginia
Charlottesville, Virginia 22901

Costas Varmazis
Physics Department
State University of New York
Stony Brook, New York 11794

Ashley Vincent
University of South Florida
166 Baltic Circle
Tampa, Florida 33606

Jiri Vrba
CTF Systems, Inc. No. 15
1750 McLean Avenue
Port Coquitlam, British Columbia
Canada V3C 1M9

Michael Waldner
Hughes Research Labs
3011 Malibu Canyon Road
Malibu, California 90265

Li-Kong Wang
Department of Physics
University of Virginia
Charlottesville, Virginia 22901

R. Han Wang
Los Alamos Scientific Lab
E-DOR, M. S. 429
Los Alamos, New Mexico 87545

Wasyl Wasylkiwskyj
Institute for Defense Analysis
400 Army Navy Drive
Arlington, Virginia 22202

Denis C. Webb
Naval Research Laboratory
Washington, D. C. 20375

Richard A. Webb
Solid State Science Division
Argonne National Laboratory
Argonne, Illinois 60439

Nancy K. Welker
Laboratory for Physical Sciences
College Park, Maryland 20740

Charles Wellemeyer
Department of Physics
University of Virginia
Charlottesville, Virginia 22901

Thomas R. Werner
Argonne National Laboratory
9700 South Cass Avenue
Argonne, Illinois 60439

Robert M. White
Xerox Research Center
Coyote Hill Road
Palo Alto, California 94304

John P. Wikswo, Jr.
Department of Physics
Vanderbilt University
Box 1807 Station B
Nashville, Tennessee 37235

Samuel J. Williamson
Physics Department
New York University
4 Washington Place
New York, New York 10003

Stuart Wolf
Code 5280
Naval Research Lab
Washington, D. C. 20375

James T. Yeh
IBM Watson Research Center
P. O. Box 218
Yorktown Heights, New York 10598

Max N. Yoder
Office of Naval Research
Arlington, Virginia 22217

Hans H. Zappe
IBM Watson Research Center
P. O. Box 218
Yorktown Heights, New York 10598

Daniel Zenatti
CENG–LETI/EPA
85 X 38041
Grenoble Cedex, France

J. E. Zimmerman
National Bureau of Standards
Boulder, Colorado 80303

Fred Zutavern
Department of Physics
University of Virginia
Charlottesville, Virginia 22901

AUTHOR INDEX

AIP Conference Proceedings

No.	Title	L.C. Number	ISBN
No.1	Feedback and Dynamic Control of Plasmas (Princeton) 1970	70-141596	0-88318-100-2
No.2	Particles and Fields - 1971 (Rochester)	71-184662	0-88318-101-0
No.3	Thermal Expansion - 1971 (Corning)	72-76970	0-88318-102-9
No.4	Superconductivity in d- and f-Band Metals (Rochester, 1971)	74-18879	0-88318-103-7
No.5	Magnetism and Magnetic Materials - 1971 (2 parts) (Chicago)	59-2468	0-88318-104-5
No.6	Particle Physics (Irvine, 1971)	72-81239	0-88318-105-3
No.7	Exploring the History of Nuclear Physics (Brookline, 1967, 1969)	72-81883	0-88318-106-1
No.8	Experimental Meson Spectroscopy - 1972 (Philadelphia)	72-88226	0-88318-107-X
No.9	Cyclotrons - 1972 (Vancouver)	72-92798	0-88318-108-8
No.10	Magnetism and Magnetic Materials - 1972 (2 parts) (Denver)	72-623469	0-88318-109-6
No.11	Transport Phenomena - 1973 (Brown University Conference)	73-80682	0-88318-110-X
No.12	Experiments on High Energy Particle Collisions - 1973 (Vanderbilt Conference)	73-81705	0-88318-111-8
No.13	π-π Scattering - 1973 (Tallahassee Conference)	73-81704	0-88318-112-6
No.14	Particles and Fields - 1973 (APS/DPF Berkeley)	73-91923	0-88318-113-4
No.15	High Energy Collisions - 1973 (Stony Brook)	73-92324	0-88318-114-2
No.16	Causality and Physical Theories (Wayne State University, 1973)	73-93420	0-88318-115-0
No.17	Thermal Expansion - 1973 (Lake of the Ozarks)	73-94415	0-88318-116-9
No.18	Magnetism and Magnetic Materials - 1973 (2 parts) (Boston)	59-2468	0-88318-117-7
No.19	Physics and the Energy Problem - 1974 (APS Chicago)	73-94416	0-88318-118-5
No.20	Tetrahedrally Bonded Amorphous Semiconductors (Yorktown Heights, 1974)	74-80145	0-88318-119-3
No.21	Experimental Meson Spectroscopy - 1974 (Boston)	74-82628	0-88318-120-7
No.22	Neutrinos - 1974 (Philadelphia)	74-82413	0-88318-121-5
No.23	Particles and Fields - 1974 (APS/DPF Williamsburg)	74-27575	0-88318-122-3
No.24	Magnetism and Magnetic Materials - 1974 (20th Annual Conference, San Francisco)	75-2647	0-88318-123-1

		L.C. Number	ISBN
No.25	Efficient Use of Energy (The APS Studies on the Technical Aspects of the More Efficient Use of Energy)	75-18227	0-88318-124-X
No.26	High-Energy Physics and Nuclear Structure - 1975 (Santa Fe and Los Alamos)	75-26411	0-88318-125-8
No.27	Topics in Statistical Mechanics and Biophysics: A Memorial to Julius L. Jackson (Wayne State University, 1975)	75-36309	0-88318-126-6
No.28	Physics and Our World: A Symposium in Honor of Victor F. Weisskopf (M.I.T., 1974)	76-7207	0-88318-127-4
No.29	Magnetism and Magnetic Materials - 1975 (21st Annual Conference, Philadelphia)	76-10931	0-88318-128-2
No.30	Particle Searches and Discoveries - 1976 (Vanderbilt Conference)	76-19949	0-88318-129-0
No.31	Structure and Excitations of Amorphous Solids (Williamsburg, Va., 1976)	76-22279	0-88318-130-4
No.32	Materials Technology - 1975 (APS New York Meeting)	76-27967	0-88318-131-2
No.33	Meson-Nuclear Physics - 1976 (Carnegie-Mellon Conference)	76-26811	0-88318-132-0
No.34	Magnetism and Magnetic Materials - 1976 (Joint MMM-Intermag Conference, Pittsburgh)	76-47106	0-88318-133-9
No.35	High Energy Physics with Polarized Beams and Targets (Argonne, 1976)	76-50181	0-88318-134-7
No.36	Momentum Wave Functions - 1976 (Indiana University)	77-82145	0-88318-135-5
No.37	Weak Interaction Physics - 1977 (Indiana University)	77-83344	0-88318-136-3
No.38	Workshop on New Directions in Mössbauer Spectroscopy (Argonne, 1977)	77-90635	0-88318-137-1
No.39	Physics Careers, Employment and Education (Penn State, 1977)	77-94053	0-88318-138-X
No.40	Electrical Transport and Optical Properties of Inhomogeneous Media (Ohio State University, 1977)	78-54319	0-88318-139-8
No.41	Nucleon-Nucleon Interactions - 1977 (Vancouver)	78-54249	0-88318-140-1
No.42	Higher Energy Polarized Proton Beams (Ann Arbor, 1977)	78-55682	0-88318-141-X
No.43	Particles and Fields - 1977 (APS/DPF, Argonne)	78-55683	0-88318-142-8
No.44	Future Trends in Superconductive Electronics (Charlottesville, 1978)	78-66638	0-88318-143-6